# Subcellular Biochemistry

## Volume 29

### Plant–Microbe Interactions

---

*Recent Volumes in This Series:*



---

A Continuation Order Plan is available for this series. A continuation order will bring delivery of each new volume immediately upon publication. Volumes are billed only upon actual shipment. For further information please contact the publisher.

# Subcellular Biochemistry

## Volume 29

## Plant–Microbe Interactions

Edited by

**B. B. Biswas**
*University of Calcutta*
*Calcutta, India*

and

**H. K. Das**
*Jawaharlal Nehru University*
*New Delhi, India*

PLENUM PRESS • NEW YORK AND LONDON

The Library of Congress cataloged the first volume of this title as follows:

Sub-cellular biochemistry.
London, New York, Plenum Press.
v. illus. 23 cm. quarterly.
Began with Sept. 1971 issue. Cf. New serial titles.
1. Cytochemistry—Periodicals. 2. Cell organelles—Periodicals.
QH611.S84 574.8'76 73-643479

ISSN 0306-0225

ISBN 0-306-45678-8

This series is a continuation of the journal *Sub-Cellular Biochemistry*, Volumes 1 to 4 of which were published quarterly from 1972 to 1975


A Division of Plenum Publishing Corporation
233 Spring Street, New York, N.Y. 10013

http://www.plenum.com

10 9 8 7 6 5 4 3 2 1



Printed in the United States of America

# Contributors

**Anke Becker** Lehrstuhl für Genetik, Fakultät für Biologie, Universität Bielefeld, 33501 Bielefeld, Germany

**Carol L. Bender** Department of Plant Pathology, 110 Noble Research Center, Oklahoma State University, Stillwater, Oklahoma 74078-3032

**Carl W. Bergmann** Complex Carbohydrate Research Center and Department of Biochemistry and Molecular Biology, University of Georgia, Athens, Georgia 30602-4712

**Ton Bisseling** Department of Molecular Biology, Agricultural University, 6703 HA Wageningen, The Netherlands

**Maria Teresa Cervera** Laboratorium voor Genetica, Department of Genetics, Flanders Interuniversity Institute for Biotechnology (VIB), Universiteit Gent, B-9000 Gent, Belgium; *present address:* Instituto Nacional de Investigación y Tecnología Agraria y Alimentaria (CIT-INIA), E-28040 Madrid, Spain

**François Côté** Complex Carbohydrate Research Center and Department of Botany, University of Georgia, Athens, Georgia 30602-4712

**Anath Das** Department of Biochemistry and Plant Molecular Genetics Institute, University of Minnesota, St. Paul, Minnesota 55108

**Keith R. Davis** Plant Biotechnology Center and Department of Plant Biology, The Ohio State University, Columbus, Ohio 43210-1002

**Koen Goethals** Laboratorium voor Genetica, Department of Genetics, Flanders Interuniversity Institute for Biotechnology (VIB), Universiteit Gent, B-9000 Gent, Belgium

**Sofie Goormachtig** Laboratorium voor Genetica, Department of Genetics, Flanders Interuniversity Institute for Biotechnology (VIB), Universiteit Gent, B-9000 Gent, Belgium

**Changhui Guan** Department of Molecular Biology, Agricultural University, 6703 HA Wageningen, The Netherlands; *present address:* Max-Planck-Institut für Züchtungsforschung, D-50829 Köln, Germany.

**Michael G. Hahn** Complex Carbohydrate Research Center and Department of Botany, University of Georgia, Athens, Georgia 30602-4712

**Kyung-Sik Ham** Complex Carbohydrate Research Center and Department of Biochemistry and Molecular Biology, University of Georgia, Athens, Georgia 30602-4712; and Department of Food Engineering, Mopko National University, Chonnam 534-729, Korea

**Marcelle Holsters** Laboratorium voor Genetica, Department of Genetics, Flanders Interuniversity Institute for Biotechnology (VIB), Universiteit Gent, B-9000 Gent, Belgium

**Dimitris Kafetzopoulos** Leiden University, Institute of Molecular Plant Sciences, Clusius Laboratory, 2333 AL Leiden, The Netherlands

**Eric Kamst** Leiden University, Institute of Molecular Plant Sciences, Clusius Laboratory, 2333 AL Leiden, The Netherlands

**Wolfgang Knogge** Department of Biochemistry, Max-Planck-Institut für Züchtungsforschung, D-50829 Cologne, Germany

**Carlos A. Malpica** Laboratorium voor Genetica, Department of Genetics, Flanders Interuniversity Institute for Biotechnology (VIB), Universiteit Gent, B-9000 Gent, Belgium; *present address:* Instituto de Medio Ambiente (IMA), Centro de Investigaciones Energéticas Medioambientales y Tecnológicas (CIEMAT), E-28040 Madrid, Spain

**Peter Mergaert** Laboratorium voor Genetica, Department of Genetics, Flanders Interuniversity Institute for Biotechnology (VIB), Universiteit Gent, B-9000 Gent, Belgium

**Karsten Niehaus** Lehrstuhl für Genetik, Fakultät für Biologie, Universität Bielefeld, 33501 Bielefeld, Germany

**David A. Palmer** Department of Plant Pathology, 110 Noble Research Center, Oklahoma State University, Stillwater, Oklahoma 74078-3032; *present address:* Zeneca Ag Products, Eastern Regional Technical Center, Whitakers, North Carolina 27891

**Katharina Pawlowski** Department of Molecular Biology, Agricultural University, 6703 HA Wageningen, The Netherlands; *present address:* Albrecht-von-Haller-Institut für Pflanzenwissenschaften, Universität Göttingen, 37073 Göttingen, Germany

**Alejandro Peñaloza-Vázquez** Department of Plant Pathology, 110 Noble Research Center, Oklahoma State University, Stillwater, Oklahoma 74078-3032

**Vidhya Rangaswamy** Department of Plant Pathology, 110 Noble Research Center, Oklahoma State University, Stillwater, Oklahoma 74078-3032

**Walt Ream** Molecular and Cellular Biology Program and Department of Microbiology, Oregon State University, Corvallis, Oregon 97331

**Chris Simoens** Laboratorium voor Genetica, Department of Genetics, Flanders Interuniversity Institute for Biotechnology (VIB), Universiteit Gent, B-9000 Gent, Belgium

**Herman P. Spaink** Leiden University, Institute of Molecular Plant Sciences, Clusius Laboratory, 2333 AL Leiden, The Netherlands

**Günter Strittmatter** Plant Genetic Systems, B-9000 Gent, Belgium; *present address:* KWS Kleinwanzlebener Saatzucht AG, Institut für Pflanzenzüchtung, D-37555 Einbeck, Germany

**Matthias Ullrich** Department of Plant Pathology, 110 Noble Research Center, Oklahoma State University, Stillwater, Oklahoma 74078-3032; *present address:* Max-Planck-Institut für Terrestrische Mikrobiologie, D-35043 Marburg, Germany

**Marc Van Montagu** Laboratorium voor Genetica, Department of Genetics, Flanders Interuniversity Institute for Biotechnology (VIB), Universiteit Gent, B-9000 Gent, Belgium

**Desh Pal S. Verma** Department of Molecular Genetics and Plant Biotechnology Center, The Ohio State University, Columbus, Ohio 43210

# Preface

Recent years have seen tremendous progress in unraveling the molecular basis of different plant–microbe interactions. Knowledge has accumulated on the mechanisms of the microbial infection of plants, which can lead to either disease or resistance. The mechanisms developed by plants to interact with microbes, whether viruses, bacteria, or fungi, involve events that can lead to symbiotic association or to disease or tumor formation. Cell death caused by pathogen infection has been of great interest for many years because of its association with plant resistance. There appear to be two types of plant cell death associated with pathogen infection, a rapid hypersensitive cell death localized at the site of infection during an incompatible interaction between a resistant plant and an avirulent pathogen, and a slow, normosensitive plant cell death that spreads beyond the site of infection during some compatible interactions involving a susceptible plant and a virulent, necrogenic pathogen. Plants possess a number of defense mechanisms against infection, such as (i) production of phytoalexin, (ii) formation of hydrolases, (iii) accumulation of hydroxyproline-rich glycoprotein and lignin deposition, (iv) production of pathogen-related proteins, (v) production of oligosaccharides, jasmonic acid, and various other phenolic substances, and (vi) production of toxin-metabolizing enzymes. Based on these observations, insertion of a single suitable gene in a particular plant has yielded promising results in imparting resistance against specific infection or disease. It appears that a signal received after microbe infection triggers different signal transduction pathways. Downstream signal transduction networks play a vital role in the recognition, reaction, and response to infection and in determining whether disease, no disease, or symbiotic association occurs, indicating the molecular complexities of the whole process. A complete understanding of the different modes of signal transduction and the genes involved in the process of symbiotic association, tumor formation, as well as susceptibility and resistance exhibited by plants after infection by microbes might help in developing a unifying concept underlying all these processes.

This volume is organized around four themes: (i) the emerging concept of symbiosis; (ii) disease and resistance of plants against viruses, bacteria, and fungi and the strategies of genetic engineering to contain the diseases; (iii) tumor induction by agrobacteria and the transfer of T-DNA to the nuclei of host cells; and (iv) the role of oligosaccharide elicitors in host–pathogen interactions and signal transduction. We asked the authors to highlight their own contributions along with emerging concepts and future research directions so that this volume might inspire further research on the issues it discusses.

In Chapter 1, D. P. S. Verma deals with the development and metabolic adaptations during symbiosis between legume hosts and rhizobia. He specifically considers the early host responses to rhizobia along with metabolic adaptation particularly for nitrogen assimilation and carbon metabolism in nodules. The key role of the induction of the *de novo* purine biosynthesis pathway in the events of symbiotic nitrogen fixation is discussed. Knowledge of rhizobial genetics has advanced to the point where it might now be possible to select host–*rhizobium* interactions for optimal symbiosis under a specific environmental condition.

In Chapter 2, E. Kamst, H. P. Spaink, and D. Kafetzopoulos elaborate on the biosynthesis and secretion of rhizobial lipochitin-oligosaccharide (LCO) signal molecules and their role in nodulation due to legume–rhizobia interaction. The formation of a root nodule involves attachment of bacteria to the root, invasion of the root by the bacteria, development of the root nodule, and differentiation of the bacteria into nitrogen-fixing bacteroids. Different aspects of LCO production by rhizobia through proteins coded by *nod* genes are discussed in detail. The molecular basis of LCO perception by plants is an interesting problem which is tackled by analyzing the functional domains of the Nod proteins synthesizing or modifying chitin-oligosaccharides. This will enable us to trace the morphogenetic aspects not only of plant development, but also those of other organisms in order to establish a unifying concept of the key role played by LCOs in developmental processes.

The role of microbial surface polysaccharides in *Rhizobium*–legume interactions is reviewed elegantly by K. Niehaus and A. Becker in Chapter 3. The establishment of the *Rhizobium*–legume symbiosis is controlled by a complex network of signal molecules that are produced by both the bacteria and the plant. Particular attention is drawn to the roles and functions of several surface carbohydrates bound to the outer membrane of rhizobial cells. Rhizobial mutants defective in the biosynthesis of exopolysaccharides (EPS), lipopolysaccharides (LPS), or capsular polysaccharides (CPS) have been found to be defective in the establishment of symbiosis. Rhizobial exopolysaccharides act as specific signals as well as suppressors of plant defense, enabling the symbiont to infect the plant.

The symbiosis between *Azorhizobium caulinodans* and *Sesbania rostrata* is an interesting beneficial plant–bacterium interaction. This is very lucidly described by S. Goormachtig, P. Mergaert, M. Van Montagu, and M. Holsters in Chapter 4. Besides the *nod* and *nif* genes of *Azorhizobium*, the participation of *Sesbania* genes is involved in the development of stem nodules. Some of the genes are homologous to chitinase III genes, extensin genes, and chalcone reductase genes. An intriguing observation is that stem nodulation is suppressed by root nodulation. Also of interest is the interaction of *Azorhizobium caulinodans* with rice root to fix nitrogen. In fact, more strains of *Azorhizobium* need to be monitored for such an interaction with cereal plants.

In Chapter 5, C. Guan, K. Pawlowski, and T. Bisseling treat the problems of interaction between *Frankia* and actinorhizal plants. *Frankia*, a genus of Actinomycetes, can induce nodules on the roots of a diverse group of dicotyledonous plants (collectively referred to as actinorhizal plants). *Frankia* can infect its hosts in two different ways, root hair infection or intracellular penetration. A comparative picture of legume–*Rhizobium* and *Alnus–Frankia* symbiosis is presented to clarify the more versatile system of actinorhizal symbiosis. Actinorhizal symbiosis emerges as a valuable system for understanding the diverse ways by which plants can establish appropriate environments for nitrogen fixation.

G. Strittmatter, K. Goethals, and M. Van Montagu review in Chapter 6 strategies for engineering plants resistant to bacterial and fungal diseases. To combat bacterial infection, the use of antibacterial enzymes and peptides, detoxification of bacterial toxins, and production of active oxygen and hydroxyl radicals inducing systemic acquired resistance (SAR) and pathogen-related protein (PR) are discussed. Similarly, in the case of fungal infection, in addition to the use of antifungal proteins, phytoalexins, and the oxidative burst, recent developments in the simulation of hypersensitive cell death in the hosts are examined. It is emphasized that molecular ecological studies on the manifold ways in which nature controls the complex interactions between plants and microorganisms have to be undertaken in order to understand these processes.

In Chapter 7, W. Knogge studies plant–fungus interactions and plant disease starting from the mode of fungal invasion, its pathogenicity, and its virulence to the resistance exhibited by host plants. He discusses the interaction of host-resistance and pathogen-virulence gene products and the subsequent signal transduction pathway leading to overall resistance. Finally, he addresses how new recognition specificities might evolve in plants. For example, is there any similarity of the process in plants to immunity in mammals? Answers to such questions may lead to the development of efficient strategies to improve disease resistance of crop plants, which is so urgently needed.

In Chapter 8, K. R. Davis describes *Arabidopsis thaliana* as a model system for studying plant–pathogen interactions. He discusses not only pathogenesis by bacteria, fungi, and viruses, but also nematodes, insects, and other pathogens of *Arabidopsis*, such as mycoplasma-like organisms (MLOs) and beet leaf hopper-transmitted virescence agent (BLTVA). The availability of several disease-resistant mutants of *A. thaliana* has aided the identification of components involved in the signal transduction pathway(s) leading ultimately to the resistance of such host plants.

In Chapter 9, C. A. Malpica, M. T. Cervera, C. Simoens, and M. Van Montagu discuss different strategies for engineering resistance against viral diseases in plants, including pathogen-derived, host-gene-derived, and imported strategies. This is a field where rDNA technology has been applied to achieve viral control through transgenesis. Recombinants only survive under appropriate selective pressures, and therefore it is believed that naturally mixed infections (occurring in nature) may provide better opportunities for RNA recombination than transgenic gene–virus genome recombination.

Phytotoxin production by *Pseudomonas syringae* is treated by C. L. Bender, D. A. Palmer, A. Peñaloza-Vázquez, V. Rangaswamy, and M. Ullrich in Chapter 10, which discusses the pathogenicity of *P. syringae* and the production of toxins in general and of coronatine in particular. With the exception of coronatine, whose gene is plasmid-borne, toxin biosynthetic gene clusters are generally chromosomally encoded. In several pathovars, toxin-producing gene clusters also contain a resistance gene to protect the strains from the biocidal effects of the toxin. The precise mode of action of coronatine is still unclear, though a receptor-mediated action may be speculated. Coronatine generally produces chlorosis and necrosis in sensitive plants. The chemical structure of coronatine consists of coronafacic acid (polyketide origin) and coronamic acid (ethylcyclopropyl amino acid). The possibility has been raised of the antifungal and antibacterial activities of polyketide derivatives. Since coronatine contains polyketide, it might be possible to deploy the specific genes for coronatine biosynthesis in other systems to develop new compounds with a wide range of activities.

*Agrobacterium tumefaciens* is the causative organism for crown gall tumor formation in most dicotyledonous plants. The necessary genes for the induction of tumor are encoded within the plasmid borne by the bacteria. The Vir region of the Ti plasmid is required for the transfer of foreign genes inserted between the right and left borders of T-DNA into the host cell. A. Das describes in Chapter 11 how DNA is transferred from *Agrobacterium* to plant cells, through the expression of the set of *vir* genes coupled with the processing of T-DNA and the formation of transport pores through which single-stranded DNA enters into a host cell and finally into the nucleus and is integrated into the genome. Interestingly, striking similarities are noted in the essential features among the *Agrobacterium irB* operon essential for T-DNA transfer to plants, the *tra* region required for conjugal transfer of plasmids between two *E. coli* strains, and the *ptl* operon essential for the secretion of pertussis toxin. The nature of the products transported by these systems suggests the existence of common macromolecule export machinery that different bacteria have adapted to carry out specialized functions.

In Chapter 12, W. Ream discusses the import of *Agrobacterium tumifaciens* virulence proteins and transfer of DNA into plant cell nuclei. This entails the stringent regulation of virulence genes as well as the protein export channel, origin of transfer, secretion of single-stranded-DNA-binding protein, and finally the entry of T-DNA into the plant cell. He also examines the different *vir* gene products and some of the homologous genes for symbiosis and nitrogen fixation in other systems to show the similarity of plant–bacteria interaction in general. Finally he raises the questions of what plant proteins are involved in T-DNA integration and how *A. tumefaciens* might be modified to permit efficient gene replacement in plants via homologous recombination.

In Chapter 13, F. Côté, K.-S. Ham, M. G. Hahn, and C. W. Bergmann discuss the generation of oligosaccharide elicitors in host–pathogen interactions. The perception by the plant cell in particular of the signal processing for the oligosaccharides apparently involved in plant–pathogen interactions through the putative receptors is elucidated, touching upon the questions of the biological relevance of oligosaccharides generated in plants. Further, they discuss (i) oligoglucoside, (ii) oligochitin and oligochitosan, and (iii) oligogalacturonide-signal transduction processes in plant defense responses and the regulation of release of oligosaccharides from the complex carbohydrate as well as the ultimate biological response.

These topics suggest that the invasion of hosts by microbes can elicit different reactions leading to symbiotic association of certain microbes with the hosts or to infection causing diseases or tumor formation. Thus symbiosis appears to resemble a refined pathogenic interaction in which the host selectively lowers its protective barriers to accommodate a controlled invasion by the microbes. In the event of disease the host protective barrier is lost and in tumor formation, microbes insert a segment of plasmid DNA into the host with the combinatorial assistance of several *vir* gene products as well as possible host factors. Initially the interaction of host and microbial factor(s) results in either a compatible or an incompatible situation. The molecular basis of these interactions is now emerging, pinpointing the involvement of different modes of signal transduction. The challenge in future is to elucidate how and to what extent these modes are convergent or divergent in situations of symbiosis, disease, or no disease. Perhaps with the emergence of new techniques and approaches the underlying principles of plant–microbe interaction will be elucidated fully in molecular terms. We hope this volume will help plant scientists in general and plant pathologists in particular in their future research.

H. K. Das
B. B. Biswas

*New Delhi and Calcutta, India*

# Contents

*Chapter 1*

**Developmental and Metabolic Adaptations during Symbiosis between Legume Hosts and Rhizobia**

Desh Pal S. Verma

*Chapter 2*

**Biosynthesis and Secretion of Rhizobial Lipochitin-Oligosaccharide Signal Molecules**

Eric Kamst, Herman P. Spaink, and Dimitris Kafetzopoulos



*Chapter 3*

**The Role of Microbial Surface Polysaccharides in the *Rhizobium*–Legume Interaction**

Karsten Niehaus and Anke Becker

*Chapter 4*

**The Symbiotic Interaction between *Azorhizobium caulinodans* and *Sesbania rostrata*: Molecular Cross-Talk in a Beneficial Plant–Bacterium Interaction**

Sofie Goormachtig, Peter Mergaert, Marc Van Montagu, and Marcelle Holsters

*Chapter 5*

**Interaction between *Frankia* and Actinorhizal Plants**

Changhui Guan, Katharina Pawlowski, and Ton Bisseling



*Chapter 6*

**Strategies to Engineer Plants Resistant to Bacterial and Fungal Diseases**

Günter Strittmatter, Koen Goethals, and Marc Van Montagu

*Chapter 7*

**Plant–Fungal Interactions and Plant Disease**

Wolfgang Knogge



*Chapter 8*

***Arabidopsis thaliana:* A Robust Model System for Studying Plant–Pathogen Interactions**

Keith R. Davis

*Chapter 9*

**Engineering Resistance against Viral Diseases in Plants**

Carlos A. Malpica, Maria Teresa Cervera, Chris Simoens, and Marc Van Montagu



*Chapter 10*

**Biosynthesis and Regulation of Coronatine, a Non-Host-Specific Phytotoxin Produced by *Pseudomonas syringae***

Carol L. Bender, David A. Palmer, Alejandro Peñaloza-Vázquez, Vidhya Rangaswamy, and Matthias Ullrich

*Chapter 11*

## DNA Transfer from *Agrobacterium* to Plant Cells in Crown Gall Tumor Disease

Anath Das



*Chapter 12*

## Import of *Agrobacterium tumefaciens* Virulence Proteins and Transferred DNA into Plant Cell Nuclei

Walt Ream

*Chapter 1*

# Developmental and Metabolic Adaptations during Symbiosis between Legume Hosts and Rhizobia

Desh Pal S. Verma

## 1. INTRODUCTION

Root nodule development is a unique process where bacteria reside intracellularly inside a host cell and contribute to the nitrogen requirement of the host plant. A molecular communication between the bacteria and host plants results in the organogenesis of nodules. The host roots exude flavonoids that act as chemotactic agents to attract the rhizobia to move toward the root surface and act as inducers for rhizobial *nod* genes. The *nod* genes encode enzymes responsible for biosynthesis and secretion of Nod factors (Carlson *et al.*, 1994). The latter are active in eliciting morphological changes in the host roots and initiate root nodule organogenesis. It has been suggested that such factors may exist in plants to regulate normal morphogenesis programs. The host induces a set of genes encoding the nodulin proteins that are required for organogenesis and function of the

**Desh Pal S. Verma** Department of Molecular Genetics and Plant Biotechnology Center, The Ohio State University, Columbus, Ohio 43210.

*Subcellular Biochemistry, Volume 29: Plant–Microbe Interactions*, edited by Biswas and Das. Plenum Press, New York, 1998.

root nodules. The metabolism of the host is adapted to meet the requirements of nitrogen fixation by bacteria and their assimilation by the host in root nodules.

## 2. EARLY HOST RESPONSES TO RHIZOBIA

Genetic studies have shown that synthesis of specific flavonoids and their secretion in the rhizosphere leads to the induction of the *Rhizobium nod* genes. Induction of *Rhizobium nod* genes is assayed by monitoring the level of the bacterial *LacZ* gene under the control of a *nod* gene promoter. Using this assay, it was first demonstrated that extracts of several legume plants are able to induce *nod* genes (Rossen *et al.*, 1985; Mulligan and Long, 1985). The first plant flavonoid compound identified to be active as inducer was luteolin from alfalfa seed extracts (Peters *et al.*, 1986). Flavonoids are often found as conjugates with sugars which may be released enzymatically by specific glycosidases (Hartwig and Phillips, 1991). It has been shown that different legume species release specific flavonoid molecules and an individual species can exude several different flavonoids. Soybean exudates contain at least two active isoflavones, daisein and genistein (Sadowsky *et al.*, 1988). Common bean roots exude at least three flavonoids, while bean seeds release several active flavonoids that act as inducers of the *nod* genes (Hungria *et al.*, 1991).

Nod factors produced by rhizobia in response to flavonoids are key to opening the door for the entry of rhizobia in their respective hosts (Relic *et al.*, 1994). They induce a rapid depolarization of membrane potential in the legume root hairs (Ehrhardt *et al.*, 1992). Additional signals are required from both host plant and rhizobia for a successful establishment of symbiosis (Verma, 1992). A body of evidence from genetic and biochemical studies on rhizobia has suggested that bacterial factors (BFs), such as diglycosyl diacylglycerol, and surface components of rhizobial cells may play important roles in the interaction between the two partners (see next section). Phytohormones, among the key regulators in plant development, are likely to act as mediators in regulating nodule organogenesis. In addition, small peptides and ethylene have been suggested to be involved in this process (van der Sande *et al.*, 1996; Heidstra *et al.*, 1997).

## 3. NODULATION SIGNALS

Over 40 nodulation (*nod*) genes have been identified in rhizobia (Schultze and Kondorosi, 1996). Since all the letters of the alphabet were used up in designating the first 26 members of this group of genes, additional nodulation genes have been designated as *nol* and *noe* genes. The *nod* genes are involved in the production of Nod factors and are clustered on a large plasmid called pSym in

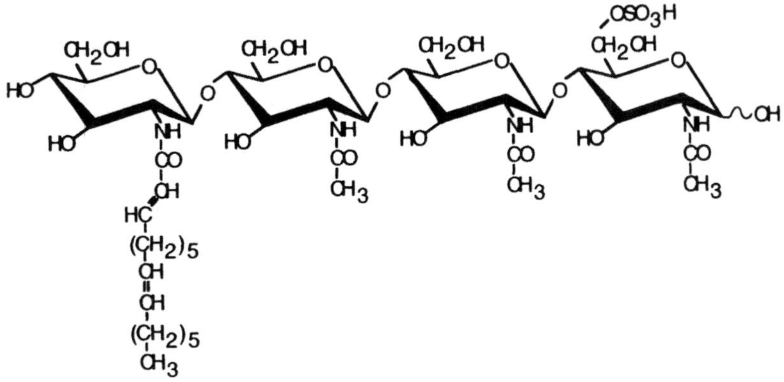


**FIGURE 1.** Structure of the Nod factor NodRm-1 from *Rhizobium meliloti*.

the fast-growing rhizobia, while in the slow-growing species such as *Bradyrhizobium japonicum* they are found on the chromosome.

Early studies indicated that the sterile supernatants of rhizobial cultures that had been treated by root extracts were able to cause changes in root or root hair morphology (Dénarié *et al.*, 1992). Using this bioassay, Nod factors were purified from a *Rhizobium meliloti* strain overexpressing *nod* genes (Lerouge *et al.*, 1990; Roche *et al.*, 1991). These compounds represent a new family of lipooligosaccharides with the basic structure being a tetra- or pentamer of β-1,4-linked *N*-acetylglucosamine residues (Fig. 1). On the terminal nonreducing sugar residue, the *N*-acetyl group is substituted with an *N*-fatty acyl (C16–18) chain. Some of the nod factors are also *O*-sulfated on the reducing end. An individual strain may produce a family of these factors with slightly different structures (Relic *et al.*, 1994). Several Nod factors have now been characterized from different rhizobial species including *Rhizobium, Bradyrhizobium,* and *Azorhizobium* spp.

### 3.1. Plant Flavonoids and the Synthesis of Nod Factors

While most of the *nod* genes are expressed only in the presence of a host-specific flavonoid, the *nodD* gene is constitutively expressed in free-living cells. The *nodD* gene encodes a transcriptional activator that, with the help of plant flavonoids, binds to the *nod* box, a highly conserved DNA sequence present in the promoter regions of the *nod* operons (Fisher *et al.*, 1988). This binding greatly promotes the expression of *nod* genes leading to the synthesis of the Nod factors. Although there is no direct evidence for an interaction of flavonoids with the NodD proteins, it is believed that they interact with the *Rhizobium* inner membrane where both NodD protein and plant flavonoids were found to accumulate (Schlaman *et al.*, 1989). This assumption has been supported by the identification of the C-terminal regions of the NodD proteins that are apparently responsible for flavonoid recognition and activation (Spaink *et al.*, 1989). Since most rhizobia possess multiple copies of *nodD* genes (Honma and Ausubel, 1987; Davis and Johnson, 1990), it is likely that different flavonoids bind to and activate a specific NodD protein. It has also been shown that some legumes secrete two types of phenolic compounds: one stimulates and the other represses *nod* gene expression, for example, in *Rhizobium trifolii* (Djordjevic *et al.*, 1987b; see also Peters and Verma, 1990).

Genetic and biochemical studies have led to the dissection of key steps in the synthesis of the Nod factors. The *nodABC* genes are responsible for the synthesis of the lipooligosaccharide backbone (Spaink *et al.*, 1991). The NodC expressed in *Escherichia coli* possesses the *N*-acetylglycosaminyl transferase activity and is responsible for the assembly of the chitooligosaccharide chain, the first step in biosynthesis of Nod factors (Geremia *et al.*, 1994). NodB is a chitooligosaccharide deacetylase that removes the *N*-acetyl group at the non-reducing end of chitooligosaccharides (John *et al.*, 1993). Finally, *N*-acylation on the deacetylated nonreducing end of chitooligosaccharides is carried out by NodA, which is an *N*-acyl-transferase (Rohrig *et al.*, 1994). The *nodIJ* genes encode an ATP-binding transport protein and a transmembrane protein, respectively (Evans and Downie, 1986; Schlaman *et al.*, 1990). Both proteins are located on the inner membrane and may form a transport complex responsible for the transport of Nod factors.

### 3.2. Host-Specific *nod* Genes and the Modification of Nod Factors

The host range is determined by the rhizobial host-specific *nod* genes, for examples, *nodFEG*, *nodH*, *nodP*, and *nodQ* in *R. meliloti* (Faucher *et al.*, 1989; Lerouge *et al.*, 1990; Schwedock and Long, 1990). Transfer of these genes from one *Rhizobium* species to another can change the host range (Cervantes *et al.*, 1989). Mutations in these genes do not totally block nodulation, but cause a delay in nodulation or a change, in some cases, in the host range. The host specificity of *R. meliloti* is primarily determined by the presence of a sulfate group on the reducing end and modification of the acyl chain on the other end of the Nod factor (Lerouge *et al.*, 1990; Spaink *et al.*, 1991).

Sulfation reaction in both prokaryotes and eukaryotes includes activation of sulfate, generating adenosine 5′-phosphosulfate (APS) and 3′-phosphoadenosine 5′-phosphosulfate (PAPS), and transfer of the activated sulfate group to a target. APS and PAPS are generated by ATP sulfurylase, which is composed of two subunits encoded by *cysN* and *cysD* in *E. coli*. NodP and NodQ have been shown to be homologs of *cysN* and *cysD* and possess ATP sulfurylase activity both *in vitro* and *in vivo* and are apparently involved in production of PAPS (Schwedock and Long, 1990). Transfer of the sulfate to Nod factor precursors is catalyzed by

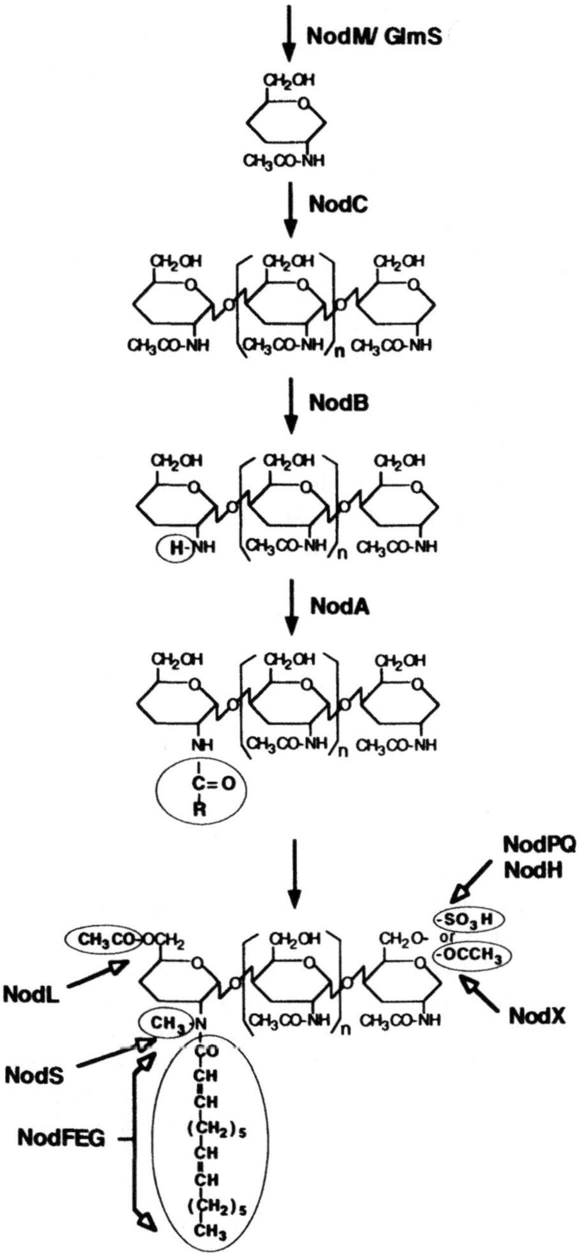


**FIGURE 2.** Synthesis of Nod factors in *Rhizobium*. Host specificity of Nod factors is often determined by appropriate modification of the chitooligosaccharide backbone (see text for detail).

sulfotransferase encoded by *nodH* (Roche *et al.*, 1991). Mutations in *nodH* or *nodQ* result in the production of an unsulfated Nod factor that can no longer induce nodule structure on the respective host (alfalfa) roots, but is still active on vetch (Roche *et al.*, 1991). Figure 2 shows the role of various *nod* genes in the synthesis of Nod factors.

Other *Rhizobium* genes, such as *nodX* from *R. leguminosarum* bv. *viciae* TOM strain, have homology to transacetylase and catalyze the addition of an acetyl group at O-6 position of the acetylglucosamine residue (Firmin *et al.*, 1993). This modification is required for nodulation on Afghanistan peas. The *nodZ* gene from *B. japonicum* encodes a fuctosyl transferase required for adding fucose at the reducing end of acetylglucosamine residue (Stacey *et al.*, 1994). NodS has homology to *S*-adenosylmethionine-dependent methyl transferase and appears to be involved in the addition of an *N*-methyl group at the nonreducing end (Geelen *et al.*, 1993). NodL shares homology with the *E. coli* acetyltransferase (Downie, 1989) and is able to add *O*-acetyl group *in vitro* to the chitin fragments and *N*-acetylglusamine (Bloemberg *et al.*, 1994). The deduced amino acid sequences of NodFEG proteins are homologous to those of acyl carrier proteins, β-ketoacyl synthase, and β-ketoacyl reductase, respectively. These proteins have been implicated in the synthesis and modification of the specific fatty acyl chains that are transferred by NodA to the nonreducing end of the oligosaccharide backbone (Horvath *et al.*, 1987; Bibb *et al.*, 1989). Double *nodF*/*nodL* mutants are unable to enter the host roots (Ardourel *et al.*, 1994).

Some host-specific *nod* genes appear to be important for nodulation on one host, but are dispensable for interaction with another host plant. The nodV and nodW genes are essential for nodulation on bean and cowpea, but contribute only marginally to the ability to nodulate soybean (Gottfert *et al.*, 1990). These genes constitute a two-component regulatory system that seems to sense the presence of host flavonoid compounds and activate the common *nod* genes (Sanjuan *et al.*, 1994). The *nodT* and *nodX* from *R. leguminosarum* and *nolYZ* from *B. japonicum* are required for efficient nodulation on a particular host (Dockendorff *et al.*, 1994). In *R. meliloti*, the *nodO* gene encodes a secreted protein that can form a $Ca^{2+}$-regulated ion channel in an artificial membrane (Sutton *et al.*, 1994). This host-specific gene has been implicated in the early recognition between rhizobia and the host. The *nodM* encodes an amidotransferase and is involved in the production of glucosamine, the precursor for the Nod factor backbone. The *R. meliloti nodM* gene can complement a glucosamine (*glmS*) auxotroph of *E. coli* (Baev *et al.*, 1991). There are two genes (*nodM* and *glmS*) coding for glucosamine synthases in *R. leguminosarum* (Marie *et al.*, 1992) which are functionally homologous; *glmS* is expressed constitutively and is required for normal growth, whereas *nodM* is induced during early plant–rhizobial interactions only.

## 4. OTHER RHIZOBIAL SIGNALS

With the exception of *Glycine soja*, in which Nod factors seem to induce complete nodule-like structures (Stokkermans and Peters, 1994), development of nodule-like tissues induced by Nod factors is limited. No infection threads are formed and no late-nodulin gene expression has been detected in these nodule-like structures. This suggests that other rhizobial factors may be required for nodule development. Among these factors, bacteria-produced phytohormones, bacterial factors (BF), and polysaccharides are important. In addition, specific proteins are known to be secreted by bacteria that may interact with the host. Moreover, physical presence of bacteria is essential for the development of complete nodule structure and for induction of most of the late nodulin genes.

### 4.1. Phytohormones

Rhizobia produce significant levels of plant hormones such as auxins, gibberellins, and cytokinins both in free-living conditions and in association with legumes. *B. japonicum* can synthesize indole-3-acetic acid (IAA) via an indoleacetamine intermediate, a pathway that is also used by *Agrobacterium tumefaciens*. IAA production by *R. meliloti* is stimulated by the addition of flavonoids (Prinsen *et al.*, 1991). Treatment of *R. meliloti* with flavonoids induces production of new cytokinins, suggesting that these phytohormones may be involved in the development of nodule structure (Sturtevant and Taller, 1989; Taller and Sturtevant, 1991). It has been shown that a *Bradyrhizobium elkanii* mutant strain with reduced IAA levels induces significantly fewer nodules on soybean roots and this deficiency can be overcome by the application of exogenous IAA (Fukuhara *et al.*, 1994). Overexpression of an *Agrobacterium trans*-zeatin secretion (*tzs*) gene in a *Rhizobium nod*⁻ mutant (Cooper and Long, 1994) was able to complement a *nod* gene mutation.

### 4.2. Bacterial Factors and Extracellular Polysaccharides

Rhizobia also produce a number of bacterial factors (BF) that are active on legume roots, causing root hair proliferation, branching or deformation, cortical cell divisions, or thick-short-root response. Of these factors, BF-5 has been identified as *N*-acetylglutamic acid (Philip-Hollingsworth *et al.*, 1991) and BF-7 as diglycosyl diacylglycerol, a membrane glycolipid (Orgambide *et al.*, 1994). The latter can induce root hair deformation and root cortical cell division at concentrations of $10^{-11}$ M. Production of these two factors is dependent on *nod* gene induction by flavonoids and their biological activity is host-specific.

Several rhizobial genes, *ndv*, *lps*, and *exo*, are involved in the formation of infection threads and affect the entry of rhizobia into plant cells. Mutations in these genes lead to the retention of bacteria in the infection threads or the formation of empty nodules devoid of bacteria, a phenotype termed nodule-development-defective (Ndv$^-$). These genes encode bacterial surface polysaccharides (Noel, 1992).

Exopolysaccharides (EPSs) are the dominant components of extracellular material. EPSs act as signals for eliciting plant response or act as suppressors of host pathogenic reactions (Halverson and Stacey, 1986). The *exo* mutants fail to synthesize acidic EPSs and induce small, white nodules (Finan *et al.*, 1985). Different *exo* mutants were isolated by selecting for nonfluorescent colonies in the presence of Calcofluor, which binds to polysaccharides. These mutants usually contain functioning *nod* genes and are able to induce root hair deformation and initiate infection thread formation (Klein *et al.*, 1988). However, the infection threads abort early, leading to the formation of empty nodules. The Ndv$^-$ phenotype in *R. leguminosarum exo* mutants can be suppressed by the addition of exogenous high-molecular-weight EPS (Djordjevic *et al.*, 1987c). Two types of EPS have been identified: EPS I, which binds Calcofluor, and EPS II, which does not bind this compound (Leigh *et al.*, 1985). A number of genetic loci related to EPS synthesis and secretion have been identified and some of these loci have been sequenced from *R. meliloti* (Leigh and Walker, 1994; Petrovics *et al.*, 1993). It is still not clear what role EPS plays during infection. The acidic EPSs may serve as a protective layer that blocks diffusion of bacterial "toxins" to the host plant cells or plant phytoalexins to the bacteria (O'Connell *et al.*, 1990) and prevent elicitation of host defense (Verma and Nadler, 1984).

Other components of the bacterial outer membrane, lipopolysaccharides (LPSs), are required for normal nodule development in most of the legume plants except alfalfa, on which *R. meliloti lps* mutants induce functioning nodules (Clover *et al.*, 1989). Defects in LPS synthesis in *R. leguminosarum* result in the induction of empty nodules in which rhizobia are not released from the infection threads (Priefer, 1989) or the infection threads abort at early stage (Noel *et al.*, 1986). A substantial variation in the structure of LPS appears to be tolerated for its role in nodule development. Moreover, the structure of LPS itself seems to change during nodule formation (Wood *et al.*, 1980) or in the presence of host flavonoids (Reuhs *et al.*, 1994). LPS does not seem to serve as a host-range determinant because replacing one type of the *O*-antigen in a bacterial strain by another type from a different rhizobial strain does not change the host specificity of the strain (Brink *et al.*, 1990).

Most *Rhizobium* species synthesize β-1,2-glucan, consisting of about 20 glucose residues, while *B. japonicum* also produces β-1,3-glucan and β-1,6-glucan. A class of rhizobial Ndv$^-$ mutants are β-1,2-glucan-deficient. Two genes, *ndvA* and *ndvB*, have been identified in *R. meliloti* to be involved in β-1,2-glucan synthesis and export. The former encodes a hemolysin homolog and may be involved in the ATP-dependent transport of β-1,2-glucan to the outer membrane (Stanfield *et al.*, 1988). The latter encodes a membrane protein that binds to UDP-glucose, an intermediate in β-1,2-glucan synthesis (Ielpi *et al.*, 1990). Mutations in *ndvA* and *ndvB* genes result in the formation of incompletely developed nodules that are largely devoid of bacteria (Dylan *et al.*, 1986).

## 5. HOST RESPONSE TO NOD FACTORS

Nodule-like structures often appear spontaneously in the absence of *Rhizobium* (Truchet *et al.*, 1989), indicating that the legume plants possess the essential components of nodule formation machinery. This nodule organogenesis program is triggered by bacterial infection (Verma, 1992). Induction of nodule-like structures on alfalfa roots by auxin-transport inhibitors (Hirsch *et al.*, 1989) suggests that the nodule-formation machinery can be activated by changing the hormone balance of the root cells. Adjustment of the levels of endogenous phytohormones may constitute the basic response of the host to rhizobial Nod factors. All five classes of phytohormones have been implicated in nodule development (Grobbelaar *et al.*, 1971). Treatment of legume roots with exogenous auxins and cytokinins induces cortical cell divisions (Libbenga *et al.*, 1973). Induction of cell-cycle-controlling genes occurs by the treatment with nod factors (D. P. S. Verma, unpublished data). Expression of *Agrobacterium rhizogenes rolABC* genes also changes hormone balance in transgenic plants. Plants expressing *nolABC* form nodules faster and produce more nodules than the control plants (Kondorosi *et al.*, 1993).

Legume roots also produce a significant amount of ethylene, and exogenous addition of ethylene strongly inhibits nodulation (Goodlass and Smith, 1979). Treatment with the ethylene inhibitor aminoethoxyvinylglycine (AVG) has been shown to stimulate nodule formation in alfalfa (Peters and Crist-Estes, 1989). AVG treatment can even overcome nitrate inhibition of nodulation, indicating that the inhibitory effect of nitrate is mediated through the phytohormone ethylene (Ligero *et al.*, 1991). The restoration of pea *sym5* mutant phenotype to wild type by AVG also provides evidence for a role of ethylene in nodulation (Fearn and LaRue, 1991).

The early nodulin genes *ENOD5* and *ENOD12* are shown to be induced in Nod factor-treated root hairs (Horvath *et al.*, 1993). Two *ENOD12* genes are differentially induced during early infection: *ENOD12b* expression is related to the action of Nod factors, whereas *ENOD12a* expression is associated with the invasion process in nodules (Bauer *et al.*, 1994). Studies on early nodulin gene expression have suggested that lithium, an inhibitor of phosphoinositide metabolism and an inducer of mitogenic activity in yeasts and animals, may also be involved in early host response to bacterial signals (T. Bisseling, personal communication). An intriguing observation is that the *enod40* gene may encode a peptide hormone able to induce mitogenic activity in the host plant (Vijn *et al.*,

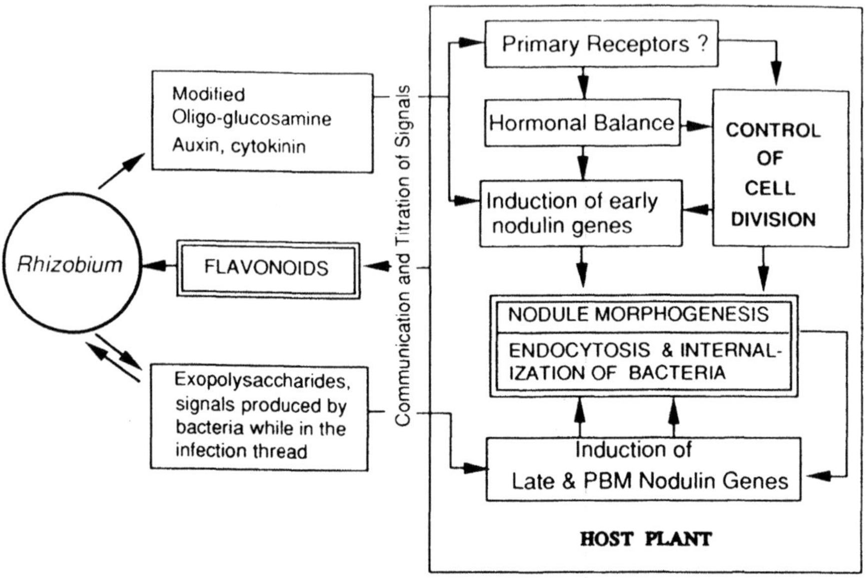


**FIGURE 3.** Communication between *Rhizobium* and legume roots during nodule organogenesis (see Verma, 1992, for details).

1995). Figure 3 summarizes host–rhizobia interactions leading to nodule organogenesis.

## 6. ENDOCYTOSIS OF RHIZOBIA AND SUBCELLULAR COMPARTMENTATION

Endocytosis of *Rhizobium* from the infection thread into the host cell is a poorly understood process (Verma, 1992). Several rhizobial or host signal molecules that mediate and/or control this process have been identified. Following endocytosis, the host plasma membrane that surrounds the infection thread becomes the peribacteroid membrane (PBM). PBM undergoes a significant change in composition during nodule development. A number of nodulins are targeted to the newly formed PBM (Fortin *et al.*, 1985, 1987; Katinakis and Verma, 1985; Cheon *et al.*, 1994; Miao *et al.*, 1992). The ATPase found in the PBM resembles that of the plasma membrane (Verma *et al.*, 1978; Day *et al.*, 1990). A vacuolar marker enzyme (α-mannosidase) and a vacuolar protease inhibitor have been found in the PBM compartment (Kinnback *et al.*, 1987; Manen *et al.*, 1991). This membrane, therefore, appears to be a mosaic membrane with properties common to both the plasma membrane and vacuolar membrane. PBM controls the transport of dicarboxylates (malate and succinate) that serve as carbon source for the bacteroids, and allows diffusion transport of oxygen, nitrogen, carbon dioxide, and ammonia (Day *et al.*, 1990). At the same time, it prevents entry of any substance from bacteria that may harm the plant cell. Failure to form PBM or any defect in this membrane may elicit pathogenic response in the host (see next section). Several *sym* mutants have been identified in pea that affect biogenesis of this membrane (Tikhonovich *et al.*, 1995).

## 7. AVOIDANCE OF HOST DEFENSE RESPONSES

The rhizobial symbiosis resembles a refined pathogenic interaction (Djordjevic *et al.*, 1987a) in which the legume host selectively lowers its protective barriers to accommodate a controlled invasion by *Rhizobium*. For example, soybean roots infected with compatible *B. japonicum* strain express a distinct subset of chalcone synthase and phenylalanine ammonia lyase enzymes which are different from the enzymes expressed as part of a defense response in pathogenic or ineffective symbiotic interactions (Estabrook and Sengupta-Gopalan, 1991). Also, a lower level of a phytoalexin, glyceollin I, is produced by soybean roots in symbiotic interactions compared to pathogenic interactions (Schmidt *et al.*, 1992). Furthermore, in ineffective symbiosis, the host plant can mount a vigorous chemical attack against the *Rhizobium* "pathogen" (Werner *et al.*, 1984, 1985), presumably mediated by the expression of genes that are normally silenced in effective symbioses. The lysis of bacteria in the vicinity of the host cell nucleus in bacteria-induced ineffective nodules (Werner *et al.*, 1985) resembles a conditional mutation in the host imposed by reducing the expression of *rab7* gene in soybean (Cheon *et al.*, 1993), indicating a gene-for-gene type of interaction observed in many plant–pathogen interactions.

It is not known how rhizobia succeed in suppressing the host defense response. The Nod factors that are evolutionarily related to chitin fragments and their chemical modifications, acylation, acetylation, sulfation, etc., of the basic oligosaccharide structure may help identify rhizobia as symbionts rather than pathogens (Fisher and Long, 1992). As discussed above, the rhizobial mutants with altered surface exo- and lipopolysaccharides (Noel, 1992) abort nodule development at different stages. It is known that the frequency of infection events that successfully result in nodule formation is extremely low, as many infection threads are aborted in the root hairs. The abortion of infection threads apparently occurs due to a hypersensitive response by the plant (Puhler *et al.*, 1991) and the production of phytoalexins. Accumulation of enzymes involved in phytoalexin production has been demonstrated histochemically (Vasse *et al.*, 1993).

Following endocytosis, rhizobia are compartmentalized within the peribacteroid membrane when they enter the infected cells from the infection threads.

Thus the bacteria are kept out of direct contact with the host-cell cytoplasm. Failure to form this membrane compartment or its premature disintegration, as seen with certain bacterial mutants, renders the association pathogenic and evokes a defense reaction from the host (Werner *et al.*, 1985).

## 8. EFFECTIVE VERSUS INEFFECTIVE NODULES AND CELLULAR ORGANIZATION

The nodules that follow normal development and achieve nitrogen-fixing activity are termed effective nodules. Those that fail to achieve symbiotic nitrogen fixation due to either host or bacterial defects are termed ineffective nodules. The ineffective nodules are generally white due to the lack of leghemoglobin, but some leghemoglobin-containing nodules are also ineffective. Cells in the central cortex contain large amounts of starch grains (Forrest *et al.*, 1990). In the ineffective indeterminate nodules, large amyloplasts are found, and the numbers of rough endoplasmic reticulum, Golgi bodies (Hirsch *et al.*, 1983), and mitochondria (Mackenzie and Jordon, 1974) are elevated, indicating the increased plant metabolic activity to meet the carbon and nitrogen demands of bacteria (Hirsch *et al.*, 1983).

Based on the types of nitrogen compounds produced following nitrogen fixation, legume plants are divided into two groups, ureide-transporters and amide-transporters. Tropical legumes such as soybean, cowpea, and bean primarily transport ureides, allantoin and allantoic acid, as fixed-nitrogen compounds, while temperate legumes such as *Lotus* and alfalfa are amide-transporters, producing L-glutamine and L-asparagine as fixed-nitrogen compounds. The ureide-producing nodules are usually determinate, but not all determinate nodules, such as those in *Lotus*, are ureide producers.

The pattern of development and organization of ureide-producing nodules is different from that in indeterminate nodules. The latter continue to grow and therefore contain dividing, nitrogen-fixing, and senescing zones of cells, while the determinate nodules stop cell division prior to commencement of nitrogen fixation. The central infected region of the nodule consists of two types of cells, those that are infected (bacteroid-containing) and those that remain uninfected. In determinate nodules, the metabolic flow of fixed nitrogen is well compartmentalized in the organelles contained in either infected or uninfected cells (Newcomb and Tandon, 1981; Shelp *et al.*, 1983). The infected cells are microaerobic, while the uninfected cells contain normal levels of oxygen (Hu, 1991). Although the infected cells occupy more space than the uninfected cells, the number of the latter cells is much higher than that of the infected cells. Both infected and uninfected cells are spatially organized in such a manner that the uninfected tissue produces a large surface area, making the transport of fixed nitrogen from the infected cells to the uninfected cells more efficient (Selker and Newcomb, 1985). In the infected cells of determinate nodules, two prominent organelles, mitochondria and plastids, congregate at the cell periphery, and the numbers of these organelles increase as the infected cells enlarge (Newcomb *et al.*, 1985). Small peroxisome-like structures are occasionally found in the periphery of the infected cell cytoplasm. Unlike in the infected cells, the plastids and mitochondria in the uninfected cells are dispersed in the cytoplasm (Newcomb *et al.*, 1985). Uninfected cells of soybean root nodules are vacuolated, but have enlarged peroxisomes as well as a prominent tubular endoplasmic reticulum (Newcomb and Tandon, 1981).

In the infected cells of indeterminate nodules, the central vacuole frequently persists in the host cells (Hirsch *et al.*, 1983), and proplastids, small amyloplasts, and the mitochondria are displaced to the cell periphery (Hirsch *et al.*, 1983). The rough endoplasmic reticulum and Golgi bodies are abundant in the infected cells at early stages, but the numbers of these organelles decrease as nodules mature (Verma *et al.*, 1978). Ineffective nodules mostly induced by mutant rhizobia (Hirsch *et al.*, 1983; VandenBosch *et al.*, 1985; Morrison and Verma, 1987; Grosskopf *et al.*, 1993) fail to fix nitrogen. Many host genes have also been identified that cause ineffectiveness of nodules. The host controls the type of nodule subcellular organization, and type of metabolism (i.e., amide vs. ureide transporting system for nitrogen assimilation) (Gresshoff and Delves, 1986; Egli *et al.*, 1989).

## 9. METABOLIC ADAPTATIONS

### 9.1. Adaptations in Nitrogen Assimilation

Symbiotically reduced nitrogen is transported as ammonia from the PBM and is assimilated in the host cell cytoplasm by glutamine sythetase (GS) and glutamate synthase. Meeks *et al.* (1978) demonstrated that $NH_4^+$ is first incorporated into the amide position of glutamine (GLN) by the GS-catalyzed reaction using [$^{13}$N]$N_2$. This incorporation is blocked by methionine sulfoximine, a GS inhibitor, and the amide group is subsequently transferred to the 2-carbon of oxoglutarate by the activity of glutamate synthase. Recently, comprehensive studies on the induction of GS have been performed using GS genes and showed that the basic nitrogen assimilation route is brought under an $NH_4$-induced control mechanism (Hirel *et al.*, 1987; Miao *et al.*, 1991). Hirel *et al.* (1987) showed that availability of ammonium ions enhanced the expression of GS in root tissue within 2 hr, reaching a level similar to that in nodules by 8 hr, while $KNO_3$ had no effect during this period. This was confirmed by isolating the GS gene promoter from soybean and fusing it with a reporter [β-glucuronidase (GUS)] gene (Miao *et al.*, 1991). The GS–GUS fusion was introduced into a legume (*Lotus corniculatus*) and a nonlegume (tobacco) plant by *Agrobacterium*-mediated transformations. This chimeric gene was expressed in a root-specific manner in both plants (Miao *et al.*, 1991). Treatment with exogenous ammonia increased the expression of this chimeric gene in the legume background, while no induction was observed in tobacco roots. These results suggested that the tissue specificity of the soybean cytosolic GS gene is conserved in both tobacco and legumes; however, this gene is directly regulated by externally provided (or symbiotically fixed) nitrogen (Miao *et al.*, 1991) only in the legume background.

In tropical legume root nodule, the GLN formed in the cytosol of infected cells is transported to the plastids, where it is used for *de novo* purine biosynthesis. GLN is one of the substrates catalyzing the first step of *de novo* purine biosynthesis. It is likely that a plastid-specific GLN transporter is operating in these nodules for the uptake of assimilated nitrogen from the cytosol. In soybean, the concentration of asparagine in nodules is increased during early nodule development (McClure and Israel, 1979; Schubert, 1981). The specific activity of aspargine synthetase (AS), which is located in plastids of tropical legume nodules (Boland *et al.*, 1982), has been shown to decline rapidly (Reynolds *et al.*, 1982) after an initial increase as nodules matured. Therefore, AS in tropical legumes plays a role only during early nodule development (Reynolds *et al.*, 1982). In amide-transporting legumes, the activity of AS increases with the onset of nitrogen fixation (Scott *et al.*, 1976).

### 9.1.1. Induction of *De Novo* Purine Biosynthesis Pathway

Production of ureides requires a large amount of purines, which are oxidatively catabolized to ureides (Atkins, 1981). Therefore, the rate of *de novo* purine biosynthesis in tropical legume nodules is highly induced (Kohl *et al.*, 1988). In all organisms thus far characterized, ten enzymatic reactions (Fig. 4) convert the activated ribose precursor phosphoribosylpyrophosphate (PRPP) to the purine nucleotide inosine 5′-monophosphate (IMP). The first step in this pathway, involving the synthesis of 5-phosphoribosyl-1-amine from PRPP and glutamine, is catalyzed by PRPP-amidotransferase (PRAT). End-product inhibition of this enzyme in bacteria (Messenger and Zalkin, 1979) as well as in plants (Reynolds *et al.*, 1984) plays a key role in regulating the metabolic flux of carbon and nitrogen through this pathway and in controlling the level of purine synthesis.

Recently, several plant genes involved in *de novo* purine biosynthesis have been cloned and characterized from plants (Senecoff and Meagher, 1993; Chapman *et al.*, 1994; Ito *et al.*, 1994; Schnorr *et al.*, 1994). Four enzymes of the pathway, PRAT, glycinamide ribonucleotide (GAR) synthetase, GAR transformylase, and adenylosuccinate-AMP lyase, have been partially purified from

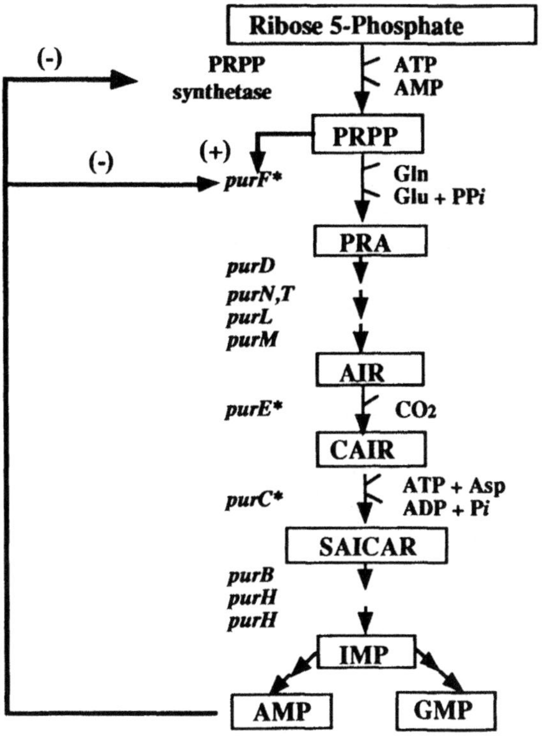


**FIGURE 4.** The *de novo* purine biosynthesis pathway and its regulation in root nodules. Enzymes of relevance are identified by their gene symbols, and the genes cloned from legume nodules are marked by asterisks; *purF*, phosphoribosylpyrophosphate (PRPP) amidotransferase; *purD*, glycineamide ribonucleotide (GAR) synthetase; *purN,T*, GAR transformylase; *purL*, formylglycinamidine ribonucleotide (FGAM) synthetase; *purM*, aminoimidazole ribonucleotide (AIR) synthetase; *purE*, AIR carboxylase; *purC*, phosphoribosylaminoimidazole carboxamide (SAICAR) synthetase; *purB*, adenylosuccinate lyase; *purH*, aminoimidazole carboxamide ribonucleotide (AICAR) transformylase/inosine monophosphate (IMP) cyclohydrolase; PRA, phosphoribosylamine; CAIR, carboxyaminoimidazole ribonucleotide; AMP, adenosine monophosphate; GMP, guanosine monophosphate.

soybean and cowpea root nodules (Reynolds *et al.*, 1984). In nodules, most of these enzymes are localized in the plastids of infected cells (Boland and Schubert, 1983; Shelp *et al.*, 1983; Atkins, 1991). The soybean PRAT cDNA encodes a protein with an N-terminal sequence resembling a plastid-targeting peptide, confirming the location of this enzyme in plastids. Downstream from this peptide is a sequence similar to the 11-amino acid propeptide found in the *Bacillus subtilis*, chicken, rat, and human PRAT proteins. The cowpea cDNA encodes the putative propeptide and efficiently complements purine auxotrophy in an *Escherichia coli purF* mutant impaired in PRAT synthesis.

The levels of PRAT mRNA in soybean and cowpea nodules were found to increase steadily as the nodule matures. PRAT gene is induced prior to the commencement of nitrogen fixation (Nguyen *et al.*, 1985; Kim *et al.*, 1995a,b), indicating a developmental control on the induction of this pathway, which is later used for assimilation of reduced nitrogen following the onset of nitrogen fixation. PRAT mRNA was not detectable in uninfected root tissue, but a low level of transcript was detected in leaves. Treatment of uninfected root with L-glutamine induced the PRAT mRNA transcript, suggesting that glutamine produced as a result of assimilation of fixed nitrogen induces *de novo* purine biosynthesis (Kim *et al.*, 1995b).

The soybean PRAT promoter was transcriptionally fused to GUS gene and introduced into both tobacco and soybean by *A. tumefaciens*- and *A. rhizogenes*-mediated transformations, respectively. The GUS expression was detected in both tobacco roots and soybean hairy roots. The transgenic roots from both plants provided with 10 mM L-glutamine showed higher GUS expression than control roots, indicating that glutamine, one of the substrates for PRAT, acts as an inducer of this gene (Kim and Verma, 1996). The results confirm our earlier studies (Kim *et al.*, 1995b), and demonstrate that ureide biosynthesis in determinate nodules is regulated by the availability of glutamine formed by symbiotically reduced nitrogen. Figure 5 shows that early and late induction of *de novo* purine biosynthesis pathways may be controlled by glutamine, first supplied by the host and later as a result of symbiotically fixed nitrogen.

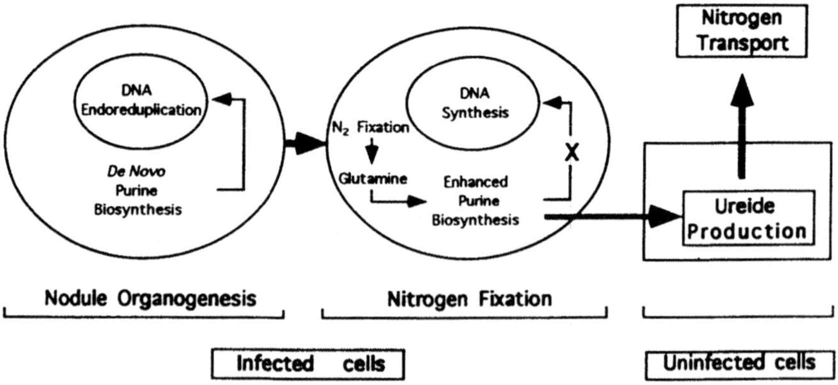


**FIGURE 5.** Purine biosynthesis during nodule organogenesis and the supply of glutamine first from the host and then from nitrogen fixation, which leads to enhanced induction of the *de novo* purine biosynthesis pathway for ureide production in tropical legume nodules.

### 9.1.2. Ureide Synthesis and Transport

Following *de novo* purine synthesis in the infected cells of tropical legume root nodules, IMP or xanthine are transported to the uninfected cells, where they are converted into ureides (Datta *et al.*, 1991). The compartmentation of the purine and ureide biosynthetic pathways in different organelles and between infected and uninfected cells is important for maintaining metabolite channeling as well as levels of key intermediates and effectors (Schubert and Boland, 1990). Oxygen seems to play an important role in the intercellular compartmentation of ureide biosynthetic enzymes. For example, PRAT is oxygen-sensitive and is located in the infected cells, where the oxygen level is low, while uricase requires oxygen and is located in the uninfected cells. Therefore, different steps of the ureide biosynthetic pathway are operating in different types of cells of nodules (Schubert, 1986; Verma, 1989). The fact that purines inhibit uricase activity (Hurst *et al.*, 1985) also indicates that these steps need to be compartmentalized. It has been speculated that plasma membrane is impermeable to nucleotides (Muller *et al.*, 1982), and therefore these nucleotides need to be converted to either nucleoside or free base to be translocated to the uninfected cells by an active transport system (Schubert and Boland, 1990). The IMP, final product of *de novo* purine biosynthesis, appears to be transported from plastids in root nodules (Schubert and Boland, 1990). Considering the location of xanthine dehydrogenase (XDH) in the uninfected cells, we can assume that xanthine monophosphate (XMP), xanthosine, or xanthine, converted from IMP, may also be transported from the infected to uninfected cells.

The purines are catabolized to uric acid by XDH in the uninfected cells. The uric acid is further oxidized to allantoin by uricase, present in peroxisomes. The soybean uricase (nodulin-35) gene has been isolated and characterized (Nguyen *et al.*, 1985). *Vigna* nodulin-35 cDNA has also been cloned. When introduced into the plant in the antisense orientation driven by CaMV-35S promoter (Lee *et al.*, 1993), plants exhibited a nitrogen deficiency phenotype (Lee *et al.*, 1993). Uricase activity in these nodules was reduced (50–60% reduction), and peroxisomes were reduced in size (Lee *et al.*, 1993). These results were similar to what was found after treatment with allopurinol (Atkins *et al.*, 1988), a uricase inhibitor, and indicate amides do not compensate for the nitrogen requirement of tropical legumes blocked in ureide biosynthesis (Lee *et al.*, 1993).

These results also indicate that leguminous plants operate two parallel nitrogen-assimilating systems. When legumes are growing on a nitrate-rich environment, they reduce the nitrate through nitrate reductase (NR) and utilize it as a major nitrogen source, while under a nitrogen-limiting environment, when root nodules are developed, they assimilate symbiotically reduced nitrogen via amide or ureide pathways. In this organ, bacteroid-excreted ammonia acts as an inducer of the GS gene, and is incorporated into glutamine. Glutamine then induces

PRAT gene expression catalyzing the first committed step of the *de novo* purine biosynthesis pathway. It is known that nitrate inhibits the synthesis of urides (McNeil land LaRue, 1984), and high levels of nitrate almost block ureide synthesis in soybean and cowpea and switch these legumes to amide metabolism (Schubert and Boland, 1990).

The regulation of GS and PRAT genes showed that the expression of these genes is tightly controlled by the availability of ammonia and glutamine, the nitrogen-fixation products, respectively. The root nodule is a well-adapted organ that not only meets the requirements of low oxygen for nitrogenase and the high demand for energy, but also has adapted well to assimilate symbiotically fixed nitrogen. Since nitrogen fixation requires large amounts of energy, this organ provides carbon flux to the bacteroids to meet the energy requirements.

### 9.2. Adaptations in Carbon Metabolism in Nodules

With the onset of nitrogen fixation, there is a rapid increase in the carbon sink activity (Streeter, 1980) as these compounds are delivered to the nodule to serve as substrates for bacteroid respiration as well as provide skeletons for assimilation of the fixed nitrogen. The primary photosynthate transported into the nodule is sucrose (Reibach and Streeter, 1983). Sucrose can be hydrolyzed by two enzymes, invertase and sucrose synthase, both of which are active in legume nodules. Sucrose syntnase is more important in sucrose hydrolysis since its activity, but not that of invertase, correlates with nitrogenase activity (Anthon and Emerich, 1990). A nodule-enhanced protein identified as nodulin-100 was shown to be a subunit of sucrose synthase (Thummler and Verma, 1987). The soybean nodule enzyme was recently shown to be concentrated mainly in the cytoplasm of uninfected cells in the central infected region of the nodules. The cellular localization of sucrose synthase and the apparent arrangement of uninfected cells in continuous files extending from the vascular bundles toward the center of the infected region (Gordon *et al.*, 1992) suggest that the uninfected cells may provide channels through which sucrose from the phloem is hydrolyzed and translocated into the central region of the nodule. This is consistent with the suggestion (Day and Copeland, 1991) that the production of $C_4$-dicarboxylic acids from sucrose occurs in the uninfected cells with the subsequent transfer of these organic acids to the infected cells. Furthermore, the infected cells isolated from soybean nodules readily take up malate, but are impermeable to sucrose (Li *et al.*, 1990).

Synthesis for dicarboxylic acids may be a major adaptation of nodule carbohydrate metabolism to the prevailing microaerobic conditions. It has been demonstrated using labeled sucrose that the major fate of sucrose transported to nodules is conversion to organic acids (Reibach and Streeter, 1983). Flow to malate and succinate under microaerobic and anaerobic conditions has been noted in a variety of plants (Vance and Heichel, 1991). Interestingly, the synthesis of $C_4$-dicarboxylic acids during anaerobic shift to reductive pathways increases in response to $NH_4^+$ in certain plants *Selenastrum minutum*.

$C_4$-dicarboxylic acids constitute the major carbon substrates used to fuel nitrogenase activity in the bacteroids. The bacteroids generally have low levels of glycolytic enzymes for the catabolism of sugars (Streeter, 1991), whereas malate is readily catabolized via bacteroidal malic enzyme and malate dehydrogenase (MDH) (Day and Copeland, 1991); consequently, mutants unable to utilize glucose and fructose still form effective nodules (Vance and Heichel, 1991; Streeter, 1991). Glycolysis may be an adaptation of host metabolism to the microaerobic environment of the nodule, and so for the preferential utilization of organic acids, rather than sugars, as respiratory substrates by the bacteroids, which represents a matching bacterial adaptation to the symbiotic state.

The high level of photosynthate requirement in the nodule raises the question of whether N fixation is limited by the supply of carbon. Vance and Heichel (1991) argued that N fixation is limited not by the delivery of photosynthate, but rather by carbon utilization within the nodule. That excess carbohydrate is delivered to the nodule is substantiated by the fact that starch accumulates at high levels (Forrest *et al.*, 1990) in nodule cells and is metabolized only when photosynthesis becomes limited. Similarly, the accumulation of large quantities of poly-β-hydroxybutyrate by bacteroids suggests that carbon substrates delivered to rhizobia exceed their respiratory capacity in a microaerobic environment (Day and Copeland, 1991). The major factor limiting carbon utilization in nodules is the low concentration of $O_2$ available, which restricts oxidative catabolism in both host mitochondria (Rawsthorne and LaRue, 1986; Suganuma *et al.*, 1987) and bacteroids (McDermott *et al.*, 1989). However, since a low $O_2$ environment is essential for the activity of the nitrogenase enzyme, the modifications to carbon utilization pathways operative in both the host cytoplasm and the bacteroids may be effective for improving the symbiotic state even if these pathways are energetically inefficient.

## 10. IMPROVEMENT OF SYMBIOTIC NITROGEN FIXATION

A variety of experimental evidence indicates that legume nodules do not generally have excess N-fixation capacity that can be activated by increasing carbon supply to the nodule. In that event, it may be feasible to alter the sink activity of the nodule by genetic manipulation of its sucrose hydrolytic activities (Sonnewald *et al.*, 1994). Although the functions of most nodulin genes remain to be ascertained, our knowledge of the hosts' contribution to symbiosis at the genetic and biochemical level has increased significantly over the last decade. In particular, a comprehensive picture of the key pathways of carbon and nitrogen metabolism in the nodule is now available. With the genes encoding several of the enzymes in these pathways having been cloned, it is now possible to contemplate genetic manipulation of the major metabolic pathways in the nodule with a view to increasing the efficiency of symbiotic nitrogen fixation. Moreover, as the genetics of *Rhizobium* has advanced, it is now possible to judicially select host/*Rhizobium* combinations for optimum symbiosis under a specific environmental condition. Thus, a three-way solution is needed for developing specific germ plasm and an inoculum that yields a high nitrogen-fixing soybean crop in a particular agroclimatic condition.

## 11. REFERENCES

Anthon, G., and Emerich, D., 1990, Developmental regulation of enzymes of sucrose and hexose metabolism in effective and ineffective soybean nodules, *Plant Physiol.* **92:**346–351.

Ardourel, M., Demont, N., Debelle, F., Maillet, F., deBilly, F., Prome, J.-C., Denarie, J., and Truchet, G., 1994, *Rhizobium meliloti* lipooligosaccharide nodulation factors: Different structural requirements for bacterial entry into target root hair cells and induction of plant symbiotic developmental responses, *Plant Cell* **6:**1357–1374.

Atkins, C. A., 1981, Metabolism of purine nucleotides to form ureides in nitrogen-fixing nodules of cowpea (*Vigna unguicilata L.* Walp), *FEBS Lett.* **125:**89–93.

Atkins, C. A., 1991, Ammonia assimilation and export of nitrogen from the legume nodule, in *Biology and Biochemistry of Nitrogen Fixation* (M. Dilworth and A. Glenn, eds.), pp. 293–319, Elsevier, Amsterdam.

Atkins, C. A., Sanford, P. J., Storer, P. J., and Pate, J. S., 1988, Inhibition of nodule functioning in cowpea by a xanthine oxido reductase inhibitor, allopurinol, *Plant Physiol.* **88:**1229–1234.

Baev, N., Endre, G., Petrovics, G., Banfalvi, Z., and Kondorosi, A., 1991, Six nodulation genes of *nod* box locus-4 in *Rhizobium meliloti* are involved in nodulation signal production—*nodM* codes for D-glucosamine synthetase, *Mol. Gen. Genet.* **228:**113–124.

Bauer, P., Crespi, M., Szecsi, J., Allison, L., Schultze, M., Ratet, P., Kondorosi, E., and Kondorosi, A., 1994, Alfalfa *Enod12* genes are differentially regulated during nodule development by Nod factors and *Rhizobium* invasion, *Plant Physiol.* **105:**585–592.

Bibb, M. J., Biro, S., Motamedi, H., Collins, J. F., and Hutchinson, C. R., 1989, Analysis of the nucleotide sequence of the *Streptomyces glaucescens tmcI* gene provides information about the enzymology of polyketide antibiotic biosynthesis, *EMBO J.* **8:**2727–2736.

Bloemberg, G. V., Thomas, O. J., Lugtenberg, B., and Spaink, H. P., 1994, Nodulation protein NodL of *Rhizobium leguminosarum O*-acetylates lipo-oligosaccharides, chitin fragments and *N*-acetylglucosamine *in vitro*, *Mol. Microbiol.* **11:**793–804.

Boland, M. J., and Schubert, K. R., 1983, Biosynthesis of purines by a proplastid fraction from soybean nodules, *Arch. Biochem. Biophys.* **220:**179–187.

Boland, M. J., Hanks, J. F., Reynolds, P. H. S., Belvins, D. G., Tolbert, N. E., and Schubert, K. R., 1982, Subcellular organization of ureide biogenesis from glycolytic intermediates and ammonium from nitrogen-fixing soybean nodules, *Planta* **155:**45–51.

Brink, B. A., Miller, J., Carlson, R. W., and Noel, K. D., 1990, Expression of *Rhizobium leguminosarum* CFN42 genes for lipo-polysaccharides in strains derived from different *R. leguminosarum* soil isolates, *J. Bacteriol.* **172:**548–555.

Carlson, R., Price, N., and Stacey, G., 1994, The biosynthesis of rhizobial lipooligosaccharide nodulation signal molecules, *Mol. Plant–Microbe Interact.* **7:**684–695.

Cervantes, E., Sharma, S. B., Maillet, F., Vasse, J., Truchet, G., and Rosenberg, C., 1989, The *Rhizobium meliloti* host range *nodQ* gene encodes a gene which share homology with translation elongation and initiation factors, *Mol. Microbiol.* **3:**745–755.

Chapman, K. A., Delauney, A. J., Kim, J. H., and Verma, D. P. S., 1994, Structural characterization of *de novo* purine biosynthesis enzymes in plants: 5-Aminoimidazole ribonucleotide carboxylase and 5-aminoimidazole-4-*N*-succinocarboxamide ribonucleotide synthetase cDNAs from *Vigna aconitifolia, Plant Mol. Biol.* **24:**389–395.

Cheon, C.-I., Lee, N.-G., Siddique, A.-B. M., Bal, A. K., and Verma, D. P. S., 1993, Roles of plant homologs of Rab1p and Rab7p in the biogenesis of the peribacteroid membrane, a subcellular compartment formed *de novo* during root nodule symbiosis, *EMBO J.* **12:**4125–4135.

Cheon, C., Hong, Z., and Verma, D. P. S., 1994, Nodulin-24 follows a novel pathway for integration into the peribacteroid membrane in soybean root nodules, *J. Biol. Chem.* **269:**6598–6602.

Clover, R. H., Kieber, J., and Signer, E. R., 1989, Lipopolysaccharide mutants of *Rhizobium meliloti* are not defective in symbiosis, *J. Bacteriol.* 171:3961–3967.

Cooper, J. B., and Long, S. R., 1994, Morphogenetic rescue of *Rhizobium meliloti* nodulation mutants by *trans*-zeatin secretion, *Plant Cell* **6:**215–225.

Datta, D. B., Triplett, E. W., and Newcomb, E. H., 1991, Localization of xanthine dehydrogenase in cowpea root nodules: Implications for the interaction between cellular compartments during ureide biogenesis, *Proc. Natl. Acad. Sci. USA* **88:**4700–4702.

Davis, E. O., and Johnston, A. W. B., 1990, Regulatory functions of the 3 *nodD* genes of *Rhizobium leguminosarum* biovar *phaseoli, Mol. Microbiol.* **4:**933–941.

Day, D., and Copeland L., 1991, Carbon metabolism and compartmentation in nitrogen-fixing legume nodules, *Plant Physiol. Biochem.* **29:**185–201.

Day, D. A., Yang, L.-J. O., and Udardi, M. K., 1990, Nutrient exchange across the peribacteroid membrane of isolated symbiosomes, in *Nitrogen Fixation: Achievements and Objectives* P. Gresshoff, J. Roth, G. Stacey, and W. E. Newton, eds., pp. 219–226, Chapman & Hall, New York.

Dénarié, J., Debellé, F., and Rosenberg, C., 1992, Signaling and host range variation in nodulation, *Annu. Rev. Microbiol.* **46:**497–531.

Djordjevic, M., Gabrielle, D., and Rolfe, B., 1987a, *Rhizobium*—the refined parasite of legumes, *Annu. Rev. Phytopathol.* **25:**145–168.

Djordjevic, M. A., Redmond, J. W., Batley, M., and Rolfe, B. G., 1987b, Clovers secrete specific phenolic compounds which either stimulate or repress *nod* gene expression in *Rhizobium trifolii, EMBO J.* **6:**1173–1179.

Djordjevic, S. P., Chen, H., Batley, M., Redmond, J. W., and Rolfe, B. G., 1987c, Nitrogen fixation ability of exopolysaccharide synthesis mutants of *Rhizobium* sp. strain NGR234 and *Rhizobium trifolii* is restored by addition of homologous exopolysaccharides, *J. Bacteriol.* **169:**53–60.

Dockendorff, T., Sharma, A., and Stacey, G., 1994, Identification and characterization of the *nolYZ* genes of *Bradyrhizobium japonicum, Mol. Plant Microbe Interact.* **7:**173–180.

Downie, J. A., 1989, The *nodL* gene from *Rhizobium leguminosarum* is homologous to the acetyl transferases encoded by *iacA* and *cysE, Mol. Microbiol.* **3:**1649–1651.

Dylan, T., Ielpi, L., Stanfield, S., Kashyap, L., Douglas, C., Yanofsky, M., Nester, E., Helinski, D. R., and Ditta, G., 1986, *Rhizobium meliloti* genes required for nodule development are related to chromosomal virulence genes in *Agrobacterium tumefaciens, Proc. Natl. Acad. Sci. USA* **83:**4403–4407.

Egli, M. A., Griffith, S. M., Miller, S. S., Anderson, M. A., and Vance, C. P., 1989, Nitrogen assimilating enzyme activities and enzyme protein during development and senescence of effective and plant gene-controlled ineffective alfalfa nodules, *Plant Physiol.* **91:**898–904.

Ehrhardt, D. W., Atkinson, E. M., and Long, S. R., 1992, Depolarization of alfalfa root hair membrane potential by *Rhizobium meliloti* Nod factors, *Science* **256:**998–1000.

Estabrook, E. M., and Sengupta-Gopalan, C., 1991, Differential expression of phenylalanine ammonia-lyase and chalcone synthase during soybean nodule development, Plant Cell **3:**299–308.

Evans, I. J., and Downie, J. A., 1986, The *nodI* gene product of *Rhizobium leguminosarum* is closely related to ATP-binding bacterial transport proteins: Nucleotide sequence analysis of the *nodI* and *nodJ* genes, *Gene* **43:**95–101.

Faucher, C., Camut, S., Dénarié, J., and Truchet, G., 1989, The *nodH* and *nodQ* host range genes of *Rhizobium meliloti* behave as avirulence genes in *R. leguminosarum* bv. *viciae* and determine changes in the production of plant-specific extracellular signals, *Mol. Plant–Microbe Interact.* **2:**291–300.

Fearn, J. C., and LaRue, T. A., 1991, A temperature-sensitive nodulation mutant (Sym-5) of *Pisum sativum* L. *Plant Cell Environ.* **14:**221–227.

Finan, T. M., Hirsch, A. M., Leigh, J. A., Johnson, E., Kuldau, G. A., Deegan, S., Walker, G. C., and Signer, E. R., 1985, Symbiotic mutants of *Rhizobium meliloti* that uncouple plant from bacterial differentiation, *Cell* **40:**869–877.

Firmin, J., Wilson, K., Carlson, R., Davies, A., and Downie, J., 1993, Resistance to nodulation of cv. Afghanistan peas is overcome by *nodX*, which mediates an *O*-acetylation of the *Rhizobium leguminosarum* lipo-oligosaccharide nodulation factor, *Mol. Microbiol.* **10:**351–360.

Fisher, R. F., and Long, S. R., 1992, *Rhizobium*–plant signal exchange, *Nature* **357:**655–660.

Fisher, R. F., Egelhoff, T. T., Mulligan, J. T., and Long, S. R., 1988, Specific binding of proteins from *Rhizobium meliloti* cell-free extracts containing NodD to DNA sequences upstream of inducible nodulation genes. *Genes Dev.* **2:**282–293.

Forrest, S. I., Verma, D. P. S., and Dhindsa, R. S., 1990, Starch content and activities of starch mobilizing enzymes in effective and ineffective root nodules of soybean, *Can. J. Bot.* **69:**697–701.

Fortin, M. G., Zelechowska, M., and Verma, D. P. S., 1985, Specific targeting of membrane nodulins to the bacteroid-enclosing compartment in soybean nodules, *EMBO J.* **4:**3041–3046.

Fortin, M., Morrison, N. A., and Verma, D. P. S., 1987, Nodulin-26, a peribacteroid membrane nodulin is expressed independently of the development of the peribacteroid compartment, *Nuc. Acids Res.* **15:**813–824.

Fukuhara, H., Minakawa, Y., Akao, S., and Minamisawa, K., 1994, The involvement of indole-3-acetic acid produced by *Bradyrhizobium elkanii* in nodule formation, *Plant Cell Physiol.* **35:**1261–1265.

Geelen, D., Mergaert, P., Geremia, R. A., Goormachtig, S., Van Montagu, M., and Holsters, M., 1993, Identification of *nodSUIJ* genes in *nod* locus 1 of *Azorhizobium caulinodans:* Evidence that *nodS* encodes a methyltransferase involved in Nod factor modification, *Mol. Microbiol.* **9:**145–154.

Geremia, R. A., Mergaert, P., Geelen, D., Van Montagu, M., Holsters, M., 1994, The NodC protein of *Azorhizobium caulinodans* is an *N*-acetylglucosaminyltransferase, *Proc. Natl. Acad. Sci. USA* **91:**2669–2673.

Goodlass, G., and Smith, K. A., 1979, Effect of ethylene on root extension and nodulation of pea (*Pisum sativum* L.) and white clover (*Trifolium repens* L.), *Plant Soil* **51:**387–395.

Gordon, A., Thomas, B., and Reynolds, P., 1992, Localization of sucrose synthase in soybean root nodules, *New Phytol.* **122:**35–44.

Gottfert, M., Hitz, S., and Hennecke, H., 1990, Identification of *nodS* and *nodU*, two inducible genes inserted between the *Bradyrhizobium japonicum nodYABC* and *nodIJ* genes, *Mol. Plant–Microbe Interact.* **3:**308–316.

Gresshoff, P. M., and Delves, A. C., 1986, Plant genetic control to symbiotic nodulation and nitrogen fixation in legumes, in *A Genetic Approach to Plant Biochemistry* (A. D. Blonstein and P. J. King, eds.), pp. 159–206, Springer-Verlag, New York.

Grobbelaar, N., Clarke, B., and Hough, M. C., 1971, The nodulation and nitrogen fixation of isolated roots of *Phaseolus vulgaris* L, *Plant Soil* (Special Issue) **1971:**215–223.

Grosskopf, E., Ha, D. T. C., Wingender, R., Rohrig, H., Szecsi, J., Kondorosi, E., Schell, J., and Kondorosi, A., 1993, Enhanced levels of chalcone synthase in alfalfa nodules induced by a Fix$^-$ mutant of *Rhizobium meliloti*, *Mol. Plant–Microbe Interact.* **6:**173–181.

Halverson, L. J., and Stacey, G., 1986, Signal exchange in plant–microbe interactions, *Microbiol. Rev.* **50:**193–225.

Hartwig, U. A., and Phillips, D. A., 1991, Release and modification of *nod* gene-inducing flavonoids from alfalfa seed, *Plant Physiol.* **95:**804–807.

Heidstra, R., Yang, W. C., Yalcin, Y., Peck, S., Emons, A. M., van Kammen, A., and Bisseling, T., 1997, Ethylene provides positional information on cortical cell division but is not involved in nod factor-induced root hair tip growth in Rhizobium-legume interaction, *Development* **124:** 1781–1787.

Hirel, B., Bouet, C., King, B., Layzell, D., Jacobs, F., and Verma, D. P. S., 1987, Glutamine synthetase genes are regulated by ammonia provided externally or by symbiotic nitrogen fixation, *EMBO J.* **6:**1167–1171.

Hirsch, A. M., Bang, M., and Ausubel, F. M., 1983, Ultrastructural analysis of ineffective nodules formed by *nif*::Tn5 mutants of *Rhizobium meliloti*, *J. Bacteriol.* **155:**367–380.

Hirsch, A., Bhuvaneswari, T., Torrey, J., and Bisseling, T., 1989, Early nodulin genes are induced in alfalfa root outgrowths elicited by auxin transport inhibitors, *Proc. Natl. Acad. Sci. USA* **86:**1244–1248.

Honma, M. A., and Ausubel, F. M., 1987, *Rhizobium meliloti* has three functional copies of the *nodD* symbiotic regulatory gene, *Proc. Natl. Acad. Sci. USA* **84:**8558–8562.

Horvath, B., Bachem, C. W. B., Schell, J., and Kondorosi, A., 1987, Host-specific regulation of nodulation genes in *Rhizobium* is mediated by a plant-signal, interacting with the *nodD* gene product, *EMBO J.* **6:**841–848.

Horvath, B., Heidstra, R., Lados, M., Moerman, M., Spaink, H. P., Promé, J. C., Van Kaman, A., and Bisseling, T., 1993, Lipo-oligosaccharides of *Rhizobium* induce infection-related early nodulin gene expression in pea root hairs, *Plant J.* **4:**727–733.

Hu, Chien-an, 1991, Aerobic status of nodule cells as revealed by ADH-GUS gene introduction in transgenic *Lotus corniculatus*, Master's thesis, Ohio State University, Columbus, Ohio.

Hungria, M., Joseph, C. M., and Phillips, D. A., 1991, Anthocyanidins and flavonols, major *nod* gene inducers from seeds of a black seeded common bean (*Phaseolus vulgaris* L), *Plant Physiol.* **97:**751–758.

Hurst, D. T., Griffiths, E., and Vayianos, C., 1985, Inhibition of uricase by pyrimidine and purine drugs, *Clin. Biochem.* **18:**247–251.

Ielpi, L., Dylan, T., Ditta, G. S., Helinski, D. R., and Stanfield, S. W., 1990, The *ndvB* locus of *Rhizobium meliloti* encodes a 319-kDa protein involved in the production of β-(1→2)-glucan, *J. Biol. Chem.* **265:**2843–2851.

Ito, T., Shiraishi, H., Okada, K., and Shimura, Y., 1994, Two amidophosphoribosyltransferase genes of *Arabidopsis thaliana* expressed in different organs, *Plant Mol. Biol.* **26:**529–533.

John, M., Rohrig, H., Schmidt, J., Wieneke, U., and Schell, J., 1993, *Rhizobium*, NodB protein involved in nodulation signal synthesis is a chitooligosaccharide deacetylase, *Proc. Natl. Acad. Sci. USA* **90:**625–629.

Katinakis, P., and Verma, D. P. S., 1985, Nodulin-24 gene of soybean codes for a peptide of the peribacteroid membrane and was generated by a tandem duplication of a sequence resembling an insertion element, *Proc. Natl. Acad. Sci. USA* **82:**4157–4161.

Kim, J. H., and Verma, D. P. S., 1996, Induction and characterization of glutamine phosphoribosylpyrophosphate amidotransferase gene promoter from soybean, *Plant Physiol. Suppl.*, in press.

Kim, J. H., Humphreys, J. M., and Verma, D. P. S., 1985a, Regulation of ureide biosynthesis genes in soybean root nodules, *Plant Physiol.* **108S:**72.

Kim, J. H., Delauney, A. J., and Verma, D. P. S., 1995b, Control of *de novo* purine biosynthesis genes in ureide-producing legumes: Induction of glutamine phosphoribosylpyrophosphate amidotransferase gene and characterization of its cDNA from soybean and *Vigna*, *Plant J.* **7:**77–86.

Kinnback, A., Mellor, R., and Werner, D., 1987, α-Mannosidase II isoenzyme in the peribacteroid space of *Glycine max* root nodules, *J. Exp. Bot.* **38:**1373–1377.

Klein, S., Walker, G. C., and Signer, E. R., 1988, All *nod* genes of *Rhizobium meliloti* are involved in alfalfa nodulation by *exo* mutants, *J. Bacteriol.* **170:**1003–1006.

Kohl, D. H., Schubert, K. R., Carter, M. B., Hagedorn, C. H., and Shearer, G., 1988, Proline metabolism in $N_2$-fixing root nodules: Energy transfer and regulation of purine synthesis, *Proc. Natl. Acad. Sci. USA* **85:**2036–2040.

Kondorosi, A., Schultze, M., Savoure, A., Hoffmann, B., Dudits, D., Pierre, M., Allison, L., Bauer, P., Kiss, G., and Kondorosi, A., 1993, Control of nodule induction and plant cell growth by Nod factors, in *Advances in Molecular Genetics of Plant–Microbe Interactions* (E. W. Nester and D. P. S. Verma, eds.), pp. 143–150, Kluwer, Dordrecht.

Lee, N. G., Stein, B., Suzuki, H., and Verma, D. P. S., 1993, Antisense expression of nodulin-35 RNA in *Vigna aconitifolia* roof nodules retards peroxisome development and the availability of nitrogen to the plant, *Plant J.* **3:**599–606.

Leigh, J. A., and Walker, G. C., 1994, Exopolysaccharides of *Rhizobium:* Synthesis, regulation and symbiotic function, *Trends Genet.* **10:**63–67.

Leigh, J. A., Signer, E. R., and Walker, G. C., 1985, Exopolysaccharide-deficient mutants of *Rhizobium meliloti* that form ineffective nodules, *Proc. Natl. Acad. Sci. USA* **82:**6231–6235.

Lerouge, P., Roche, P., Faucher, C., Maillet, F., Truchet, G., Promé, J. C., and Dénarié, J., 1990, Symbiotic host-specificity of *Rhizobium meliloti* is determined by a sulphated and acylated glucosamine oligosaccharide signal, *Nature* **344:**781–784.

Li, Y., Ou Yang, L.-J., Quinnel, R. G., Udvardi, M. K., and Day, D. A., 1990, Malate transport and metabolism in infected cells of soybean nodules, in *Nitrogen Fixation: Achievements and Objectives*, P. M. Gresshoff, L. E. Roth, G. Stacey, and W. E. Newton, eds.), p. 753, Chapman and Hall, New York.

Libbenga, K. R., Van Iren, F., Bogers, R. J., and Schraag-Lamers, M. F., 1973, The role of hormones and gradients in the initiation of cortex proliferation and nodule formation in *Pisum sativum* L, *Planta* **114:**29–39.

Ligero, F., Caba, J. M., Lluch, C., and Olivares, J., 1991, Nitrate inhibition of nodulation can be overcome by the ethylene inhibitor aminoethoxyvinylglycine, *Plant Physiol.* **97:**1221–1225.

Mackenzie, C. R., and Jordon, D. C., 1974, Ultrastructure of root nodules formed by ineffective strains of *Rhizobium meliloti*, *Can. J. Microbiol.* **20:**755–758.

Manen, J. F., Simon, P., Vanslooten, J. C., Osteras, M., Frutiger, S., and Hughes, G. J., 1991, A nodulin specifically expressed in senescent nodules of winged bean is a protease inhibitor, *Plant Cell* **3:**259–270.

Marie, C., Barny, M. A., and Downie, J. A., 1992, *Rhizobium leguminosarum* has two glucosamine synthases, *glmS* and *nodM*, required for nodulation and development of nitrogen-fixing nodules, *Mol. Microbiol.* **6:**843–851.

McClure, P. R., and Israel, D. W., 1979, Transport of nitrogen in the xylem of soybean plants, *Plant Physiol.* **64:**411–416.

McDermott, T., Griffith, S., Vance, C., and Graham, P., 1989, Carbon metabolism in *Bradyrhizobium japonicum* bacteroids, *FEBS Microbiol. Rev.* **63:**327–340.

McNeil, D. L., and LaRue, T. A., 1984, Effect of nitrogen source on ureides in soybean, *Plant Physiol.* **74:**227–232.

Meeks, J. C., Wolk, C. P., Schilling, N., Shaffer, P. W., Avissar, Y., and Chien, W. S., 1978, Initial organic products of fixation of [$^{13}$N]dinitrogen by root nodules of soybean (*Glycine max*), *Plant Physiol.* **61:**980–983.

Messenger, L. J., and Zalkin, H., 1979, Glutamine phosphoribosylpyrophosphate amidotransferase from *Escherichia coli*, *J. Biol. Chem.* **254:**3382–3392.

Miao, G. H., Hirel, B., Marsolier, M. C., Ridge, R. W., and Verma, D. P. S., 1991, Ammonia-regulated expression of a soybean gene encoding cytosolic glutamine synthetase in transgenic *Lotus corniculatus*, *Plant Cell* **3:**11–22.

Miao, G. H., Hong, Z., and Verma, D. P. S., 1992, Topology and phosphorylation of soybean nodulin-26, an intrinsic protein of the peribacteroid membrane, *J. Cell Biol.* **118:**481–490.

Morrison, N., and Verma, D. P. S., 1987, A block in the endocytosis of *Rhizobium* allows cellular differentiation in nodules but affects the expression of some peribacteroid membrane nodulins, *Plant Mol. Biol.* **9:**185–196.

Muller, M., Kraupp, M., Chiba, P., and Rumpold, H., 1982, Regulation of purine uptake in normal and neoplastic cells, *Adv. Enzyme Regul.* **21:**239–256.

Mulligan, J., and Long, S., 1985, Induction of *Rhizobium meliloti nodC* expression by plant exudate requires *nodD*, *Proc. Natl. Acad. Sci. USA* **82:**1–5.

Newcomb, E. H., and Tandon, S. R., 1981, Uninfected cells of soybean root nodules: Ultrastructure suggests key role in ureide production, *Science* **212:**1394–1396.

Newcomb, E. H., Tandon, Sh. R., and Kowal, R. R., 1985, Ultrastructural specialization for ureide production in uninfected cells of soybean root nodules, *Protoplasma* **125:**1–12.

Nguyen, T., Zelechowska, M., Foster, V., Bergmann, H., and Verma, D. P. S., 1985, Primary structure of the soybean nodulin-35 gene encoding uricase II localized in the peroxisomes of uninfected cells of nodules, *Proc. Natl. Acad. Sci. USA* **82:**5040–5044.

Noel, K. D., 1992, Rhizobial polysaccharides required in symbioses with legumes, in *Molecular Signals in Plant–Microbe Communications* (D. P. S. Verma, ed.), pp. 341–357, CRC Press, Boca Raton, Florida.

Noel, K. D., Vandenbosch, K. A., and Kalpaca, B., 1986, Mutations in *Rhizobium phaseoli* that lead to arrested development of infection threads, *J. Bacteriol.* **168:**1392–1401.

O'Connell, D. P., Araujo, R. S., and Handelsman, J., 1990, Exopolysaccharide-deficient mutants of *Rhizobium* spp strain CIAT899 induce chlorosis in common bean (*Phaseolus vulgaris*), *Mol. Plant–Microbe Interact.* **3:**424–428.

Orgambide, G. G., Philip-Hollingsworth, S., Hollingsworth, R. I., and Dazzo, F. B., 1994, Flavone-enhanced accumulation and symbiosis-related biological activity of a diglycosyl diacylglycerol membrane glycolipid from *Rhizobium leguminosarum* biovar *trifolii*, *J. Bacteriol.* **176:**4338–4347.

Peters, N. K., and Crist-Estes, D. K., 1989, Nodule formation is stimulated by the ethylene inhibitor aminoethoxyvinylglycine, *Plant Physiol.* **91:**690–693.

Peters, N., and Verma, D., 1990, Phenolic compounds as regulators of gene expression in plant–microbe interaction, *Mol. Plant–Microbe Interact.* **3:**4–8.

Petrovics, G., Putnoky, P., Reuhs, B., Kim, J., Thorp, T. A., Noel, K. D., Carlson, R. W., and Kondorosi, A., 1993, The presence of a novel type of surface polysaccharide in *Rhizobium meliloti* requires a new fatty acid synthase-like gene cluster involved in symbiotic nodule development, *Mol. Microbiol.* **8:**1083–1094.

Philip-Hollingsworth, S., Hollingsworth, R. I., and Dazzo, F. B., 1991, *N*-Acetylglutamic acid: An extracellular nod signal of *Rhizobium trifolii* ANU843 that induces root hair branching and nodule-like primordia in white clover roots, *J. Biol. Chem.* **266:**16854–16858.

Priefer, U. B., 1989, Genes involved in lipopolysaccharide production and symbiosis are clustered on the chromosome of *Rhizobium leguminosarum* biovar *viciae* Vf39, *J. Bacteriol.* **171:**6161–6168.

Prinsen, E., Chauvaux, N., Schmidt, J., John, M., Wieneke, U., Degreef, J., Schell, J., and Vanonckelen, H., 1991, Stimulation of indole-3-acetic acid production in *Rhizobium* by flavonoids, *Ind. J. Exp. Biol.* **29:**184–186.

Puhler, A., Arnold, W., Buendia-Claveria, A., Kapp, D., Keller, M., Niehaus, L., Quant, J., Roxlau, A., and Weng, W., 1991, The role of *Rhizobium meliloti* exopolysaccharides EPSI and EPSII in the infection process of alfalfa nodules, in *Advances in Molecular Genetics of Plant-Microbe Interactions* (H. Hennecke and D. P. S. Verma, eds.), pp. 189–194, Kluwer, Dordrecht.

Rawsthorne, S., and LaRue, T., 1986, Preparation and properties of mitochondria from cowpea nodules, *Plant Physiol.* **81:**1092–1096.

Reibach, P., and Streeter, J., 1983, Metabolism of $^{14}C$-labeled photosynthate and distribution of enzymes of glucose metabolism in soybean nodules, *Plant Physiol.* **72:**634–640.

Relic, B., Perret, X., Estrada-Garcia, M. T., Kopcinska, J., Golinowski, W., Krishnan, H. B., Pueppke, S. G., and Broughton, W. J., 1994, Nod-factors of *Rhizobium* are a key to the legume door, *Mol. Microbiol.* **13:**171–178.

Reuhs, B., Kim, J., Badgett, A., and Carlson, R., 1994, Production of cell-associated polysaccharides of *Rhizobium fresii* USDA205 is modulated by apigenin and host root extract. *Mol. Plant–Microbe Interact.* **7:**240–247.

Reynolds, P. H. S., Boland, M. J., Blevins, D. G., Schubert, K. P., and Randall, D. D., 1982, Enzymes of amide and ureide biogenesis in developing soybean nodules, *Plant Physiol.* **69:**1334–1338.

Reynolds, P. H. S., Blevins, D. G., and Randall, D. D., 1984, 5-Phosphoribosylpyrophosphate amidotransferase from soybean root nodules: Kinetic and regulatory properties, *Arch. Biochem. Biophys.* **229:**623–631.

Roche, P., Debelle, F., Maillet, F., Lerouge, P., Faucher, C., Truchet, G., Dénarié, J., and Promé, J. C., 1991, Molecular basis of symbiotic host specificity in *Rhizobium meliloti—nodH*and *nodPQ* genes encode the sulfation of lipo-oligosaccharide signals, *Cell* **67:**1131–1143.

Rohrig, H., Schmidt, J., Wieneke, U., Kondorosi, E., Barlier, I., Schell, J., and John, M., 1994, Biosynthesis of lipooligosaccharide nodulation factors: *Rhizobium* NodA protein is involved in *N*-acylation of the chitooligosaccharide backbone, *Proc. Natl. Acad. Sci. USA* **91:**3122–3126.

Rossen, L., Shearman, C., Johnston, A., and Downie, J., 1985, The *nodD* gene of *Rhizobium leguminosarum* is autoregulatory and in the presence of plant exudate induces the *nodABC* genes, *EMBO J.* **4:**3369–3373.

Sadowsky, M., Olson, E., Foster, V., Kosslak, R., and Verma, D. P. S., 1988, Two host-inducible genes of *Rhizobium fredii* and characterization of the inducing compound, *J. Bacteriol.* **170:**171–178.

Sanjuan, J., Grob, P., Göttfert, M., Hennecke, H., and Stacey, G., 1994, NodW is essential for full expression of the common nodulation genes in *Bradyrhizobium japonicum*, *Mol. Plant–Microbe Interact.* **7:**364–369.

Schlaman, H. R., Spaink, H. P., Okker, R. J., and Lugtenberg, B. J., 1989, Subcellular localization of the *nodD* gene product in *Rhizobium leguminosarum*, *J. Bacteriol.* **171:**4686–4693.

Schlaman, H. R. M., Okker, R. J. H., and Lugtenberg, B. J. J., 1990, Subcellular localization of the *Rhizobium leguminosarum nodI* gene product, *J. Bacteriol.* **172:**5486–5489.

Schmidt, P. E., Parniske, M., and Werner, D., 1992, Production of the phytoalexin glyceollin-I by soybean roots in response to symbiotic and pathogenic infection, *Botan. Acta* **105:**18–25.

Schnorr, K. M., Nygaard, P., and Laloue, M., 1994, Molecular characterization of *Arabidopsis thaliana* cDNAs encoding three purine biosynthetic enzymes, *Plant J.* **6:**113–121.

Schubert, K. R., 1981, Enzymes of purine biosynthesis and catabolism in *Glycine max*, *Plant Physiol.* **68:**1115–1122.

Schubert, K. R., 1986, Products of biological nitrogen fixation in higher plants: Synthesis, transport, and metabolism, *Annu. Rev. Plant Physiol.* **37**:539–574.

Schubert, K. R., and Boland, M. J., 1990, The ureides, in *The Biochemistry of Plants*, Vol. 16 (B. J. Miflin and P. J. Lea, eds.), pp. 197–281, Academic Press, New York.

Schultze, M., and Kondorosi, A., 1996, The role of lipochitooligosaccarides in root nodule organogenesis and plant cell growth, *Current Opin. Genet. Dev.* **6**:631–638.

Schwedock, J., and Long, S. R., 1990, ATP sulphurylase activity of the *nodP* and *nodQ* gene products of *Rhizobium meliloti*, *Nature* **348**;644–647.

Scott, D. B., Farnden, K. J. F., and Robertson, J. G., 1976, Ammonia assimilation in lupin nodules, *Nature* **263**:703–705.

Selker, J. M. L., and Newcomb, E. H., 1985, Spatial relationships between uninfected and infected cells in root nodules of soybean, *Planta* **165**:446–454.

Senecoff, J. F., and Meagher, R. B., 1993, Isolating the *Arabidopsis Thaliana* genes for *de novo* purine synthesis by suppression of *Escherichia coli* mutants, *Plant Physiol.* **102**:387–399.

Shelp, B. J., Atkins, C. A., Storer, P. J., and Canvin, D. T., 1983, Cellular and subcellular organization of pathways of ammonia assimilation and ureide synthesis in nodules of cowpea [*Vigna unguiculata* (L.) Walp.] *Arch. Biochem. Biophysiol.* **224**:429–441.

Sonnewald, U., Lerchl, J., Zrenner, R., and Frommer, W., 1994, Manipulation of sinksource relations in transgenic plants, *Plant Cell Environ.* **17**:649–658.

Spaink, H. P., Wijffelman, C. A., Pees, E., Okker, R. J. H., and Lugtenberg, B. J. J., 1989, Localization of functional regions of the *Rhizobium nodD* product using hybrid *nodD* genes, *Plant Mol. Biol.* **12**:59–73.

Spaink, H. P., Sheeley, D. M., vanBrussel, A. A. N., Glushka, J., York, V. N., and Lugtenberg, B. J. J., 1991, A novel highly unsaturated fatty acid moiety of lipo-oligosaccharide signals determines host specificity of *Rhizobium*, *Nature* **354**:125–129.

Stacey, G., Luka, S., Sanjuan, J., Banfalvi, Z., Niewkoop, A. J., Chun, J. Y., Forsberg, L. S., and Carlson, R., 1994, *nodZ*, a unique host-specific nodulation gene, is involved in the fucosylation of the lipooligosaccharide nodulation signal of *Bradyrhizobium japonicum*, *J. Bacteriol.* **176**:620–633.

Stanfield, S. W., Ielpi, L., O'Brochta, D., Helinski, D. R., and Ditta, G. S., 1988, The *ndvA* gene product of *Rhizobium meliloti* is required for β-(1→2) glucan production and has homology to the ATP-binding export protein HlyB, *J. Bacteriol.* **170**:3523–3530.

Stokkermans, T. J. W., and Peters, N. K., 1994, *Bradyrhizobium elkanii* lipooligosaccharide signals induce complete nodule structures on *Glycine soja*, *Planta* **193**:413–420.

Streeter, J. G., 1980, Carbohydrates in soybean nodules. II Distribution of compounds in seedlings during the onset of nitrogen fixation, *Plant Physiol.* **66**:471–476.

Streeter, J. G., 1991, Transport and metabolism of carbon and nitrogen in legume nodules, *Adv. Bot. Research* **18**:129–187.

Sturtevant, D., and Taller, B., 1989, Cytokinin production by *Bradyrhizobium japonicum*, *Plant Physiol.* **89**:1247 1252.

Suganuma, N., Kitou, M., and Yamaoto, Y., 1987, Carbon metabolism in relation to cellular organization of soybean root nodules and respiration of mitochondria aided by leghemoglobin, *Plant Cell Physiol.* **28**:113–122.

Sutton, J., Lea, E., and Downie, J., 1994, The nodulation-signaling protein NodO from *Rhizobium leguminosarum* biovar *viciae* forms ion channels in membranes, *Proc. Natl. Acad. Sci. USA* **91**:9990–9994.

Taller, B. J., and Sturtevant, D. B., 1991, Cytokinin production by rhizobia, in *Advances in Molecular Genetics of Plant–Microbe Interactions* (H. Hennecke and D. P. S. Verma, eds.), pp. 215–221, Kluwer, Dordrecht.

Thummler, F., and Verma, D., 1987, Nodulin-100 of soybean is the subunit of sucrose synthase regulated by the availability of free heme in nodules, *J. Biol. Chem.* **262**:14730–14736.

Tikhonovich, I. A., Borisov, A. Y., Chvabauskene, I. A., Kamardin, N. N., Kravchenko, L. V., Lebsky, V. K., Minchin, F., Morzhina, E. V., Romanov, V. I., Skot, L., Tchetokova, S. A., and Tsyganov, V. E., 1995, Plant genetical control at the crucial steps of symbiosis, *Nitrogen Fixation: Fundamentals and Applications* (I. A. Tikhonovich *et al.*, eds.), pp. 461–466, Dordrecht, The Netherlands.

Truchet, G., Barker, D. G., Camut, S., Debilly, F., Vasse, J., and Huguet, T., 1989, Alfalfa nodulation in the absence of *Rhizobium*, *Mol. Gen. Genet.* **219:**65–68.

Vance, C., and Heichel, G., 1991, Carbon in $N_2$ fixation limitations or exquisite adaptation, *Annu. Rev. Plant Physiol. Plant Mol. Biol.* **42:**373–392.

van de Sande, K., Pawlowski, K., Czaja, I., Weineke, U., Schell, J., Schmidt, R., Walden, R., Matvienko, M., Wellink, J., van Kammen, A., Franssen, H., and Bisseling, T., 1996, Modification of phytohormone response by a peptide encoded by ENOD40 of legumes and a nonlegume, *Science* **273:**370–373.

Vasse, J., de Billy, F., Camut, S., and Truchet, G., 1993, Correlation between ultrastructural differentiation of bacteroids and nitrogen fixation in alfalfa nodules, *J. Bacteriol.* **172:**4296–4306.

Verma, D. P. S., 1989, Plant genes involved in carbon and nitrogen assimilation in root nodules, in *Plant Nitrogen Metabolism* (J. E. Poulton, J. T. Romeo, and E. E. Conn, eds.), pp. 43–63, Plenum Press, New York.

Verma, D. P. S., and Nadler, K., 1984, Legume–*Rhizobium* symbiosis: Host's point of view, in *Genes Involved in Microbe–Plant Interactions* (D. P. S. Verma and T. Hohn, eds.), pp. 57–93, Springer-Verlag, New York.

Verma, D. P. S., 1992, Signals in root nodule organogenesis and endocytosis of *Rhizobium*, *Plant Cell* **4:**373–382.

Verma, D. P. S., Kazazian, V., Zogbi, V., and Bal, A. K., 1978, Isolation and characterization of the membrane envelope enclosing the bacteroids in soybean root nodules, *J. Cell Biol.* **78:**919–936.

Vijn, I., Yang, W. C., Pallisgard, N., Ostergaard-Jensen, E., van Kammen, A., and Bisseling, T., 1995, *VsENOD5*, *VsENOD12* and *VsENOD40* expression during *Rhizobium*-induced nodule formation on *Vicia sativa* roots, *Plant Mol. Biol.* **28:**1111–1119.

Werner, D., Morschel, E., Korat, R., Meilor, R., and Bassarab, S., 1984, Lysis of bacteroids in the vicinity of the host cell nucleus in an ineffective ($fix^-$) root nodule of soybean (*Glycine max*), *Planta* **16:**8–16.

Werner, D., Mellor, R., Hahan, M., and Grisbach, H., 1985, Soybean root response to symbiotic infection glyceollin I accumulation in an ineffective type of soybean nodules with an early loss of the peribacteroid membrane, *Z. Naturforsch.* **40:**179–181.

Wood, E. A., Butcher, G. W., Brewin, N. J., and Kannenberg, E. L., 1980, Genetic derepression of a developmentally regulated lipopolysaccharide antigen from *Rhizobium leguminosarum* 3841, *J. Bacteriol* **171:**4545–4555.

*Chapter 2*

# Biosynthesis and Secretion of Rhizobial Lipochitin-Oligosaccharide Signal Molecules

Eric Kamst, Herman P. Spaink,
and Dimitris Kafetzopoulos

## 1. INTRODUCTION

Plants that belong to the *Leguminosae* family are able to establish a symbiosis with soil bacteria belonging to the genera *Rhizobium, Bradyrhizobium,* and *Azorhizobium* (collectively called rhizobia). The result of this symbiosis is the formation of a new organ, the root nodule, in which differentiated bacteria convert atmospheric nitrogen into ammonia. The formation of a root nodule involves attachment of the bacteria to the root, invasion of the root by bacteria, development of the root nodule, and differentiation of the bacteria into nitrogen-fixing bacteroids. The exchange of molecular signals between plant and bacterium during nodule formation has been the subject of much research. This chapter will focus on the early stages of nodulation, and will present an overview of the present knowledge on the synthesis of nodulation (Nod) factors: rhizobial lipochitin oligosaccharide signal molecules which are essential for nodulation.

---

**Eric Kamst, Herman P. Spaink, and Dimitris Kafetzopoulos** Leiden University, Institute of Molecular Plant Sciences, Clusius Laboratory, 2333 AL Leiden, The Netherlands.
*Subcellular Biochemistry, Volume 29: Plant–Microbe Interactions,* edited by Biswas and Das. Plenum Press, New York, 1998.

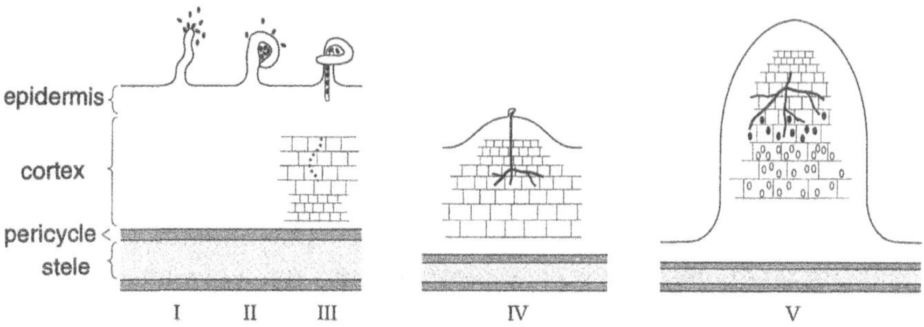


**FIGURE 1.** Schematic representation of root nodule formation. Formation of an indeterminate root nodule. Successive stages in root nodule formation are (I) attachment of rhizobia and induction of root hair deformations, (II) curling of root hairs and subsequent entrapment of bacteria, (III) formation of an infection thread, a pre-infection thread in the cortical cells in front of the infection thread, and induction of cell divisions in the inner cortex, (IV) development of the nodule primordium and branching of the infection thread as it reaches the basal cells of the primordium, and (V) development of the apical primordium cells into a persistent meristem, release of bacteria from the infection thread into plant cells, and differentiation of bacteria inside the plant cells into nitrogen-fixing bacteroids.

## 2. THE *RHIZOBIUM*–LEGUME SYMBIOSIS

### 2.1. The Nodulation Process

Nodulation is a complex process, outlined in Fig. 1; for reviews which describe the early events in nodulation in detail we refer to Spaink (1995), Mylona *et al.* (1995), Göttfert (1993), Kijne *et al.* (1992), and Franssen *et al.* (1992). Rhizobia are attracted to the roots of leguminous plants by metabolic compounds and secondary metabolites secreted by the plant (Barbour *et al.*, 1991; Caetano-Anolles *et al.*, 1988; Gaworzewska and Carlile, 1982). The bacteria subsequently attach to the tip of young root hairs. This attachment is likely to be a two-step process (Smit *et al.*, 1992, and references therein). A bacterial $Ca^{2+}$-binding protein, rhicadhesin, is involved in the first step. Accumulation and anchoring of bacteria to the root surface then leads to a firmer attachment. Plant lectins and/or bacterial structures such as cellulose fibrils and fimbriae can be involved in this latter step, depending on the growth conditions of the bacteria. Nodule development involves the induction of cortical cell divisions in the plant roots. In some plant species this eventually leads to the formation of nodules with a persistent meristem (indeterminate nodules). Other leguminous plant species form determinate nodules which have a round shape and lack such a meristem. The events in the development of indeterminate nodules have been studied in detail and will be briefly discussed. Rhizobia can induce curling of the root hairs. In this way, the attached bacteria get trapped in the so-called Shepherd's crook. Bacteria then enter the roots through a special tubular structure, the infection thread. Formation of the infection thread starts with the localized hydrolysis of plant cell wall structures, after which the plant membrane invaginates. The growing infection thread contains bacteria, and a layer of plant-derived wall material is deposited around it. In root cells, in front of the growing infection thread, cytoskeletal rearrangements, movement of the nuclei, and the formation of cytoplasmic bridges are observed and result in formation of so-called pre-infection threads (van Brussel *et al.*, 1992). During the infection process, plant cells located in the inner cortex and opposite to the protoxylem poles of the central cylinder start to proliferate. This results in the formation of the nodule primordium. The continuous cell divisions result in the growth of the nodule. The outer (apical) cells of the primordium will develop into a persistent meristem. This results in the formation of a nodule of the indeterminate type, which continues to consist of several developmental layers. In contrast, nodules of the determinate type start to develop by cell division in the outer cortex and lack a persistent meristem. When the infection thread reaches the nodule primordium it branches, and bacteria, surrounded by a plant-derived membrane, are released inside the plant cells. Bacteria lose their cell wall after being released inside plant cells and subsequently differentiate into bacteroids. In the bacteroids, nitrogen fixation is catalyzed by the nitrogenase complex (reviewed by Mylona *et al.*, 1995). This results in the formation of ammonia, which is taken up by the plant.

The synthesis of rhizobial signal molecules known as Nod factors is essential for root hair deformation, infection, and induction of cell division. Rhizobia which are defective in the formation of Nod factors are unable to elicit any of the responses in plants described above. The production of exopolysaccharides (EPS), lipopolysaccharides (LPS), capsular polysaccharides (CPS), and cyclic β-glucans by the rhizobia is also important for a successful infection. For a detailed discussion of the role of these polysaccharides in nodulation, the reader is referred to other review articles (Breedveld and Miller, 1994; Kannenberg and Brewin, 1994; Leigh and Walker, 1994). There are a number of reports which indicate that LPS and EPS may be involved in preventing a defense response of the plant against the bacteria (van Workum *et al.*, 1995; Parniske *et al.*, 1994; Perotto *et al.*, 1994).

### 2.2. Host Specificity of Nodulation

Each rhizobial strain is only capable of nodulating a limited group of plant genera. There are rhizobia with a narrow host range which only nodulate plants of one or a few genera, such as the *Rhizobium leguminosarum* biovar *trifolii* strains, which only nodulate clover (*Trifolium*) plants. Other rhizobia have a larger group of host plants and are so-called broad-host strains. The most pronounced example of such a strain is *Rhizobium* sp. strain NGR234, which is capable of nodulating over 70 genera (S. G. Pueppke and W. J. Broughton, unpublished results). The only non-legume that can be nodulated is *Parasponia* (Trinick and Galbraith, 1980). The rhizobia are divided into several cross-inoculation groups on the basis of their host range. A major determinant of host range is the signal exchange between plants and bacteria that results in the induction of nodule primordia and infection thread formation. These signals will be discussed in more detail in the following paragraphs. The production of EPS by rhizobia also seems to contribute to host specificity, since the infection-deficient phenotype of EPS mutants can be restored by the exogenous addition of EPS from a wild-type nodulating strain, but not by EPS from rhizobia which do not nodulate the plant species used in these studies (Battisti *et al.*, 1992; Djordjevic *et al.*, 1987).

## 3. STRUCTURES OF NOD FACTORS: LIPOCHITIN-OLIGOSACCHARIDE SIGNAL MOLECULES

In response to flavonoids present in the exudate of host plants, rhizobia secrete metabolites which were initially detected due to their ability to elicit root hair deformation on host plants. These metabolites (named Nod factors) have been purified from a large number of rhizobial strains. Their chemical structure has been elucidated using nuclear magnetic resonance (NMR) and mass spectrometry (MS). An overview of reported Nod factor structures is given in Fig. 2. Most Nod factors characterized so far have a common structure which consists of an oligosaccharide backbone of $\beta 1 \rightarrow 4$-linked *N*-acetyl-D-glucosamine (GlcNAc) residues, with a fatty acid group attached to the nitrogen of the nonreducing terminus. Due to the structural resemblance of the oligosaccharide backbone to chitin, the polymer of $\beta 1 \rightarrow 4$-linked GlcNAc, the Nod factors have been designated as lipochitin oligosaccharides (LCOs). All rhizobial strains examined produce a mixture of different LCOs. Recently two LCOs which are produced in relatively low abundance have been reported to contain an unusual oligosaccharide backbone. One of these LCOs is produced by *Rhizobium fredii* and contains a glucose residue in the middle of the oligosaccharide (Bec-Ferté *et al.*, 1996). The other unusual oligosaccharide backbone is present in an LCO produced by a *Rhizobium tropici* strain, and contains a mannose as the reducing terminal residue (Folch-Mallol *et al.*, 1996). Purified LCOs, applied in nanomolar concentrations, can elicit several responses in plants which are indistinguishable from responses during nodulation [recently reviewed by Spaink (1996)]. In the case of alfalfa plants, purified LCOs from the natural symbiont *Rhizobium meliloti* can even elicit the formation of complete nodules (Truchet *et al.*, 1991).

The structures of LCOs produced by different rhizobia vary in (i) the presence of additional groups on either the reducing or nonreducing terminus of the chitin oligosaccharide, (ii) the type of acyl chain present on the nonreducing terminus, and (iii) the length of the oligosaccharide backbone (Fig. 2). These variations are major determinants of host specificity. The presence of special polyunsaturated fatty acids in the LCOs of *R. leguminosarum* and *R. meliloti*, for instance, is essential for these strains to nodulate their natural host plants (Spaink *et al.*, 1991, 1995a; Demont *et al.*, 1993). The *R. leguminosarum* biovars *viciae* and *trifolii* are closely related. Their host range, however, does not overlap. Acyl chains in the LCOs produced by these *R. leguminosarum* biovars differ in length and the degree of unsaturation. These differences in LCO structure fully determine the inability of *R. l.* biovar *viciae* strains to nodulate *Trifolium* plants, the natural host plants of *R. l.* biovar *trifolii* (Bloemberg *et al.*, 1995a; Spaink *et al.*, 1989). Another example of a host-range-determining modification is the sulfate group in LCOs of *R. meliloti*. Mutant strains that lack this sulfate on their LCOs no longer nodulate alfalfa plants (Roche *et al.*, 1991; Lerouge *et al.*, 1990). Chemically desulfated *R. meliloti* LCOs or LCOs isolated from sulfation-deficient mutants are no longer capable of inducing cell division and the formation of nodule primordia in alfalfa roots, but have acquired the ability to do so in the non-host-plant *Vicia* (Roche *et al.*, 1991; Truchet *et al.*, 1991).

Although LCOs with oligosaccharide backbones ranging in length between three and six GlcNAc residues have been reported, the majority of LCOs produced by most rhizobia contain a chitinpentaose backbone. Exceptions are the *R. leguminosarum* biovars and *R. meliloti*. Approximately 30% of the LCOs produced by *R. l.* bv. *viciae*, 60% of the LCOs produced by *R. l.* by *trifolii*, and 75% of the LCOs produced by *R. meliloti* contain a chitintetraose backbone (Spaink *et al.*, 1991, 1995a; Schultze *et al.*, 1992). Several reports suggest that the length of the oligosaccharide moiety in LCOs has a role in determining specificity (Bec-Ferté *et al.*, 1994; Felle *et al.*, 1995; Heidstra *et al.*, 1994; Schultze *et al.*, 1992; Stokkermans *et al.*, 1995). *Rhizobium fredii* LCOs containing a tetrasaccharide backbone, for instance, are three orders of magnitude more active in inducing root hair deformations on natural host plants than LCOs containing a trisaccharide backbone. Lipochitin-trisaccharides constitute only a minor fraction of the LCOs produced by *R. fredii*. Surprisingly, these lipochitin-trisaccharides induce root hair deformations at 100-fold lower concentrations than *R. fredii* LCOs with a pentasaccharide backbone (Bec-Ferté *et al.*, 1994). *Rhizobium meliloti* LCOs with a tetrasaccharide backbone are between 10- and 100-fold more active in inducing root hair deformation, membrane depolarization, and meristematic activity than those containing a pentasaccharide backbone (Felle *et al.*, 1995; Schultz *et al.*, 1992). For recent reviews on the role of LCO structures in host specificity, see Dénarié *et al.* (1996), Spaink (1996), and Schultze and Kondorosi (1996).

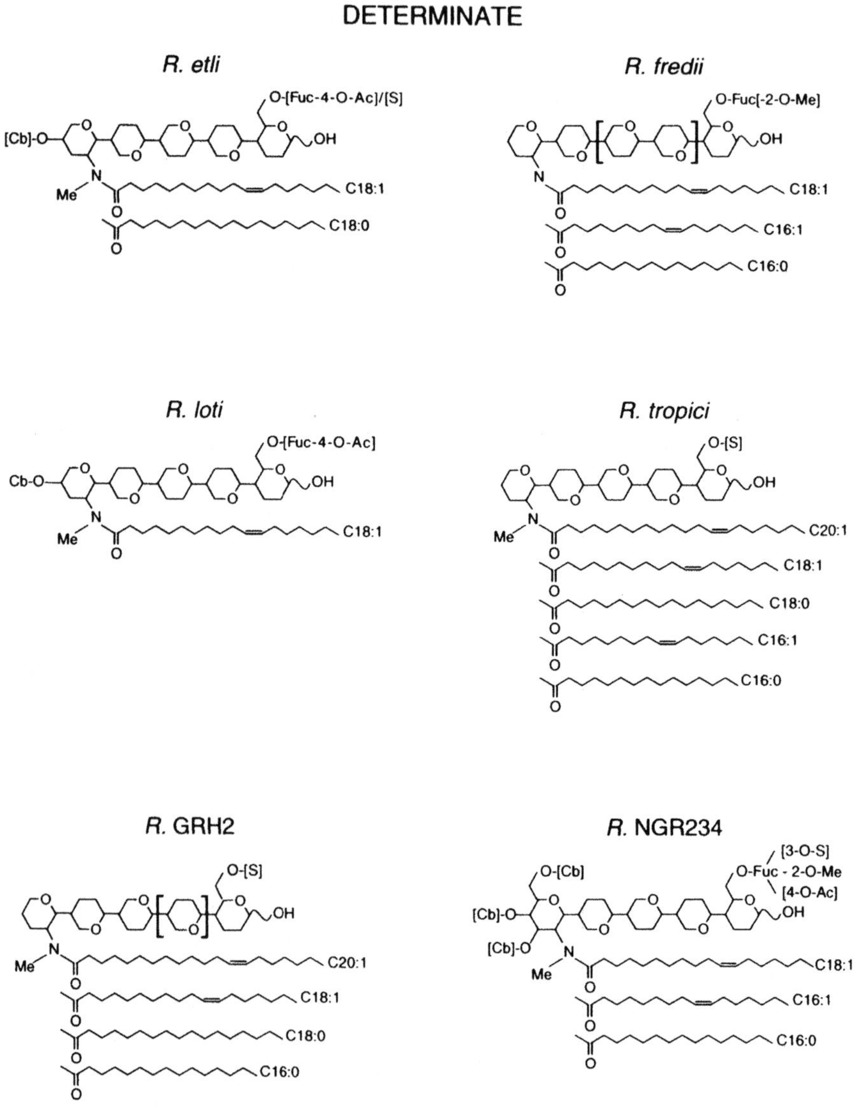


**FIGURE 2.** Nod factor structures. Structures of lipochitin oligosaccharides produced by rhizobia. Hexagons represent GlcNAc, and the various substitutions on the oligosaccharide are indicated with the following abbreviations: Me, methyl; Ac, acetyl; Cb, carbamoyl; Ara, D-arabinose; Fuc, L-fucose. Brackets indicate partial substitution or variation in oligosaccharide length. The *nodFE*-dependent fatty acids are indicated with asterisks. Besides oligosaccharide backbones of five GlcNAc residues, several rhizobial species also produce LCOs with a tetraose backbone, which is indicated by a single GlcNAc surrounded by brackets. In addition, some strains produce chitintriose-containing LCOs as well, which is indicated by brackets around a pair of GlcNAcs. The structures have been

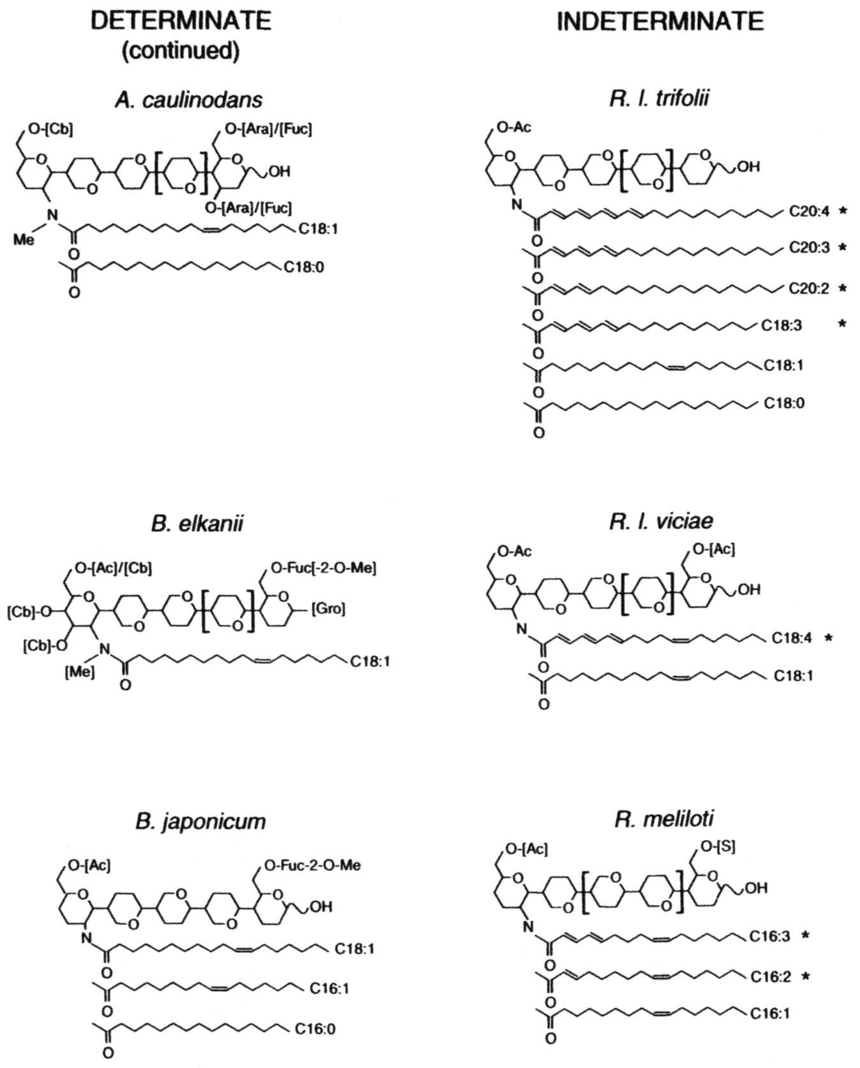


**FIGURE 2.** (*Continued*) derived from the publications listed below. Unless only LCOs containing a chitinpentaose backbone have been detected, the major length of the oligosaccharide moiety is given by a roman numeral, followed by the estimated percentage of LCOs which contain a backbone of this length. *R. etli* (Poupot *et al.*, 1995; Cardenas *et al.*, 1995), *R. fredii* (V, 75%) (Bec-Ferté *et al.*, 1994, 1996), *R. loti* (López-Lara *et al.*, 1995a), *R. tropici* (Folch-Mallol *et al.*, 1996; Poupot *et al.*, 1993), *R. l.* bv. *viciae* (V, 70%) (Spaink *et al.*, 1991), *R. l.* bv. *trifolii* (IV, 60%) (van der Drift *et al.*, 1996; Spaink *et al.*, 1995a), *R. meliloti* (IV, 75%) (Demont *et al.*, 1993; Schultze *et al.*, 1992; Roche *et al.*, 1991; Lerouge *et al.*, 1990), *Rhizobium* sp. NGR234 (Price *et al.*, 1992), *Rhizobium* sp. GRH2 (V, 90%) (López-Lara *et al.*, 1995b), *B. japonicum* (Carlson *et al.*, 1993; Luka *et al.*, 1993), *B. elkanii* (V, 95%) (Stokkermans *et al.*, 1996; Carlson *et al.*, 1993), *A. caulinodans* (V, 95%) (Mergaert *et al.*, 1993, 1996).

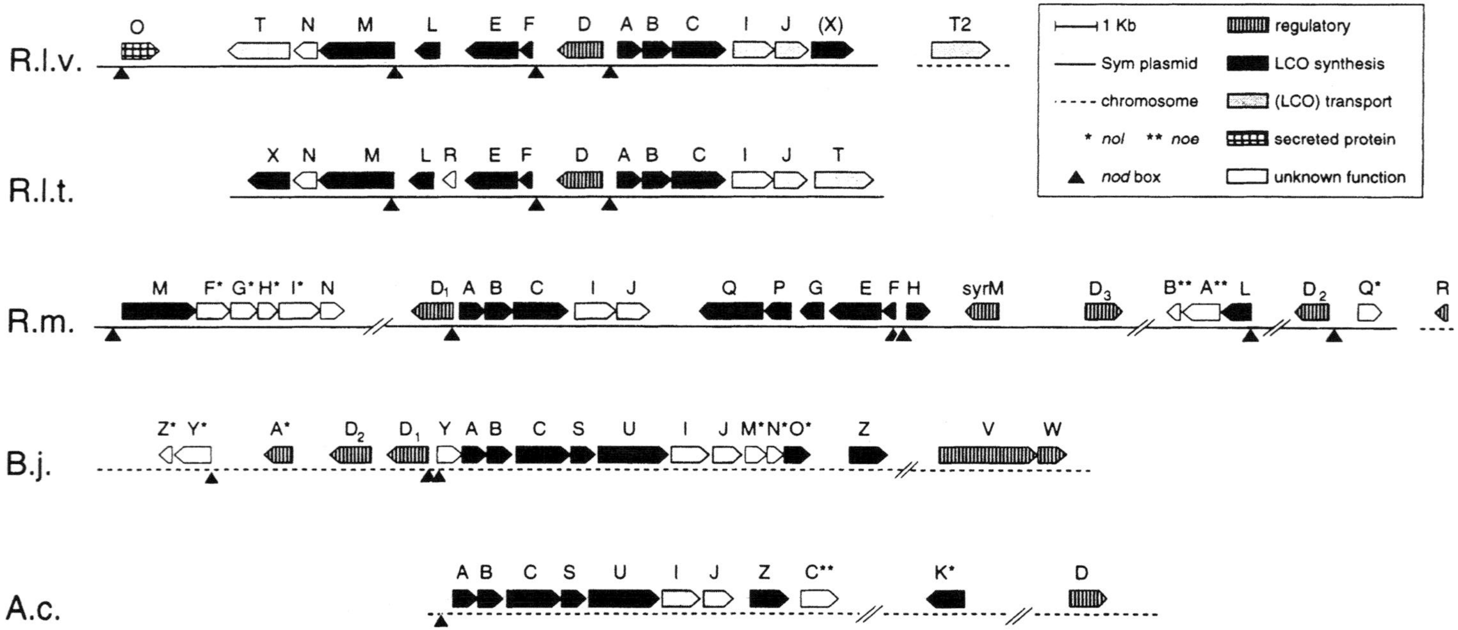


**FIGURE 3.** Examples of nodulation gene organization. Arrows indicate the open reading frames of the *nod*, *nol*, and *noe* genes of *R. leguminosarum* biovar *viciae* (R.l.v.) strain 248 [except for *nodX*, which is only present in strain TOM (Davis *et al.*, 1988)], *R. leguminosarum* biovar *trifolii* (R.l.t) strain ANU843, *R. meliloti* (R.m.) strains AK41 and 2011, *B. japonicum* (B.j.) strain USDA110, and *A. caulinodans* (A.c.) strain ORS571. Maps are based on a large number of reports. References are given by Schultze *et al.* (1994) and Göttfert (1993). Additional references are Plazanet *et al.* (1995), for *nolQ* and Ardourel *et al.* (1995) for *noeA* and *noeB*. Other known nodulation genes, not included in this figure, are *nolS* in *R. meliloti* (Plazanet *et al.*, 1995), *nodK* of *B. elkanii* [homologous to *nodY* (Dobert *et al.*, 1994)], *nolE* and *nolP* of *R. etli* (Vázquez *et al.*, 1991), *nolB*, *nolC*, *nolJ*, *nolT*, *nolU*, *nolV*, *nolW*, *nolX* of *R. fredii* (Boundy-Mills *et al.*, 1994; Meinhardt *et al.*, 1993), and ORFA, ORFC, ORFD, and ORFE of *R.* sp. NGR234 (Fellay *et al.*, 1995). The *nolXWBTUV* locus is also present in *Rhizobium* sp. NGR234 (Balatti *et al.*, 1995).

## 4. ORGANIZATION AND REGULATION OF THE NODULATION GENES

Bacterial genes required for nodulation have been identified mainly by the analysis of rhizobial mutant strains obtained by transposon mutagenesis. Examples of how the nodulation genes are organized in several rhizobia are shown in Fig. 3. The nodulation genes (indicated by *nod, nol,* or *noe*) are located on large symbiotic plasmids in most *Rhizobium* species (with the exception of *R. loti*) or at several loci of the chromosome such as in *Rhizobium* sp. NGR234, *Bradyrhizobium* strains, and *Azorhizobium* strains. The *nodDABC* genes are present in all rhizobia investigated so far. Other genes are clearly strain-specific, such as *nodFEL, nodH, nodSU,* and *nodZ.*

All rhizobia contain one or more *nodD* genes. The NodD proteins are transcriptional activators of the LysR family (reviewed by Pueppke, 1996; Schultze *et al.*, 1994; Schlaman *et al.*, 1992). Most nodulation genes are organized in operons which are preceded by a promotor containing a consensus sequence, the so-called *nod* box, which is recognized by NodD (Goethals *et al.*, 1992; Fisher *et al.*, 1988). Specific flavonoids in the exudates of host plants activate the NodD protein, which in turn promotes the transcription of the nodulation genes. NodD proteins of different rhizobia respond to different flavonoids (Peters and Verma, 1990; Spaink *et al.*, 1987), and this flavonoid specificity of NodD is also a determinant of host specificity (Spaink *et al.*, 1987). *Rhizobium meliloti* contains the *syrM* gene, which also encodes a LysR-type regulator and shows homology with *nodD* (Barnett and Long, 1990; Mulligan and Long, 1989). The products of the *R. meliloti nolR* and the *B. japonicum nolA* genes repress nodulation gene expression and may be involved in active downregulation of *nod* gene expression after infection (Cren *et al.*, 1995; Sadowsky *et al.*, 1991). In *B. japonicum,* gene regulation by isoflavonoids also involves the *nodVW* genes, which are members of the two-component regulatory family (Sanjuan *et al.*, 1994). Sequence similarities indicate that NodV is a sensor kinase and that NodW is a transcription activator. Recent data of Loh *et al.* (1997) have confirmed this. The NodVW regulatory system is essential for nodulation of several host plants, but not for nodulation of soybean (Göttfert *et al.*, 1990). NodV may therefore recognize flavonoids which do not interact with or activate NodD. For a more detailed overview of transcriptional regulation of nodulation gene expression, see Pueppke (1996), Stacey (1995), and Göttfert (1993).

## 5. FUNCTION OF NODULATION GENE PRODUCTS IN LCO BIOSYNTHESIS

About 10 years ago it was recognized that the synthesis of Nod factors depends on the expression of nodulation genes (van Brussel *et al.*, 1986). Since then, most of the nodulation genes have been shown to be directly or indirectly involved in the biosynthesis of LCOs. Information on the biochemical function of nodulation proteins involved in LCO synthesis has mainly come from structural analysis of LCOs produced by strains carrying mutations in nodulation genes, and by analysis of LCOs after introduction of cloned nodulation genes of one rhizobial strain into another. Insight in the biochemical activities of nodulation proteins has also been obtained by expression of cloned nodulations genes in *Escherchia coli,* followed by the characterization of *nod* gene-dependent metabolites produced by such strains. Five nodulation proteins have been purified to homogeneity. *In vitro* analysis of their substrate specificities has provided information about their position in the LCO biosynthesis pathway. Table I presents the biosynthetic Nod proteins for which an enzymatic activity is known or has been proposed. A model which shows how the Nod proteins act together in the synthesis of LCOs in *R. meliloti* and *R. loti* is shown in Fig. 4.

### 5.1. Synthesis of the Chitin Oligosaccharide Backbone

Biosynthesis of the precursor for chitin oligosaccharide synthesis, UDP-GlcNAc, involves several proteins encoded by the *glm* genes. The first step in this pathway is the conversion of fructose-6-P into glucosamine-6-P by the *glmS* gene product (Dutka-Malen *et al.*, 1988; Walker *et al.*, 1984). The NodM protein has sequence homology to several amido phosphoribosyl transferases; the highest homology is observed with *glmS* of *E. coli* (Baev *et al.*, 1991). The *nodM* gene complements a *glmS* mutant of *E. coli* (Baev *et al.*, 1991) and a mutation in the chromosomally encoded *glmS* of *R. l.* bv. *viciae* (Marie *et al.*, 1992). Mutations in *nodM* lead to reduction in LCO production (Spaink *et al.*, 1992) and a delayed-nodulation phenotype (Marie *et al.*, 1992). In conclusion, expression of NodM seems to be necessary for an increase in the cellular UDP-GlcNAc supply to allow a maximal rate of LCO synthesis upon *nod* gene induction. No other nodulation genes with homology to *glm* genes have been reported.

The backbone of LCOs consists of a linear oligosaccharide of $\beta 1 \rightarrow 4$-linked GlcNAc residues. The sequence similarity between *Saccharomyces cerevisiae* chitin synthases and the nodulation protein NodC suggested that synthesis of the chitin oligosaccharide backbone of LCOs was performed by NodC (Bulawa and Wasco, 1991). Membranes isolated from *E. coli* or *Rhizobium* strains expressing *nodC* produce molecules, using UDP-GlcNAc as precursor, which are chitinase-degradable and migrate as chitin oligosaccharides in TLC and HPLC systems (Kamst *et al.*, 1997; Mergaert *et al.*, 1995; Iñón de Iannino *et al.*, 1995; Geremia *et al.*, 1994). Structural analysis, using mass spectrometry, of these compounds produced in *E. coli in vivo* has shown that these *nodC*-dependent metabolites are indeed linear oligosaccharides of $\beta 1 \rightarrow 4$-linked GlcNAc residues with a degree of polymerization ranging from 2 to 5 (Kamst *et al.*, 1995; Mergaert *et al.*, 1995).

**Table I**
**Function of Nodulation Proteins Involved in Nod Factor Biosynthesis**

| Nodulation protein | Function | Evidence for function[a] | References |
|---|---|---|---|
| *Nodulation proteins involved in the biosynthesis of the chitin oligosaccharide backbone* | | | |
| NodM | Glucosamine synthase | seq, vivo | Marie *et al.*, 1992; Baev *et al.*, 1991 |
| NodC | *N*-Acetyl-glucosamine transferase | seq, vivo, vitro | Kamst *et al.*, 1995; Mergaert *et al.*, 1995; Spaink *et al.*, 1994; Geremia *at al.*, 1994; Bulawa and Wasco, 1991 |
| NodB* | Chitin oligosaccharide deacetylase | seq, vivo, vitro | Spaink *et al.*, 1994; John *et al.*, 1993; Kafetzopoulos *et al.*, 1993 |
| *Nodulation proteins involved in the biosynthesis and transfer of the fatty acid moiety* | | | |
| NodF* | Acyl carrier protein | seq, vivo | Ritsema *et al.*, 1994; Geiger *et al.*, 1991; Shearman *et al.*, 1986 |
| NodE | β-Keto acyl synthase | seq, vivo | Bloemberg *et al.*, 1995a; Demont *et al.*, 1993; Spaink *et al.*, 1991; Bibb *et al.*, 1989 |
| NodA | Acyl transferase | vitro | Atkinson *et al.*, 1994; Röhrig *et al.*, 1994 |
| *Nodulation proteins involved in the modification of the nonreducing terminus* | | | |
| NodL* | Acetyl transferase | seq, vivo, vitro | Bloemberg *et al.*, 1994; Spaink *et al.*, 1991; Downie, 1989 |
| NodS* | Methyl transferase | seq, vivo, vitro | Geelen *et al.*, 1995; Jabbouri *et al.*, 1995 |
| NodU | Carbamoyl transferase | vivo | Jabbouri *et al.*, 1995 |
| NolO | Carbamoyl transferase | seq | Luka *et al.*, 1993 |
| *Nodulation proteins involved in the modification of the reducing terminus* | | | |
| NodP | ATP sulfurylase subunit | seq, vivo, vitro | Schwedock *et al.*, 1994; Schwedock and Long, 1990 |
| NodQ | ATP sulfurylase subunit/ APS kinase | seq, vivo, vitro | Schwedock *et al.*, 1994; Schwedock and Long, 1990 |
| NodH | Sulfo transferase | vivo, vitro | Ehrhardt *et al.*, 1995; Schultze *et al.*, 1995 |
| NoeE | Sulfo transferase | vivo | Hanin *et al.*, 1997 |
| NolK | Epimerase/reductase | seq, vivo | Mergaert *et al.*, 1996, 1997 |
| NodZ* | Fucosyl transferase | vivo, vitro | Quinto *et al.*, 1997; Stacey *et al.*, 1994 |
| NolL | Acetyl transferase | seq, vivo | Ovtsyna *et al.*, 1997; Scott *et al.*, 1996 |
| NodX | Acetyl transferase | seq, vivo | Firmin *et al.*, 1993; Clark *et al.*, 1991 |

(*continued*)

**Table I (Continued)**

| Nodulation protein | Function | Evidence for function[a] | References |
|---|---|---|---|
| *Nodulation proteins involved in the secretion of LCOs* | | | |
| NodI | ATPase | seq, vivo | Fernández-López *et al.*, 1996; Spaink *et al.*, 1995b; Mckay and Djordjevic, 1993; Vázquez *et al.*, 1993; Evans and Downie, 1986 |
| NodJ | ATP-dependent transporter | seq, vivo | Fernández-López *et al.*, 1996; Spaink *et al.*, 1995b; Mckay and Djordjevic, 1993; Vázquez *et al.*, 1993; Evans and Downie, 1986 |

[a]seq: sequence similarity with genes or proteins of known function; vivo: structural analysis of compounds produced by recombinant strains or functional complementation of mutant strains; vitro: biochemical demonstration of function.
*Nod proteins that have been purified to homogeneity.

The synthesis of polysaccharides and oligosaccharides often involves lipid-linked intermediates (Whitfield and Valvano, 1993). The synthesis of chitin oligosaccharides by NodC has been studied *in vitro* using membranes from *E. coli* or *Rhizobium* strains expressing *nodC*. Several compounds which inhibit the synthesis of lipid-linked oligosaccharides or recycling of prenyl carriers do not affect the synthesis of chitin oligosaccharides by NodC (Kamst *et al.*, 1997b; Mergaert *et al.*, 1995). It is therefore likely that lipid-linked intermediates are not involved in the synthesis of chitin oligosaccharides in rhizobia. Since free chitin oligosaccharides are not used as substrates for elongation, chitin oligosaccharide synthesis by NodC occurs by a processive mechanism. In such a mechanism, addition of monosaccharides to the growing chain only proceeds while the enzyme continuously remains in contact with the oligosaccharide. Elongation of chitin oligosaccharides by NodC has been suggested to take place at the reducing end (Saxena *et al.*, 1994). However, we recently observed that NodC, in addition to GlcNAc from UDP-GlcNAc, also directly incorporates free GlcNAc into chitin oligosaccharides (Kamst *et al.*, 1997b). Further analysis showed that the incorporated GlcNAc had been used as a primer. Therefore elongation must have taken place at the non-reducing end. This conclusion was confirmed by the observation that the synthetic glycoside p-nitrophenyl-β-N-acetylglucosaminide, presenting a free hydroxyl group at C4 but not at C1 could also be used as acceptor by NodC. A proposed model for the synthesis of chitin oligosaccharides in rhizobia based on these results is shown in Fig. 5.

The length of the major products of the NodC proteins of *R. loti*, *R. meliloti*, and *R. l.* bv. *viciae* in *E. coli* corresponds to the oligosaccharide length

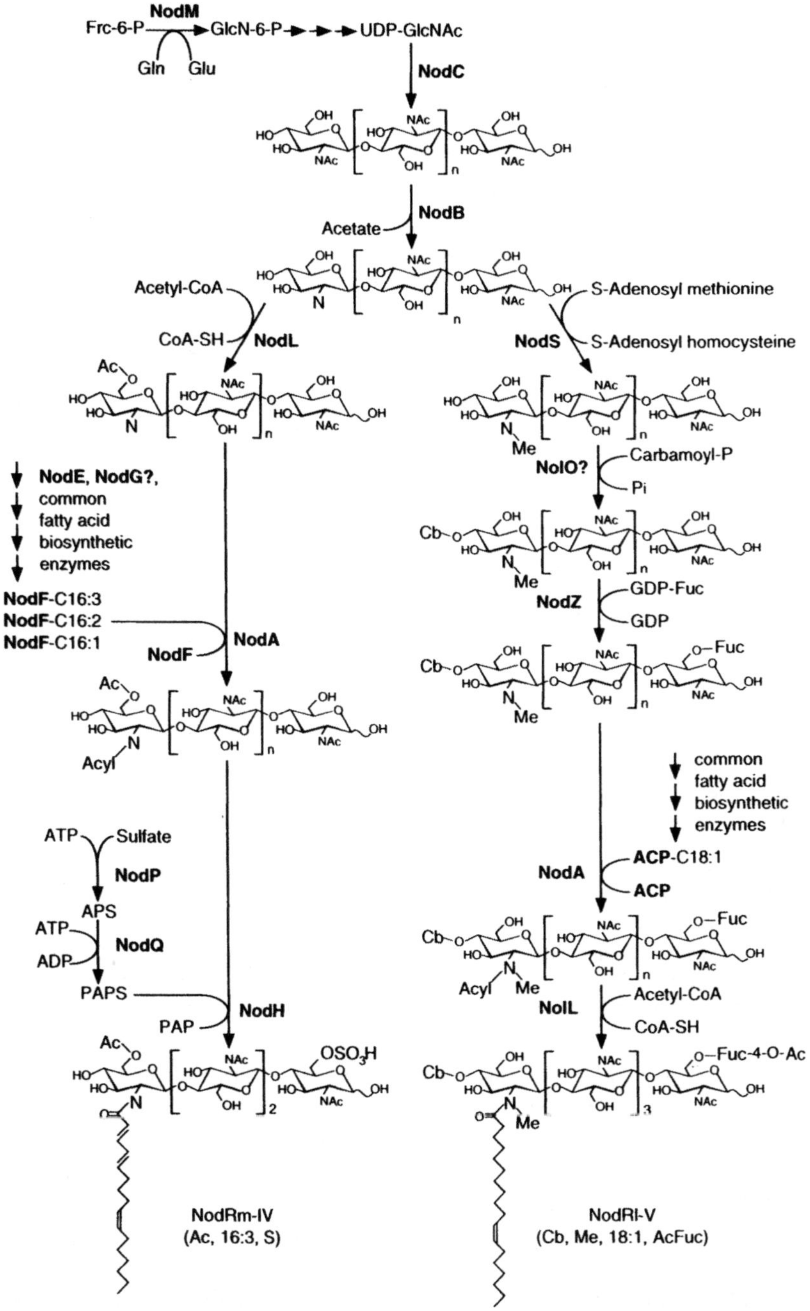


**FIGURE 4.** Proposed LCO biosynthesis pathways in *R. meliloti* and *R. loti*. The figure shows the synthesis of the *R. meliloti O*-acetylated, sulfated lipochitin tetrasaccharide containing the *nodFE*-dependent polyunsaturated C16:3 fatty acid [NodRM-IV(Ac, C16:3, S)] and the carbamoylated, methylated, acetyl-fucosylated lipochitin pentasaccharide [NodRl-V(Cb, Me, C18:1, AcFuc)] produced by *R. loti*.

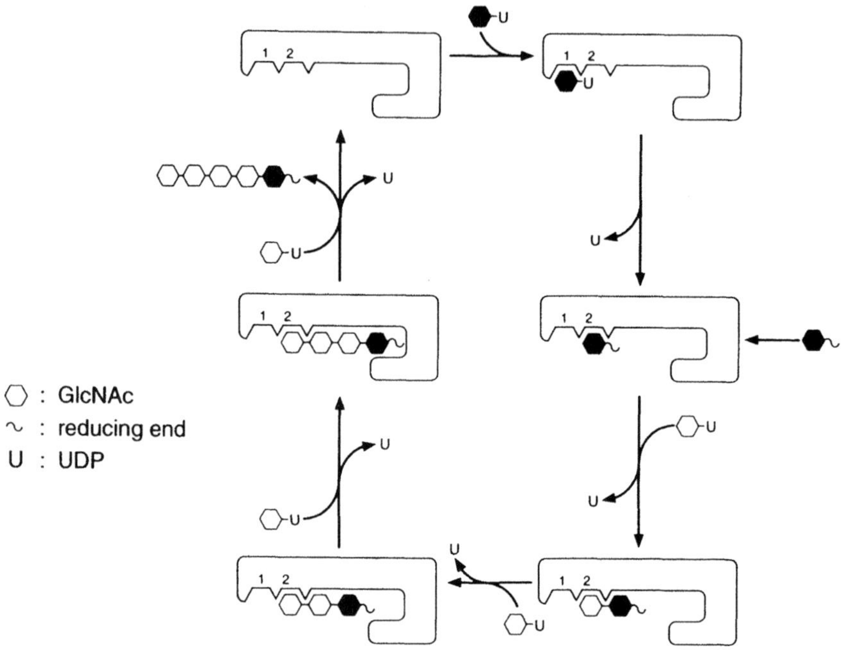


**FIGURE 5.** Proposed model for chitin oligosaccharide synthesis by the NodC protein. Synthesis of chitin oligosaccharides is likely to start with the binding of UDP-GlcNAc to NodC, followed by its hydrolysis, yielding free UDP and GlcNAc. The resulting GlcNAc residue may then be translocated to a second binding site to allow binding of another UDP-GlcNAc. In each following step, UDP-GlcNAc binding at site 1 is followed by hydrolysis of UDP-GlcNAc, transfer of GlcNAc to the nonreducing end of the growing oligosaccharide, and immediate translocation of the elongated chain. When the oligosaccharide reaches the maximal length, controlled by NodC, the saccharide dissociates from NodC, and NodC becomes available for the synthesis of another chitin oligosaccharide molecule. The observed priming and stimulation of chitin oligosaccharide synthesis may result from the direct binding of GlcNAc to site 2. Limitation of oligosaccharide chain length may result from a direct interaction between the oligosaccharide and NodC, but this is still speculative.

present in the major LCOs produced by the *Rhizobium* strains from which these *nodC* genes originate (Kamst *et al.*, 1997a). Complementation of *nodC* mutants of *R. meliloti* (Debellé *et al.*, 1995) and *R. l.* bv. *viciae* (Kamst *et al.*, 1997a) with *nodC* genes of different origins confirms that *nodC* is a major determinant of oligosaccharide chain length in LCO biosynthesis. This may seem to be in contrast to a report of Iñón de Iannino (1995), who showed that the length of chitin oligosaccharides produced *in vitro* by *R. fredii* membranes depends on the UDP-GlcNAc concentration. However, a more detailed analysis *in vitro* using *R. loti* and *R. meliloti* NodC has shown that this reduction in oligosaccharide chain length is probably due to premature chain termination, since it is only observed at UDP-GlcNAc concentrations that are rate-limiting (Kamst *et al.*, 1997a). The identification of the *nodC* gene as a major determinant of chitin oligosaccharide chain length provides a tool to further investigate the proposed role of variations in the length of the oligosaccharide backbone of LCOs in host specificity.

NodC has been shown to be located in the inner cytoplasmic membrane (Barny and Downie, 1993). More recently, a membrane topology was proposed, based on the analysis of several NodC–PhoA fusion proteins (Barny *et al.*, 1996). In the predicted intracellular domain, NodC shows a sequence similarity to other glycosyltransferases involved in the synthesis of the linear polysaccharides chitin, alginate, and hyaluronic acid, which all contain β1→4-linked sugar residues (Fig. 6). Several of the conserved amino acids have been shown to

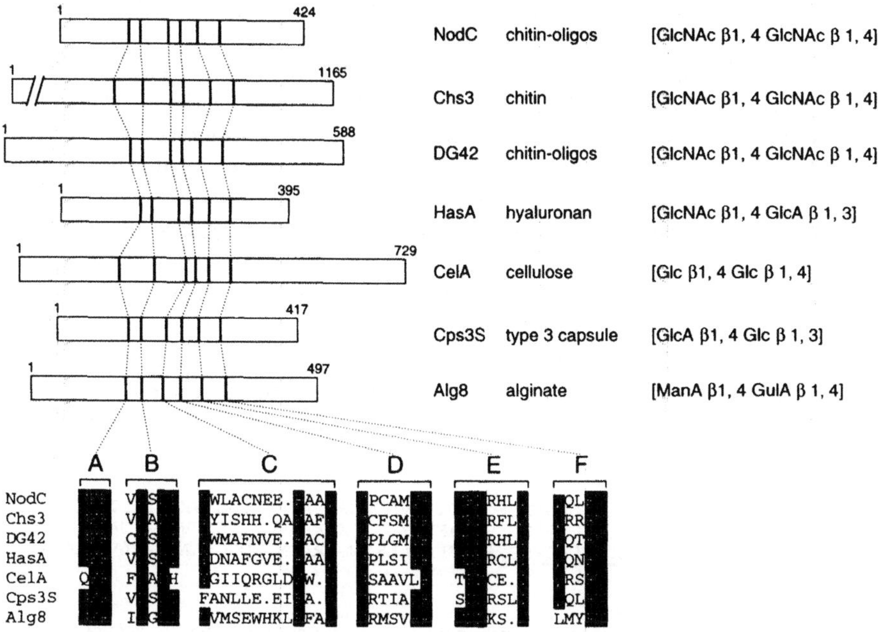


**FIGURE 6.** Conserved regions in processive β-glycosyl transferases. Horizontal bars in the upper left represent the various proteins, and the number of amino acids in each protein is given at the right end of each bar. The black vertical bars indicate the positions of the sequences that are conserved throughout the entire family. These sequences are shown at the bottom of the figure. The names and the repeating units of the oligosaccharides or polysaccharides which are produced by the different enzymes are listed on the left. The proteins shown are: NodC, *Rhizobium loti* nodulation protein (Accession No. X01650; Rossen *et al.*, 1984); Chs3, *Saccharomyces cereviciae* chitin synthase CSD2/CAL1 (Accession No. P29465; Valdivieso *et al.*, 1991); DG42, *Xenopus laevis* developmental protein (Accession No. P13563; Rosa *et al.*, 1988); HasA, *Streptococcus pyogenes* hyaluronan synthase (Accession No. L20853; DeAngelis *et al.*, 1993); CelA, *Agrobacterium tumefaciens* cellulose synthase (Accession No. L38609; Matthysse *et al.*, 1995); Cps3S, *Streptococcus pneumoniae* capsular polysaccharide synthase (Accession No. U15171; Dillard *et al.*, 1995); Alg8, *Pseudomonas auruginosa* alginate synthase (Accession No. 122611; Maharaj *et al.*, 1993). Abbreviations for monosaccharides are: GlcNAc, *N*-acetylglucosamine; Glc, glucose; GlcA, glucuronic acid; ManA, mannuronic acid; and GulA, guluronic acid.

be essential for enzyme activity of *Saccharomyces cerevisiae* chitin synthase 2 (Nagahashi *et al.*, 1995). Even conservative substitutions for the amino acids corresponding to $Asp^{140}$ (first Asp in region B), $Asp^{242}$ (region E), $Gln^{278}$, $Arg^{281}$, and $Trp^{282}$ (region F) in *R. loti* NodC drastically decreased chitin synthase activity. Similar data were recently also reported for *A-xylinum* cellulose synthase mutants (Saxena and Brown, 1997). On the basis of these results it is expected that these conserved amino acids are catalytic residues or residues involved in substrate binding. It is not known which amino acids are responsible for the substrate specificity of the transferases involved in the synthesis of polysaccharides. In this respect, it is interesting to note that the tyrosine preceding the conserved arginine in region D is conserved in the GlcNAc transferases NodC, hyaluronan synthase, DG42, and chitin synthase, but not in the cellulose synthase and the alginate synthase. In addition, the GlcNAc transferases shown in Fig. 6 all contain the G(E/D)DR motif (region E), whereas TEDA is found in the cellulose synthase and GDDK in alginate synthase. These differences may be the basis for the different substrate specificities of these enzymes, but experimental evidence for such a role has not been reported.

The nonreducing terminal GlcNAc residue in chitin oligosaccharides needs to be deacetylated to allow a fatty acid to be coupled to the resulting free amino group. After cloning and sequence analysis of the gene encoding the chitin deacetylase of *Mucor rouxii* a sequence similarity with NodB was observed which suggests that NodB functions as a chitin-oligosaccharide deacetylase (Fig. 7) (Kafetzopoulos *et al.*, 1993). The similarity between NodB and the acetyl xylan esterase of *Streptomyces lividans* (Fig. 7) indicates that the observed se-

```
                   ** ****           *      *  *  *      *      *  *
RlenodB   19 EARCVYLTFDDGPNPFCTPQILDVLAEHRVPATFFAIGSYVKDHPELIRRLVAEGH
 MroCDA  157 EPETWGLTYDDGPNC.SHNAFYDYLQEQKLKASMFYIGSNVVDWPYGAMRGVVDGH
 SliAXE   46 CNGYVGLTFDDGPSG.STQSLLNALRQNGLRATMFNQGQYAAQNPSLVRAQVDAGM
                      I

               * *   **            *      **  *
RlenodB      DVANHTMTHPDLATCDPKDVKREIDEAHQAIVSACPQALVRHLRAPYGVWTEDVLSASVR
 MroCDA      HIASHTWSHPQMTTKTNQEVLAEFYYTQKAIKLATGLT.PRYWRPPYGDIDDRVRWIASQ
 SliAXE      WVANHSYTHPHMTQLGQAQMDSEISRTQQAIAGAGGG..RQAVPAAVRQTNATLRSVEAK
                  II

              **  * *  *  **        *  *                       *  *
RlenodB      AGLGAVHWSADPRDWSCPGVDVIVDEVLAA.............ARPGAIVLLHDGC 178
 MroCDA      LGLTAVIWNLDTDDWS.AGVTTTVEAVEQSYSDYIAMGTNGTFANSGNIVLTHEIN 324
 SliAXE      YGLTEVIWDVDSQDWNNASTDAIVQAVSRL..............GNGQVILMHDWP 199
                   III                                          IV
```

**FIGURE 7.** Sequence homology between NodB and related deacetylases. RlenodB: *Rhizobium leguminosarum* bv. *viciae* NodB protein (216 amino acid residues, Accession No. X01650; Rossen *et al.*, 1984); MroCDA: *Mucor rouxii* chitin deacetylase (401 amino acid residues, Accession No. Z19109; Kafetzopoulos *et al.*, 1993); SliAXE: *Streptomyces lividans* acetyl xylan esterase (334 amino acid residues, Accession No. M64552; Shareck *et al.*, 1995). Invariant residues among all proteins are indicated by asterisks. Highly conserved regions are underlined and numbered by Roman numerals.

quence homologies are related to enzymatic activity and the β1→4 configuration in the substrate. Expression of *nodB* together with *nodC* in either *E. coli* or a *Rhizobium* strain lacking its symbiotic plasmid leads to the synthesis of chitin oligosaccharides which are *N*-deacetylated at the nonreducing-end residue (Spaink *et al.*, 1994). Purified recombinant NodB protein is able to deacetylate chitin oligosaccharides, but not GlcNAc monomers (John *et al.*, 1993). Nevertheless, NodB of *R. l.* bv. *viciae* seems to preferentially deacetylate chitin oligosaccharides with a minimal length of four GlcNAc residues (Spaink *et al.*, 1994). De-*N*-acetylation of the nonreducing terminal GlcNAc residue of chitin oligosaccharides by NodB is most likely to occur directly after the synthesis of these oligosaccharides by NodC, since the other modifying Nod proteins whose substrate specificity has been investigated *in vitro* show a preference for terminally de-*N*-acetylated chitin oligosaccharides or LCOs, instead of fully *N*-acetylated chitin oligosaccharides (Bloemberg *et al.*, 1995b; Geelen *et al.*, 1995; Schultze *et al.*, 1995; Quinto *et al.*, 1997).

## 5.2. Modifications at the Nonreducing Terminus

### 5.2.1. *O*-Acetylation

The presence of an *O*-acetyl group at C6 of the nonreducing GlcNAc residue of the oligosaccharide backbone of *R. leguminosarum* and *R. meliloti* LCOs is due to the enzymatic activity of NodL. The function of NodL was initially proposed on the basis of its sequence similarity with the two *E. coli* acetyl transferases encoded by *lacA* (thio galactosyl acetyl transferase) and *cysE* (serine acetyl transferase) (Downie, 1989). Furthermore, it has been shown that *nodL* mutants produce LCOs which lack the 6-*O*-acetyl group (Ardourel *et al.*, 1995; Spaink *et al.*, 1991). The NodL protein has been purified after overexpression in *E. coli* and has been demonstrated to show acetyltransferase activity, using acetyl-CoA as acetyl donor (Bloemberg *et al.*, 1994). Acceptors for acetylation by NodL are GlcNAc, chitin oligosaccharides, chitosan oligosaccharides (oligomers of glucosamine), and LCOs, but the lowest $K_M$ values were observed with chitosan oligosaccharides (Bloemberg *et al.*, 1995a). The efficiency of acetylation increases when oligosaccharides with increasing length are used as acceptor. The most efficient acceptors, however, are chitin oligosaccharides which are de-*N*-acetylated only at the nonreducing-end position. These results suggest that the terminal de-*N*-acetylated saccharides produced by NodC and NodB are the *in vivo* substrates of NodL.

### 5.2.2. *N*-Methylation

Most rhizobial strains that form determinate nodules produce LCOs which carry an *N*-methyl group at the nonreducing-end GlcNAc residue of the chitin-oligosaccharide backbone (Fig. 2). *N*-Methylation in these strains is catalyzed by the NodS protein. Analysis of LCO intermediates produced *in vivo* after expression of various combinations of the *A. caulinodans nodABCS* genes in *E. coli* suggests that NodS methylates the de-*N*-acetylated chitin oligosaccharides produced by NodC and NodB (Mergaert *et al.*, 1995). The NodS protein has been purified after overexpression as a NodS–glutathion-*S*-transferase fusion protein in *E. coli,* and *in vitro* experiments show that NodS uses *S*-adenosyl-methionine (SAM) as methyl donor to methylate chitin oligosaccharides which are de-*N*-acetylated at the nonreducing terminus (Geelen *et al.*, 1995). Terminally de-*N*-acetylated oligosaccharides with a chain length of 2–6 sugar residues have been tested as substrate, and can all be used by NodS as methyl acceptor. However, GlcNAc, fully *N*-acetylated chitin oligosaccharides, and LCOs cannot be methylated by NodS, clearly showing that methylation indeed takes place after de-*N*-acetylation and prior to acylation. Introduction of the *nodS* gene of *Rhizobium* sp. NGR234 into *R. fredii* strain USDA257 results in the synthesis of *N*-methylated LCOs, which are not produced by *R. fredii* USDA257 wild-type strains (Jabbouri *et al.*, 1995). Interestingly, introduction of *Rhizobium* sp. NGR234 *nodS* into *R. fredii* USDA 257 leads to the production of LCOs which only contain chitinpentaose backbones, whereas the wild-type *R. fredii* USDA257 strain also produces LCOs with chitintriose and chitintetraose backbones. This suggests a preferential acylation of methylated chitinpentaose in this strain. Expression of *nodS* results in a 20-fold increase in the overall LCO production of *R. fredii* USDA257, whereas LCO production in *Rhizobium* sp. NGR234 *nodSU* mutants is strongly reduced (Jabbouri *et al.*, 1995). A similar effect on overall LCO production has been observed with the *O*-acetyl transferase NodL (Spaink *et al.*, 1992), again suggesting that acylation of the backbone may be enhanced by the presence of additional groups on the chitin oligosaccharide.

### 5.2.3. *O*-Carbamoylation

Carbamoylation has been observed at C6 and C4 (Fig. 2). However, it cannot be excluded that carbamoylation of LCOs in *Rhizobium* sp. NGR234 occurs at C6 and C3 (Jabbouri *et al.*, 1995). There are several indications that carbamoylation at C6 is performed by the NodU protein. Carbamoylation at C6 is absent in a *nodU* mutant of *Rhizobium* sp. NGR234, whereas carbamoylation at C3/C4 in this strain is unaffected (Jabbouri *et al.*, 1995). Moreover, introduction of *Rhizobium* NGR234 *nodU* into *R. fredii* strain USDA257 results in the synthesis of LCOs which only carry a 6-*O*-carbamoyl group, strongly suggesting that carbamoylation at C3 or C4 in *Rhizobium* sp. NGR234 is not due to *nodU*. It is not clear which protein is responsible for carbamoylation at C3/C4 in *Rhizobium* sp. NGR234. Luka *et al.* (1993) reported a homology between the NolO protein of *japonicum* strain USDA110 and NodU, and showed that the *nolO* gene is also present in *Bradyrhizobium elkanii*. The LCOs produced by *B. elkanii* can be carbamoylated and some LCOs carry two carbamoyl groups at the nonreducing GlcNAc, just as in *Rhizobium* sp. NGR234, but the position of these carbamoyl groups has not been determined (Stokkermans *et al.*, 1996). Since an *nolO* mutant of *B. elkanii* has not been reported, the possible role of NolO in carbamoylation at C3/C4 is still unclear.

### 5.3. Synthesis and Transfer of the *N*-Acyl Moiety

#### 5.3.1. Synthesis of α,β-Polyunsaturated Fatty Acids

Rhizobia that nodulate plant species which form indeterminate nodules produce LCOs with unique, polyunsaturated fatty acids in addition to the LCOs containing acyl chains from the common fatty acid synthesis pathway (Fig. 2) (van der Drift *et al.*, 1996; Spaink *et al.*, 1991, 1995a; Demont *et al.*, 1993; Schultze *et al.*, 1992; Roche *et al.*, 1991; Lerouge *et al.*, 1990). The synthesis of these polyunsaturated fatty acids requires the presence of functional *nodF* and *nodE* genes (Ritsema *et al.*, 1994; Demont *et al.*, 1993; Spaink *et al.*, 1991). Fatty acid biosynthesis occurs by the repetitive, sequential action of β-keto-acylsynthase, acylreductase, dehydrogenase, and enoylreductase [recently reviewed by Magnuson *et al.* (1993), and schematically presented in Fig. 8]. The

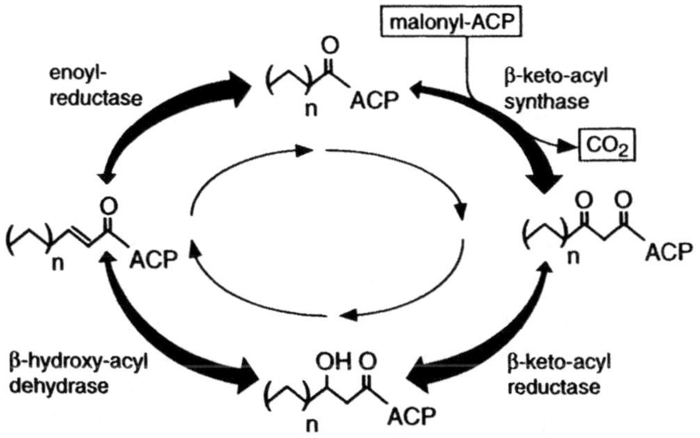


**FIGURE 8.** The elongation cycle in fatty acid biosynthesis. The model is according to Heath and Rock (1995). Each cycle in fatty acid elongation consists of four steps. The first reaction is the transfer of an acyl chain from acyl-ACP to malonyl-ACP, catalyzed by β-keto-acyl-synthases (FabH, FabB, or FabF). The next step is the reduction of the resulting keto group by β-keto-acyl-reductase (FabG), followed by the action of a dehydrogenase (FabA or FabZ). The last step is the reduction of the unsaturated enoyl-ACP, resulting in the formation of an elongated acyl-ACP. The outer arrows indicate equilibrium positions as reported by Heath and Rock (1995) and the inner arrows show the overall direction of the cycle.

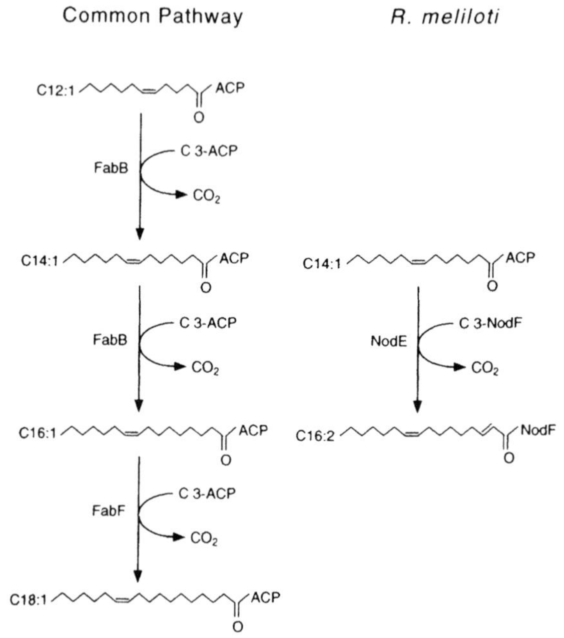


**FIGURE 9.** Proposed pathway for the synthesis of *nodFE*-dependent fatty acids in *Rhizobium*. As examples of the synthesis of α,β-polyunsaturated fatty acids in *Rhizobium*, the proposed pathways for the synthesis of the *nodFE*-dependent fatty acids C16:2 in *R. meliloti*, C18:4 in *R. l.* bv. *viciae*, and C20:4 in *R. l.* bv. *trifolii* are shown. Also shown is the synthesis of the most abundant common fatty acid C18:1. For a model describing common fatty acid biosynthesis, see Fig. 8. NodF of *Rhizobium* is likely to function as a specialized ACP, possibly preventing reduction of enoyl intermediates. The rhizobial NodE protein is homologous to FabB and FabF and, unlike the regular β-keto-acyl-synthases, probably elongates enoyl-NodF yielding α,β-polyunsaturated fatty acids.

acyl intermediates during fatty acid biosynthesis are covalently linked to an acyl carrier protein (ACP), and a 4′-phosphopantetheine prosthetic group functions as a linker between ACP and the acyl chain. The NodF protein has sequence similarity to ACPs of *E. coli* and *Streptomyces* and it carries a 4′-phosphopantetheine prosthetic group (Geiger *et al.*, 1991; Shearman *et al.*, 1986) which is essential for the synthesis of host-specific LCOs carrying the C18:4 fatty acid in *R. l.* bv. *viciae* (Ritsema *et al.*, 1994). LCOs containing polyunsaturated fatty acids are also no longer produced in *R. meliloti*, *R. l.* bv. *viciae*, and *R. l.* bv. *trifolii* strains carrying deletions or Tn*5* insertions in the *nodE* gene (Spaink *et al.*, 1991, 1995a; Demont *et al.*, 1993). Since NodE has sequence homology with β-keto acyl synthases of *Streptomyces* and *E. coli*, in particular FabB (Bibb *et al.*, 1989), it is most likely that NodE functions as the condensing enzyme in the synthesis of the polyunsaturated fatty acids in *Rhizobium* LCO biosynthesis.

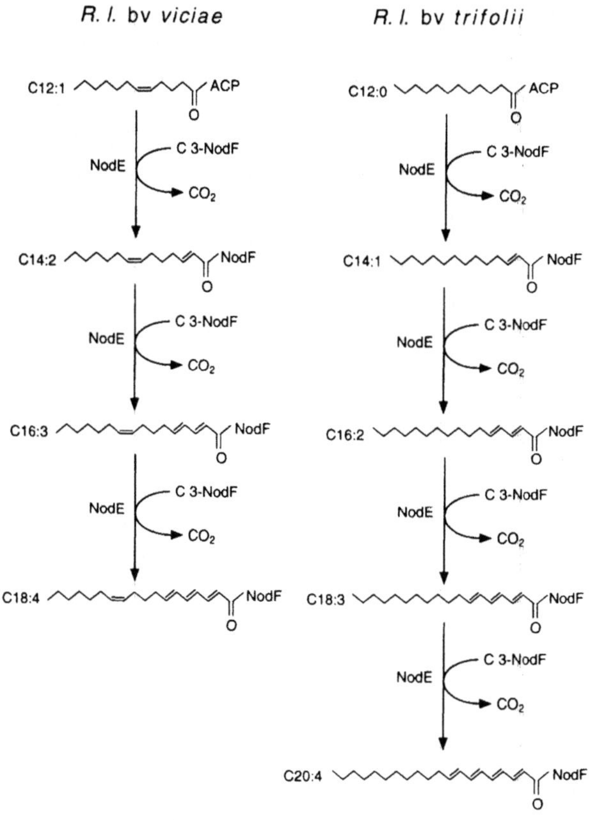


**FIGURE 9.** *(Continued).*

A model illustrating the most likely pathway for the synthesis of *nodFE*-dependent polyunsaturated fatty acids in *R. meliloti* and in *R. leguminosarum* biovars *viciae* and *trifolii* is shown in Fig. 9. The structure of *nodFE*-dependent fatty acids indicates that their synthesis is probably due to the inability of enoyl-reductase to reduce enoyl-NodF intermediates, and the unique ability of NodE to elongate α,β-unsaturated fatty acids. It seems likely that NodE preferentially interacts with NodF instead of ACP, but experiments confirming this hypothesis have not been reported.

NodF is the first nodulation protein for which the tertiary structure has been analyzed. Using NMR, Ghose *et al.* (1996) recently showed that NodF consists of three helices which fold in an antiparallel fashion, with the prosthetic group attachment site located near the beginning of the second helix. This is similar to the three-dimensional structure of *E. coli* ACP (Kim and Prestegard, 1990; Holak *et al.*, 1988). FabD (malonyl-CoA-acyltransferase) of *E. coli* can form malonyl-

NodF *in vitro* (Ritsema *et al.*, 1997). However, malonyl-NodF is not elongated by FabH, the enzyme which catalyzes the first committed step in fatty acid biosynthesis, namely the condensation of two malony-ACPs. It therefore seems most likely that the first specific reaction in the synthesis of the *nodFE*-dependent polyunsaturated fatty acids is the condensation of an acyl-ACP from the common fatty acid synthesis pathway with a malonyl-NodF by NodE, resulting in the formation of β-keto-acyl-NodF (Fig. 9).

The polyunsaturated fatty acids in the LCOs of *R. meliloti, R. l.* bv. *viciae,* and *R. l.* bv. *trifolii* differ in length. The *nodFE*-dependent fatty acids in LCOs produced by these rhizobia are C16, C18, and a mixture of C18 and C20 fatty acids, respectively. In a *R. meliloti nodE* mutant, a combination of NodE of *R. l.* bv. *viciae* and NodF of *R. meliloti* results in the synthesis of polyunsaturated C18 fatty acids (Demont *et al.*, 1993), which is the same length as that of the *nodFE*-dependent fatty acids in *R. l.* bv. *viciae* (Spaink *et al.*, 1991). This observation indicates that NodF is not likely to be involved in the determination of the length of *nodFE*-dependent fatty acids. It is therefore most likely that the length of the *nodFE*-dependent fatty acids is mainly determined by NodE. This is not unusual since the condensing enzymes involved in the common fatty acid biosynthesis are known to display distinct specificities for certain acyl chain lengths (Magnuson *et al.*, 1993). The absorption maxima of the polyunsaturated acyl chains produced in *R. l.* bv. *trifolii* indicate that all the *trans* double bonds are conjugated with the carbonyl group (van der Drift *et al.*, 1996; Spaink *et al.*, 1995a). Therefore these fatty acids apparently lack a *cis* double bond. This lack of *cis* double bonds in these fatty acids seems to be due to the special characteristics of *R. l.* bv. *trifolii* NodE, since the polyunsaturated fatty acids produced after expression of *R. l.* bv. *trifolii nodE* in a *R. l.* bv. *viciae nodE* mutant lack a *cis* double bond as well (Bloemberg *et al.*, 1995a).

A puzzling observation is that the synthesis of LCOs with C16:1 acyl chains in *R. meliloti* largely depends on the expression of *nodF* and *nodE*, even though C16:1 fatty acids can also be produced by the common fatty acid biosynthesis pathway (Demont *et al.*, 1993). However, *nodFE*-dependent synthesis of C16:1-containing LCOs can easily be explained by postulating enoyl-NodF reduction by enoyl-reductase in *R. meliloti*, since reduction of the *nodFE*-dependent C16:2-NodF results in the formation of C16:1-NodF.

*Rhizobium meliloti* is the only species in which the *nodG* gene has been found. NodG shows homology to β-keto-acyl reductases and dehydrogenases (Slabas *et al.*, 1992; Sherman *et al.*, 1989; Debellé *et al.*, 1986). An increase in β-keto-acyl reductase or a dehydrase is expected to result in an increased synthesis of enoyl-NodF, providing that the dehydrase would shift the equilibrium between β-hydroxy-acyl-NodF and enoyl-NodF toward enoyl-NodF (see Fig. 8). However, mutations in *nodG* hardly change the relative amounts of C16:1, C16:2, and C16:3 fatty acids in *R. meliloti* LCOs, showing that *nodG* is not essential for the synthesis of the *nodFE*-dependent polyunsaturated fatty acids in *R. meliloti* under laboratory conditions (Demont *et al.*, 1993). Interestingly, Demont *et al.* (1993) described that under culture conditions which lead to low oxygen concentrations, the relative amount of *R. meliloti* LCOs containing C16:1 fatty acids increases. This suggests that at low oxygen concentrations, enoyl-NodF reduction increases relative to elongation by NodE. The function of NodG may be to increase enoyl-NodF formation at low oxygen concentrations, thereby counteracting increased reduction of enoyl-NodF and ensuring the synthesis of C16:2 and C16:3 fatty acids under these conditions. Such a function of NodG would explain why inactivation of *nodG* has no effect on the synthesis of the *nodFE*-dependent polyunsaturated fatty acids in routinely used, well-aerated cultures.

### 5.3.2. *N*-Acylation of Chitin Oligosaccharides

The addition of the acyl chain to the terminally de-*N*-acetylated chitin oligosaccharides produced by NodC and NodB is mediated by NodA. Expression of *nodABC* in *Rhizobium* is sufficient for the synthesis of a basic LCO, containing a common acyl chain and lacking other modifications of the oligosaccharide backbone (Spaink *et al.*, 1991). NodA-dependent acylation of oligosaccharide backbones has been shown *in vitro* with extracts of *E. coli* strains expressing *R. meliloti nodA* and *nodB*, and extracts of *R. meliloti* cells (Atkinson *et al.*, 1994; Röhrig *et al.*, 1994). Acylation of sulfated chitin oligosaccharides was tested with di-, tri-, and tetrasaccharides, and could only be detected using the tetraose (Atkinson *et al.*, 1994). It therefore seems that oligosaccharides with a length of more than three sugar residues are the preferred acceptors for acylation by NodA. The *nodA* genes of *Bradyrhizobium* and *R. fredii* strains, which only produce LCOs containing common fatty acids, were introduced into *nodA* mutants of *R. l.* bv. *viciae* and *R. meliloti,* and are unable to direct the transfer of the *nodFE*-dependent polyunsaturated fatty acids to chitin oligosaccharides (Ritsema *et al.*, 1996; Debellé *et al.*, 1996). Whether this specificity is due to a specialized interaction between NodA and NodF, between NodA and the acyl chain, or a combination of both is not known.

## 5.4. Modifications at the Reducing Terminus

### 5.4.1. *O*-Sulfation

The enzymes involved in the synthesis of 3′-phosphoadenosine 5′-phosphosulfate (PAPS), the activated form of sulfate, are CysD and CysN (subunits of ATP sulfurylase) and CysC (APS kinase) (reviewed by Kredich, 1996). The *R. meliloti* nodulation proteins NodP and NodQ are homologous to the *E. coli* CysDNC proteins (Schwedock and Long, 1990). NodQ is a single polypeptide that appears to consist of two domains, one of which is similar to CysN, the other to CysC. NodQ is therefore expected to function both as an ATP sulfurylase subunit and as APS kinase. The C-terminal part of NodQ has a GTP-binding consensus, just like CysC. NodP and NodQ have been partially purified as a complex (Schwedock *et al.*, 1994). This complex synthesizes PAPS *in vitro* and is stimulated more than 50-fold by the addition of GTP. The *nodQ* gene can complement a *cysC* mutant, but complementation of a *cysNC* mutant requires both *nodP* and *nodQ*. This suggests that NodQ cannot act together with CysD, but specifically interacts with NodP. The *nodPQ* genes are present in two copies in *R. meliloti,* which are both required for complete sulfation of LCOs *in vivo* (Roche *et al.*, 1991). Only a part of the LCOs produced by *R. tropici* are sulfated (Fig. 2). This has been shown to be due to limiting NodPQ activity (Poupot *et al.*, 1995). These data show that the increased synthesis of the sulfate donor PAPS in *R. meliloti* and *R. tropici* by NodP and NodQ is essential for the production of sulfated LCOs.

Several reports have shown a *nodH*-dependent sulfotransferase activity *in vitro* using cell extracts or partially purified protein (Bourdineaud *et al.*, 1995; Schultze *et al.*, 1995; Ehrhardt *et al.*, 1995). NodH mediates the transfer of a sulfate group from PAPS to C6 of the reducing-end sugar of chitin oligosaccharides with a length ranging from two to six GlcNAc residues (longer oligosaccharides have not been tested). Schultze *et al.* (1995) showed that LCOs are the preferred substrate for NodH, and LCOs with a backbone of four GlcNAc residues are a better substrate than LCOs with a chitinpentaose backbone. These results indicate that sulfation of the chitin-oligosaccharide backbone by NodH occurs after acylation of the oligosaccharide by NodA.

### 5.4.2. *O*-Acetylation

*Rhizobium l.* bv. *viciae* strain TOM produces LCOs which contain an *O*-acetyl group at C6 of the reducing GlcNAc residue (Firmin *et al.*, 1993). The presence of this modification depends on the expression of the *nodX* gene. Insertion of a transposon in *nodX* leads to the production of LCOs lacking the *O*-acetyl group at the reducing end. Expression of *nodX* in a *R. l.* bv. *viciae* strain lacking *nodX* results in the production of LCOs with an *O*-acetyl group at C6 of the reducing GlcNAc residue (Firmin *et al.*, 1993). These data indicate that *nodX* encodes an *O*-acetyl transferase. Only LCOs containing a chitinpentaose backbone carry the *nodX*-dependent *O*-acetyl group. At present there are no experimental data which indicate at which position in the LCO synthesis pathway acetylation of LCOs by NodX occurs. However, NodX only has some homology with acetyl transferases involved in the acetylation of LPS and EPS and not with NodL (Fig. 10) (Buendia *et al.*, 1991; Clark *et al.*, 1991), suggesting that NodX-

**A**

```
                         **          *             *   *
RlvnodL   71  DYGFNIRIGAWVYINYNCVILDVAAVTIGDGTAIGPAVQIYTADHPHDPEQRQAGL
EcolacA   71  SYGSNIHIGRNFYANFNLTIVDDYTVTIGDNVLIAPNVTLSVTGHPVHHELRKNGE
EcocysE  133  DIHPAAKIGRGIMLDHATGIVVGETAVIENDVSILQSVTLGGTGKSGGDRHPK...

                           **    *      *    ***** *    *    * * * *
RlvnodL       QLGRPVSIGRHAWIGGGAIILPGVTIGDHAVIGAGSVVTRDVPAGSTAMGNPARVK 182
EcolacA       MYSFPITIGNNVWIGSHVVINPGVTIGDNSVIGAGSIVTKDIPPNVVAAGVPCRVI 182
EcocysE       .......IREGVMIGAGAKILGNIEVGRGAKIGAGSVVLQPVPPHTTAAGVPARIV 234
```

**B**

```
                             *                              *    *
RlenodX   15  FDLLRLFAACQVMFSHAWNWLHLGDPLNGTWVFDLLFSAPGVA..IFFLISGFLVT  66
RmeexoZ    1              MLFHAAEKT............GHHFTIGAAGVDVFFVISGFIMW  30
Sf6OAAT    7  FDTARLVAAMMVLVSHHYALSGQPEP....YLFGFE.SAGGIAVIIFFSISGYLIS  55
SenOAAT   16  MDYVKAFGMIFVLVGHI........N.NDIFNV.YY..AYLFHMPLFFFIGGVLYK  57
XcaOAT    32  IDATKAIAILLVVFCHA........K.GVPHGMTLF..AYSFHVPLFFLVSGWLAA  74
RlonolL   26  FDFAKAILITLVIIGHLLQYLIYQGT.DAFWLSPYFKSIYMFHMPLFMAISGYLSS  78
```

**FIGURE 10.** Multiple sequence alignments of acetyl transferases. (A) Multiple sequence alignment of cytosolic acetyl transferases. RlvnodL: *Rhizobium leguminosarum* bv. *viciae* NodL protein (190 amino acid residues, Accession No. P08632; Surin and Downie, 1988); EcolacA: *Escherichia coli* galactoside acetyl transferase (203 amino acid residues, Accession No. P07464; Hediger *et al.*, 1985); EcocysE: *Escherichia coli* serine acetyl transferase (273 amino acid residues, Accession No. P05796; Denk and Bock, 1987). These proteins belong to the CYSE/LACA/LPXA/NODL family of acetyl transferases and are composed of multiple repeats of the [LIV]-G-X motif. Invariant residues among all proteins are indicated by asterisks. (B) Multiple sequence alignment of membrane acetyl transferases. RlenodX: *Rhizobium leguminosarum* bv. *viciae* NodX protein (367 amino acid residues, Accession No. P08888; Davis *et al.*, 1988); RmeexoZ: *Rhizobium meliloti* exopolysaccharide production protein ExoZ (317 amino acid residues, Accession No. P26502; Buendia *et al.*, 1991); Sf6OAAT: bacteriophage SF6 *O*-antigen acetylase (333 amino acid residues, Accession No. P23214; Verma *et al.*, 1991); SenOAAT: *Salmonella enterica O*-antigen acetyltransferase (367 amino acid residues, Accession No. X60666; Wang *et al.*, 1992); XcaOAT: *Xanthomonas campestris* LPS, xanthan *O*-acetyltransferase (364 amino acid residues, Accession No. X78451); RlonolL: *Rhizobium loti* NolL protein (373 amino acid residues, Accession No. U22899; Scott *et al.*, 1996). Invariant residues among all proteins are indicated by asterisks.

mediated acetylation is more likely to take place with LCOs as acceptor than with free chitin oligosaccharides.

### 5.4.3. *O*-Fucosylation

The presence of an *O*-fucosyl residue at C6 of the reducing GlcNAc in LCOs is due to the activity of NodZ. NodZ has recently been purified to homogeneity after overexpression in *E. coli*, and was shown to have fucosyl transferase activity using GDP-fucose as fucosyl donor (Quinto *et al.*, 1997). These results are in agreement with the previous finding that a *nodZ* mutant produces LCOs lacking the 2-*O*-Me-fucose moiety (Stacey *et al.*, 1994). Moreover, introduction of *nodZ* into *R. l.* bv. *viciae* results in the synthesis of LCOs carrying an *O*-fucose at the reducing GlcNAc residue of the sugar backbone (López-Lara *et al.*, 1996). The *in vitro* study by Quinto *et al.* (1997) has shown that chitin oligosaccharides are more efficiently fucosylated by NodZ than LCOs, indicating that fucosylation occurs prior to acylation.

Insertional mutagenesis and *in vivo* studies have shown that *nolK* is essential for the synthesis of fucosylated LCOs in *A. caulinodans in vivo* (Mergaert *et al.*, 1996, 1997). Extracts of this mutant, however, are capable of catalyzing the transfer of fucose to chitin oligosaccharides or LCOs when radiolabeled GDP-fucose is added to the reaction mixture. These results indicate that NolK is likely to be involved in the synthesis of GDP-fucose, rather than being a fucosyltransferase itself. NolK is homologous to the *E. coli* proteins encoded by the *orf0.9* and *orf6.7* genes (Mergaert *et al.*, 1996, 1997). The *orf0.9* gene encodes a protein which is likely to be involved in the conversion of GDP-4-keto-6-deoxy-D-mannose into GDP-fucose (Aoyama *et al.*, 1994), whereas the *orf6.7* gene product is involved in the synthesis of GDP-colitose, which has some steps in common with the GDP-fucose pathway (Bastin and Reeves, 1995). Two different genes on the symbiotic plasmid of *Rhizobium* sp. NGR 234, which also produces fucosylated LCOs, show homology to *nolK* and other genes involved in GDP-fucose synthesis as well (Mergaert *et al.*, 1996, 1997; Fellay *et al.*, 1995).

**5.4.3a. 4-*O*-Sulfation of Fucose.** Recently, Hanin *et al.* (1997) have shown that *Rhizobium* sp. NGR234 carrying a mutation in *noeE* no longer produces LCOs which are sulfated on the fucose residue present on the reducing-end GlcNAc of the chitin-oligosaccharide backbone. Introduction of the *noeE* gene of *Rhizobium* sp. NGR234 into *R. fredii* USDA 257, a strain producing fucosylated LCOs which lack a sulphate group, results in the synthesis of sulfated LCOs. The *noeE* and *nodH* gene products show a clear sequence homology but these sulphotransferases have distinct substrate specificities. NoeE specifically transfers a sulphate group to fucosylated LCOs, whereas the sulphotransferase NodH is capable of using both fucosylated LCOs and nonfucosylated LCOs as substrate.

**5.4.3b. 2-*O*-Methylation of Fucose.** On the basis of insertional mutagenesis, *nolO* has been suggested to be involved in the synthesis of the 2-*O*-fucose moiety in *B. japonicum* LCOs, since the synthesis of 2-*O*-Me-fucosylated LCOs is apparently strongly reduced in a *nolO* mutant (Luka *et al.*, 1993). NolO, however, shows sequence similarity with carbamoyltransferases, as was discussed in a previous paragraph of this section. A likely explanation for this discrepancy is a possible negative effect of the transposon insertion in *nolO* on the expression of the *nodZ* gene, which encodes the fucosyltransferase responsible for the synthesis of fucosylated LCOs, and is located immediately downstream of *nolO* (see Fig. 3).

**5.4.3c. 3,4-*O*-Acetylation of Fucose.** There are no experimental indications for the identity of the gene encoding the acetyl transferase responsible for the synthesis of *O*-acetyl-fucose LCOs. However, Scott *et al.* (1996) recently reported the identification of a gene in *Rhizobium* sp. NGR234, *nolL*, which is essential for nodulation of two *Lotus* species. NolL interestingly has sequence homology with an *O*-acetyltransferase from *Xanthomonas campestris* and an *O*-antigen acetylase from *Shigella flexneri* (Clark *et al.*, 1991). NolL also shows some homology with NodX, which is likely to be an acetyl transferase and is required for the presence of an *O*-acetyl group at the reducing-end GlcNAc of the LCO oligosaccharide backbone (Fig. 10). These data suggest that NolL may acetylate fucosylated LCOs at the fucose residue. A preliminary characterization of LCOs produced by *nolL* mutants and *nolL* transconjugants is in agreement with this conclusion (Ovtsyna *et al.*, 1997).

### 5.4.4. *O*-Arabinosylation

An arabinosyl moiety has been detected in LCOs produced by *A. caulinodans* (Mergaert *et al.*, 1993). The *O*-arabinose group can be present at C6 or C3 of the reducing GlcNAc of the LCO sugar backbone (Mergaert *et al.*, 1996). A mutation in *noeC* of *A. caulinodans* prevents the synthesis of arabinose-containing LCOs, but whether this is due to inactivation of *noeC* itself or of downstream genes is still under investigation.

## 5.5. Topology of LCO Biosynthesis

The NodC protein, directing the synthesis of chitin oligosaccharides, is located in the inner membrane (Barny and Downie, 1993). The chitin oligosaccharide deacetylase NodB, however, has been localized in the cytosol (Schmidt *et al.*, 1988). This suggests that chitin oligosaccharides, synthesized at the inner membrane, may be released into the cytosol. The Nod proteins NodL and NodS, which are also cytosolic (Bloemberg *et al.*, 1994; Geelen *et al.*, 1995), modify the chitin oligosaccharide before it is acylated by NodA. The proposed acyltransferase NodA has been localized in the cytosol (Schmidt *et al.*, 1986), but has also been reported to be associated with the inner membrane (Johnson *et al.*, 1989). It is therefore not clear where acylation of the chitin oligosaccharides occurs. Due to the hydrophobic nature of the acyl chain, LCOs are expected to insert into the membrane, where they could be further modified by membrane-localized Nod proteins. NodH, for instance, mediates sulfation of LCOs, and has been reported to be membrane-associated (Schmidt *et al.*, 1991). The hydrophobicity of NodX suggests a membrane localization of this protein (Firmin *et al.*, 1993). It is, however, not known whether *O*-acetylation of the reducing GlcNAc in LCOs by NodX takes place before or after the transfer of a fatty acid to the chitin oligosaccharide backbone. In the model described above, a membrane localization would suggest that NodX acts after acylation of the oligosaccharide backbone. At present there are no experimental indications for the presence of Nod protein-enzyme complexes. Such complexes, however, could very well exist, since the experimental conditions used so far are not suited to demonstrate such protein interactions.

## 6. LCO SECRETION

LCOs produced in laboratory cultures are secreted in the medium, indicating the presence of an LCO secretion system in rhizobia. The *nodIJ* genes have been found in most rhizobial strains, and show a sequence similarity to a family of ATPases and membrane proteins known as the ATP-binding cassette transport systems (ABC transporters) (Vázquez *et al.*, 1993; Reizer *et al.*, 1992; Higgins *et al.*, 1986; Evans and Downie, 1986). Recent studies with *R. leguminosarum, Azorhizobium,* and *E. coli* strains containing cloned nodulation genes have shown that *nodI* and *nodJ* together are involved in the efficient secretion of LCOs (Fernández-López *et al.*, 1996; Spaink *et al.*, 1995b; Mckay and Djordjevic, 1993). In *R. l.* bv. *viciae, R. l.* bv. *trifolii,* and *Azorhizobium,* mutations in *nodIJ* only result in a reduction in LCO transport during the first few hours after *nod* gene induction, and do not seem to affect LCO transport after prolonged incubations, suggesting the existence of alternative LCO transport systems (Fernández-López *et al.*, 1996; Spaink *et al.*, 1995b). In contrast, Mckay and Djordjevic (1993) reported that *nodI* and *nodJ* are essential for LCO secretion in *R. l.* bv. *trifolii* strains after incubations overnight. The reason for this apparent discrepancy is not clear, but may be due to the differences in experimental conditions.

The NodIJ proteins are structurally most similar to the capsular polysaccharide transport systems KpsMT and CtrDBC of Gram-negative bacteria and the drug resistance system DrrAB of *Streptomyces* (Vázquez *et al.*, 1993; Reizer *et al.*, 1992; Higgins *et al.*, 1986; Evans and Downie, 1986). ABC export systems generally consist of an outer membrane or periplasmic protein (accessory proteins), a hydrophobic inner membrane protein (membrane-spanning proteins), and an ATP-binding protein [reviewed by Fath and Kolter (1993)]. A detailed model for polysialic acid transport by the Kps system in *E. coli* has recently been proposed by Bliss and Silver (1996). In this system a transient fusion or close contact between the inner and outer membranes has been proposed to be an important step in polysialic acid transport [also discussed by Whitfield and Valvano (1993)], and the accessory proteins are thought to play a role in this process. The membrane-spanning protein KpsM has been shown to act as a dimer (Pigeon and Silver, 1994). NodI clearly has ATP-binding motifs which are well conserved among all ATP-binding proteins. The NodJ proteins show sequence homology with the membrane-spanning proteins KpsT, BexB, and CtrD. NodJ proteins furthermore contain a series of leucines, each separated by six other amino acid residues in their N-terminus. This motif is known as a "leucine zipper," which has been identified as a dimerization domain in several other proteins (Busch and Sassone Corsi, 1990). If NodI and NodJ function in a way that is similar to the CPS transporters, accessory proteins may also be required for LCO transport. The NodT protein has similarity with outer membrane proteins that function with ABC transporters (Rivilla *et al.*, 1995). The highest sequence homology, however, is observed with proteins involved in protein export rather than export of lipid-linked saccharides. Moreover, a mutation in *nodT* does not seem to affect LCO transport in *R. l.* bv. *viciae* (Spaink *et al.*, 1995b). Another possible accessory protein in LCO transport may be a chromosomally located homolog of *nodT* which has been identified in *R. l.* bv. *viciae* (Rivilla and Downie, 1994). Other genes that may encode membrane proteins involved in LCO export are *nolF*, which resembles a putative membrane fusion protein, and *nolGHI*, which show homology to several outer membrane proteins involved in the efflux of cations (Saier *et al.*, 1994; Paulson *et al.*, 1996). These considerations are obviously rather speculative, since no experimental data have been reported on a possible involvement of these genes in LCO secretion.

*Escherichia coli* strains containing *nodABCSU* of *A. caulinodans*, and wild-type *A. caulinodans* strains produce a mixture of LCOs. Fernández-López *et al.* (1996) have shown that these LCOs are not all secreted with the same efficiency. This raises the possibility that NodIJ export systems not only may be specific for LCOs, but may even discriminate between different LCOs.

## 7. CONCLUDING REMARKS AND FUTURE PROSPECTS

### 7.1. LCO Biosynthesis

Although much progress has been made in assigning biochemical functions to nodulation proteins involved in the biosynthesis of LCOs, our knowledge about the precise order in which they act is still quite limited. The presence of decorations in LCOs does not appear to be a prerequisite for the addition of more modifications. This eliminates the possibility of elucidating the biosynthesis pathway by studying selective increases in LCO intermediates in rhizobial nodulation gene mutants. Consequently, detailed study on the substrate specificities of the individual enzymes in LCO biosynthesis has been the approach in several investigations. It will therefore be necessary to obtain the various LCO intermediates in large quantities to allow more detailed studies. Recently the complete nucleotide sequence of the symbiotic plasmid of *Rhizobium* sp. NGR234 was reported (Freiberg *et al.*, 1997). Many new open reading frames were found which may play a role in nodulation. The functional analysis of these new genes is expected to provide more insight in LCO biosynthesis and transport.

The increasing knowledge about nodulation gene functions and LCO structures provides explanations for observations which were not understood in the past. In contrast to *R. l.* bv. *viciae* strain TOM, for instance, other strains of *R. l.* bv. *viciae* are unable to nodulate *Pisum sativa* cultivar Afghanistan (Götz *et al.*, 1985; Lie, 1978). Canter Cremers *et al.* (1988) constructed *R. l.* bv. *trifolii* strains containing, in addition to their own symbiotic plasmids, the symbiotic plasmid of such a *R. l.* bv. *viciae* strain. Surprisingly, these recombinant *R. l.* bv. *trifolii* strains had acquired the ability to nodulate Afghanistan pea plants. Recent results have shown that *O*-acetylation at the reducing GlcNAc of LCOs by NodX is essential for nodulation of the Afghanistan cultivar by *R. l.* bv. *viciae* strain TOM (Firmin *et al.*, 1993). The *nodX* gene has been shown also to be present in *R. l.* bv. *trifolii* strains (Firmin *et al.*, 1993; Lewis-Henderson and Djordjevic, 1991). It is therefore expected that in the recombinant strain of Canter Cremers *et al.* NodX, encoded by the *nodX* gene on the *R. l.* bv. *trifolii* symbiotic plasmid, directed the acetylation of *R. l.* bv. *viciae* LCOs produced by the Nod proteins encoded by the *nod* genes on the introduced *R. l.* bv. *viciae* symbiotic plasmid. This would result in the synthesis of LCOs with a structure identical to that of strain TOM, and explains the nodulation of this recombinant strain on Afghanistan pea plants. No *O*-acetyl group on the reducing GlcNAc in *R. l.* bv. *trifolii* LCOs has been reported, but preliminary results obtained in our laboratory show that some *R. l.* bv. *trifolii* strains indeed produce such LCOs (M. Olsthoorn, personal communication).

Culture conditions can influence the level of LCO production and the synthesis of specific LCOs (Theberge *et al.*, 1996; Spaink *et al.*, 1995a; Demont *et al.*, 1993; Mckay and Djordjevic, 1993). It can therefore not be excluded that rhizobia in the soil produce LCOs with different characteristics. In addition, it is possible that LCOs that are produced as minor components under laboratory conditions may be synthesized more efficiently in the soil. The observation that several newly isolated strains of *B. japonicum* produce LCOs which are not produced by the related laboratory strains (J. Streeter, personal communication) also illustrates that the LCOs produced by laboratory strains are not necessarily representative for an entire rhizobial species.

### 7.2. LCO Perception by Plants

Several possible models for LCO perception by plants have been proposed. LCOs may interact with a classical transmembrane signaling receptor which allows some variation in LCO structure (Schultze and Kondorosi, 1996). Alternatively, multiple receptors could be present which may recognize different LCOs and may be responsible for different effects induced by LCOs (Heidstra and Bisseling, 1996; Ardourel *et al.*, 1994). A third possibility is that there is no specialized LCO receptor, but that LCOs interact with receptors or regulatory proteins which normally bind endogenous plant signal molecules controlling differentiation and development. LCOs may structurally resemble such plant molecules (Spaink *et al.*, 1993). Studies with *R. meliloti* LCOs have shown that modifications of the oligosaccharide backbone of LCOs can protect them against hydrolysis by root chitinases (Staehelin *et al.*, 1994, 1995). A lipochitin oligosaccharide hydrolase which does not degrade chitin has been detected in the exudate of intact plant roots. Interestingly, the production of this hydrolytic activity is induced by LCOs (Staehelin *et al.*, 1995). The biological role of this specific LCO hydrolysis, however, is unclear.

### 7.3. Do Plants Produce LCO-like Signals Themselves?

There are several indications that LCO-like molecules are involved in plant developmental processes other than nodulation. These indications have come from studies with nonleguminous plants. A carrot temperature-sensitive somatic embryogenesis mutant can be rescued by the addition of a chitinase purified from wild-type cultures (De Jong *et al.*, 1992). Interestingly, the same mutant could also be complemented by the addition of purified LCOs in nanomolar concentrations, suggesting that the chitinase releases a signal with structural homology with LCOs (De Jong *et al.*, 1993). Expression of the *Rhizobium nodA* and *nodB* genes in tobacco plants, either separately or in combination, leads to a disturbed development (Schmidt *et al.*, 1993). This indicates that NodA and NodB interfere with the synthesis of plant molecules which are important in plant morphogenesis, and which are structurally homologous with the chitin oligosaccharide backbone of rhizobial LCOs. Finally, LCOs have been shown to induce cell division and expression of an auxin-responsive gene in tobacco cells (Röhrig *et al.*, 1995). In this study synthetic LCOs were used and were active at concentrations as low as $10^{-13}$ M.

Since the studies described above were performed with nonleguminous plants, the ability to recognize LCOs as such appears not to be unique for the *Leguminosae*. The results furthermore support the hypothesis that LCO-like signal molecules may be produced by plants themselves. The NodC protein, responsible for the synthesis of chitin oligosaccharides, shows a significant similarity with the *Xenopus* DG42 protein which is transiently expressed during embryogenesis (Bulawa and Wasco, 1991; Rosa *et al.*, 1988). The DG42 protein, produced in an *in vitro* transcription/translation system, synthesizes chitin oligosaccharides *in vitro* (Semino and Robbins, 1995). Recently, it was reported that the development of zebrafish embryos is severely disturbed by microinjection of either (i) an antiserum that specifically recognizes DG42 and inhibits its chitin oligosaccharide synthase activity *in vitro* (Bakkers *et al.*, 1997), (ii) the fucosyltransferase NodZ which fucosylates chitin oligosaccharides (Bakkers *et al.*, 1997), and (iii) a chitinase which degrades chitin oligosaccharides (C. E. Semino, personal communication). Together these data indicate that chitin oligosaccharides or chitin oligosaccharide derivatives even play an important role during vertebrate development. A detailed characterization of the substrate specificity of the rhizobial enzymes involved in LCO biosynthesis will enable us to use these enzymes as valuable tools in the search for LCO-like compounds in plants and animals. Moreover, analysis of conserved functional domains in the Nod proteins which synthesize or modify chitin oligosaccharides will facilitate the cloning of genes which encode similar enzymes in other organisms.

ACKNOWLEDGMENTS. The authors wish to thank Drs. M. Hanin, C. Quinto, and T. Ritsema for making available unpublished results and helpful discussions.

## 8. REFERENCES

Aoyama, K., Haase, A. M., and Reeves, P. R., 1994, Evidence for effect of random genetic drift on G+C content after lateral transfer of fucose pathway genes to *Escherichia coli* K-12, *Mol. Biol. Evol.* **11**:829–838.

Ardourel, M., Demont, N., Debellé, F., Maillet, F., Debilly, F., Promé, J. C., Dénarié, J., and Truchet, G., 1994, *Rhizobium meliloti* lipooligosaccharide nodulation factors: Different structural requirements for bacterial entry into target root hair cells and induction of plant symbiotic developmental responses, *Plant Cell* **6**:1357–1374.

Ardourel, M., Lortet, G., Maillet, F., Roche, P., Truchet, G., Promé, J. C., and Rosenberg, C., 1995, In *Rhizobium meliloti*, the operon associated with the *nod* box n5 comprises *nodL*, *noeA* and *noeB*, three host-range genes specifically required for the nodulation of particular *Medicago* species, *Mol. Microbiol.* **17**:687–699.

Atkinson, E. M., Palcic, M. M., Hindsgaul, O., and Long, S. R., 1994, Biosynthesis of *Rhizobium meliloti* lipooligosaccharide Nod factors: NodA is required for an *N*-acyltransferase activity, *Proc. Natl. Acad. Sci. USA* **91**:8418–8422.

Baev, N., Endre, G., Petrovics, G., Banfalvi, Z., and Kondorosi, A., 1991, Six nodulation genes of *nod* box locus 4 in *Rhizobium meliloti* are involved in nodulation signal production: *nodM* codes for *D*-glucosamine synthetase, *Mol. Gen. Genet.* **228**:113–124.

Bakkers, J., Semino, C. E., Stroband, H., Kijne, J. W., Robbins, P. W., and Spaink, H. P., 1997, An important developmental role for oligosaccharides during early embryogenesis of cyprinid fish, *Proc. Natl. Acad. Sci. USA* **94**:7982–7986.

Balatti, P. A., Kovacs, L. G., Krishnan, H. B., and Pueppke, S. G., 1995, *Rhizobium* sp. NGR234 contains a functional copy of the soybean cultivar specificity locus, *nolXWBTUV*, *Mol. Plant–Microbe Interact.* **8**:693–699.

Barbour, W. M., Hattermann, D. R., and Stacey, G., 1991, Chemotaxis of *Bradyrhizobium japonicum* to soybean exudates, *Appl. Environ. Microbiol.* **57**:2635–2639.

Barnett, M. J., and Long, S. R., 1990, DNA sequence and translational product of a new nodulation regulatory locus: *syrM* has sequence similarity to NodD proteins, *J. Bacteriol.* **172**:3695–3700.

Barny, M. A., and Downie, J. A., 1993, Identification of the NodC protein in the inner but not the outer membrane of *Rhizobium leguminosarum*, *Mol. Plant–Microbe Interact.* **6:**669–672.

Barny, M. A., Schoonejans, E., Economou, E. S. A., Johnston, A. W. B., and Downie, J. A., 1996, The C-terminal domain of the *Rhizobium leguminosarum* chitin synthase NodC is important for function and determines the orientation of the N-terminal region in the inner membrane, *Mol. Microbiol.* **19:**443–453.

Bastin, D. A., and Reeves, P. R., 1995, Sequence and analysis of the *O* antigen gene (rfb) cluster of *Escherichia coli* O111, *Gene* **164:**17–23.

Battisti, L., Lara, J. C., and Leigh, J. A., 1992, Specific oligosaccharide form of the *Rhizobium meliloti* exopolysaccharide promotes nodule invasion in alfalfa, *Proc. Natl. Acad. Sci. USA* **89:**5625–5629.

Bec-Ferté, M. P., Krishnan, H. B., Promé, D., Savagnac, A., Pueppke, S. G., and Promé, J. C., 1994, Structures of nodulation factors from the nitrogen-fixing soybean symbiont *Rhizobium fredii* USDA257, *Biochemistry* **33:**11782–11788.

Bec-Ferté, M. P., Krishnan, H. B., Savagnac, A., Pueppke, S. G., and Promé, J. C., 1996, *Rhizobium fredii* synthesizes an array of lipooligosaccharides, including a novel compound with glucose inserted into the backbone of the molecule, *FEBS Lett.* **393:**273–279.

Bibb, M. J., Biro, S., Montamedi, H., Collins, J. F., and Hutchinson, C. R., 1989, Analysis of the nucleotide sequence of the *Streptomyces glaucescens tcmI* genes provides key information about the enzymology of polyketide antibiotic biosynthesis, *EMBO J.* **9:**2727–2736.

Bliss, J. M., and Silver, R. P., 1996, Coating the surface: A model for expression of capsular polysialic acid in *Escherichia coli* K1, *Mol. Microbiol.* **21:**221–231.

Bloemberg, G. V., Thomas-Oates, J. E., Lugtenberg, B. J. J., and Spaink, H. P., 1994, Nodulation protein NodL of *Rhizobium leguminosarum* *O*-acetylates lipo-oligosaccharides, chitin fragments and *N*-acetylglucosamine *in vitro*, *Mol. Microbiol.* **11:**793–804.

Bloemberg, G. V., Kamst, E., Harteveld, M., van der Drift, K. M. G. M., Haverkamp, J., Thomas-Oates, J. E., Lugtenberg, B. J. J., and Spaink, H. P., 1995a, A central domain of *Rhizobium* NodE protein mediates host specificity by determining the hydrophobicity of fatty acyl moieties of nodulation factors, *Mol. Microbiol.* **16:**1123–1136.

Bloemberg, G. V., Lagas, R., van Leeuwen, S., van der Marel, G., van Boom, J. H., Lugtenberg, B. J. J., and Spaink, H. P., 1995b, Substrate specificity and kinetic studies of nodulation protein NodL of *Rhizobium leguminosarum*, *Biochemistry* **34:**12712–12720.

Boundy-Mills, K. L., Kosslak, R. M., Tully, R. E., Pueppke, S. G., Lohrke, S., and Sadowsky, M. J., 1994, Induction of the *Rhizobium fredii nod* box-independent nodulation gene *nolJ* requires a functional *nodD1* gene, *Mol. Plant–Microbe Interact.* **7:**305–308.

Bourdineaud, J. P., Bono, J. J., Ranjeva, R., and Cullimore, J. V., 1995, Enzymatic radiolabelling to a high specific activity of legume lipo-oligosaccharide nodulation factors from *Rhizobium meliloti*, *Biochem. J.* **306:**259–264.

Breedveld, M. W., and Miller, K. J., 1994, Cyclic β-glucans of members of the family Rhizobiaceae, *Microbiol. Rev.* **58:**145–161.

Buendia, A. M., Enenkel, B., Köplin, R., Niehaus, K., Arnold, W., and Pühler, A., 1991, The *Rhizobium meliloti exoZ/exoB* fragment of megaplasmid2: ExoB functions as a UDP-glucose 4-epimerase and ExoZ shows homology to NodX of *Rhizobium leguminosarum* biovar *viciae* strain TOM, *Mol. Microbiol.* **5:**1519–1530.

Bulawa, C. E., and Wasco, W., 1991, Chitin and nodulation, *Nature* **353:**710.

Busch, S. J., and Sassone Corsi, P., 1990, Dimers, leucine zippers and DNA-binding domains, *Trends Genet.* **6:**36–40.

Caetano-Anolles, G., Wall, L. G., De Micheli, A. T., Macchi, E. M., Bauer, W. D., and Favelukas, G., 1988, Role of motility and chemotaxis in efficiency of nodulation by *Rhizobium meliloti*, *Plant Physiol.* **86:**1228–1235.

Canter Cremers, H. C. J., Wijffelman, C. A., Pees, E., Rolfe, B. G., Djordjevic, M. A., and

Lugtenberg, B. J. J., 1988, Host specific nodulation of plants of the pea cross-inoculation group is influenced by genes in fast growing *Rhizobium* downstream *nodC*, *J. Plant Physiol.* **132:**398–404.

Cardenas, L., Dominquez, J., Quinto, C., López-Lara, I. M., Lugtenberg, B. J. J., Spaink, H. P., Rademaker, G. J., Haverkamp, J., and Thomas-Oates, J. E., 1995, Isolation, chemical structure and biological activity of the lipo-chitin oligosaccharide nodulation signals from *Rhizobium etli*, *Plant Mol. Biol.* **29:**453–464.

Carlson, R. W., Sanjuan, J., Bhat, R., Glushka, J., Spaink, H. P., Wijfjes, A. H. M., van Brussel, A. A. N., Stokkermans, T. J. W., Peters, K., and Stacey, G., 1993, The structures and biological activities of the lipo-oligosaccharide nodulation signals produced by type I and type II strains of *Bradyrhizobium japonicum*, *J. Biol. Chem.* **268:**18372–18381.

Clark, C. A., Beltrame, J., and Manning, P. A., 1991, The *oac* gene encoding a lipopolysaccharide O-antigen acetylase maps adjacent to the integrase-encoding gene on the genome of *Shigella flexneri* bacteriophage Sf6, *Gene* **107:**43–52.

Cren, M., Kondorosi, A., and Kondorosi, E., 1995, NolR controls expression of the *Rhizobium meliloti* nodulation genes involved in the core Nod factor synthesis, *Mol. Microbiol.* **15:**733–747.

Davis, E. O., Evans, I. J., and Johnston, A. W. B., 1988, Identification of *nodX*, a gene that allows *Rhizobium leguminosarum* biovar *viciae* strain TOM to nodulate Afghanistan peas, *Mol. Gen. Genet.* **212:**531–535.

DeAngelis, P. L., Papaconstantinou, J., and Weigel, P. H., 1993, Molecular cloning, identification, and sequence of the hyaluronan synthase gene from group A *Streptococcus pyogenes*, *J. Biol. Chem.* **268:**19181–19184.

Debellé, F., Rosenberg, C., Vasse, J., Maillet, F., Martinez, E., Dénarié, J., and Truchet, G., 1986, Assignment of symbiotic developmental phenotypes to common and specific nodulation (*nod*) genetic loci of *Rhizobium meliloti*, *J. Bacteriol.* **168:**1075–1086.

Debellé, F., Roche, P., Plazanet, C., Maillet, F., Pujol, C., Ardourel, M., Demont, N., Rosenberg, C., Truchet, G., Promé, J. C., and Dénarié, J., 1995, The genetics of *Rhizobium* host range control: Allelic and non-allelic variation, in *Nitrogen Fixation: Fundamentals and Applications* (I. A. Tikhonovich, ed.), pp. 275–280. Kluwer, The Netherlands.

Debellé, F., Plazanet, C., Roche, P., Pujol, C., Savagnac, A., Rosenberg, C., Promé, J. C., and Dénarié, J., 1996, The NodA proteins of *Rhizobium meliloti* and *Rhizobium tropici* specify the *N*-acylation of Nod factors by different fatty acids, *Mol. Microbiol.* **22:**303–314.

De Jong, A. J., Cordewener, J. H. G., Lo Schavio, F., Terzi, M., Vandekerckhove, J., van Kammen, A., and de Vries, S., 1992, A carrot somatic embryo mutant is rescued by chitinase, *Plant Cell* **4:**425–433.

De Jong, A. J., Heidstra, R., Spaink, H. P., Hartog, M. V., Hendriks, T., Lo Schavio, F., Terzi, M., Bisseling, T., van Kammen, A., and de Vries, S., 1993, A plant somatic embryo mutant is rescued by rhizobial lipo-oligosaccharides, *Plant Cell* **5:**615–620.

Demont, N., Debellé, F., Aurelle, H., Dénarié, J., and Promé, J. C., 1993, Role of the *Rhizobium meliloti nodF* and *nodE* genes in the biosynthesis of lipo-oligosaccharidic nodulation factors, *J. Biol. Chem.* **268:**20134–20142.

Dénarié, J., Debellé, F., and Promé, J. C., 1996, *Rhizobium* lipo chitooligosaccharide nodulation factors: Signaling molecules mediating recognition and morphogenesis, *Annu. Rev. Biochem.* **65:**503–535.

Denk, D., and Bock, A., 1987, L-Cysteine biosynthesis in *Escherichia coli:* Nucleotide sequence and expression of the serine acetyltransferase (*cysE*) gene from the wild-type and a cysteine-excreting mutant, *J. Gen. Microbiol.* **133:**3–25.

Dillard, J. P., Vandersea, M. W., and Yother, J., 1995, Characterization of the cassette containing genes for type 3 capsular polysaccharide biosynthesis in *Streptococcus pneumoniae*, *J. Exp. Med.* **181:**973–983.

Djordjevic, S. P., Chen, H., Batley, M., Redmond, J. W., and Rolfe, B. G., 1987, Nitrogen fixation ability of exopolysaccharide synthesis mutants of *Rhizobium* sp. strain NGR234 and *Rhizobium trifolii* is restored by the addition of homologous exopolysaccharides, *J. Bacteriol.* **169:**53–60.

Dobert, R. C., Brei, B. T., and Triplett, E. W., 1994, DNA sequence of the common nodulation genes of *Bradyrhizobium elkanii* and their phylogenetic relationship to those of other nodulating bacteria, *Mol. Plant–Microbe Interact.* **7:**564–572.

Downie, J. A., 1989, The *nodL* gene from *Rhizobium leguminosarum* is homologous to the acetyl transferases encoded by *lacA* and *cysE*, *Mol. Microbiol.* **3:**1649–1651.

Dutka-Malen, S., Mazodier, P., and Badet, B., 1988, Molecular cloning and overexpression of the glucosamine synthase gene from *Escherichia coli*, *Biochimie* **70:**287–290.

Ehrhardt, D. W., Atkinson, E. M., Faull, K. F., Freedberg, D. I., Sutherlin, D. P., Armstrong, R., and Long, S. R., 1995, *In vitro* sulfotransferase activity of NodH, a nodulation protein of *Rhizobium meliloti* required for host-specific nodulation, *J. Bacteriol.* **177:**6237–6245.

Evans, I. J., and Downie, J. A., 1986, The *nodI* gene product of *Rhizobium leguminosarum* is closely related to ATP-binding bacterial transport proteins: Nucleotide sequence analysis of the *nodI* and *nodJ* genes, *Gene* **43:**95–101.

Fath, M. J., and Kolter, R., 1993, ABC transporters: Bacterial exporters, *Microbiol. Rev.* **57:**995–1017.

Fellay, R., Perret, X., Viprey, V., Broughton, W. J., and Brenner, S., 1995, Organization of host-inducible transcripts on the symbiotic plasmid of *Rhizobium* sp NGR234, *Mol. Microbiol.* **16:**657–667.

Felle, H. H., Kondorosi, E., Kondorosi, A., and Schultze, M., 1995, Nod signal-induced plasma membrane potential changes in alfalfa root hairs are differentially sensitive to structural modifications of the lipochitooligosaccharide, *Plant J.* **7:**939–947.

Fernández-López, M., D'Haeze, W., Mergaert, P., Verplancke, C., Promé, J. C., van Montagu, M., and Holsters, M., 1996, Role of *nodI* and *nodJ* in lipo chitooligosaccharide secretion in *Azorhizobium caulinodans* and *Escherichia coli*, *Mol. Microbiol.* **20:**993–1000.

Firmin, J. L., Wilson, K. E., Carlson, R. W., Davies, A. E., and Downie, J. A., 1993, Resistance to nodulation of cv Afghanistan peas is overcome by *nodX* which mediates an *O*-acetylation of the *Rhizobium leguminosarum* lipo-oligosaccharide nodulation factor, *Mol. Microbiol.* **10:**351–360.

Fisher, R. F., Egelhoff, T., Mulligan, J. T., and Long, S. R., 1988, Specific binding of proteins from *Rhizobium meliloti* cell-free extracts containing NodD to DNA sequences upstream of inducible nodulation genes, *Genes Dev.* **2:**282–293.

Folch-Mallol, J. L., Marroqui, S., Sousa, S., Manyani, H., López-Lara, I. M., van der Drift, K. M. G. M., Haverkamp, J., Quinto, C., Gil-Serrano, A., Thomas-Oates, J. E., Spaink, H. P., and Megias, M., 1996, Characterization of *Rhizobium tropicii* CIAT899 nodulation factors: The role of *nodH* and *nodPQ* genes in their sulphation, *Mol. Plant–Microbe Interact.* **9:**151–163.

Franssen, H. J., Vijn, I., Yang, W. C., and Bisseling, T., 1992, Developmental aspects of the *Rhizobium*–legume symbiosis, *Plant Mol. Biol.* **19:**89–107.

Freiberg, C., Fellay, R., Bairoch, A., Broughton, W. J., Rosenthal, A., and Perret, X., 1997, Molecular basis of symbiosis between Rhizobium and Legumes, *Nature* **387:**394–401.

Gaworzewska, E. T., and Carlile, M. J., 1982, Positive chemotaxis of *Rhizobium leguminosarum* and other bacteria towards root exudates from legumes and other plants, *J. Gen. Microbiol.* **128:**1179–1188.

Geelen, D., Leyman, B., Mergaert, P., Klarskov, K., van Montagu, M., Geremia, R., and Holsters, M., 1995, NodS is an *S*-adenosyl-*L*-methionine-dependent methyltransferase that methylates chitooligosaccharides deacetylated at the non-reducing end, *Mol. Microbiol.* **17:**387–397.

Geiger, O., Spaink, H. P., and Kennedy, E. P., 1991, Isolation of the *Rhizobium leguminosarum* NodF nodulation protein: NodF carries a 4′-phosphopantetheine prosthetic group, *J. Bacteriol.* **173:**2872–2878.

Geremia, R. A., Mergaert, P., Geelen, D., van Montagu, M., and Holsters, M., 1994, The NodC protein of *Azorhizobium caulinodans* is an *N*-acetylglucosaminyltransferase, *Proc. Natl. Acad. Sci. USA* **91**:2669–2673.

Ghose, R., Geiger, O., and Prestegard, J. H., 1996, NMR investigations of the structural properties of the nodulation protein, NodF, from *Rhizobium leguminosarum* and its homology with *Escherichia coli* acyl carrier protein, *FEBS Lett.* **388**:66–72.

Goethals, K., Mergaert, P., Gao, M., Geelen, D., van Montagu, M., and Holsters, M., 1992, Identification of a new inducible nodulation gene in *Azorhizobium caulinodans*, *Mol. Plant–Microbe Interact.* **5**:405–411.

Göttfert, M., 1993, Regulation and function of rhizobial nodulation genes, *FEMS Microbiol Rev.* **104**:39–64.

Göttfert, M., Grob, P., and Hennecke, H., 1990, Proposed regulatory pathway encoded by the *nodV* and *nodW* genes, determinants of host specificity in *Bradyrhizobium japonicum*, *Proc. Natl. Acad. Sci. USA* **87**:2680–2684.

Götz, R., Evans, I. J., Downie, J. A., and Johnston, A. W. B., 1985, Identification of the host-range DNA which allows *Rhizobium leguminosarum* strain TOM to nodulate cv. Afghanistan peas, *Mol Gen. Genet.* **201**:296–300.

Hanin, M., Jabbouri, S., Quesadavincens, D., Freiberg, C., Perret, X., Promé, J. C., Broughton, W. J., and Fellay, R., 1997, Sulphation of Rhizobium sp. NGR234 Nod factors is dependent on noeE, a new host specificity gene, *Mol. Microbiol.* **24**:1119–1129.

Heath, R. J., and Rock, C. O., 1995, Regulation of malonyl-CoA metabolism by acyl–acyl carrier protein and beta-ketoacyl–acyl carrier protein synthases in *Escherichia coli*, *J. Biol. Chem.* **270**:15531–15538.

Hediger, M. A., Johnson, D. F., Nierlich, D. P., and Zabin, I., 1985, DNA sequence of the lactose operon: The *lacA* gene and the transcriptional termination region, *Proc. Natl. Acad. Sci. USA* **82**:6414–6418.

Heidstra, R., and Bisseling, T., 1996, Nod factor induced host responses and mechanisms of Nod factor perception, *New Phytol.* **133**:25–43.

Heidstra, R., Geurts, R., Franssen, H., Spaink, H. P., van Kammen, A., and Bisseling, T., 1994, Root hair deformation activity of nodulation factors and their fate on *Vicia sativa*, *Plant Physiol.* **105**:787–797.

Higgins, C. F., Hiles, I. D., Salmond, G. P. C., Gill, D. R., Downie, J. A., Evans, I. J., Holland, I. B., Gray, L., Buckel, S. D., Bell, A. W., and Hermodson, M. A., 1986, A family of related ATP-binding subunits coupled to many distinct biological processes in bacteria, *Nature* **323**:448–450.

Holak, T. A., Nilges, M., Prestegard, J. H., Gronenborn, A. M., and Clore, G. M., 1988, Three-dimensional structure of acyl carrier protein in solution determined by nuclear magnetic resonance and the combined use of dynamical simulated annealing and distance geometry, *Eur. J. Biochem.* **175**:9–15.

Iñón de Iannino, N., Pueppke, S. G., and Ugalde, R. A., 1995, Biosynthesis of the Nod factor chitooligosaccharide backbone in *Rhizobium fredii* is controlled by the concentration of UDP-*N*-acetyl-*D*-glucosamine, *Mol. Plant–Microbe Interact.* **8**:292–301.

Jabbouri, S., Fellay, R., Talmont, F., Kamalaprija, P., Burger, U., Relic, B., Promé, J. C., and Broughton, W. J., 1995, Involvement of *nodS* in *N*-methylation and *nodU* in 6-*O*-carbamoylation of *Rhizobium* sp NGR234 nod factors, *J. Biol. Chem.* **270**:22968–22973.

John, M., Röhrig, H., Schmidt, J., Wieneke, U., and Schell, J., 1993, *Rhizobium* NodB protein involved in nodulation signal synthesis is a chitooligosaccharide deacetylase, *Proc. Natl. Acad. Sci. USA* **90**:625–629.

Johnson, D., Evans Roth, L., and Stacey, G., 1989, Immunogold localization of the NodC and NodA proteins of *Rhizobium meliloti*, *J. Bacteriol.* **171**:4583–4588.

Kafetzopoulos, D., Thireos, G., Vournakis, J. N., and Bouriotis, V., 1993, The primary structure of a fungal chitin deacetylase reveals the function for two bacterial gene products, *Proc. Natl. Acad. Sci. USA* **90:**8005–8008.

Kamst, E., van der Drift, K. M. G. M., Thomas-Oates, J. E., Lugtenberg, B. J. J., and Spaink, H. P., 1995, Mass spectrometric analysis of chitin oligosaccharides produced by *Rhizobium* NodC protein in *Escherichia coli, J. Bacteriol.* **177:**6282–6285.

Kamst, E., Lugtenberg, B. J. J., and Spaink, H. P., 1996, Chitin-oligosaccharide synthesis by the *Rhizobium* NodC protein, in *Chitin Enzymology* (R. A. A. Muzzarelli, ed.), pp. 329–338, Atec Edizioni, Italy.

Kamst, E., Pilling, J., Raamsdonk, L. M., Lugtenberg, B. J. J., and Spaink, H. P., 1997a, *Rhizobium* nodulation protein NodC is an important determinant of chitin oligosaccharide chain length in Nod factor biosynthesis, *J. Bacteriol.* **179:**2103–2108.

Kamst, E., Bakkers, J., Quaedvlieg, N. E. M., Pilling, J., Kijne, J. W., Lugtenberg, B. J. J., and Spaink, H. P., 1997b, Chitin oligosaccharide synthesis by the bacterial NodC protein and zebrafish DG42 starts at the reducing terminal residue, submitted.

Kannenberg, E. L., and Brewin, N. J., 1994, Host-plant invasion by *Rhizobium:* The role of cell surface components, *Trends Microbiol.* **2:**277–283.

Kijne, J. W., Bakhuizen, R., van Brussel, A. A. N., Caner Cremers, H. C. J., Diaz, C. L., de Pater, B. S., Smit, G., Spaink, H. P., Swart, S., Wijffelman, C. A., and Lugtenberg, B. J. J., 1992, The *Rhizobium* trap: Root hair curling in root-nodule symbiosis, in *Perspectives in Plant Cell Recognition* (J. A. Callow and J. R. Green, eds.), pp. 267–284, Cambridge University Press, Cambridge.

Kim, Y., and Prestegard, J. H., 1990, Refinement of the NMR structures for acyl carrier protein with scalar coupling data, *Proteins* **8:**377–385.

Kredich, N. M., 1996, Biosynthesis of cysteine, in *Escherichia coli and Salmonella. Cellular and Molecular Biology* Vol. 2 (F. C. Neidhardt, ed.), pp. 514–527, ASM Press, Washington, D.C.

Leigh, J. A., and Walker, G. C., 1994, Exopolysaccharides of *Rhizobium*—Synthesis, regulation and symbiotic function, *Trends Genet.* **10:**63–67.

Lerouge, P., Roche, P., Faucher, C., Maillet, F., Truchet, G., Promé, J. C., and Dénarié, J., 1990, Symbiotic host-specificity of *Rhizobium meliloti* is determined by a sulphated and acylated glucosamine oligosaccharide signal, *Nature* **344:**781–784.

Lewis-Henderson, W. R., and Djordjevic, M. A., 1991, *nodT*, a positively acting cultivar specificity determinant controlling nodulation of *Trifolium subterraneum* by *Rhizobium leguminosarum* biovar *trifolii, Plant Mol. Biol.* **16:**515–526.

Lie, T. A., 1978, Symbiotic specialization in pea plants: The requirement of specific *Rhizobium* strains for peas from Afghanistan, *Ann. Appl. Biol.* **88:**462–465.

Loh, J., Garcia, M., and Stacey, G., 1997, NodV and NodW, a second flavonoid recognition system regulating nod gene expression in Bradyrhizobium japonicum, *J. Bacteriol.* **179:**3013–3020.

López-Lara, I. M., van den Berg, J. D. J., Thomas-Oates, J. E., Glushka, J., Lugtenberg, B. J. J., and Spaink, H. P., 1995a, Structural identification of the lipo-chitin oligosaccharide nodulation signals of *Rhizobium loti, Mol. Microbiol.* **15:**627–638.

López-Lara, I. M., van der Drift, K. M. G. M., van Brussel, A. A. N., Haverkamp, J., Lugtenberg, B. J. J., Thomas-Oates, J. E., and Spaink, H. P., 1995b, Induction of nodule primordia on *Phaseolus* and *Acacia* by lipo-chitin oligosaccharide nodulation signals from broad host range *Rhizobium* strain GRH2, *Plant Mol. Biol.* **29:**465–477.

López-Lara, I. M., Blok-Tip, L., Quinto, C., Garcia, M. L., Stacey, G., Bloemberg, G. V., Lamers, G. E. M., Lugtenberg, B. J. J., Thomas-Oates, J. E., and Spaink, H. P., 1996, NodZ of *Bradyrhizobium* extends the nodulation host range of *Rhizobium* by adding a fucosyl residue to nodulation signals, *Mol. Microbiol.* **21:**397–408.

Luka, S., Sanjuan, J., Carlson, R. W., and Stacey, G., 1993, *nolMNO* genes of *Bradyrhizobium japonicum* are co-transcribed with *nodYABCSUIJ*, and *nolO* is involved in the synthesis of the lipo-oligosaccharide nodulation signals, *J. Biol. Chem.* **268:**27053–27059.

Magnuson, K., Jackowski, S., Rock, C. O., and Cronan, J. E., 1993, Regulation of fatty acid biosynthesis in *Escherichia coli*, *Microbiol. Rev.* **57:**522–542.

Maharaj, R., May, T. B., Wang, S. K., and Chakrabarty, A. M., 1993, Sequence of the *alg8* and *alg44* genes involved in the synthesis of alginate by *Pseudomonas aeruginosa*, *Gene* **136:**267–269.

Marie, C., Barny, M. A., and Downie, J. A., 1992, *Rhizobium leguminosarum* has two glucosamine synthases, GlmS and NodM, required for nodulation and development of nitrogen-fixing nodules, *Mol. Microbiol.* **6:**843–851.

Matthysse, A. G., White, S., and Lightfoot, R., 1995, Genes required for cellulose synthesis in *Agrobacterium tumefaciens*, *J. Bacteriol.* **177:**1069–1075.

Mckay, I. A., and Djordjevic, M. A., 1993, Production and excretion of Nod metabolites by *Rhizobium leguminosarum* bv. *trifolii* are disrupted by the same environmental factors that reduce nodulation in the field, *Appl. Env. Nicrobiol.* **59:**3385–3392.

Meinhardt, L. W., Krishnan, H. B., Balatti, P. A., and Pueppke, S. G., 1993, Molecular cloning and characterization of a sym plasmid locus that regulates cultivar-specific nodulation of soybean of *Rhizobium fredii* USDA257, *Mol. Microbiol.* **9:**17–29.

Mergaert, P., van Montagu, M., Promé, J. C., and Holsters, M., 1993, Three unusual modifications, a D-arabinosyl, a *N*-methyl, and carbamoyl group, are present on the Nod factors of *Azorhizobium caulinodans* strain ORS571, *Proc. Natl. Acad. Sci. USA* **90:**1551–1555.

Mergaert, P., D'Haeze, W., Geelen, D., Promé, D., van Montagu, M., Geremia, R., Promé, J. C., and Holsters, M., 1995, Biosynthesis of *Azorhizobium caulinodans* Nod factors: Study of the activity of the NodABC proteins by expression of the genes in *Escherichia coli*, *J. Biol. Chem.* **270:**29217–29223.

Mergaert, P., D'Haeze, W., Fernández-López, M., Geelen, D., Goethals, K., Promé, J. C., van Montagu, M., and Holsters, M., 1996, Fucosylation and arabinosylation of Nod factors in *Azorhizobium caulinodans:* Involvement of *nolK*, *nodZ* as well as *noeC* and/or downstream genes, *Mol. Microbiol.* **21:**409–419.

Mergaert, P., van Montagu, M., and Holsters, M., 1997, The nodulation gene nolK of Azorhizobium caulinodans is involved in the formation of GDP fucose from GDP mannose, *FEBS Lett.* **409:**312–316.

Mulligan, J. T., and Long, S. R., 1989, A family of activator genes regulates expression of *Rhizobium meliloti* nodulation genes, *Genetics* **127:**7–18.

Mylona, P., Pawlowski, K., and Bisseling, T., 1995, Symbiotic nitrogen fixation, *Plant Cell* **7:**869–885.

Nagahashi, S., Sudoh, M., Ono, N., Sawada, R., Yamaguchi, E., Uchida, Y., Mio, T., Takagi, M., Arisawa, M., and Yamadaokabe, H., 1995, Characterization of chitin synthase 2 of *Saccharomyces cerevisiae*—Implication of two highly conserved domains as possible catalytic sites, *J. Biol. Chem.* **270:**13961–13967.

Ovtsyna, A. O., Veldhuis, A., López-Lara, I. M., Wijfjes, A. H. M., Quinto, C., Geurts, R., Bisseling, T., Scott, D. B., Tikhonovich, I. A., Lugtenberg, B. J. J., and Spaink, H. P., 1997, The host-specific role of chemical modifications at the reducing terminus of lipo-chitin oligosaccharides, in press.

Parniske, M., Schmidt, P. E., Kosch, K., and Muller, P., 1994, Plant defense responses of host plants with determinate nodules induced by EPS-defective *exoB* mutants of *Bradyrhizobium japonicum*, *Mol. Plant–Microbe Interact.* **7:**631–638.

Paulson, I. T., Brown, M. H., and Skurray, R. A., 1996, Proton-dependent multidrug efflux systems, *Microbiol. Rev.* **60:**575–608.

Perotto, S., Brewin, N. J., and Kannenberg, E. L., 1994, Cytological evidence for a host defense response that reduces cell and tissue invasion in pea nodules by lipopolysaccharide-defective mutants of *Rhizobium leguminosarum* strain 3841, *Mol. Plant–Microbe Interact.* **7**:99–112.

Peters, N. K., and Verma, D. P. S., 1990, Phenolic compounds as regulators of gene expression in plant–microbe interactions, *Mol. Plant–Microbe Interact.* **3**:4–8.

Pigeon, R. P., and Silver, R. P., 1994, Topological and mutational analysis of KpsM, the hydrophobic component of the ABC-transporter involved in the export of polysialic acid in *Escherichia coli* K1, *Mol. Microbiol.* **14**:871–881.

Plazanet, C., Refregier, G., Demont, N., Truchet, G., and Rosenberg, C., 1995, The *Rhizobium meliloti* region located downstream of the *nod* box n6 is involved in the specific nodulation of *Medicago lupulina*, *FEMS Microbiol. Lett.* **133**:285–291.

Poupot, R., Martinez-Romero, E., and Promé, J. C., 1993, Nodulation factors from *Rhizobium tropici* are sulphated or non-sulphated chitopentasaccharides containing an *N*-methyl-*N*-acylglucosaminyl terminus, *Biochemistry,* **32**:10430–10435.

Poupot, R., Martinez-Romero, E., Gautier, N., and Promé, J. C., 1995, Wild type *Rhizobium etli,* a bean symbiont, produces acetyl-fucosylated, *N*-methylated, and carbamoylated nodulation factors, *J. Biol. Chem.* **270**:6050–6055.

Price, N. P. J., Relic, B., Taimont, F., Lewin, A., Promé, D., Pueppke, S. G., Maillet, F., Dénarié, J., Promé, J. C., and Broughton, W. J., 1992, Broad-host-range *Rhizobium* species strain NGR234 secretes a family of carbamoylated, and fucosylated, nodulation signals that are *O*-acetylated or sulphated, *Mol. Microbiol.* **6**:3575–3584.

Pueppke, S. G., 1996, The genetic and biochemical basis for nodulation of legumes by rhizobia, *Crit. Rev. Biotech.* **16**:1–51.

Quinto, C., Wijfjes, A. H. M., Bloemberg, G. V., Blok-Tip, L., López-Lara, I. M., Lugtenberg, B. J. J., Thomas-Oates, J. E., and Spaink, H. P., 1997, Bacterial nodulation protein NolZ is a chitin oligosaccharide fucosyltransferase which can also recognize related substrates of animal origin, *Proc. Natl. Acad. Sci. USA* **94**:4336–4341.

Reizer, J., Reizer, A., and Saier, M. H., 1992, A new subfamily of bacterial ABC-type transport systems catalyzing export of drugs and carbohydrates, *Protein Sci.* **1**:1326–1332.

Ritsema, T., Geiger, O., van Dillewijn, P., Lugtenberg, B. J. J., and Spaink, H. P., 1994, Serine residue 45 of nodulation protein NodF from *Rhizobium leguminosarum* bv. *viciae* is essential for its biological function, *J. Bacteriol.* **176**:7740–7743.

Ritsema, T., Wijfjes, A. H. M., Lugtenberg, B. J. J., and Spaink, H. P., 1996, *Rhizobium* nodulation protein NodA is a host-specific determinant of the transfer of fatty acids in Nod factor biosynthesis, *Mol. Gen. Genet.* **251**:44–51.

Ritsema, T., Gehring, A. M., Stuitje, A. R., van der Drift, K. M. G. M., Dandal, I., Lambalot, R. H., Walsh, C. T., Thomas-Oates, J. E., Lugtenberg, B. J. J., and Spaink, H. P., 1997, Functional analysis of interspecies chimera of acyl carrier proteins indicates a specialized domain for protein recognition, *Mol. Gen. Genet.*, in press.

Rivilla R., and Downie, J. A., 1994, Identification of a *Rhizobium leguminosarum* gene homologous to *nodT* but located outside the symbiotic plasmid, *Gene* **144**:87–91.

Rivilla, R., Sutton, J. M., and Downie, J. A., 1995, *Rhizobium leguminosarum* NodT is related to a family of outer-membrane transport proteins that includes TolC, PrtF, CyaE and AprF, *Gene* **161**:27–31.

Robyt, J. F., 1979, Mechanisms involved in the biosynthesis of polysaccharides, *Trends Biochem. Sci.* **4**:47–49.

Roche, P., Debellé, F., Maillet, F., Lerouge, P., Faucher, C., Truchet, G., Dénarié, J., and Promé, J. C., 1991, Molecular basis of symbiotic host specificity in *Rhizobium meliloti: nodH* and *nodPQ* genes encode the sulfation of lipooligosaccharides signals, *Cell* **67**:1131–1143.

Röhrig, H., Schmidt, J., Wieneke, U., Kondorosi, E., Barlier, I., Schell, J., and John, M., 1994,

Biosynthesis of lipooligosaccharide nodulation factors—*Rhizobium* NodA protein is involved in *N*-acylation of the chitooligosaccharide backbone, *Proc. Natl. Acad. Sci. USA* **91:**3122–3126.

Röhrig, H., Schmidt, J., Walden, R., Czaja, I., Miklasevics, E., Wieneke, U., Schell, J., and John, M., 1995, Growth of tobacco protoplasts stimulated by synthetic lipo-chitooligosaccharides, *Science* **269:**841–843.

Rosa, F., Sargent, T. D., Rebbert, M. L., Michaels, G. S., Jamrich, M., Grunz, H., Jonas, E., Winkles, J. A., and Dawid, I. B., 1988, Accumulation and decay of DG42 gene products follow a gradient pattern during *Xenopus* embryogenesis, *Dev. Biol.* **129:**114–123.

Rossen, L., Johnston, A. W. B., and Downie, J. A., 1984, DNA sequence of the *Rhizobium leguminosarum* nodulation genes *nodAB* and *C* required for root hair induction, *Nucleic Acids Res.* **12:**9497–9508.

Sadowsky, M. J., Cregan, P. B., Gottfert, M., Sharma, A., Gerhold, D., Rodriguez Quinones, F., Keyser, H. H., Hennecke, H., and Stacey, G., 1991, The *Bradyrhizobium japonicum nolA* gene and its involvement in the genotype-specific nodulation of soybeans, *Proc. Natl. Acad. Sci. USA* **88:**637–641.

Saier, M. H., Tam, R., Reizer, A., and Reizer, J., 1994, Two novel families of bacterial membrane proteins concerned with nodulation, cell division and transport, *Mol. Microbiol.* **11:**841–847.

Sanjuan, J., Grob, P., Göttfert, M., Hennecke, H., and Stacey, G., 1994, NodW is essential for full expression of the common nodulation genes in *Bradyrhizobium japonicum*, *Mol. Plant–Microbe Interact.* **7:**364–369.

Saxena, I. M., and Brown, R. M., 1996, Identification of cellulose synthase(s) in higher plants: Sequence analysis of processive glycosyltransferases with the common motif D,D,D 35 Q(R/Q)XRW, *Cellulose.*

Saxena, I. M., Kudlicka, K., Okuda, K., and Brown, R. M., 1994, Characterization of genes in the cellulose-synthesizing operon, *J. Bacteriol.* **176:**5735–5752.

Schlaman, H. R. M., 1992, Regulation of nodulation gene expression in *Rhizobium leguminosarum* biovar *viciae*, PhD thesis, Leiden University.

Schlaman, H. R. M., Okker, R. J. H., and Lugtenberg, B. J. J., 1992, Regulation of nodulation gene expression by NodD in rhizobia, *J. Bacteriol.* **174:**5177–5182.

Schmidt, J., John, M., Wieneke, U., Krussmann, H. D., and Schell, J., 1986, Expression of the nodulation gene *nodA* in *Rhizobium meliloti* and localization of the gene product in the cytosol, *Proc. Natl. Acad. Sci. USA* **83:**9581–9585.

Schmidt, J., Wingender, R., John, M., Wieneke, U., and Schell, J., 1988, *Rhizobium meliloti nodA* and *nodB* genes are involved in generating compounds that stimulate mitosis of plant cells, *Proc. Natl. Acad. Sci. USA* **85:**8578–8582.

Schmidt, J., John, M., Wieneke, U., Stacey, G., Röhrig, H., and Schell, J., 1991, Studies of the function *Rhizobium meliloti* nodulation genes, in *Advances in Molecular Genetics of Plant–Microbe Interactions* (H. Hennecke and D. P. S. Verma, eds.), pp. 150–155, Kluwer, Dordrecht, The Netherlands.

Schmidt, J., Röhrig, H., John, M., Wieneke, U., Stacey, G., Koncz, C., and Schell, J., 1993, Alteration of plants growth and development by *Rhizobium nodA* and *nodB* genes involved in the synthesis of oligosaccharide signal molecules, *Plant J.* **4:**651–658.

Schultze, M., and Kondorosi, A., 1996, The role of Nod signal structures in the determination of host specificity in the *Rhizobium*–legume symbiosis, *World J. Microbiol. Biotechnol.* **12:**137–149.

Schultze, M., Quiclet-Sire, B., Kondorosi, E., Virelizier, H., Glushka, J. N., Endre, G., Géro, S. D., and Kondorosi, A., 1992, *Rhizobium meliloti* produces a family of sulphated lipo-oligosaccharides exhibiting different degrees of plant host specificity, *Proc. Natl. Acad. Sci. USA* **89:**192–196.

Schultze, M., Kondorosi, E., Ratet, P., Buire, M., and Kondorosi, A., 1994, Cell and molecular biology of *Rhizobium*–plant interactions, *Int. Rev. Cytol.* **156:**1–75.

Schultze, M., Staehelin, C., Röhrig, H., John, M., Schmidt, J., Kondorosi, E., Schell, J., and Kondorosi, A., 1995, *In vitro* sulfotransferase activity of *Rhizobium meliloti* NodH protein: Lipochitooligosaccharide nodulation signals are sulfated after synthesis of the core structure, *Proc. Natl. Acad. Sci. USA* **92:**2706–2709.

Schwedock, J. S., and Long, S. R., 1990, ATP sulphurylase activity of the *nodP* and *nodQ* gene products of *Rhizobium meliloti, Nature* **348:**644–647.

Schwedock, J. S., Liu, C. X., Leyh, T. S., and Long, S. R., 1994, *Rhizobium meliloti* NodP and NodQ form a multifunctional sulfate-activating complex requiring GTP for activity, *J. Bacteriol.* **176:**7055–7064.

Scott, D. B., Young, C. A., Collins-Emerson, J. M., Terzaghi, E. A., Rockman, E. S., Lewis, P. E., and Pankhurst, C. E., 1996, Novel and complex chromosomal arrangement of *Rhizobium loti* nodulation genes, *Mol. Plant–Microbe Interact.* **9:**187–197.

Semino, C. E., and Robbins, P. W., 1995, Synthesis of "Nod"-like chitin oligosaccharides by the Xenopus developmental protein DG42, *Proc. Natl. Acad. Sci. USA* **92:**3498–3501.

Shareck, F., Biely, P., Morosoli, R., and Kluepfel, D., 1995, Analysis of DNA flanking the *xlnB* locus of *Streptomyces lividans* reveals genes encoding acetyl xylan esterase and the RNA component of ribonuclease P, *Gene* **153:**105–109.

Shearman, C. A., Rossen, L., Johnston, A. W. B., and Downie, J. A., 1986, The *Rhizobium leguminosarum* nodulation gene *nodF* encodes a polypeptide similar to acyl-carrier protein and is regulated by *nodD* plus a factor in pea root exudate, *EMBO J.* **5:**647–652.

Sherman, D. H., Malpartida, F., Bibb, M. J., Kieser, H. M., and Hopwood, D. A., 1989, Structure and function of the granatacin-producing polyketide synthase cluster of *Streptomyces violaceoruber, EMBO J.* **8:**2717–2725.

Slabas, A. R., Chase, D., Nishida, I., Murata, N., Sidebottom, C., Safford, R., Sheldon, P. S., Kekwick, R. G., Hardie, D. G., and Mackintosh, R. W., 1992, Molecular cloning of higher-plant 3-oxoacyl-(acyl carrier protein) reductase, *Biochem J.* **283:**2–6.

Smit, G., Swart, S., Lugtenberg, B. J. J., and Kijne, J. W., 1992, Molecular mechanisms of attachment of *Rhizobium* bacteria to plant roots, *Mol. Microbiol.* **6:**2897–2903.

Spaink, H. P., 1995, The molecular basis of infection and nodulation by rhizobia: The ins and outs of sympathogenesis, *Annu. Rev. Phytopathol.* **33:**345–368.

Spaink, H. P., 1996, Regulation of plant morphogenesis by lipo chitin oligosaccharides, *Crit. Rev. Plant Sci.* **15:**559–582.

Spaink, H. P., Wijffelman, C. A., Pees, E., Okker, R. J. H., and Lugtenberg, B. J. J., 1987, *Rhizobium* nodulation gene *nodD* as a determinant of host specificity, *Nature* **328:**337–340.

Spaink, H. P., Weinman, J., Djordjevic, M. A., Wijffelman, C. A., Okker, R. J. H., and Lugtenberg, B. J. J., 1989, Genetic analysis and cellular localization of the *Rhizobium* host specificity-determining NodE protein, *EMBO J.* **8:**2811–2818.

Spaink, H. P., Sheeley, D. M., van Brussel, A. A. N., Glushka, J., York, W. S., Tak, T., Geiger, O., Kennedy, E. P., Reinhold, V. N., and Lugtenberg, B. J. J., 1991, A novel highly unsaturated fatty acid moiety of lipo-oligosaccharide signals determines host specificity of *Rhizobium, Nature* **354:**125–130.

Spaink, H. P., Aarts, A., Stacey, G., Bloemberg, G. V., Lugtenberg, B. J. J., and Kennedy, E. P., 1992, Detection and separation of *Rhizobium* and *Bradyrhizobium* Nod metabolites using thin layer chromatography, *Mol. Plant–Microbe Interact.* **5:**72–80.

Spaink, H. P., Wijfjes, A. H. M., van Vliet, T. B., Kijne, J. W., and Lugtenberg, B. J. J., 1993, Rhizobial lipo-oligosaccharide signals and their role in plant morphogenesis: Are analogous lipophilic chitin derivatives produced by the plant? *Aust. J. Plant Physiol.* **20:**381–392.

Spaink, H. P., Wijfjes, A. H. M., van der Drift, K. M. G. M., Haverkamp, J., Thomas-Oates, J. E., and Lugtenberg, B. J. J., 1994, Structural identification of metabolites produced by the NodB and NodC proteins of *Rhizobium leguminosarum*, *Mol. Microbiol.* **13**:821–831.

Spaink, H. P., Bloemberg, G. V., van Brussel, A. A. N., Lugtenberg, B. J. J., van der Drift, K. M. G. M., Haverkamp, J., and Thomas-Oates, J. E., 1995a, Host specificity of *Rhizobium leguminosarum* is determined by the hydrophobicity of highly unsaturated fatty acyl moieties of the nodulation factors, *Mol. Plant–Microbe Interact.* **8**:155–164.

Spaink, H. P., Wijfjes, A. H. M., and Lugtenberg, B. J. J., 1995b, *Rhizobium* NodI and NodJ proteins play a role in the efficiency of secretion of lipochitin oligosaccharides, *J. Bacteriol.* **177**:6276–6281.

Stacey, G., 1995, *Bradyrhizobium japonicum* nodulation genetics, *FEMS Microbiol. Lett.* **127**:1–9.

Stacey, G., Luka, S., Sanjuan, J., Banfalvi, Z., Nieuwkoop, A. J., Chun, J. Y., Forsberg, L. S., and Carlson, R., 1994, NodZ, a unique host-specific nodulation gene, is involved in the fucosylation of the lipooligosaccharide nodulation signal of *Bradyrhizobium japonicum*, *J. Bacteriol.* **176**:620–633.

Staehelin, C., Granado, J., Muller, J., Wiemken, A., Mellor, R. B., Felix, G., Regenass, M., Broughton, W. J., and Boller, T., 1994, Perception of *Rhizobium* nodulation factors by tomato cells and inactivation by root chitinases, *Proc. Natl. Acad. Sci. USA* **91**:2196–2200.

Staehelin, C., Schultze, M., Kondorosi, E., and Kondorosi, A., 1995, Lipo-chitooligosaccharide nodulation signals from *Rhizobium meliloti* induce their rapid degradation by the host plant alfalfa, *Plant Physiol.* **108**:1607–1614.

Stokkermans, T. J. W., Ikeshita, S., Cohn, J., Carlson, R. W., Stacey, G., Ogawa, T., and Peters, N. K., 1995, Structural requirements of synthetic and natural product lipo-chitin oligosaccharides for induction of nodule primordia on *Glycine soja*, *Plant Physiol.* **108**:1587–1595.

Stokkermans, T. J. W., Orlando, R., Kolli, V. S. K., Carlson, R. W., and Peters, N. K., 1996, Biological activities and structures of *Bradyrhizobium elkanii* low abundance lipo chitin oligosaccharides, *Mol. Plant–Microbe Interact.* **9**:298–304.

Surin, B. P., and Downie, J. A., 1988, Characterization of the *Rhizobium leguminosarum* genes *nodLMN* involved in efficient host nodulation, *Mol. Microbiol.* **2**:173–184.

Theberge, M. C., Prevost, D., and Chalifour, F. P., 1996, The effect of different temperatures on the fatty acid composition of Rhizobium leguminosarum bv. viciae in the Faba bean symbiosis, *New Phytol.* **134**:657–664.

Trinick, M. J., and Galbraith, J., 1980, The *Rhizobium* requirements of the non-legume *Parasponia* in relationship to the cross inoculation concept of legumes, *New Phytol.* **86**:17–26.

Truchet, G., Roche, P., Lerouge, P., Vasse, J., Camut, S., de Billy, F., Promé, J. C., and Dénarié, J., 1991, Sulphated lipo-oligosaccharide signals of *Rhizobium meliloti* elicit root nodule organogenesis in alfalfa, *Nature* **351**:670–673.

Valdivieso, M. H., Mol, P. C., Shaw, J. A., Cabib, E., and Duran, A., 1991, *CAL1*, a gene required for activity of chitin synthase 3 in *Saccharomyces cerevisiae*, *J. Cell. Biol.* **114**:101–109.

Van Brussel, A. A. N., Zaat, S. A. J., Canter Cremers, H. C. J., Wijffelman, C. A., Pees, E., Tak, T., and Lugtenberg, B. J. J., 1986, Role of plant root exudate and sym plasmid-localized nodulation genes in the synthesis by *Rhizobium leguminosarum* of Tsr factor which causes thick and short roots on common vetch, *J. Bacteriol.* **165**:517–522.

Van Brussel, A. A. N., Bakhuizen, R., van Spronsen, P. C., Spaink, H. P., Tak, T., Lugtenberg, B. J. J., and Kijne, J. W., 1992, Induction of pre-infection thread structures in the leguminous host plant by mitogenic lipo-oligosaccharides of *Rhizobium*, *Science* **257**:70–72.

Van der Drift, K. M. G. M., Spaink, H. P., Bloemberg, G. V., Van Brussel, A. A. N., Lugtenberg, B. J. J., Haverkamp, J., and Thomas-Oates, J. E., 1996, *Rhizobium leguminosarum* bv. *trifolii* produces lipo chitin oligosaccharides with *nodE*-dependent highly unsaturated fatty acyl moieties: An electrospray ionization and collision induced dissociation tandem mass spectrometric study, *J. Biol. Chem.* **271**:22563–22569.

Van Workum, W. A. T., Van Brussel, A. A. N., Tak, T., Wijffelman, C. A., and Kijne, J. W., 1995, Ethylene prevents nodulation of *Vicia sativa* ssp. *nigra* by exopolysaccharide-deficient mutants of *Rhizobium leguminosarum* bv. *viciae*, *Mol. Plant–Microbe Interact.* **8:**278–285.

Vázquez, M., Davalos, A., de las Penas, A., Sanchez, F., and Quinto, C., 1991, Novel organization of the common nodulation genes in *Rhizobium leguminosarum* bv. *phaseoli* strains, *J. Bacteriol.* **173:**1250–1258.

Vázquez, M., Santana, O., and Quinto, C., 1993, The NodI and NodJ proteins from *Rhizobium* and *Bradyrhizobium* strains are similar to capsular polysaccharide secretion proteins from Gram-negative bacteria, *Mol. Microbiol.* **8:**369–377.

Verma, N. K., Brandt, J. M., Verma, D. J., and Lindberg, A. A., 1991, Molecular characterization of the *O*-acetyl transferase gene of converting bacteriophage SF6 that adds group antigen 6 to *Shigella flexneri*, *Mol. Microbiol.* **5:**71–75.

Walker, J. E., Gay, N. J., Saraste, M., and Eberle, A. N., 1984, DNA sequence around the *Escherichia coli unc* operon. Completion of the sequence of a 17 kilobase segment containing *asnA*, *oriC*, *unc*, *glmS*, and phoS, Biochem. J. **224:**799–815.

Wang, L., Romana, L. K., and Reeves, P. R., 1992, Molecular analysis of a *Salmonella enterica* group E1 *rfb* gene cluster: *O* antigen and the genetic basis of the major polymorphism, *Genetics* **130:**429–443.

Whitfield, C., and Valvano, M. A., 1993, Biosynthesis and expression of cell-surface polysaccharides in Gram-negative bacteria, in *Advances in Microbial Physiology* (A. H. Rose, ed.), pp. 135–246, Academic Press, London.

Wong, H. C., Fear, A. L., Calhoon, R. D., Eichinger, G. H., Mayer, R., Amikam, D., Benziman, M., Gelfand, D. H., Meade, J. H., Emerick, A. W., *et al.*, 1990, Genetic organization of the cellulose synthase operon in *Acetobacter xylinum*, *Proc. Natl. Acad. Sci. USA* **87:**8130–8134.

*Chapter 3*

# The Role of Microbial Surface Polysaccharides in the *Rhizobium*–Legume Interaction

Karsten Niehaus and Anke Becker

## 1. INTRODUCTION TO THE *RHIZOBIUM*-LEGUME SYMBIOSIS

The symbiotic interaction between the soil bacteria classified as *Rhizobium* or *Bradyrhizobium* and members of the plant family *Leguminosae* enables the study of cell-to-cell recognition, signal exchange, and induced morphogenesis. Rhizobia are able to enter the plant root via a sophisticated infection mechanism, induce a new plant organ, the root nodule, and differentiate within the infected plant cells into nitrogen-fixing bacteroids which have properties of an endosymbiont (Hirsch, 1992). The establishment of this symbiosis is controlled by a complex network of signals which are produced by both the bacteria and the plant. In this review we would like to draw attention to the role and functions of rhizobial surface carbohydrates during the establishment of the symbiosis. Rhizobial cells are able to synthesize several surface carbohydrates: lipopolysaccharides (LPS) bound to the outer membrane of the cell, capsular polysaccharides (CPS) and 3-deoxy-D-manno-2-octulosonic acid (KDO)-rich poly-

**Karsten Niehaus and Anke Becker** Lehrstuhl für Genetik, Fakultät für Biologie, Universität Bielefeld, 33501 Bielefeld, Germany.

*Subcellular Biochemistry, Volume 29: Plant–Microbe Interactions*, edited by Biswas and Das. Plenum Press, New York, 1998.

saccharides (KPS) associated with the cell surface, and exopolysaccharides (EPS) secreted to the medium.

Surface carbohydrates are the interface between the bacterial cell and their environment. For *Rhizobium* these surroundings could mean the soil, the host plant surface, the infection threat matrix, or the peribacteroid space. Several questions can be asked regarding the functions of rhizobial surface polysaccharides in the *Rhizobium*–legume interaction. Surface carbohydrates can function (i) in the soil, (ii) in attachment to the plant roots, (iii) in the infection of the host plant, (iv) in the release of bacteria from the infection thread, and (v) in the establishment of nitrogen-fixing bacteroids. Surface carbohydrates may be active in more than one environment or developmental stage of the plant–microbe interaction process. In addition, different carbohydrates may play the same role in a particular step of the interaction, complicating the mutational analysis of proposed activities. The environment can be complex and nonhomogenous. Due to soil types and other exobiotic factors (e.g water supply and temperature) many microenvironments may be present, leading to the need for polysaccharides adapted to these conditions. It can be considered that the most dramatic change in the environment for a rhizobial cell is the transition from the free-living stage in the soil to a nitrogen-fixing bacteroid within a plant cell. There is growing evidence that surface carbohydrates play a highly active role in this transition. One of the emerging questions to be answered in this circumstance is whether surface carbohydrates are signal molecules specifically recognized by the host plant.

### 1.1. Infection of the Host Plant by *Rhizobium*

The interaction of the soil bacteria and its host plants starts with a chemotactic attraction of rhizobia to the plant root hairs (Fig. 1). The attraction is mainly mediated by aromatic compounds secreted by the plant (Currier and Strobel, 1976; Parke *et al.*, 1985). The aromatic compounds, namely flavonoids, are not uniformly secreted, resulting in an attraction of rhizobia to localized sites of legume roots (Gulash *et al.*, 1984). Rhizobia attach mainly to emerging root hairs close to the tip of the growing root (Bhuvaneswari *et al.*, 1980). These root hairs are highly susceptible to the binding of rhizobial cells by a two-step attachment process (Dazzo *et al.*, 1984) (Fig. 1). Initially, binding is mediated by the calcium-dependent protein rhicadesin, which appears to be common among different *Rhizobium* species (Smith *et al.*, 1987, 1989). A more effective contact is then established by cellulose fibrils or fimbriae (Vesper and Bauer, 1986; Vesper *et al.*, 1987). The contribution of legume lectins and rhizobial surface carbohydrates to the attachment process and specificity of the symbiosis has been discussed for more than two decades (Albersheim and Anderson-Prouty, 1975; Dazzo, 1980; Bauer, 1981; Smith *et al.*, 1992; Ho *et al.*, 1994; Brewin and

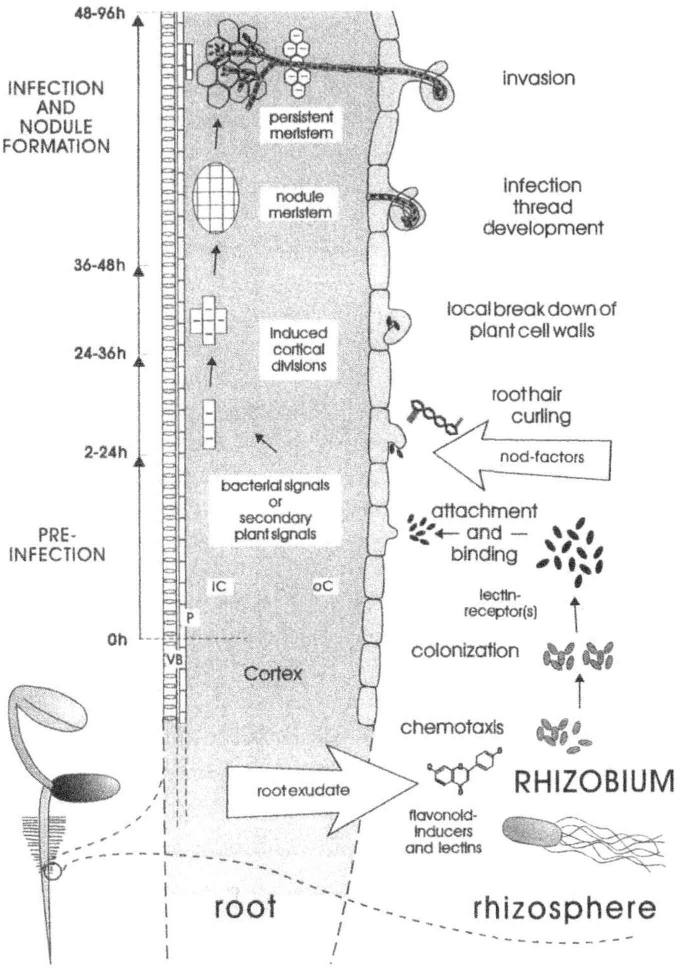


**FIGURE 1.** Early steps in the establishment of the *Rhizobium*–legume symbiosis. Rhizobia become attracted chemotactically by root exudates and colonize the plant host's rhizosphere. Phenolic signal molecules (*nod*-inducers) contained in the exudate activate the transcription of the rhizobial *nod* genes. As a result, host-specific chitooligosaccharide signal molecules called Nod factors are synthesized by the bacterial symbiont. Rhizobia become attached to emerging root hairs—possibly with the help of lectins—forming microcolonies. As a consequence of Nod factor activity the root tips bend, entrapping the rhizobial colony within pockets formed by the walls of the curled root hairs. The hair walls adjacent to the microcolonies become degraded and rhizobia invade the underlying periplasmic space of the hair. Then a new wall is synthesized at the infection site, forming a tunnel-shaped structure, the infection thread. Rhizobia pass via infection threads through the root hair and enter the underlying cortex by repetition of wall degradation and initiation of infection threads. Simultaneously, cells in the root cortex (inner cortex [iC] for indeterminate, outer cortex [oC] for determinate nodules) divide under the influence of Nod-factor signal molecules or possibly a secondary plant factor. When these young meristematic cells are reached by infection threads, rhizobia are released into the host cells.

Kardalsky, 1997). Nevertheless, the function(s) remain unclear. Bohlool and Schmidt (1974) found that soybean lectin specifically combines with all but three of the 25 strains of *Bradyrhizobium japonicum* that were tested. In addition, it was shown that soybean lectin does not bind to any of 23 heterologous *Rhizobium* strains. These results indicated a specific binding of rhizobia to their host plant via microbial cell surface carbohydrates and plant lectins. In accordance with this finding, Dazzo and Brill (1977) reported that 2-deoxyglucose specifically inhibits agglutination of *Rhizobium trifolii* by the clover lectin trifoliin and the binding of rhizobial cells to clover root hairs. In contrast, soybean lines lacking the *N*-acetylgalactosamine-specific lectin which binds to surface carbohydrates of the microsymbiont are still nodulated by *B. japonicum* (Pull *et al.*, 1978). Strong evidence for the involvement of lectins in the recognition of Rhizobia by the host plant came from a transgenic plant expressing a foreign lectin gene. Introduction of the *Pisum sativum* lectin gene *psl* into white clover hairy roots enables heterologous infection and nodulation by the pea symbiont *R. leguminosarum* bv. *viciae* (van Eijsden *et al.*, 1995). This result strongly suggests that a surface carbohydrate of the microsymbiont is specifically recognized, enabling the bacteria to infect the plant. The first microscopically visible reaction of the plant to the attached rhizobia is the deformation of young root hairs (Fahraeus, 1957). A few susceptible root hairs coil 360° at the tip and form so-called shepherd's crooks (Newcomb, 1981). Rhizobia locked in by the curled root hair form a microcolony of bacteria and start the infection through an invagination of the plant cell wall (Napoli and Hubbell, 1975; Turgeon and Bauer, 1985). It is not absolutely clear whether the penetration of rhizobia is mediated by a hydrolytic activity softening the cell wall of the root hair (Turgeon and Bauer, 1985; Hirsch, 1992). Recent results also indicate a direct effect of isolated nodulation factors on the cell wall crystallinity and growth dynamics (Dazzo *et al.*, 1996). The hydrolytic activity of rhizobial enzymes and the effect of nodulation factors is combined in the two-step model by van Spronsen and coworkers (1994). The authors propose a two-step cell wall degradation process, preceding the formation of the infection thread. The first step is a cell wall modification by plant enzymes, induced by the nodulation factors. The second step is complete cell wall degradation in the presence of rhizobia. In susceptible root hairs this initial stage is followed by the generation of infection threads that grow through the root hair cell into the root cortex (Fig. 1). As discussed later, this step is highly dependent on the presence of rhizobial surface carbohydrates. The infection threads are surrounded by the plasma membrane of the plant cell followed by a cell-wall-like layer building the infection thread wall (Callaham and Torrey, 1981; Newcomb, 1981; Turgeon and Bauer, 1985). Within the infection thread, bacteria are surrounded by the infection thread matrix containing glycoproteins secreted by the plant (VandenBosch *et al.*, 1989a) and most probably rhizobial exopolysaccharides (Niehaus *et al.*, 1993). The mode of infection may differ, especially in the case of tropical legumes. Infection of the aquatic mimosoid legume *Neptunia* occurs via rhizobia spread between epidermal cells and intercellular spaces of the root (Schaede, 1940; James *et al.*, 1992). In ground nut and *Stylosanthes* species rhizobial infection takes place at points of lateral root emergence (Chandler, 1978; Chandler *et al.*, 1982), while infection along the stem of *Sesbania rostrata* involves a "crack entry"-type system.

## 1.2. Establishment of the Root Nodule

Simultaneously with the initiation of the infection thread, cell divisions are induced in the cortex of the plant root (Bond, 1948; Libbenga and Harkes, 1973; Dudley *et al.*, 1987). Root hair curling (Hirsch and Spokes, 1994), infection thread formation, and the initiation of cell divisions are the consequences of a molecular cross-talk that guarantees the positive recognition of the specific *Rhizobium* species and the accessory host plant (Denarie *et al.*, 1992). As summarized by Brewin (1991) and Hirsch (1992), the induced meristem gives rise to a highly organized new plant organ, the root nodule (Fig. 2). Root nodules on temperate legumes like alfalfa, clover, or vetch are of the indeterminate type (Fig. 2). The meristematic activity is typically initiated within the inner root cortex followed by cell divisions in the pericycle. The cells of the persistent meristem divide constantly, forming the indeterminate nodule. While the meristem itself is not infected, the proximal cells are invaded by infection threads. Bacteria are released within this zone from the infection threads into the plant cells. Infection droplets lacking cell walls are formed at the tip of the infection threads. From these infection droplets the bacteria are endocytosed in the cytoplasm of the plant cell. During this process, the bacteria become surrounded by a membrane of plant origin, the peribacteroid membrane. Bacteria multiply within the cytoplasm of the plant cell and differentiate into bacteroids, still surrounded by the peribacteroid membrane. During this process different stages of bacteroid differentiation can be defined by electron microscopy (Vasse *et al.*, 1990). To emphasize the similarity with plant organelles the bacteroids enclosed by the peribacteroid membrane are designated as symbiosomes (Roth *et al.*, 1988). In this stage bacteroids are fed by the plant with assimilates fix atmospheric nitrogen (Fischer, 1994).

Beside the indeterminate nodules of temperate legumes such as peas, beans, clovers, and medics, determinated nodules are formed on tropical legumes such as soybeans, common beans, or cowpeas (Fig. 2). The main difference between the two nodules is the nonpersistent meristem. The meristem in the determinated nodules is induced in the outer root cortex in comparison to that in the inner root cortex in the case of indeterminate nodules. In soybean the infection also occurs within curled root hairs. The plant cells built by the nodule meristem get infected by rhizobia which differentiate into nitrogen-fixing bacteroids. The developmen-

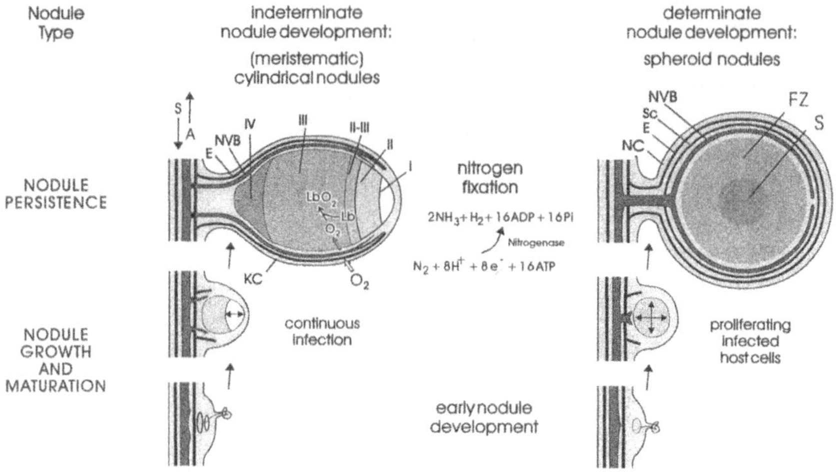


**FIGURE 2.** Sketch of the symbiotic development of indeterminate and determinate types of root nodules. While in the case of indeterminate nodule development, infected cells lose their ability to divide and a new meristematic zone is induced in the distal cortex, in the case of determinate-type nodules, infected cells continue divisions along with their internalized rhizobial symbionts for several days, thus forming the infected nodule tissue. In contrast, the newly induced meristems of indeterminate nodules never contain infection threads or rhizobia and continue to divide. Host cells proliferated proximally result in cylindrically shaped nodules, which are continuously infected by infection threads. Rhizobia taken up into the host cytoplasm by a phagocytotic process multiply and differentiate into mature bacteroids which actively fix atmospheric nitrogen. In order to protect the sensitive nitrogenase enzyme complex from destructive oxygen concentrations, microaerobic conditions are maintained in the central nodule tissue. This is accomplished both by the specialized morphology of nodules and a high metabolic activity in the symbiotic tissue. Leghemoglobin (Lb), a hemoprotein synthesized by the host, acts in the provision of oxygen to the aerobic bacteroids. Ammonia produced by the bacteroids becomes incorporated into appropriate transport molecules, such as amino acids or amides (A), which consequently are exported via phloem vessels into the whole plant body. Sucrose (S) from the green parts of the plant is imported into the nodule for the supply of energy and carbon units. Nodule meristem (I); prefixation zone (II); interzone (II–III); fixation zone (III; FZ); senescence zone (IV; S); endodermis (E); nodule cortex (NC); nodule vascular bundle (NVB); sclerenchyme (SC); pericycle (P).

tal stages of meristematic activity are, sequentially, building up of the nodule tissue, infection of the plant cells, multiplication of rhizobia within the plant tissue, differentiation into nitrogen-fixing bacteroids, and senescence of the old root nodule (Fig. 2). These developmental stages are represented by differentiation zones along the apical–lateral axis of the indeterminate root nodule. Nodulation is a highly developmentally regulated phenomenon controlled by a regulated expression of specific genes coding for proteins which are found exclusively, or highly amplified, in root nodules. These so-called nodulins are proposed to be responsible for structural, physiological, and regulatory functions within the root nodule, reflecting the highly differentiated nature of this plant organ (Legocki and Verma, 1980). Up to now the regulation and function of nodulins has been poorly understood. Some nodulins provide basic physiological functions like uricase, glutamine synthetase, sucrose synthase, and the oxygen binding leghaemoglobins. For the majority of nodulins the functions are unknown. Plant nodulation mutants will help reveal the functions of these nodulins (Caetano Anolles and Gresshoff, 1991). Nevertheless, the understanding of specific nodulin functions will not sufficiently explain the "biology" of a root nodule, rather than a holistic view of the molecular biology of the plant cell. The search for plant receptors, signal-transduction elements, cytoskeleton compounds, or GTP-binding proteins is just opening the door to this exciting field.

## 1.3. Genes Involved in the *Rhizobium*–Legume Interaction

Beside microscopic analysis of the infection process, one of the most powerful tools to analyze the *Rhizobium*–legume interaction is transposon-based mutagenesis of the microsymbiont. The mutagenesis of *Rhizobium* resulted in the definition of developmental steps involved in the establishment of symbiosis that are controlled by defined rhizobial genes. Vincent (1980) proposed several phenotypic codes to allow the classification of symbiotic mutants in plants and bacteria. With the growing data on the nodulation process, especially with the growing number of plant mutants, new phenotypic codes were added. As pointed out by Caetano Anolles and Gresshoff (1991), blockage at any one of the phenotypic points may be impossible because of multiple phenotypes and developmental pleiotropy. However, the genes responsible for the synthesis and reception of the initial signals have been identified in *Rhizobium*. The genes responsible for the nodulation, termed *nod* genes, are located on the bacterial chromosome in the case of *Bradyrhizobium* and on a large plasmid, called the Sym plasmid, in the case of *Rhizobium*. Flavonoids secreted by the plant bind to the constitutively expressed NodD protein. The NodD so activated then promotes the transcription of the *nod* genes leading to the biosynthesis of the nodulation factors (Fig. 1). These nodulation factors are responsible for the root hair curling and the induction of the nodule meristem by *Rhizobium* (see review in Chapter 1 of this book). Of particular interest is that this molecule determines the host specificity and induces the root hair curling and cortical cell divisions when present at pico- to nanomolar levels (Truchet *et al.*, 1991; Denarie *et al.*, 1994). In particular, substitutions at the C6 of the terminal sugar at the reducing end of the oligosaccharide are important for this function (Schultze and Kondorosi, 1995). Besides the decipherment of the nodulation enigma, mutagenesis of the microsymbiont showed the importance of surface carbohydrates for the establishment of symbiosis. Rhizobial mutants defective in the biosynthesis of surface carbohydrates such as exopolysaccharides (EPS), lipopolysaccharides (LPS), or capsular polysaccharides (CPS) were defective in the establishment of symbiosis. Unfortunately, the situation is complex because more than one surface carbohydrate is involved in this process. In addition, the observed function of a particular surface carbohydrate may be different according to the *Rhizobium*–legume combination.

## 2. GENETICS, BIOSYNTHESIS, AND FUNCTIONS OF RHIZOBIAL EXOPOLYSACCHARIDES

### 2.1. Structures of Rhizobial Exopolysaccharides

Rhizobial cells often produce large quantities of extracellular polysaccharides (EPS) as indicated by the very mucoid phenotype when cultivated on media rich in sugars. The EPS are loosely attached to the cell surface and secreted into the growth medium. The chemical and cytological relationships between a more tightly bound capsule and a secreted EPS are uncertain. Microscopically detectable capsules may be present or absent within one species (Mutaftschiev *et al.*, 1982). One of the most extensively studied rhizobial EPS is the succinoglycan produced by several *Rhizobium meliloti*, *Agrobacterium*, and *Alcaligenes* strains. The principal structure of the succinoglycan repeating unit was demonstrated by sequential degradation of the EPS, through gas chromatography-mass spectrometry and NMR analysis (Jansson *et al.*, 1977). The *R. meliloti* SU47 derivative strain Rm1021 (Aman iet al., 1981; Reinhold *et al.*, 1994) and *R. meliloti* YE-2 (Matulová *et al.*, 1994) were studied in detail (Fig. 3). Succinoglycan is composed of octasaccharide repeating units containing one galactose and seven glucose residues joined by $\beta$-1→3, $\beta$-1→4, and $\beta$-1→6 glycosidic linkages. The repeating unit can be associated with additional acetyl, pyruvyl, and succinyl groups. Apart from succinoglycan, often referred to as EPS I, these *R. meliloti* strains have the ability to produce a second EPS, termed galactoglucan or EPS II (Her *et al.*, 1990; Zevenhuizen and Faleschini, 1991) (Fig. 3). EPS II contains glucose and galactose residues in the ratio of 1:1 joined by $\alpha$-1→3 and $\beta$-1→3 glycosidic bonds. The repeating unit is substituted by acetyl and pyruvyl groups. The EPS produced by broad-host-range *Rhizobium* sp. NGR234 is structurally related to succinoglycan (Djordjevic *et al.*, 1986) (Fig. 3). Concerning the five sugars of the reducing terminal part of the repeating unit, the structures of succinoglycan and the *Rhizobium* sp. NGR234 EPS are identical. Acetyl and pyruvyl groups were also identified in the EPS of *Rhizo-*

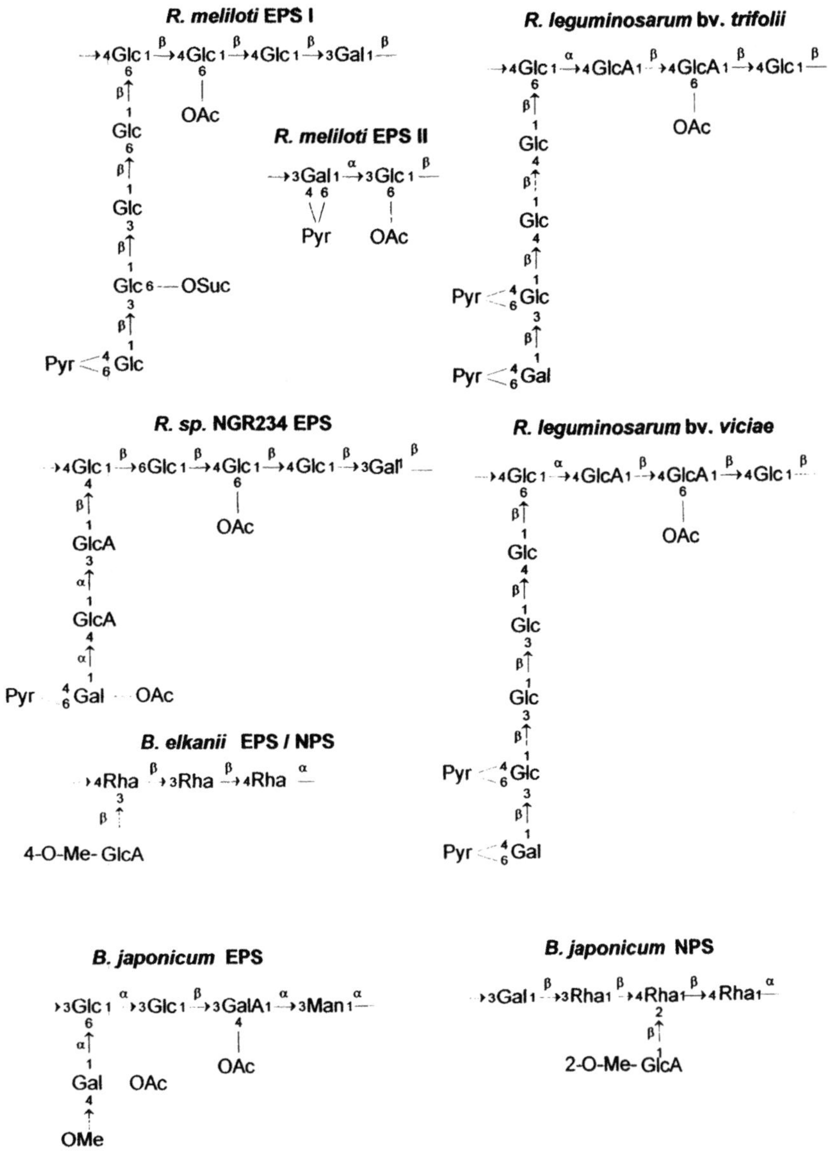


**FIGURE 3.** Structures of some surface polysaccharides from various *Rhizobium* and *Bradyrhizobium strains*. Proposed structures of *R. meliloti* EPS I (succinoglycan) (Jansson *et al.*, 1977; Aman *et al.*, 1981; Reinhold *et al.*, 1994), EPS II (galactoglucan) (Her *et al.*, 1990), *Rhizobium sp.* NGR234 EPS (Djordjevic *et al.*, 1986), *R. leguminosarum* bv. *viciae* EPS (Robertsen *et al.*, 1981), *B. elkanii* EPS/NPS (An *et al.*, 1995), *B. japonicum* EPS (Mort and Bauer, 1982), and the nodule-specific *B. japonicum* NPS (An *et al.*, 1995). Glc, glucose; Gal, galactose; Man, mannose; Rha, rhamnose; GlcA, glucuronic acids; GalA, galacturonic acid; Pyr, pyruvate, OAc, *o*-acetyl; OMe, *o*-methyl; OSuc, *o*-succinyl.

*bium* sp. NGR234. The *R. leguminosarum* group consists of the two biovars, *R. leguminosarum* bv. *trifolii* and bv. *viciae*. Strains of the different biovars nodulating different hosts have a similar basic EPS structure (Fig. 3), although the pattern of substitution and the length of the side chain may vary (Hollingsworth and Dazzo, 1988; Canter Cremers *et al.*, 1991). Several strains of *Rhizobium etli* (formerly *R. leguminosarum* bv. *phaseoli*) produce EPS which differ from the basic structure shown in Fig. 3 (Gil-Serrano *et al.*, 1990).

The slow-growing *Bradyrhizobium japonicum* strains produce EPS containing glucose, mannose, galactose, galacturonic acid, and 4-*O*-methylglucose, whereas *B. elkanii* (former type II *B. japonicum*) strains produce EPS composed of rhamnose and 4-*O*-methylglucoronic acid (Dudman, 1978; Mort and Bauer, 1982). Both strains nodulate soybean. As indicated by electron-microscopic analysis, certain strains of *B. japonicum* and *B. elkanii* produce a polysaccharide within the root nodules of their legume host (Newcomb, 1981). These nodule polysaccharides (NPS) were isolated and characterized (An *et al.*, 1995). The *B. elkanii* EPS from free-living bacteria and those isolated from root nodules have the same structure with a repeating unit consisting of rhamnose linked $\alpha$-1→4, $\beta$-1→3, and $\beta$-1→4 with a $\beta$-1→3 linked 4-*o*-methyl-glucoronic acid as side chain (Fig. 3). The structure of the *B. japonicum* EPS isolated from free-living cells shows no obvious similarity (Mort and Bauer, 1982) (Fig. 3). Interestingly, the structure of the NPS from *B. japonicum* proved to be quite different from that of its EPS, indicating that genes responsible for the biosynthesis of NPS are induced by conditions or factors unique to the host root nodule (An *et al.*, 1995) (Fig. 3). The repeating unit of the *B. japonicum* NPS is composed of galactose linked $\alpha$-1→3 to rhamnose linked $\beta$-1→3, $\beta$-1→4, and $\beta$-1→4 to rhamnose with a $\beta$-1→3 linked 4-*o*-methyl-glucoronic acid as side chain. The authors speculate that a general NPS structural motif, i.e., a rhamnose-rich backbone with a single branching methylated glucuronosyl residue, has an important function in nodule physiology (An *et al.*, 1995).

Besides the monosaccharide structure, several rhizobial strains produce EPS fractions differing in molecular weight. Two major fractions, high-molecular-weight (HMW) polymers and low-molecular-weight (LMW) oligosaccharides, were identified in the culture supernatants of *R. meliloti* (Amemura *et al.*, 1983; Leigh and Lee, 1988; Golzalez *et al.*, 1996), *Rhizobium* sp. NGR234 (Djordjevic *et al.*, 1987), *R. leguminosarum* bv. *trifolii* (Zevenhuizen and van Neerven, 1983; Djordjevic *et al.*, 1987), *Rhizobium* sp. GRH2 (Lopez Lara *et al.*, 1993), and *B. japonicum* strains (Kosch *et al.*, 1994).

Cellulose, one of the most abundant biologically produced polymers, is synthesized by many *Rhizobium* (Ross *et al.*, 1991; Mateos *et al.*, 1995) and *Agrobacterium* strains (Matthysse, 1983). Generally cellulose does not contribute much to the overall polysaccharide production of rhizobia. Nevertheless, cellulose fibrils are involved in the adhesion of rhizobial cells to the root hairs (Mateos *et al.*, 1995).

## 2.2. Organization of EPS Biosynthesis Genes and EPS Biosynthetic Pathways

Bacterial EPS can be either homopolymers, composed of a single type of monosaccharide, or heteropolymers, containing different monosaccharides. Homopolymers composed of only one monosaccharide may have a relatively simple biosynthetic pathway. An example is levan, one of the EPSs produced by the phytopathogenic bacterium *Erwinia amylovora*. Levan is synthesized by a single enzyme, the levansucrase, from sucrose (Gross *et al.*, 1992). Another example is the cellulose synthesis by *Acetobacter xylinum* via a single polymerization step, utilizing UDP-glucose as the substrate (Ross *et al.*, 1991). The biosynthesis is carried out by the cellulose synthetase with the cellulose remaining attached to the membrane-bound enzyme during polymerization (Wong *et al.*, 1990). The extracellular ribbon of cellulose microfibrils is exported, with the help of the participating protein AcsC and the crystallization-promoting protein AcsD, to the bacterial surface (Saxena *et al.*, 1994). Generally the export of polysaccharides to the cell surface is poorly understood. ATP-binding cassette (ABC) transporters seem to be involved in the transport of homopolymers, polymerized in a processive manner across the inner membrane (Whitfield, 1995).

The biosynthesis of heteropolysaccharides seems to be more complex. Mutations in the EPS-biosynthesis genes in combination with the biochemical analysis of intermediates were valuable for the understanding of the sequential steps of the bacterial EPS production. EPS biosynthesis was intensively studied in cell-free enzyme systems in the phytopathogenic bacterium *Xanthomonas campestris* (Ielpi *et al.*, 1981a,b) and in *R. meliloti* (Tolmasky *et al.*, 1982; Reuber and Walker, 1993b). The biosynthetic pathway can be subdivided into four steps: (i) The synthesis of nucleotide diphosphosugars which normally serve as precursors for the formation of heteropolysaccharides; (ii) stepwise assembly of the repeating unit of the EPS by transfer of the monosaccharides to a membrane-bound lipid carrier (C-55 undecaprenyl phosphate) by specific glycosyl transferases; (iii) substitution of the repeating unit with nonsugar groups like pyruvate, succinate, hydroxybutyrate, or acetate; and (iv) the transfer of the subunit to the growing polysaccharide chain (Coplin and Cook, 1990).

## 2.3. *Rhizobium meliloti* as a Model for the Analysis of Rhizobial Exopolysaccharide Biosynthesis

The EPS I biosynthesis of *R. meliloti* represents one of the best-studied examples of rhizobial EPS production and therefore serves as a model system to discuss the genetics and the biosynthetic pathway of a complex rhizobial heteropolysaccharide. In most cases genes directing the biosynthesis of bacterial EPS are situated in large gene clusters, but examples where such genes are organized as single loci distributed in the genome have been reported. Genes involved in EPS production by *R. meliloti* were mapped on the chromosome and on large plasmids (Finan *et al.*, 1985; Leigh *et al.*, 1985; Hynes *et al.*, 1986).

*Rhizobium meliloti* mutants deficient in EPS I production were initially isolated by transposon mutagenesis (Finan *et al.*, 1985; Leigh *et al.*, 1985; Müller *et al.*, 1988). Loci affected by transposon insertions were mapped on the chromosome and on the second megaplasmid (Finan *et al.*, 1986; Long *et al.*, 1988; Müller *et al.*, 1988). A large gene cluster involved in EPS I biosynthesis comprising at least 27 kb was identified on megaplasmid 2 of the *R. meliloti* SU47 derivative strains Rm1021 and Rm2011. Analysis of the nucleotide sequence revealed the presence of 21 genes, termed *exo* or *exs* genes (Becker *et al.*, 1993a–c, 1995b; Glucksmann *et al.*, 1993a,b; Müller *et al.*, 1993; Buendia *et al.*, 1991; Reed *et al.*, 1991a). In addition, five loci affecting EPS I biosynthesis were identified on the chromosome. These include the regulatory genes *exoR* (Doherty *et al.*, 1988; Reed *et al.*, 1991b), *exoS* (Doherty *et al.*, 1988), and *mucR* (Zhan *et al.*, 1989; Keller *et al.*, 1995), as well as *exoC* (Finan *et al.*, 1986), coding for a phosphoglucomutase (Uttaro *et al.*, 1990), and *exoD* (Reed and Walker, 1991b), with unknown molecular functions. Homologies of the deduced *exo* and *exs* gene products to proteins of known function suggested that the *exo* and *exs* genes encode proteins required for the synthesis of precursors, assembly of the repeating unit, substitution, polymerization, and export of EPS I, as well as for the regulation of the polysaccharide production. A detailed genetic analysis revealed that the *exo/exs* gene cluster is organized in several operons (Fig. 4) (Becker *et al.*, 1993a–c, 1995b; Müller *et al.*, 1993; Buendia *et al.*, 1991). In all, at least 18 promoters were identified by plasmid integration

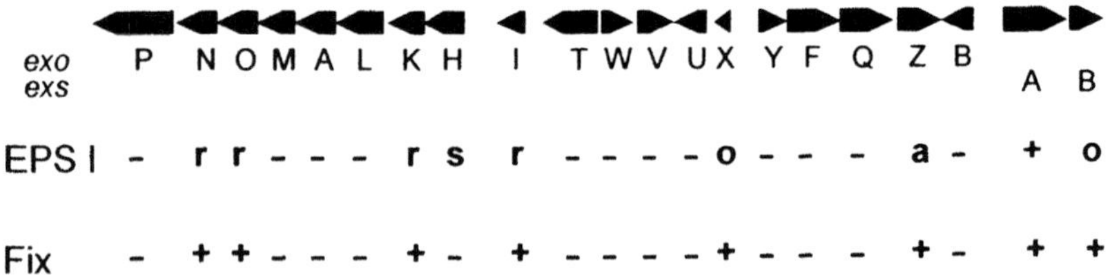


**FIGURE 4.** *Rhizobium meliloti* EPS I biosynthesis genes and their requirement for EPS production and symbiotic functions. The organization of the 27-kb *exo/exs* gene cluster of *R. meliloti* 2011. The effect of nonpolar mutations on exopolysaccharide production and the symbiotic phenotype is indicated. (+) Nitrogen fixation of the wild type; (−) no production of EPS I and no nitrogen fixation of the root nodule; (a) production of nonacetylated EPS I; (o) overproduction of EPS I; (r) reduced production of EPS I; (s) production of nonsuccinylated EPS I.

mutagenesis of the *R. meliloti* Rm2011 wild-type strain or mutants carrying *exo–lacZ* and *exs–lacZ* transcriptional fusions. The transcription directed by several promoters results in the formation of overlapping transcripts. This is particularly obvious in the *exoHKLAMONP*, *exoWV*, and *exoYFQ* regions.

The biosynthetic pathway of EPS I resembles that of other heteropolysaccharides composed of repeating units. Tolmasky *et al.* (1982) demonstrated that the repeating unit is assembled on a lipid carrier resulting in a polyprenyl diphosphate octasaccharide intermediate. UDP-galactose and UDP-glucose were identified as the cytoplasmic precursors for the formation of the lipid-linked repeating unit (Glucksmann *et al.*, 1993b; Reuber and Walker, 1993b; Buendia *et al.*, 1991). ExoC, ExoB, and ExoN are involved in the biosynthesis of the sugar nucleotide precursors. Mutations in ExoA and ExoB have an influence on more than one surface carbohydrate. The pleiotropic phenotype of such mutants complicates the analysis of proposed functions of defined surface molecules in the plant–microbe interaction (Fig. 4). The phosphoglucomutase encoded by *exoC* converts glucose-6-phosphate to glucose-1-phosphate (Uttaro *et al.*, 1990). This enzymatic activity is essential for the biosynthesis of EPS I, EPS II, LPS, and β-(1-2)glucans (Uttaro *et al.*, 1990; Zhan *et al.*, 1989).

Functions were assigned to most of the *exo* genes by the analysis of lipid-linked biosynthetic intermediates that accumulated in *exo* mutants (Reuber and Walker, 1993b). The sequential activities of eight glycosyl transferases and three enzymes for nonsugar substitutions result in the formation of the lipid-linked octasaccharide repeating unit. ExoY and ExoF were identified to be involved in the initial transfer of the galactosyl residue to the lipid carrier. ExoA, ExoL, ExoM, ExoO, ExoU, and ExoW are involved in the sequential addition of the subsequent sugars (Table I). The glycosyl transferase responsible for the addition of the terminal sugar of the repeating unit remains to be identified. Analysis of EPS I produced by *exoH* and *exoZ* mutants revealed that ExoH is involved in succinylation, whereas ExoZ is implicated in acetylation (Leigh *et al.*, 1987; Reuber and Walker, 1993a). In addition, *exoP*, *exoQ*, or *exoT* mutants involved in polymerization or export of EPS I accumulated lipid-linked octasaccharide subunits that were acetylated, succinylated, and pyruvylated to various extents. This indicates that substitution might occur prior to polymerization of the repeating units. Nonacetylated and nonsuccinylated EPS I can still be polymerized and exported. Recently, Gonzáles *et al.* (1996b) reported evidence for the ExoQ protein to be involved in the synthesis of HMW EPS I, whereas ExoT was proposed to be involved in the synthesis of EPS I octasaccharide trimers and tetramers. Since octasaccharide monomers were exclusively synthesized by *exoP* mutants, the ExoP protein seems to be essential for the synthesis of HMW and LMW EPS I polymerization products. *Rhizobium meliloti* SU47 mutants carrying a truncated *exoP** gene exclusively encoding the N-terminal domain produced reduced amounts of EPS I (Becker *et al.*, 1995a). This reduction could be

**Table I**
**Properties of Genes and Gene Products Involved in EPS I Biosynthesis of *R. meliloti* SU47**

| Protein | Protein localization[a] | Proposed function | Reference |
|---|---|---|---|
| ExoC | CY | phosphoglucomutase | Uttaro *et al.* (1990) |
| ExoB | CY | UDP-glucose-4-epimerase | Buendia *et al.* (1991) |
| ExoN | CY | UDP-glucose pyrophosphorylase | Becker *et al.* (1993b), Glucksmann *et al.* (1993b) |
| ExoY | CMa | galactosyl transferase | Müller *et al.* (1993), Reuber and Walker (1993a), Reed *et al.* (1991a) |
| ExoA | CMa | glucosyl transferase | Becker *et al.* (1993b), Glucksmann *et al.* (1993a), Reuber and Walker (1993b) |
| ExoL | CY | glucosyl transferase | Becker *et al.* (1993a), Glucksmann *et al.* (1993a), Reuber and Walker (1993b) |
| ExoM | CY | glucosyl transferase | Becker *et al.* (1993b), Glucksmann *et al.* (1993a), Reuber and Walker (1993b) |
| ExoO | CY | glucosyl transferase | Becker *et al.* (1993b), Glucksmann *et al.* (1993a), Reuber and Walker (1993b) |
| ExoU | CMa | glucosyl transferase | Becker *et al.* (1993c), Glucksmann *et al.* (1993a), Reuber and Walker (1993b) |
| ExoW | CY | glucosyl transferase | Becker *et al.* (1993c), Glucksmann *et al.* (1993a), Reuber and Walker (1993b) |
| ExoZ | CMi | acetyl transferase | Buendia *et al.* (1991), Reuber and Walker (1993a) |
| ExoH | CMi | succinyl transferase | Becker *et al.* (1993a), Leigh *et al.* (1987) |
| ExoV | CY | pyruvyl transferase | Becker *et al.* (1993c), Glucksmann *et al.* (1993a), Reuber and Walker (1993b) |
| ExoF | PP | transfer or galactose | Müller *et al.* (1993), Reuber and Walker (1993b) |
| ExoQ | CMi | export and polymerization | Müller *et al.* (1993), Reuber and Walker (1993b) |
| ExoT | CMi | export and polymerization | Becker *et al.* (1993c), Glucksmann *et al.* (1993b), Reuber and Walker (1993b) |
| ExoP | CMi | export and polymerization | Becker *et al.* (1993b), Glucksmann *et al.* (1993b), Reuber and Walker (1993b) |
| ExoA | CMi | export | Becker *et al.* (1993b) |
| ExoK | PP or EC | endo-β-1,3-1,4-glucanase | Becker *et al.* (1993a) |
| ExoX | CMa | negative regulator | Zahn and Leigh (1990), Reed *et al.* (1991a), Müller *et al.* (1993) |
| ExoI | PP or EC | unknown | Becker *et al.* (1993c) |
| ExsA | CMi | ABC transporter | Becker *et al.* (1995b) |
| ExsB | CY | negative regulator | Becker *et al.* (1995b) |
| ExsH | EC | endoglycanase | York and Walker (1997) |
| ExoR | CY | negative regulator | Doherty *et al.* (1988), Reed *et al.* (1991b), Ozga *et al.* (1994) |
| ExoS | CMa | negative regulator | Doherty *et al.* (1988), Ozga *et al.* (1994), Østeras *et al.* (1995) |
| MucR | CY | positive regulator | Zahn *et al.* (1989), Keller *et al.* (1995) |
| ExoD | CY | unknown | Reed and Walker (1991a,b) |

[a] Subcellular localization of the proteins as predicted from the hydrophobicity of the amino acid sequence or from gene fusion experiments. CY, cytoplasm; CM, cytoplasmic membrane (i, integral membrane protein; a, attached to or anchored in the membrane); PP, periplasm; EC, extracellular.

suppressed by a mutation in the negative regulatory gene *exoR*. The ratio of HMW to LMW EPS I was significantly decreased in *exoP** mutants (Becker *et al.*, 1995a). It is tempting to speculate that ExoP proteins might be implicated in processes determining the ratio of HMW to LMW EPS I. A similar function, but via a different mechanism, could be performed by the ExoK and ExoH proteins (Becker *et al.*, 1995a). These two proteins are able to specifically degrade EPS I and contribute to the production of LMW EPS I (York and Walker, 1997). While ExoP might control the distribution of HMW EPS I and LMW EPS I at the site of synthesis, ExoK and ExoH could break down HMW EPS I outside the bacterial cell yielding LMW EPS I. Both HMW and LMW EPS I may have discrete functions in the establishment of the symbiosis, as discussed in this section.

Besides the examination of the EPS I biosynthesis pathway, mutational analysis of *R. meliloti* Rm2011 revealed that this strain has the cryptic ability to synthesize an alternative exopolysaccharide EPS II (Fig. 3) also designated galactoglucan (Her *et al.*, 1990; Levery *et al.*, 1991). Significant amounts of EPS II are only produced in the presence of a mutation in either of the chromosomally located genes *expR* (Glazebrook and Walker, 1989) or *mucR* (Zahn *et al.*, 1989). In addition, regulatory genes responsible for the biosynthesis of EPS II were identified on megaplasmid 2 (Glazebrook and Walker, 1989). The cluster of *exp* genes is separated from the *exo*/*exs* genes by approximately 200 kb (Charles and Finan, 1991). Initially the *exp* gene cluster was characterized by transposon mutagenesis and complementation, revealing that at least 23 kb containing six complementation groups are required for EPS II biosynthesis (Glazebrook and Walker, 1989). Nucleotide sequence analysis of a 32-kb DNA region carrying the complete *exp* gene cluster resulted in the identification of 22 *exp* genes (Becker *et al.*, 1997). The biosynthetic pathway of EPS II has not yet been elucidated.

## 2.4. Regulation of Exopolysaccharide Biosynthesis

The regulation of rhizobial EPS biosynthesis is complex. Seven potentially regulatory genes influencing the EPS I/EPS II ratio or the amount of EPS have been identified in *R. meliloti*. Among the main regulatory factors are the nutrient composition, osmolarity, and oxygen supply of the growth medium (Zevenhuizen, 1986; Zevenhuizen and Faleschini, 1991). Generally, high amounts of sugars or sugar alcohols and limitations in the nitrogen, phosphate, and sulfur source stimulate EPS I production in *R. meliloti* (Leigh *et al.*, 1985; Doherty *et al.*, 1988). In the wild type, EPS II is synthesized under phosphate limitation (Zhan *et al.*, 1991). The osmolarity of the medium also has an influence on the EPS I/EPS II ratio. *Rhizobium meliloti* YE produced HMW EPS II and LMW EPS I in the presence of low concentrations of NaCl. Increasing the NaCl concentration of the medium resulted in increased production of EPS I at the expense of EPS II and LMW EPS I (Zevenhuizen and Faleschini, 1991). During the infection of the plant all these parameters, including nutrient supply, osmolarity of the surrounding medium, and oxygen supply, may change and influence the amount and quality of the secreted exopolysaccharides. That such a regulation occurs is indicated by electron microscopic data showing no significant polysaccharide secreted into the peribacteroid space (Vasse *et al.*, 1990). In addition, translational fusions of *exoA*, *exoP*, and *exoF* with the reporter gene *phoA* were mainly expressed up to the infection zone of alfalfa nodules, suggesting that *exo* gene expression is downregulated in the symbiotic zone (Reuber *et al.*, 1991). On the molecular level the regulation of EPS biosynthesis in *R. meliloti* is poorly understood. *Rhizobium meliloti* SU47 mutated in the chromosomally located genes *exoR* or *exoS* showed an increased production of EPS I (Doherty *et al.*, 1988). Mutations in one of these genes affected the expression of *exo* genes (Reed *et al.*, 1991b; Ozga *et al.*, 1994). Whereas the regulatory protein ExoR displayed no homologies to known regulators (Reed *et al.*, 1991b), ExoS was found to be identical to the ChvG protein (Østeras *et al.*, 1995). The ChvG protein is homologous to sensor proteins of two-component regulatory systems. Therefore, ExoS might be involved in sensing environmental factors that influence EPS I production of *R. meliloti* SU47. The ratio of *exoX* to *exoY*, located within the *exo*/*exs* gene cluster, affects the level of EPS I production of *R. meliloti* SU47 (Zahn and Leigh, 1990; Reed *et al.*, 1991a; Müller *et al.*, 1993). Both proteins are probably associated with the inner membrane (Müller *et al.*, 1993), suggesting that protein–protein interaction might occur. Moreover, the *exsB* gene also located in the *exo*/*exs* gene cluster might encode a negative regulator affecting a posttranscriptional level of *exo* gene expression (Becker *et al.*, 1995b). With the exception of strict phosphate limitation, *R. meliloti* SU47 exclusively produces EPS I. Two *R. meliloti* SU47 mutants were isolated which constitutively produced EPS II. Mutations in *mucR* resulted in the constitutive biosynthesis of EPS II and in a reduction of EPS I production (Zhan *et al.*, 1989; Keller *et al.*, 1995). In addition, only LMW EPS I was identified in the culture supernatants of *mucR* mutants (Keller *et al.*, 1995; Bertram-Drogatz *et al.*, 1998).

A further transcriptional regulator might be encoded by the *expG* gene, which is located in the *exp* gene cluster on megaplasmid 2 (Becker *et al.*, 1997). In *R. meliloti* SU47 the *mucS* gene, which is identical to *expG*, was identified by Astete and Leigh (1996). This gene was found to be required for the induction of EPS II production by low phosphate concentrations and by extra copies of the *exp* gene cluster. Finally, overexpression of *syrM* in *R. meliloti* SU47, a gene involved in regulation of *nod* gene expression, caused the enhanced production of EPS, although the regulatory mechanism is unknown (Mulligan and Long, 1989).

### 2.5. Rhizobial Exopolysaccharide Mutants Are Defective in the Infection of the Host Plant

The ability to establish an effective symbiosis is severely affected in mutants of many rhizobia deficient in EPS production. In particular, the acidic EPS turned out to be essential for the establishment of effective nitrogen-fixing symbiosis on legumes that develop an indeterminate type of nodule. Tn*5*-induced mutants of *R. leguminosarum* bv. *trifolii* defective in the EPS biosynthesis fail to establish effective root nodules on white and subterranean clovers (Chakravorty *et al.*, 1982). The induced root nodules are small and poorly infected. The bacteroids fail to differentiate (Rolfe *et al.*, 1996). Nonmucoid mutants of the broad-host-range strain NGR234 are noninfective on *Leucaena leucocephala* (Djordjevic *et al.*, 1987). Nonmucoid mutants of *R. meliloti* SU47, induce non-infected root nodules on *Medicago, Melilotus,* and *Trigonella* lacking infection threads (Finan *et al.*, 1985; Leigh *et al.*, 1985; Long *et al.*, 1988; Müller *et al.*, 1988). A plant test with more than 20 *Medicago* species revealed that the inability of EPS I-deficient *R. meliloti* mutants to establish an effective symbiosis is conserved in all tested species (Niehaus *et al.*, 1994). Severe underproduction or overproduction of EPS I has in general no significant influence on the symbiosis (Doherty *et al.*, 1988; Becker *et al.*, 1995b), while mutants in which EPS I lacks the succinyl moiety are defective in the infection process (Leigh *et al.*, 1987). This shows that not the quantity, but the quality of EPS I is essential for the infection of the host plant. The symbiotic defect of EPS I mutants can be suppressed by specific second-site mutations in the regulatory *expR* gene that lead to the synthesis of the alternative EPS II (Glazebrook and Walker, 1989; Zhan *et al.*, 1989). In contrast, Keller *et al.* (1995) found that a *R. meliloti mucR*/*exoY* mutant producing exclusively EPS II failed to establish an effective symbiosis with alfalfa. The different interaction of these mutants with the host plant may correlate with the quality of the secreted EPS II. Gonzáles *et al.* (1996a) reported that the *expR* mutation caused the production of HMW and LMW EPS II fractions, whereas the *mucR* mutant exclusively produced HMW EPS II. It seems that only the LMW EPS II is biologically active. Besides EPS I and EPS II, other surface carbohydrates can have a function in the establishment of the *R. meliloti*–*Medicago sativa* symbiosis. Rm41 is a separate isolate of *R. meliloti* that is able to produce both EPS I and EPS II and exhibits a normal symbiosis with alfalfa (Putnoky *et al.*, 1988). The nonmucoid derivative AK631, which carries a mutation in the *exoB* gene producing neither EPS I nor EPS II, shows no obvious defect in the infection process (Putnoky *et al.*, 1990). Further studies showed that mutations in the *lpsZ* gene of AK631 lead to an altered surface carbohydrate pattern and a defect in the symbiosis typical for EPS-defective mutants of *R. meliloti* (Putnoky *et al.*, 1990). *R. meliloti* AK631 has the ability to produce 3-deoxy-D-manno-2-octulosonic acid (KDO)-containing polysaccharides that are structurally analogous to group K antigens found in *Escherichia coli* (Reuhs *et al.*, 1993). The synthesis and function of KDO-containing polysaccharides will be discussed later in this section.

In spite of the pronounced effects in some systems, it seems that there is no general requirement for EPS in all rhizobial–legume interactions. A mutation in the EPS-biosynthesis gene *pssA* in *R. leguminosarum* bv. *viciae* abolished both the EPS production and the nodulation on vetch and pea (Borthakur *et al.*, 1986). *Rhizobium leguminosarum* bv. *phaseoli* strains carrying the same mutation in the *pssA* gene are also defective in EPS biosynthesis, but form effective nodules on determinate nodulating host plants like bean (Borthakur *et al.*, 1986; Diebold and Noel, 1989). On the other hand, *R. leguminosarum* bv. *phaseoli* strains carrying a multicopy polysaccharide restoration (*psr*) gene induced nonfixing nodules on bean (Borthakur and Johnston, 1987). It seems that for the effective nodulation of bean an equimolar expression of the polysaccharide inhibition (*psi*) gene and *psr*, which regulates *psi*, is needed (Borthakur and Johnston, 1987). Only the negative regulator of EPS biosynthesis, *psi*, is expressed in the symbiotic zone of the root nodule (Latchford *et al.*, 1991). This could indicate that the EPS is not needed for the establishment of this symbiosis, but the biosynthesis has to be stopped during the bacteroid development. A similar host-specific effect was also observed in other host–microbe combinations, such as in EPS-defective mutants of *R. loti*. These mutants were fully effective on the determinate nodulating host *Lotus*, but ineffective on the indeterminate nodulating host *Leucaena* (Hotter and Scott, 1991). EPS-defective mutants of the broad-host-range strain *Rhizobium* sp. NGR234 formed ineffective nodules on *Leucaena*, but effective nodules on tropical legumes forming nodules of the determinate type (Chen *et al.*, 1985; Djordjevic *et al.*, 1987). It was pointed out that the rhizobial LPS and not the EPS has a major function in the *Bradyrhizobium japonicum–Glycine max* symbiosis (Stacey *et al.*, 1991). Therefore, it was proposed that EPS is not essential for the nodulation of plants forming determinate nodules (Diebold and Noel, 1989; Hotter and Scott, 1991). A correlation of the different symbiotic phenotypes of EPS mutants with the differences in infection thread formation of determinate and indeterminate nodulating plants was suggested (Leigh and Coplin, 1992). Although EPS seems not to be essential for the formation of determinate nitrogen-fixing nodules on *Glycine max*, early stages of the development of nodules infected with *B. japonicum* EPS mutants were disturbed. Delayed nodulation and reduced competitiveness were reported (Kosch *et al.*, 1994; Parniske *et al.*, 1994). On *Glycine soja*, the effects were even more pronounced. *Bradyrhizobium japonicum exoB* mutants lost their infectivity and induced the formation of white, uninfected, and ineffective nodulelike structures at the base of lateral roots (Parniske *et al.*, 1994). Nevertheless, the deficiencies of EPS mutants in the establishment of the symbiosis with an indeterminate nodulating plant seem to be generally more severe than with a determinate nodulating host. LPS-defective mutants of rhizobial strains inducing nodules of the determinate type were reported to be defective in the establishment of the symbiosis. Examples are the LPS mutants of *B. japonicum* (Carlson and Yadav, 1985; Puvanesarajah *et al.*, 1987; Stacey *et al.*, 1991) and *R. leguminosarum* bv. *phaseoli* (Noel *et al.*, 1986). It remains to be determined whether LPS replaces the functions of EPS in these systems.

### 2.6. Rhizobial Exopolysaccharides Are Specific Signals in the Infection Process

The best-examined model system with respect to the function of exopolysaccharides is the *R. meliloti*–alfalfa symbiosis. *R. meliloti* SU47 mutants that fail to synthesize the acidic EPS I form small, spheroid root nodules which fail to fix nitrogen (Finan *et al.*, 1985; Leigh *et al.*, 1985; Müller *et al.*, 1988). Similar noninfected nodules were induced by EPS-deficient mutants of *R. leguminosarum* bv. *trifolii* on clover (Chakravorty *et al.*, 1982; Carlson *et al.*, 1986; Djordjevic *et al.*, 1987), *Rhizobium* sp. NGR234 on *Leucaena leucocephala* (Djordjevic *et al.*, 1987), and *R. leguminosarum* bv. *viciae* on peas (Borthakur *et al.*, 1986). Microscopic analysis of alfalfa nodules induced by EPS-deficient *R. meliloti* mutants revealed that the central nodule tissue was not infected. No infection threads or bacteroids were found (Finan *et al.*, 1985; Müller *et al.*, 1988). Infection threads were initiated in curled root hairs by *R. meliloti* mutants but these aborted within the peripheral cells of the developing nodule (Yang, 1992; Yang *et al.*, 1992). The infection threads induced by EPS mutants are often enlarged and covered by layers of cell-wall-like material (Niehaus *et al.*, 1993). Within these infection threads numerous vesicle-like structures become visible (Yang *et al.*, 1992). Intracellular invasion of the cortical cells on the surface of these root nodules was observed (Yang *et al.*, 1992; Niehaus *et al.*, 1993). *Exo* mutants of *R. meliloti* SU47 penetrate the cortical cell layer along the middle lamella and enter parenchymatic cells close to the nodule meristem. Within these cells large infection threads are initiated which can lead to partially infected pseudonodules. Interestingly, the infection threads contain no obvious matrix, indicating that the matrix found in wild-type infection threads consists mainly of rhizobial exopolysaccharide (Niehaus *et al.*, 1993). In contrast to the wild-type *R. meliloti*-induced nodules, the noninfected ones lack a well-defined, persistent apical meristem (Yang *et al.*, 1992). Anatomically, the noninfected nodules are comparable to some extent to those induced by hormone treatment (Hirsch and Fang, 1994), spontaneously arising in a specific alfalfa line (Truchet *et al.*, 1989; Joshi *et al.*, 1991) or induced by isolated nodulation factors (Truchet *et al.*, 1991), supplying evidence that the initiation of a root nodule is a legume plant program triggered by rhizobia. Meristem formation is independent of bacterial invasion. In the reverse case, infection of the plant is dependent on the synthesis of the nodulation factor (Ardourel *et al.*, 1994). The nodulation factor is needed for root hair curling, meristem induction, and infection of the root hair. The root hair curling itself is not an absolute need for the infection since in rare cases straight root hairs can get infected (Fahraeus, 1957; Napoli and Hubbell, 1975). The induction of changes in the organization of the plant cytoskeleton and the induction of pre-infection threads could be the most important function of nodulation factors in this developmental step (van Spronsen *et al.*, 1994). In the case of *R. meliloti* SU47 it needs at least two prerequisites, the nodulation factor and EPS I, to enter the alfalfa root hair. Alfalfa seedlings inoculated with *R. meliloti exoH* mutants, which fail to succinylate EPS I, also form ineffective nodules that do not contain intracellular bacteria or bacteroids (Leigh *et al.*, 1987). Non- or poorly infected root nodules were also induced by *mucR* mutants (Keller *et al.*, 1995), which produce only HMW EPS II (Gonzáles *et al.*, 1996). In contrast, mutants that produce LMW EPS II are fully infective, at least on most host plants (Gonzáles *et al.*, 1996a), as is *R. meliloti* AK631, which produces neither LMW EPS I nor LMW EPS II, but does produce the KDO-rich polysaccharide (KPS) (Putnoky *et al.*, 1990). The LMW fractions of EPS I, EPS II, and KPS can substitute for each other in the infection process.

### 2.7. Are Rhizobial Exopolysaccharides Suppressors of the Plant Defense System?

Are these LMW oligosaccharides a structural requirement or a specific signal for the infection process? *R. meliloti* mutants in the *ndvB* gene are defective in the biosynthesis of cyclic β-1,2-glucans and induce noninfected pseudonodules on alfalfa (Quandt *et al.*, 1992). The sensitivity of these mutants to low osmolarity is drastically increased (Ielpi *et al.*, 1990). A second-site mutation results in an osmorevertant of a *R. meliloti ndvB* deletion mutant that forms infection threads and, depending on the second-site mutation, effective root nodules (Dylan *et al.*, 1990; Quandt *et al.*, 1992). These results show that cyclic β-1,2-glucans are not essential requirements and therefore are very unlikely to be signals. It seems that the infection process puts an osmotic stress on the bacteria which can be tolerated by the wild type but not by the osmotically sensitive *ndv* mutants of *R. meliloti*. A similar situation may exist in the case of *R. meliloti exoD* mutants. *ExoD* mutants produce altered amounts of EPS I and fail to infect alfalfa nodules (Reed and Walker, 1991b). The authors showed that *exoD* mutants are sensitive to alkaline conditions. Interestingly, buffering of the plant medium at slightly acidic pH restores the ability of *exoD* mutants to invade the nodules (Reed and Walker, 1991a). Also in *exoD* mutants it is unlikely that a signal molecule is missing.

But what is the causal function of the exopolysaccharides? One important point is that the function(s) of LMW EPS I can be provided in *trans*. Alfalfa plants coinoculated with *R. meliloti* nodulation (e.g., *nodC*::Tn*5*) and infection mutants deficient in exopolysaccharide production formed mixed infected nodules that were capable of fixing atmospheric nitrogen (Klein *et al.*, 1988; Kapp *et al.*, 1990). The formation of infected nodules was dependent on close contact between the inoculum partners. When the partners were separated by a filter, noninfected nodules were formed, suggesting that infection thread formation in alfalfa is dependent on the nodulation factor and EPS synthesized during the infection process. In mixed infected nodules, both nodulation and infection mutants colonized the plant cells and differentiated into bacteroids (Klein *et al.*, 1988). Only the infection of the plant, but not the formation of bacteroids, was dependent on cell-to-cell contact between the mutants. Immunogold staining revealed that the ratio of the two mutants varied considerably in colonized plant cells following mixed inoculation (Kapp *et al.*, 1990). Colonization of the plant cell, bacteroid differentiation, and nitrogen fixation seem to be independent of *nod* and *exo* gene functions (Rolfe *et al.*, 1980; Klein *et al.*, 1988; Kapp *et al.*, 1990). In some symbiotic systems addition of purified EPS restored the ability of EPS mutants to infect nodules. Djordjevic *et al.* (1987) reported that the exogenous supplementation with purified EPS- or EPS-derived oligosaccharides restored the ability of *Rhizobium* sp. NGR234 EPS-deficient mutants to induce nitrogen-fixing nodules on *L. leucocephala.* Purified EPS or oligosaccharides also promoted invasion of *Trifolium repens* nodules by EPS mutants of *R. leguminosarum* bv. *trifolii* (Djordjevic *et al.*, 1987). Heterologous EPS did not promote nodule invasion by EPS mutants, indicating that EPS fulfills a specific function (Djordjevic *et al.*, 1987). The symbiotic deficiency of *R. meliloti* SU47 EPS I mutants was overcome by supplementation with LMW EPS I (Battisti *et al.*, 1992; Urzainqui and Walker, 1992). HMW EPS I did not promote nodule invasion of *R. meliloti* SU47 *exo* mutants. The most charged tetramer of the EPS I octasaccharide repeating unit was identified as the active compound in the LMW fraction (Battisti *et al.*, 1992). Since induction of nitrogen-fixing nodules by *exo* mutants exogenously supplemented with LMW EPS I was delayed in comparison to the formation of fixing nodules by the wild type, it was suggested that EPS I is needed within the nodule during nodule invasion or HMW EPS I might also contribute to the invasion process (Battisti *et al.*, 1992). Similar results were obtained for LMW EPS II. A LMW EPS II fraction consisting of 15–20 EPS II disaccharide subunits efficiently allows nodule invasion by noninfective strains when present in amounts as low as 7 pmol per plant (Gonzáles *et al.*, 1996a). Exopolysaccharide-defective mutants, coinoculation experiments, and the possibility to supplement with very low amounts of LMW EPS I/II suggest that these molecules represent a second signal molecule, beside the nodulation factor, needed to establish the symbiosis.

A further hint that rhizobial oligosaccharides represent signals in the plant–microbe interaction comes from observations of plant defense symptoms in noninfected root nodules induced by EPS-deficient rhizobial mutants. The symbiotic phenotype of *exoB* mutants of *B. japonicum* was analyzed on the host plants *Glycine max* and *G. soja*. While effective nodules were induced on *G. max*, the symbiosis turned out to be defective on *G. soja* (Parniske *et al.*, 1994). Microscopic data and the accumulation of the phytoalexin glyceollin proved the induction of plant defense reactions by the mutated symbiont. Similar observations were made for EPS-defective mutants of *R. leguminosarum* bv. *trifolii*. The outermost layers of plant cell walls in the empty nodules induced on clovers had thickened cell walls and deposited material (Rolfe *et al.*, 1996). The cortical cell walls of the pseudonodules induced by EPS-defective *R. meliloti* mutants were abnormally thick and incrusted with an autofluorescent material (Niehaus *et al.*, 1993). Parts of these cell walls and wall appositions contained callose. Defense-related genes of the phenylpropanoide pathway were overexpressed in the noninfected nodules (Grosskopf *et al.*, 1993; Niehaus *et al.*, 1994). Biochemical analysis of pseudonodules revealed an increase of phenolic compounds bound to the nodule cell walls when compared with the wild-type-induced nodules (Niehaus *et al.*, 1993). From these observations we proposed that LMW EPS I acts as a suppressor of the plant defense system, enabling the symbiont *R. meliloti* to infect the host plant (Niehaus *et al.*, 1993). In order to test this hypothesis, elicitor-responsive cell cultures of the host plant *M. sativa* (alfalfa) and as a control the non-host plants *Nicotiana tabacum* (tobacco) and *Lycopersicum esculentum* (tomato) were established. Besides other defense-related reactions, all cell cultures reacted to the addition of small amounts of the nonspecific yeast elicitor with a strong transient alkalinization of their culture medium (Niehaus *et al.*, 1997a). Using this assay system, purified homologous and heterologous EPS were analyzed for possible suppressor functions in the three plant suspension cultures. In alfalfa cell cultures the elicitor-induced alkalinization could be suppressed by the simultaneous application of LMW EPS I. Neither HMW EPS I, HMW EPS II, nor the heterologous EPS xanthan from *Xanthomonas campestris* provoked a reduction of the elicitor response. None of the carbohydrate preparations were able to suppress the elicitor-induced alkalinization in the cell cultures of the non-host plants tobacco and tomato. These data provide evidence for a specific recognition of LMW EPS I by the host plant as a suppressor of the plant defense system enabling the symbiont to infect the plant (Niehaus *et al.*, 1997a).

## 3. GENETICS, BIOSYNTHESIS, AND FUNCTIONS OF RHIZOBIAL LIPOPOLYSACCHARIDES

### 3.1. Structures of Rhizobial Lipopolysaccharides

Lipopolysaccharides (LPS) are an essential component of the outer leaflet of the outer membranes of Gram-negative bacteria. LPS are important virulence factors of bacteria pathogenic for animals or human beings, e.g., *Salmonella*

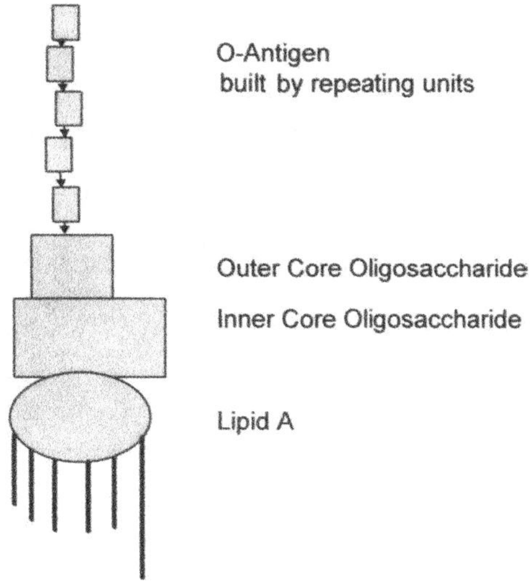


**FIGURE 5.** Schematic structure of lipopolysaccharides (LPS).

spp. (Rietschel *et al.*, 1996). Since these molecules are highly immunogenic, they are also termed endotoxins. Lipopolysaccharides can be subdivided into two principal parts, the lipid and the polysaccharide portion, leading to the amphiphilic nature of these molecules. Monosaccharide and lipid compositions are highly variable; nevertheless, the principal architecture of the molecule is conserved (Fig. 5). Lipid A consists in general of a disaccharide to which fatty acids are attached. Lipid A, the chemically most uniform portion of LPS, builds the membrane anchor of the molecule. The core oligosaccharide bound to the lipid A anchor can be subdivided into the less-variable inner and the more-variable outer core part (Fig. 5). The *O*-specific chain bound to the core, also termed *O*-antigen, is usually composed of a polymer of repeating oligosaccharide units. The total structure of LPS is known from enterobacteria of relevance to medical microbiology, e.g., *Salmonella enterica* or *Escherichia coli*. In contrast, no total structure of LPS is known for any *Rhizobium* species. Nevertheless, partial LPS structures have been recognized, exhibiting unique features among the Gram-negative bacteria. Besides saturated 3-hydroxy fatty acids with 14–18 carbon atoms, some very long acyl chains (27-OH-C:28) are found in the LPS isolated from *R. meliloti, R. fredii, R. leguminosarum* biovars *trifolii, viciae* and *phaseoli*, and *B. japonicum* (Bhat *et al.*, 1991b, 1994). The distribution of the 27-OH-C:28 fatty acid as well as analysis of lipid A backbone sugars revealed the taxonomic relatedness of various strains of the *Rhizobiaceae* (Bhat *et al.*, 1991b). Three major lipid A types were found in the different *Rhizobium* species: type 1, represented by *R. meliloti* and *R. fredii*, have a lipid A that contains only glucosamine; type 2, represented by *R. leguminosarum*, contains glucosamine and galacturonic acid; and type 3, represented by *B. japonicum*, contains either

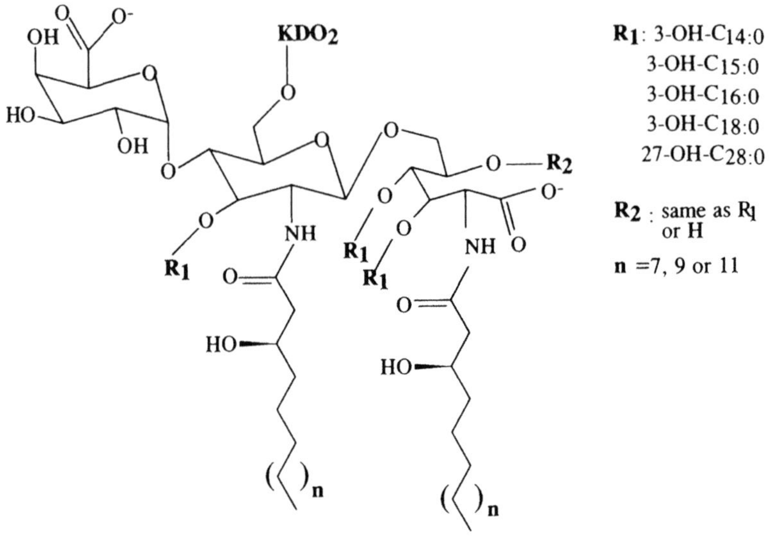


**FIGURE 6.** Structure of *R. leguminosarum* lipid A (Price *et al.*, 1995). KDO, 2-Keto-3-deoxy-octulosonate.

only 2,3-diamino-2,3-dideoxy-D-glucose (DAG) or a lipid A backbone containing DAG and glucosamine (Bhat *et al.*, 1991b). A total structure of the *R. leguminosarum* bv. *phaseoli* LPS, which is most probably identical to those of the biovars *viciae* and *trifolii*, was established (Bhat *et al.*, 1994). Most remarkably, there is a galacturonosyl residue, rather than phosphate, at the 4′ position of the glucosamine residue (Fig. 6). In all other LPS from Gram-negative bacteria examined so far a phosphate is found in lipid A (Rietschel *et al.*, 1996). Interestingly, the glucosamine residue is phosphorylated during LPS biosynthesis, but the 4′-phosphate group is enzymatically removed and replaced by an α-linked galacturonic acid (Price *et al.*, 1995). Also, the glycosidic phosphate is cleaved off. The first seven enzymes of *E. coli* lipid A biosynthesis, leading from UDP-GlcNAc to the phosphorylated intermediate 2-keto-3-deoxyoctulosonate ($KDO_2$)-lipid $IV_A$, are present in *R. leguminosarum* (Price *et al.*, 1995). After oxidation of the terminal glucosamine at $C_2$, 2-amino-2-deoxygluconic acid carrying the acyl residues in amide and ester linkages is formed (Fig. 6). There is no clue why *R. leguminosarum* would incorporate the 4′-phosphate group in the first place, only to remove it and replace it with the galacturonic acid residue at a later stage of LPS biosynthesis. Since *R. meliloti* LPS of aerophilic grown cells contains a 4′-phosphate moiety on lipid A, it seems that the replacement by galacturonic acid is not a general feature of the *Rhizobiaceae*. Nevertheless, it was speculated (Price *et al.*, 1995) that dephosphorylation is important during the microaerophilic state, and perhaps during bacteroid development within the root nodule.

The core oligosaccharide of the LPS is linked via 2-keto-3-deoxyoctulosonic acid (KDO) to lipid A. The structures of core oligosaccharides from different rhizobia are known. Partial structures have been established for *R. meliloti* (Russa *et al.*, 1996), *R. loti* (Russa *et al.*, 1995), and *B. japonicum* (Carlson and Krishnaiah, 1992). The best-examined example is again the group of *R. leguminosarum* biovars (Bhat *et al.*, 1991a, 1994; Kannenberg *et al.*, 1992) and *R. etli*, the former *R. leguminosarum* bv. *phaseoli* (Carlson *et al.*, 1995). The core oligosaccharide structures from *R. etli* CE3 and *R. leguminosarum* bv. *trifolii* and bv. *viciae* contain the α-Gal-(1→6)-[α-GalA-(1→4)]-α-Man-(1→5)-KDO and α-GalA-(1→4)-[α-GalA-(1→5)]-KDO oligosaccharides (Carlson and Garci, 1989; Hollingsworth *et al.*, 1989, 1994; Bhat *et al.*, 1991a). A precise structure for an entire core region is not known. From the data it can be concluded that the *R. leguminosarum* and *R. etli* core oligosaccharides are structurally very different from those of the enteric bacteria (Fig. 7). Besides the missing phosphorylation

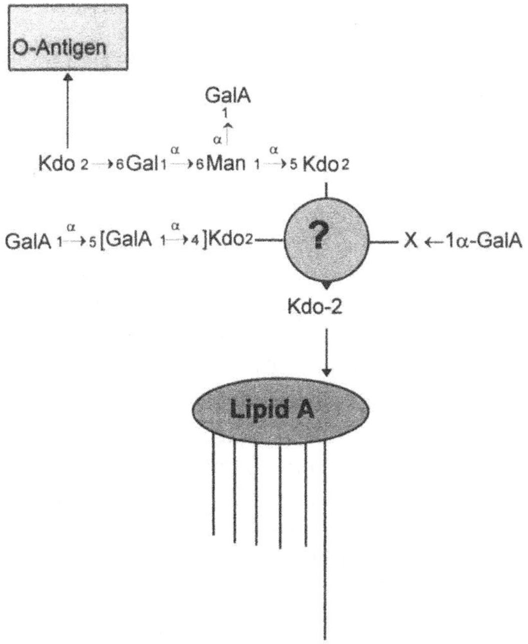


**FIGURE 7.** Schematic representation of the possible structure of the LPS from *R. etli* CE3 (Carlson *et al.*, 1995). It is not yet known how the various core oligosaccharides are linked together, as indicated by the shaded circle. The X represents a moiety or chemical environment which renders the GalA glycosidic bond labile to mild acid.

in lipid A, the core oligosaccharide contains uronosyl residues (Carlson *et al.*, 1995). A simple method to analyze the heterogeneity among LPS chemotypes is sodium dodecylsulfate–polyacrylamide gel electrophoresis (SDS–PAGE) followed by the sensitive silver-staining method (Hitchcock and Brown, 1983). Kim *et al.* (1996) adapted this method for the separation of bacterial CPS and LPS by preparative electrophoresis. Total rhizobial LPS can be separated with this technique into two classes, fast-migrating LPS II, and the often microheterogenic LPS I (Carlson, 1984). Subsequent studies revealed that LPS II represents the lipid A core lipooligosaccharide, while LPS I represents LPS II with the attached *O*-antigen polysaccharide. The *O*-antigen is more strain-specific. The *O*-antigen released by mild acid hydrolysis from the LPS I fraction contains KDO at the reducing ends (Carlson, 1984; Carlson *et al.*, 1987). Aside from the monosaccharide composition of rhizobial *O*-antigens there is a significant lack of structural information. Monosaccharide composition was reported for *B. japonicum* (Carrion *et al.*, 1990), *R. leguminosarum* (Carlson, 1984; Zhang *et al.*, 1992), and *R. meliloti* (Zevenhuizen *et al.*, 1980; Russa *et al.*, 1996). For the *R. leguminosarum O*-antigen some structural information is available (Kannenberg and Brewin, 1989; Wang and Hollingsworth, 1994).

### 3.2. Genetics and Biosynthesis of Rhizobial Lipopolysaccharides

The genes and gene functions of rhizobial LPS biosynthesis are largely unknown. In *R. meliloti* the genes *exoB*, coding for UDP-glucose-4-epimerase (Buendia *et al.*, 1991), and *exoC*, coding for phosphoglucomutase (Uttaro *et al.*, 1990), were involved in EPS and LPS biosynthesis (Leigh and Lee, 1988). Activated sugar precursors are needed for both EPS and LPS. Clover *et al.* (1989) reported four chromosomal regions of *R. meliloti* involved in LPS biosynthesis. The best-characterized region is the *lpsB* gene. Mutants in *lpsB* show a faster migrating LPS I and LPS II in SDS–PAGE (Clover *et al.*, 1989; Lagares *et al.*, 1992; Niehaus *et al.*, 1997b). This may indicate that the core region is affected by the mutation. Since the *O*-antigen seems to be present, a side chain of the core could be truncated. DNA sequencing of *lpsB* revealed no obvious homologies to known LPS-biosynthesis genes of enterobacteria (A. Lagares, unpublished). The *lpsZ* gene analyzed in *R. meliloti* turned out to be involved in the biosynthesis of a KDO-rich capsular polysaccharide rather than in LPS biosynthesis (Putnoky *et al.*, 1990; Brzoska and Signer, 1991; Reuhs *et al.*, 1995). Mutants of *B. japonicum* (Stacey *et al.*, 1991) and *R. leguminosarum* bv. *viciae* (DeMaagd *et al.*, 1989; Priefer, 1989; Kannenberg *et al.*, 1992) altered in LPS biosynthesis were isolated upon random mutagenesis. The results indicate that a cluster of genes necessary for LPS production and symbiotic efficiency is located within a defined region of 20 kb on the bacterial chromosome of *R. leguminosarum* bv. *viciae* (Priefer, 1989). Symbiotic mutants of *R. etli*, the former

*R. leguminosarum* bv. *phaseoli*, indicate that in this microbe mainly three DNA regions are involved in LPS biosynthesis (Cava *et al.*, 1989). Many genes required for the *O*-antigen synthesis, and also some genes for the core oligosaccharide synthesis, are located in the *lps* α region of the CE3 genome (Carlson *et al.*, 1995). Specific mutants could be correlated with defined truncated LPS structures (Carlson *et al.*, 1995).

### 3.3. Rhizobial Lipopolysaccharide Mutants Are Defective in the Establishment of the Symbiosis

In determinate-nodulating hosts like *P. vulgaris*, LPS mutants fail to form normal infection threads or the bacteria are not released from the infection threads (Noel *et al.*, 1986; Cava *et al.*, 1989). Also, *B. japonicum* mutants defective in LPS biosynthesis uncouple plant from bacterial differentiation. These mutants fail to establish effective root nodules on soybean (Stacey *et al.*, 1991). A host-specific defect of a Tn*5*-insertion mutant of the *Leucaena*-nodulating *Rhizobium* sp. strain TAL1145 defective in EPS and LPS synthesis was described (Praveen *et al.*, 1996). This mutant forms noninfected pseudonodules on *P. vulgaris* and *Gliricidia sepium*, while the infected nodules induced on *L. leucocephala* appear to be initially defective in nitrogen fixation (Praveen *et al.*, 1996). The nodules induced on *L. leucocephala* were defective in nitrogen fixation 4 weeks after inoculation and turned out to be effective after 8 weeks. A similar host-plant-dependent defect of LPS mutants was observed for the broad-host-range *Rhizobium* sp. GRH2. The LPS mutant GRH2-14 induced a few noninfected nodules on the legume tree *Acacia* and lost the ability for effective nodulation of *Lotus corniculatus*, *Vicia hirsuta*, *Trifolium subterraneum*, *T. repens*, and *T. incarnatum* (Lopez-Lara *et al.*, 1995). On the other hand, the mutant is able to nodulate *P. vulgaris*, but not *L. corniculatus*. The LPS-dependent defect is therefore not correlated with the determinate or indeterminate type of root nodule. LPS-defective mutants of *R. leguminosarum* bv. *trifolii* (Carlson *et al.*, 1987b; Lorkiewicz *et al.*, 1993) or *R. leguminosarum* bv. *viciae* (DeMaagd *et al.*, 1989; Priefer, 1989) are defective in infection or bacteroid development. *R. meliloti* mutants defective in the LPS biosynthesis form infected, nitrogen-fixing nodules on *M. sativa* (Clover *et al.*, 1989). A detailed analysis of an *R. meliloti* strain mutated in *lpsB* shows that this strain is altered in its competitiveness for nodulation of alfalfa (Lagares *et al.*, 1992). When alfalfa plants were inoculated with a mixed inoculum consisting of the *R. meliloti* wild-type Rm2011 and the *lpsB* mutant a strain ratio of approximately 1:100 favoring the mutant was necessary to obtain an equal ratio (1:1) of nodule occupancy (Lagares *et al.*, 1992). Most interestingly, this mutant is defective in symbiosis, depending on the host plant. When inoculated on *M. truncatula*, only small, white nodules were observed, lacking any nitrogenase activity (Niehaus *et al.*,

1997b). In general we can conclude that, depending on the particular plant–*Rhizobium* combination, EPS and LPS can have a function in both the determinate and the indeterminate type of nodule (Parniske *et al.*, 1994; Lopez-Lara *et al.*, 1995; Niehaus *et al.*, 1997b).

Light and electron microscopy showed that the *M. truncatula* nodule cells were only poorly infected by the LPS mutant. Using fluorescence microscopy, a strong autofluorescence of infected plant cells became visible, indicating a plant defense response (Niehaus *et al.*, 1997b). This phenotype is very similar to that induced by *R. leguminosarum* bv. *viciae* LPS mutants on pea root nodules (Perotto *et al.*, 1994). The host-plant-specific defect of the *R. meliloti lpsB* mutant indicates a specific recognition of the rhizobial LPS rather than a nonspecific or structural function. This hypothesis is strengthened by the observation that a crude LPS fraction obtained from *R. meliloti* suppresses the elicitor-induced alkalinization of alfalfa, but not of tobacco or tomato cell cultures (Niehaus *et al.*, 1997a).

In addition, the finding that *R. leguminosarum* bv. *trifolii* LPS binds rapidly to root hair tips of clover, and the infection-promoting effect of the homologous LPS, point to a specific recognition of this molecule by the host plant (Dazzo *et al.*, 1991). Nevertheless, a specific and high-affinity binding of rhizobial LPS by defined proteins of the host plant has to be shown. The search for LPS-binding proteins of the host plant is complicated by the fact that there is a complex change in the LPS structure during the establishment of the symbiosis. A sensitive detection of these changes could be achieved by the use of monoclonal antibodies (Brewin *et al.*, 1986). The expression of cell surface antigens, represented by the LPS, from *R. leguminosarum* is regulated by oxygen and pH (Kannenberg and Brewin, 1989; Bhat and Carlson, 1992). In addition, there is a change in the surface antigens during infection of the plant and bacteroid development (VandenBosch *et al.*, 1989b; Sindhu *et al.*, 1990). The LPS-epitope expression of *R. leguminosarum* bacteroids has been analyzed by *in situ* immunolabeling of pea root nodule sections (Kannenberg *et al.*, 1994). Many LPS epitopes were expressed in the free-living and symbiotic stages. Other epitopes were only expressed in particular regions of the nodule (Kannenberg *et al.*, 1994). These changes may reflect a physiological adaptation of rhizobia to the conditions of the root nodule. Antigenic changes in the rhizobial LPS can also be induced by exudates of the host plant (Noel *et al.*, 1996). In addition, the host-specific defect of some LPS-defective mutants indicates a more active role of LPS, perhaps in the specific recognition of the molecules by the plant.

### 3.4. The K-Antigen of *Rhizobium*

EPS-deficient mutants of the fast growing Rhizobia are generally defective in the establishment of symbiosis with their host plants. An interesting exception is the *R. meliloti* strain AK631, carrying a mutation in the *exoB* gene. This derivative of the parental strain Rm41 can no longer produce EPS I or EPS II due to the defect in UDP-galactose synthesis. Surprisingly this mutant can still establish a fully effective symbiosis with all tested host plants (Putnoky *et al.*, 1988). The mutational analysis of AK631 led to the identification of two gene regions responsible for the effective symbiosis in the absence of EPS. These regions were termed *lpsZ* and *fix*-23 (Putnoky *et al.*, 1990; Williams *et al.*, 1990). The *lpsZ* gene is not present in *R. meliloti* SU47 and its derivatives Rm1021 and Rm2011. The symbiotic defect of *exoB* mutants in Rm1021 could be corrected by the introduction of the *lpsZ* gene from *R. meliloti* AK631 (Williams *et al.*, 1990). Further it was recognized that the presence of a functional *lpsZ* gene altered the surface properties of this particular strain. Alterations were detected in the LPS and in cell bound capsular polysaccharides (CPS) containing 3-deoxy-D-manno-2-octulosonic acid (KDO) or variants of this carbohydrate. These capsular polysaccharides were classified according to the presence of KDO as K-antigens or K-polysaccharides (KPS). The complete chemical structure of the *R. meliloti* KPS is yet not determined. Biochemical analysis indicate a disaccharide repeating unit of α-keto-3,5,7,9-tetradeoxy-7,7-diamino nonulonosic acid, a variant of KDO, and aminohexuronic acid (Reuhs *et al.*, 1995; Kiss *et al.*, 1997). In case of *R. fredii* the repeating unit of the KPS was determined as [→3)-α-D-Galp-(1→5)-β-D-KDOp-(2→ ] (Reuhs *et al.*, 1993). Most interestingly the production of KPS (and LPS) of *R. fredii* is modulated by apigenin, which has been shown to induce the nodulation genes of this rhizobial strain. Root extracts had a similar effect (Reuhs *et al.*, 1994). Reuhs *et al.* (1995) also showed that the suppression of the Fix$^-$ phenotype of *R. meliloti* Rm1021 mutants by *lpsZ* correlates with a modified biosynthesis of the KPS. In other words, the KPS can replace LWW-EPS I and LMW-EPS II in the infection process. The *lpsZ* gene itself is not responsible for the synthesis of the KPS, instead it seems to be involved in the determination of the degree of polymerization (Reuhs *et al.*, 1995). The expression of *lpsZ* leads to the biosynthesis of LMW-KPS, which might be the specific compound required for the interaction with the host plant. Due to the sequence homology of *lpsZ* to the *E. coli kpcC*, involved in the expression of the group II K-antigens of *E. coli*, the *R. meliloti* gene was redesignated *rkpZ* (rhizobial KPS) (Reuhs *et al.*, 1995). Beside *rkpZ* the *fix-23* region of *R. meliloti* is involved in the synthesis of KPS. This region is organized in four complementation groups comprising of ten genes, *rkpA* to *rkpJ* (Petrovics *et al.*, 1993; Kiss *et al.*, 1997). The genes *rkpA-F*, building the first complementation group, exhibit similarities to fatty acid synthase genes (Petrovic *et al.*, 1993). This homology indicates their involvement in the biosynthesis of a lipid molecule which might serve as a lipid carrier for the synthesis or as an anchor for the produced KPS. Recently Kiss *et al.* (1997) published the analysis of the remaining complementation groups. Sequence analysis indicated homologies of *rkpG* to acyltransferases, of *rkpH* to short-chain alcohol dehydrogenases, and of *rkpJ* to bacterial polysaccharide transport proteins. Biochemical and immunological analysis of mutants in these genes demonstrated a partial or total loss of KPS from the cell surface. Reuhs *et al.* (1995) proposed that initially an oligosaccharide of 10 to 15 repeating units is synthesized. This oligosaccharide represents the LMW-KPS. This LMW-KPS is polymerized to HMW-KPS and exported to the cell surface. Purified HMW-KPS contains no fatty acid which indicates that the lipid molecules are only needed for the biosynthesis and/or export of the HMW-KPS. The fact that KPS can i. replace LMW-EPS in the infection process of the plant, ii. is modified by plant exudates, iii. and is species specific (Forsbery and Reuhs, 1997), indicates that this surface polysacchaccharide is specifically recognized by the host plant. This idea is supported by the induction of the isoflavonoid biosynthetic pathway in alfalfa leaves after infiltration with *R. meliloti* expressing KPS. A similar effect could be induced by isolated KPS, but not by KPS deficient *R. meliloti* mutants (Becquart-de Kozak *et al.*, 1997).

## 4. RHIZOBIAL SURFACE CARBOHYDRATES—THE KEY TO SUPPRESSING THE PLANT DEFENSE SYSTEM?

All rhizobial surface carbohydrates examined so far have a function in the establishment of the symbiosis in at least one *Rhizobium*–legume combination. There is no example of a *Rhizobium*–legume combination where neither EPS nor LPS nor capsular polysaccharides are needed to establish the symbiotic interaction. Surface carbohydrates are complex structures, which may change during the infection of the host plant, making the functional analysis even more complicated. Molecular genetic studies of polysaccharide biosynthesis using advanced biochemical and chemical techniques provide powerful tools for understanding the biosynthesis and the elucidation of the chemical structures of rhizobial surface carbohydrates. Nevertheless, the characterization of symbiotic defects of EPS and LPS mutants is only the first step toward understanding the probably different functions of these molecules.

Rhizobial mutants altered in EPS biosynthesis are defective in the infection of the host plant in different *Rhizobium*–legume systems. Mutants altered in LPS biosynthesis can be defective in infection, release of bacteria from the infection thread, or bacteroid development. This emphasizes the general importance of these molecules for this type of plant–microbe interaction. In this respect the situation is clearer than in the case of plant–pathogen interaction. It seems that surface carbohydrates are important in some but not all pathogen–plant combinations (Leigh and Coplin, 1992). The rhizobial EPS and LPS may have (i) structural, (ii) barrier, (iii) morphogen, or (iv) defense-suppressor functions. A structural function in stabilizing the outer membrane of the bacterium is evident especially for LPS. Nevertheless, specific EPS or LPS mutants can establish an effective symbiosis with one plant species but fail on another, indicating that the structural function is not the *specific* function in the symbiosis. A variation on this theme is the idea that EPS or LPS may be a barrier against toxic compounds, e.g., phytoalexins, of the plant. Since EPS-defective mutants are more sensitive to phytoalexins the barrier effect may contribute to the establishment of the symbiosis. On the other hand, there are a lot of arguments for a specific recognition of rhizobial surface carbohydrates by the plant. Minor changes in the chemical structure, like nonsuccinylated EPS I in case of *R. meliloti*, lead to a nonfunctional EPS, while severe over- or underproduction can have no significant influence. In addition, the defect in infection of *R. meliloti* EPS mutants can be restored by trace amounts of LMW EPS. But how does this recognition function on the molecular level? At least two possibilities are obvious, a positive recognition as morphogens or a negative one as suppressors of the plant defense system. Examples can be found for both possibilities. Oligosaccharides can induce flower formation in tobacco explants (Marfá *et al.*, 1991), inhibit auxin-induced elongation of pea stem segments (Branca *et al.*, 1988), or induce rapid responses at the plasma membrane, as reviewed by Darvill *et al.* (1992). The acceleration of cytoplasmic streaming and production of novel proteins in clover root hairs specifically induced by isolated LPS of *R. leguminosarum* bv. *trifolii* (Dazzo *et al.*, 1991) indicates a positive recognition of the molecules.

The identification of plant genes specifically induced by the LPS of homologous rhizobia would help to reveal possible functions. Positive recognition via surface carbohydrates is well documented in different animal cell systems (Varik, 1993). For example, primary adhesion of lymphocytes to activated platelets or the endothelium of venules is mediated by surface carbohydrates and selectins, specific lectins (Varik, 1993). The release of bacteria from the infection thread and their enclosure by the peribacteroid membrane could involve similar mechanisms. But there is also evidence for a suppressor function of rhizobial EPS and LPS in suppressing the host-plant defense system in order to allow infection by rhizobium. EPS- or LPS-defective strains of *Rhizobium* can induce typical symptoms of plant defense connected with the loss of infectivity. This effect can be host-plant-specific, resembling a gene-for-gene relationship found in plant–pathogen interactions (Albersheim and Anderson-Prouty, 1975). In plant–pathogen interactions oligosaccharides can be involved in both the induction and suppression of the defense reactions (Cote and Hahn, 1994). The elicitor can be a pathogen-derived oligosaccharide, e.g., a fragment of a fungal cell wall, or an endogenous elicitor, released by hydrolytic enzymes from the plant cell wall (Albersheim and Anderson-Prouty, 1975; Aldington *et al.*, 1991; Darvill *et al.*, 1992; Cote and Hahn, 1994). From the analysis of EPS- or LPS-defective rhizobial mutants it seems obvious that *Rhizobium* has to deal with the plant defense system. In host-plant cell cultures oligosaccharides derived from the EPS I biosynthesis of *R. meliloti* are able to suppress the elicitor-induced alkalinization of the culture medium. Only LMW EPS is active in this assay (Niehaus *et al.*, 1997a) or in the supplementation of rhizobial mutants (Battisti *et al.*, 1992; Gonzáles *et al.*, 1996a). This may indicate that the oligosaccharide has to cross the plant cell wall to reach a hypothetical receptor on the plasma membrane. Degradation of rhizobial polysaccharides could be involved in the regulation of the plant response (Mellor and Collinge, 1995). A specific enzymatic degradation of rhizobial polysaccharides by the host plant has been reported (Solheim and Fjellheim, 1984). Host-plant-specific suppressors have been described for plant–pathogen interactions (Shiraishi *et al.*, 1994). A glycopeptide suppressor secreted by the pathogenic fungus *Mycosphaerella pinodes* inhibits the recycling of the second messenger inositoltriphosphate in pea membranes, indicating that the elicitor-induced signal transduction is influenced by the pathogen (Shiraishi *et al.*, 1994). The host plants of *Rhizobium* have to weigh the different signals (Fig. 8). One signal is the nodulation factor, others might be suppressors represented by rhizobial oligosaccharides. A third class of as-yet-hypothetical signals could be represented by a rhizobial elicitor. In the case of ectomycorrhizal interactions, chitin-like elicitors have been identified (Salzer *et al.*, 1996). But what is the rhizobial elicitor, if it exists? Recognition and signal transduction of the different signals by the host plant may involve common compounds such as second messengers e.g., calcium or inositolphosphates. Also, the reactions of the plant, establishment of the symbiosis or plant defense, may share some common processes. Nevertheless, this seems to be a minor part (Fig. 8). A

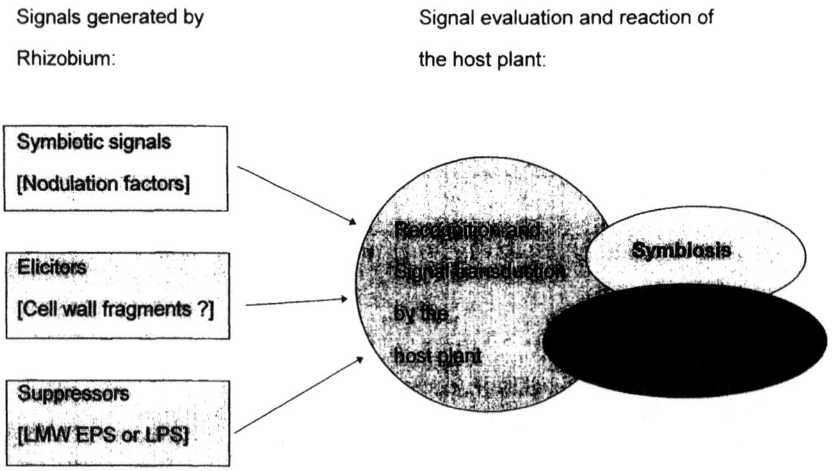


**FIGURE 8.** Hypothetical working model summarizing signals generated by the microsymbiont and their recognition and signal transduction by the host plant.

central question for further work will be the understanding of the signal transduction network making the decision for or against plant defense.

ACKNOWLEDGMENTS. The authors thank A. Pühler for his helpful discussions, D. Kapp for preparing the illustrations, and B. Subramaniam-Niehaus and A. Meyer for a critical reading of the manuscript.

## 5. REFERENCES


Albersheim, P., and Anderson-Prouty, A. J., 1975, Carbohydrates, proteins, cell surfaces, and the biochemistry of pathogenesis, *Ann. Rev. Plant Physiol.* **26**:31–52.

Aldington, S., McDougall, G. J., and Fry, S. C., 1991, Structure–activity relationships of biologically active oligosaccharides, *Plant Cell Environ.* **14**:625–636.

Aman, P., McNeil, M., Franzen, L.-E., Darvill, A. G., and Albersheim, P., 1981, Structural elucidation, using HPLC-MS and GLC-MS, of the acidic exopolysaccharide secreted by *Rhizobium meliloti* strain Rm2011, *Carbohydr. Res.* **95**:263–282.

Amemura, A., Hisamatsu, M., Mitani, H., and Harada, T., 1983, Cyclic (1-2)-β-D-glucan and the octasaccharide repeating units of extracellular acidic polysaccharides produced by *Rhizobium*, *Carbohydr. Res.* **114**:277–285.

An, J., Carlson, R. W., Glushka, J., and Streeter, J. G., 1995, The structure of a novel polysaccharide produced by *Bradyrhizobium* species with soybean nodules, *Carbohydr. Res.* **269**:303–317.

Ardourel, M., Demont, N., Debellé, F., Maillet, F., de Billy, F., Promé, J.-C., Dénarié, J., and Truchet, G., 1994, *Rhizobium meliloti* lipooligosaccharide nodulation factors: Different structural requirements for bacterial entry into target root hair cells and induction of plant symbiotic developmental responses, *Plant Cell* **6**:1357–1374.

Astete, S. G., and Leigh, J. A., 1996, *mucS*, a gene involved in activation of galactoglucan (EPS II) synthesis gene expression in *Rhizobium meliloti*, *Mol. Plant–Microbe Interact.* **9**:395–400.

Battisti, L., Lara, J. C., and Leigh, J. A., 1992, Specific oligosaccharide form of the *Rhizobium meliloti* exopolysaccharide promotes nodule invasion in alfalfa, *Proc. Natl. Acad. Sci. USA* **89**:5625–5629.

Bauer, W. D., 1981, Infection of legumes by rhizobia, *Annu. Rev. Plant Physiol.* **32**:407–449.

Becker, A., Kleickmann, A., Arnold, W., and Pühler, A., 1993a, Analysis of the *Rhizobium meliloti exoH/exoK/exoL* fragment: ExoK shows homology to excreted endo-beta-1,3–1,4-glucanases and ExoH resembles membrane proteins, *Mol. Gen. Genet.* **238**:145–154.

Becker, A., Kleickmann, A., Arnold, W., Keller, M., and Pühler, A., 1993b, Identification and analysis of the *Rhizobium meliloti exoAMONP* genes involved in exopolysaccharide biosynthesis and mapping of promoters located on the *exoHKLAMONP* fragment, *Mol. Gen. Genet.* **241**:367–379.

Becker, A., Kleickmann, A., Küster, H., Keller, M., Arnold, W., and Pühler, A., 1993c, Analysis of the *Rhizobium meliloti* genes *exoU*, *exoV*, *exoW*, *exoT*, and *exoI* involved in exopolysaccharide biosynthesis and nodule invasion, *Mol. Plant–Microbe Interact.* **6**:735–744.

Becker, A., Niehaus, K., and Pühler, A., 1995a, Low-molecular-weight succinoglycan is predominantly produced by *Rhizobium meliloti* strains carrying a mutated ExoP protein characterized by a periplasmic N-terminal domain and a missing C-terminal domain, *Mol. Microbiol.* **16**:191–203.

Becker, A., Küster, H., and Pühler, A., 1995b, Extension of the *Rhizobium meliloti* succinoglycan biosynthesis gene cluster: identification of the *exs* A gene encoding an ABC transporter protein, and the *exs* B gene which probably codes for a regulator of succinoglycan biosynthesis, *Mol. Gen. Genet.* **249:**487–497.

Becker, A., Küster, H., Roxlau, A., Keller, M., Ivashina, T., Cheng, H.-P., Walker, G. C., and Pühler, A., 1997, A 32 kb *exp* gene cluster of *Rhizobium meliloti* directing the biosynthesis of galactoglucan: Genetic organization and properties of the encoded gene products, *J. Bacteriol.* **179:**1375–1384.

Becquart-de Kozak, I., Reuhs, B. L., Buffard, D., Breda, C., Kim, J. S., Esnault, R., and Kendorosi, A., 1997, Role of the K-antigen subgroup of capsular polysaccharides in the early recognition process between *Rhizobium meliloti* and alfalfa leaves, *MPMI* **10:**114–123.

Bertram-Drogatz, P. A., Quester, I., Becker, A., and Pühler, A., 1998, The *Sinorhizobium meliloti* MucR protein is essential for the production of high molecular weight succinoglycan exopolysaccharide binds to short DNA regions upstream of exoH and exoY, *Mol. Gen. Genet.*, in press.

Bhat, U. R., and Carlson, R. W., 1992, Chemical characterization of pH-dependent structural epitopes of lipopolysaccharides from *Rhizobium leguminosarum* biovar *phaseoli, J. Bacteriol.* **174:**2230–2235.

Bhat, U. R., Krishnaiah, B. S., and Carlson, R. W., 1991a, Re-examination of the structures of the lipopolysaccharide core oligosaccharides from *Rhizobium leguminosarum* biovar *phaseoli, Carbohydr. Res.* **220:**219–227.

Bhat, U. R., Mayer, H., Yokota, A., Hollingsworth, R. I., and Carlson, R. W., 1991b, Occurrence of lipid A variants with 27-hydroxyoctacosanoic acid in lipopolysaccharides from members of the family *Rhizobiaceae, J. Bacteriol.* **173:**2155–2159.

Bhat, U. R., Forsberg, L. S., and Carlson, R. W., 1994, Structure of lipid A component of *Rhizobium leguminosarum* bv. *phaseoli* lipopolysaccharide: Unique nonphosphorylated lipid A containing 2-amino-2-deoxygluconate, galacturonate and glucosamine, *J. Biol. Chem.* **269:**14402–14410.

Bhuvaneswari, T. V., Turgeon, G. B., and Bauer, W. D., 1980, Early events in the infection of soybean by *Rhizobium japonicum*. I. Localization of infectable root cells, *Plant Physiol.* **66:**1027–1031.

Bohlool, B. B., and Schmidt, E. L., 1974, A possible basis for specificity in the *Rhizobium*–legume root nodule symbiosis, *Science* **185:**269–271.

Bond, L., 1948, Origin and developmental morphology of root nodules of *Pisum sativum, Bot. Gaz.* **109:**411–434.

Borthakur, D., and Johnston, A. W. B., 1987, Sequence of *psi*, a gene of the symbiotic plasmid of *Rhizobium phaseoli* which inhibits exopolysaccharide synthesis and nodulation and demonstration that its transcription is inhibited by *psr*, another gene on the symbiotic plasmid, *Mol. Gen. Genet.* **207:**149–154.

Borthakur, D., Barber, C. E., Lamb, J. W., Daniels, M. J., Downie, J. A., and Johnston, A. W. B., 1986, A mutation that blocks exopolysaccharide synthesis prevents nodulation of peas by *Rhizobium leguminosarum* but not by beans by *Rhizobium phaseoli* and is corrected by cloned DNA from *Rhizobium* or the phytopathogen *Xanthomonas, Mol. Gen. Genet.* **203:**320–323.

Branca, C., De Lorenzo, G., and Cervone, F., 1988, Competitive inhibition of the auxin-induced elongation by a-D-oligogalacturonides in pea stem segments, *Physiol. Plant.* **72:**499–504.

Brewin, N. J., 1991, Development of the legume root nodule, *Annu. Rev. Cell Biol.* **7:**191–226.

Brewin, N., and Kardalsky, I. V., 1997, Legume lectins and nodulation by *Rhizobium, Trends in Plant Sci.* **2:**92–98.

Brewin, N. J., Wood, E. A., Larkins, A. P., Gafaré, G., and Butcher, G. W., 1986, Analysis of lipopolysaccharide from root nodule bacteroids of *Rhizobium leguminosarum* using monoclonal antibodies, *J. Gen. Microbiol.* **132:**1959–1968.

Brzoska, P. M., and Signer, E. R., 1991, *lpsZ*, a lipopolysaccharide gene involved in symbiosis of *Rhizobium meliloti*, *J. Bacteriol.* **173:**3235–3237.

Buendia, A. M., Enenkel, B., Koplin, R., Niehaus, K., Arnold, W., and Pühler, A., 1991, The *Rhizobium meliloti exoZl exoB* fragment of megaplasmid 2: ExoB functions as a UDP-glucose 4-epimerase and ExoZ shows homology to NodX of *Rhizobium leguminosarum* biovar *viciae* strain TOM, *Mol. Microbiol.* **5:**1519–1530.

Caetano Anolles, G., and Gresshoff, P. M., 1991, Plant genetic control of nodulation, *Annu. Rev. Microbiol.* **45:**345–382.

Callaham, D. A., and Torrey, J. G., 1981, The structural basis for infection of root hairs of *Trifolium repens* by *Rhizobium*, *Can. J. Bot.* **59:**1647–1664.

Canter Cremers, H. C., Stevens, K., Lugtenberg, B. J., Wijffelman, C. A., Batley, M., Redmond, J. W., Breedveld, M. W., and Zevenhuizen, L. P., 1991, Unusual structure of the exopolysaccharide of *Rhizobium leguminosarum* bv. *viciae* strain 248, *Carbohydr. Res.* **218:**185–200.

Carlson, R. W., 1984, Heterogenicity of *Rhizobium* lipopolysaccharides, *J. Bacteriol.* **158:**548–555.

Carlson, R. W., and Garci, F., 1989, The structures of the lipopolysaccharide core components from *Rhizobium leguminosarum* biovar *phaseoli* CE3 and two of its symbiotic mutants, CE109 and CE309, *Carbohydr. Res.* **195:**101–110.

Carlson, R. W., and Krishnaiah, B. S., 1992, Structures of the oligosaccharides obtained from the core regions of the lipopolysaccharides of *Bradyrhizobium japonicum* 61A101c and its symbiotically defective lipopolysaccharide mutant, JS314, *Carbohydr. Res.* **231:**205–219.

Carlson, R. W., and Yadav, M., 1985, Isolation and partial characterization of the extracellular polysaccharides and lipopolysaccharides from fast-growing *Rhizobium japonicum* USDA205 and its nod- mutant, HC205, which lacks the symbiotic plasmid, *Appl. Environ. Microbiol.* **50:**1219–1224.

Carlson, R. W., Hanley, B., Rolfe, B. G., and Djordjevic, M. A., 1986, A structural comparison of the acidic extracellular polysaccharides from *Rhizobium trifolii* mutants affected in root hair infection, *Plant Physiol.* **80:**134–137.

Carlson, R. W., Kalembasa, S., Turowski, D., Pachori, P., and Noel, K. D., 1987a, Characterization of the lipopolysaccharide from a *Rhizobium phaseoli* mutant that is defective in infection thread development, J. Bacteriol. **169:**4923–4928.

Carlson, R. W., Shatters, R., Duh, J.-L., Turnbull, E., Hanley, B., Rolfe, B. G., and Djordjevic, M. A., 1987b, The isolation and partial characterization of the lipopolysaccharides from several *Rhizobium trifolii* mutants affected in root hair infection, *Plant Physiol.* **84:**421–427.

Carlson, R. W., Reuhs, B., Chen, T. B., Bhat, U. R., and Noel, K. D., 1995, Lipopolysaccharide core structures in *Rhizobium etli* and mutants deficient in O-antigen, *J. Biol. Chem.* **270:**11783–11788.

Carrion, M., Bhat, R., Rehus, B., and Carlson, R. W., 1990, Isolation and characterization of the lipopolysaccharides from *Bradyrhizobium japonicum*, *J. Bacteriol.* **172:**1725–1731.

Cava, J. R., Elias, P. M., Turowski, D. A., and Noel, K. D., 1989, *Rhizobium leguminosarum* CNF42 genetic regions encoding lipopolysaccharide structures essential for complete nodule development on bean plants, *J. Bacteriol.* **171:**8–15.

Chakravorty, A. K., Zurkowski, W., Shine, J., and Rolfe, B. G., 1982, Symbiotic nitrogen fixation: Molecular cloning of *Rhizobium* genes involved in exopolysaccharide synthesis and effective nodulation, *J. Mol. Appl. Gen.* **1:**585–596.

Chandler, M. R., 1978, Some observations on the infection of *Arachis hypogaea* L. by *Rhizobium*, *J. Exp. Bot.* **29:**749–755.

Chandler, M. R., Date, R. A., and Roughley, R. J., 1982, Infection and root nodule development in *Stylosanthes* species by *Rhizobium*, *J. Exp. Bot.* **33:**47–57.

Charles, T. C., and Finan, T. M., 1991, Analysis of a 1600-kilobase *Rhizobium meliloti* megaplasmid using defined deletions generated *in vivo*, *Genetics* **127:**5–20.

Chen, H., Batley, M., Redmond, J., and Rolfe, B. G., 1985, Alteration of the effective nodulation properties of a fast growing broad host range *Rhizobium* due to changes in exopolysaccharide synthesis, *J. Plant Physiol.* **120**:331–349.

Clover, R. H., Kieber, J., and Signer, E. R., 1989, Lipopolysaccharide mutants of *Rhizobium meliloti* are not defective in symbiosis, *J. Bacteriol.* **171**:3961–3967.

Coplin, D. L., and Cook, D., 1990, Molecular genetics of extracellular polysaccharide biosynthesis in vascular phytopathogenic bacteria, *Mol. Plant–Microbe Interact.* **3**:271–279.

Cote, F., and Hahn, M. G., 1994, Oligosaccharins: Structures and signal transduction, *Plant Mol. Biol.* **26**:1379–1411.

Currier, W. W., and Strobel, G. A., 1976, Chemotaxis of *Rhizobium* spp. to plant root exudates, *Plant Physiol.* **57**:820–823.

Darvill, A., Augur, C., Bergmann, C., Carlson, R. W., Cheong, J. J., Eberhard, S., Hahn, M. G., Ló, V. M., Marfà, V., Meyer, B., Mohnen, D., O'Neil, M. A., Spiro, M. D., Halbeek, H., York, W. S., and Albersheim, P., 1992, Oligosaccharins—Oligosaccharides that regulate growth, development and defence responses in plants, *Glycobiology* **2**:181–198.

Dazzo, F. B., 1980, Lectins and their saccharide receptors as determinants of specificity in the *Rhizobium*–legume symbiosis, in *The Cell Surface: Mediator of Developmental Processes* (S. Subtelny and N. K. Wessells, eds.), pp. 81–104, Academic Press, New York.

Dazzo, F. B., and Brill, W. J., 1977, Receptor site on clover and alfalfa roots for *Rhizobium*, *Appl. Environ. Microbiol.* **33**:132–136.

Dazzo, F. B., Truchet, G. L., Sherwood, J. E., Hrabak, E. M., Abe, M., and Pankratz, S. H., 1984, Specific phases of root hair attachment in the *Rhizobium trifolii*–clover symbiosis, *Appl. Environ. Microbiol.* **48**:87–94.

Dazzo, F. B., Truchet, G. L., Hollingsworth, R. I., Hrabak, E. M., Pankratz, H. S., Philip-Hollingsworth, S., Salzwedel, J. L., Chapman, K., Appenzeller, L., Squartini, A., *et al.*, 1991, *Rhizobium* lipopolysaccharide modulates infection thread development in white clover root hairs, *J. Bacteriol.* **173**:5371–5384.

Dazzo, F. B., Orgambide, G. G., Philip-Hollingsworth, J., Hollingsworth, R. I., Nike, K. O., and Salzwedel, J. L., 1996, Modulation of development, growth dynamics, wall crystallinity, and infection sites in white clover root hairs by membrane chitolipooligosaccharides from *Rhizobium leguminosarum* bv. *trifolii*, *J. Bacteriol.* **178**:3621–3627.

DeMaagd, R. A., Rao, A. S., Mulders, I. H. M., Goosen-de Roo, L., VanLoosdrecht, M. C. M., Wijffelman, C. A., and Lugtenberg, B. J. J., 1989, Isolation and characterization of mutants of *Rhizobium leguminosarum* bv. *viciae* 248 with altered lipopolysaccharides: Possible role of surface charge or hydrophobicity in bacterial release from the infection thread, *J. Bacteriol.* **171**:1143–1150.

Denarie, J., Debelle, F., and Rosenberg, C., 1992, Signaling and host range variation in nodulation, *Annu. Rev. Microbiol. 46*:497–531.

Denarie, J., Truchet, G., and Prome, J. C., 1994, Lipo-oligosacchaaride signalling: The mediation of recognition and nodule organogenesis induction in the legume–*Rhizobium* symbiosis, *Biochem. Soc. symp.* **60**:51–60.

Diebold, R., and Noel, K. D., 1989, *Rhizobium leguminosarum* exopolysaccharide mutants: Biochemical and genetic analyses and symbiotic behaviour on three hosts, *J. Bacteriol.* **171**:4821–4830.

Djordjevic, S. P., Rolfe, B. G., Batley, M., and Redmond, J. W., 1986, The structure of the exopolysaccharide from *Rhizobium* sp. strain NGR234, *Carbohydr. Res.* **148**:87–99.

Djordjevic, S. P., Chen, H., Batley, M., Redmond, J. W., and Rolfe, B. G., 1987, Nitrogen fixation ability of exopolysaccharide synthesis mutants of *Rhizobium* sp. NGR234 and *Rhizobium trifolii* is restored by the addition of homologous exopolysaccharides, *J. Bacteriol.* **169**:114–126.

Doherty, D., Leigh, J. A., Glazebrook, J., and Walker, G. C., 1988, *Rhizobium meliloti* mutants that overproduce the *R. meliloti* acidic Calcofluor-binding exopolysaccharide, *J. Bacteriol.* **170:**4249–4256.

Dudley, M. E., Jacobs, T. W., and Long, S. R., 1987, Microscopic studies of cell divisions induced in alfalfa roots by *Rhizobium meliloti, Planta* **171:**289–301.

Dudman, W. F., 1978, Structural studies of the extracellular polysaccharides of *Rhizobium japonicum* strains 71a, CC708 and CB1795, *Carbohydr. Res.* **66:**9–23.

Dylan, T., Nagpal, P., Helsinski, D. R., and Ditta, G., 1990, Symbiotic pseudorevertants of *R. meliloti ndv* mutants, *J. Bacteriol.* **172:**1409–1417.

Fahraeus, G., 1957, The infection of clover root hairs by nodule bacteria studied by a simple glass slide technique, *J. Gen. Microbiol.* **16:**374–381.

Finan, T. M., Hirsch, A. M., Leigh, J. A., Johansen, E., Kuldau, G. A., Deegan, S., Walker, G. C., and Signer, E. R., 1985, Symbiotic mutants of *Rhizobium meliloti* that uncouple plant from bacterial differentiation, *Cell* **40:**869–877.

Finan, T. M., Kunkel, B., De Vos, G. F., and Signer, E. R., 1986, Second symbiotic megaplasmid in *Rhizobium meliloti* carrying exopolysaccharide and thiamine synthesis genes, *J. Bacteriol.* **167:**66–72.

Fischer, H. M., 1994, Genetic regulation of nitrogen fixation in rhizobia, *Microbiol. Rev.* **58:**352–386.

Forsbery, L. S., and Reuhs, B. L., 1997, Structural characterization of the K antigens from *Rhizobium fredii* USDA257: Evidence for a common structural motiv, with strain specific variation, within capsular polysaccharides of *Rhizobium spp.*, *J. Bacteriol.* **179:**5366–5371.

Gil-Serrano, A., del Junco, A. S., Tejero-Mateo, P., Megias, M., and Caviedes, M. A., 1990, Structure of the extracellular polysaccharide secreted by *Rhizobium leguminosarum* var. *phaseoli* CIAT 899, *Carbohydr. Res.* **204:**103–107.

Glazebrook, J., and Walker, G. C., 1989, A novel exopolysaccharide can function in place of the Calcofluor-binding exopolysaccharide in nodulation of alfalfa by *Rhizobium meliloti, Cell* **56:**661–672.

Glucksmann, M. A., Reuber, L., and Walker, G. C., 1993a, Genes needed for the modification, polymerization, export, and processing of succinoglycan by *Rhizobium meliloti:* A model for succinoglycan biosynthesis, *J. Bacteriol.* **175:**7045–7055.

Glucksmann, M. A., Reuber, T. L., and Walker, G. C., 1993b, Family of glycosyl transferases needed for the synthesis of succinoglycan by *Rhizobium meliloti, J. Bacteriol.* **175:**7033–7044.

Gonzáles, J. E., Rehus, B. L., and Walker, G. C., 1996a, Low molecular weight EPS II of *Rhizobium meliloti* allows nodule invasion in *Medicago sativa, Proc. Natl. Acad. Sci. USA* **93:**8636–8641.

Gonzáles, J. E., York, G. M., and Walker, G. C., 1996b, *Rhizobium meliloti* exopolysaccharides: synthesis and symbiotic function, *Gene* **179:**141–146.

Gross, M., Geier, G., Geider, K., and Rodolph, K., 1992, Levan and levan sucrase synthesized by the fireblight pathogen *Erwinia amylovora, Physiol. Mol. Plant Pathol.* **40:**371–381.

Grosskopf, E., Ha, D. T. C., Wingender, R., Roehrig, H., Szesci, J., Kondorosi, E., Schell, J., and Kondorosi, A., 1993, Enhanced levels of chalcone synthase in alfalfa nodules induced by a Fix negative mutant of *Rhizobium meliloti, Mol. Plant–Microbe Interact.* **6:**173–181.

Gulash, M., Ames, P., LaRosiliere, R. C., and Bergman, K., 1984, Rhizobia are attracted to localized sites in legume roots, *Appl. Environ. Microbiol.* **48:**148–152.

Her, G. R., Glazebrook, J., Walker, G. C., and Reinhold, V. N., 1990, Structural studies of a novel exopolysaccharide produced by a mutant of *Rhizobium meliloti* strain Rm2011, *Carbohydr. Res.* **198:**305–312.

Hirsch, A. M., and Fang, Y., 1994, Plant hormones and nodulation: What's the connection? *Plant Mol. Biol.* **26:**5–9.

Hirsch, P. R., and Spokes, J. D., 1994, Survival and dispersion of genetically modified rhizobia in the field and genetic interactions with native strains, *FEMS Microbiol. Ecol.* **15:**147–159.

Hitchcock, P. J., and Brown, T. M., 1983, Morphological hetrogenicity among *Salmonella* lipopolysaccharide chemotypes in silver-stained polyacrylamide gels, *J. Bacteriol.* **154:**269–277.

Ho, J. C., Wang, J. L., Schindler, M., and Loh, S. T., 1994, Carbohydrate binding activities of *Bradyrhizobium japonicum:* III Lectin expression, bacterial binding, and nodulation efficiency, *Plant J.* **5:**873–884.

Hollingsworth, R. I., and Dazzo, F. B., 1988, The complete structure of the trifoliin A lectin-binding capsular polysaccharaide of *Rhizobium trifolii* 843, *Carbohydr. Res.* **172:**92–112.

Hollingsworth, R. I., Carlson, R. W., Garcia, F., and Gage, D. A., 1989, A new core tetrasaccharide component from the lipopolysaccharide of *Rhizobium trifolii* ANU843, *J. Biol. Chem.* **264:**9294–9299.

Hollingsworth, R. I., Zhang, Y., and Priefer, U. B., 1994, Structure of the unusual trisaccharide lipopolysaccharide component produced by a symbiotically defective mutant of *Rhizobium leguminosarum* biovar *viciae*, *Carbohydr. Res.* **264:**271–280.

Hotter, G. S., and Scott, D. B., 1991, Exopolysaccharide mutants of *Rhizobium loti* are fully effective on a determinate nodulating host but are ineffective on an indeterminate nodulating host, *J. Bacteriol.* **173:**851–859.

Hynes, M. F., Simon, R., Müller, P., Niehaus, K., Labes, M., and Pühler, A., 1986, The two megaplasmids of *Rhizobium meliloti* are involved in effective nodulation of alfalfa, *Mol. Gen. Genet.* **1986:**356–362.

Ielpi, L., Couso, R., and Dankert, M., 1981a, Lipid linked intermediates in the biosynthesis of xanthan gum, *FEBS Lett.* **130:**253–256.

Ielpi, L., Couso, R. O., and Dankert, M. A., 1981b, Xanthan gum biosynthesis: Pyruvic avid acetal residues are transferred from phosphoenolpyruvate to the pentasaccharide-P-P-lipid, *Biochem. Biophys. Res. Commun.* **102:**1400–1408.

Ielpi, L., Dylan, T., Ditta, G., Helinski, D. R., and Stanfield, S. W., 1990, The *ndvB* locus of *Rhizobium meliloti* encodes a 319 KDa protein involved in the production of β-1,2-glucan, *J. Biol. Chem.* **265:**2843–2851.

James, E. K., Sprent, J. L., Sutherland, J. M., Mcinroy, S. G., and Minchin, F. R., 1992, The structure of nitrogen fixing root nodules on the aquatic mimosoid legume *Neptunia plena*, *Annu. Bot.* **69:**173–180.

Jansson, P.-E., Kenne, L., Lindberg, B., Ljunggren, H., Lönngren, J., Rudén, U., and Svensson, S., 1977, Demonstration of an octasaccharide repeating unit in the extracellular polysaccharide of *Rhizobium meliloti* by sequential degradation, *J. Am. Chem. Soc.* **99:**3812–3815.

Joshi, P. A., Caetano-Anollés, G., Graham, E. T., and Gresshoff, P. M. (1991, Ontogeny and ultrastructure of spontaneous nodules in alfalfa (*Medicago sativa*), *Protoplasma* **162:**1–11.

Kannenberg, E. L., and Brewin, N. J., 1989, Expression of a cell surface antigen from *Rhizobium leguminosarum* 3841 is regulated by oxygen and pH, *J. Bacteriol.* **171:**4543–4548.

Kannenberg, E. L., and Brewin, N. J., 1994, Host-plant invasion by *Rhizobium:* The role of cell-surface components, *Trends Microbiol.* **2:**277–283.

Kannenberg, E. L., Rathbun, E. A., and Brewin, N. J., 1992, Molecular dissection of structure and function in the lipopolysaccharide of *Rhizobium leguminosarum* strain 3841 using monoclonal antibodies and genetic analysis, *Mol. Microbiol.* **6:**2477–2487.

Kannenberg, E. L., Perotto, S., Bianciotto, V., Rathbun, E. A., and Brewin, N. J., 1994, Lipopolysaccharide epitope expression of *Rhizobium* bacteroids as revealed by *in situ* immunolabelling of pea root nodule sections, *J. Bacteriol.* **176:**2021–2032.

Kapp, D., Niehaus, K., Quandt, J., Müller, P., and Pühler, A., 1990, Cooperative action of *Rhizobium meliloti* nodulation and infection mutants during the process of forming mixed infected alfalfa nodules, *Plant Cell* **2:**139–151.

Keller, M., Roxlau, A., Weng, W. M., Schmidt, M., Quandt, J., Niehaus, K., Jording, D., Arnold, W., and Pühler, A., 1995, Molecular analysis of the *Rhizobium meliloti mucR* gene regulating the biosynthesis of the exopolysaccharides succinoglycan and galactoglucan, *Mol. Plant-Microb Interact.* **8**:267–277.

Kim, J. S., Reuhs, B. L., Rahman, M. M., Ridley, B., and Carlson, R. W., 1996. Separation of bacterial capsular and lipopolysaccharides by preparative electrophoresis, *Glycobiology* **6**:433–437.

Kiss, E., Reuhs, B. L., Kim, J. S., Kereszt, A., Petrovics, G., Putnoky, P., Dusha, I., Carlson, R. W., and Kandorosin, A., 1997, The *rkp* GHI and -J genes are involved in capsular polysaccharide production by *Rhizobium meliloti*, *J. Bacteriol.* **179**:2132–2140.

Klein, S., Hirsch, A. M., Smith, C. A., and Signer, E. R., 1988, Interaction of nod and exo *Rhizobium meliloti* in alfalfa nodulation, *Mol. Plant-Microbe Interact.* **1**:94–100.

Kosch, K., Jacobi, A., Parniske, M., Werner, D., and Mueller, P., 1994, The impairment of the nodulation process induced by a *Bradyrhizobium japonicum* exopolysaccharide mutant is determined by the genotype of the host plant, *Z. Naturforsch. C Biosci.* **49**:727–736.

Lagares, A., Caetano Anolles, G., Niehaus, K., Lorenzen, J., Ljunggren, H. D., Pühler, A., and Favelukes, G., 1992, A *Rhizobium meliloti* lipopolysaccharide mutant altered in competitiveness for nodulation of alfalfa, *J. Bacteriol.* **174**:5941–5952.

Latchford, J. W., Borthakur, D., and Johnston, A. W., 1991, The products of *Rhizobium* genes, *psi* and *pss*, which affect exopolysaccharide production, are associated with the bacterial cell surface, *Mol. Microbiol.* **5**:2107–2114.

Legocki, R. P., and Verma, D. P., 1980, Identification of "nodule specific" host proteins (nodulins) involved in the development of rhizobium-legume symbiosis, *Cell* **20**:153–163.

Leigh, J. A., and Coplin, D. L., 1992, Exopolysaccharides in plant-bacterial interactions, *Annu. Rev. Microbiol.* **46**:307–346.

Leigh, J. A., and Lee, C. C., 1988, Characterization of polysaccharides of *Rhizobium meliloti exo* mutants that form ineffective nodules, *J. Bacteriol.* **170**:3327–3332.

Leigh, J. A., Signer, E. R., and Walker, G. C., 1985, Exopolysaccharide-deficient mutants of *Rhizobium meliloti* that form ineffective nodules, *Proc. Natl. Acad. Sci. USA* **85**:6231–6235.

Leigh, J. A., Reed, J. W., Hanks, J. F., Hirsch, A. M., and Walker, G. C., 1987, *Rhizobium meliloti* mutants that fail to succinylate their Calcofluor-binding exopolysaccharide are defective in nodule invasion, *Cell* **51**:579–587.

Levery, S. B., Zhan, H., Lee, C. C., Leigh, J. A., and Hakomori, S., 1991, Structural analysis of a second acidic exopolysaccharide of *Rhizobium meliloti* that can function in alfalfa root nodule invasion, *Carbohydr. Res.* **210**:339–347.

Libbenga, K. R., and Harkes, P. A. A., 1973, Initial proliferation of cortical cells in the formation of root nodules in *Pisum sativum* L., *Planta* **114**:17–28.

Long, S., Reed, J. W., Himawan, J., and Walker, G. C., 1988, Genetic analysis of a cluster of genes required for synthesis of the Calcofluor-binding exopolysaccharide of *Rhizobium meliloti*, *J. Bacteriol.* **170**:4239–4248.

Lopez Lara, I. M., Orgambide, G., Dazzo, F. B., Olivares, J., and Toro, N., 1993, Characterization and symbiotic importance of acidic extracellular polysaccharides of *Rhizobium* sp. strain GRH2 isolated from acacia nodules, *J. Bacteriol.* **175**:2826–2832.

Lopez-Lara, I. M., Orgambide, G., Dazzo, F. B., Olivares, J., and Toro, N., 1995, Surface polysaccharide mutants of *Rhizobium* sp. (Acacia) strain GRH2: Major requirement of lipopolysaccharide for successful invasion of Acacia nodules and host range determination, *Microbiology* **141**:573–581.

Lorkiewicz, Z., Derylo, M., Russa, R., Skorupska, A., and Urbanik Sypniewska, T., 1993, *Rhizobium leguminosarum* biovar *trifolii* mutants altered in surface structures that are defective in nodulation or nitrogen fixation, *Acta Microbiol. Polon.* **42**:219–234.

Marfá, V., Gollin, D. J., Eberhard, S., Mohnen, D., Darvill, A., and Albersheim, P., 1991, Oligogalacturonides in suspension-cultured tobacco cells, *Plant J. Cell Mol. Biol.* **1**:217–225.

Mateos, P. F., Baker, D. L., Philip Hollingsworth, S., Squartini, A., Peruffo, A. D. B., Nuti, M. P., and Dazzo, F. B., 1995, Direct *in situ* identification of cellulose microfibrils associated with *Rhizobium leguminosarum* biovar *trifolii* attached to the root epidermis of white clover, *Can. J. Microbiol.* **41**:202–207.

Matthysse, A. G., 1983, Role of bacterial cellulose fibrils in *Agrobacterium tumefaciens* infections, *J. Bacteriol.* **154**:906–915.

Matulová, M., Toffanin, R., Navarini, L., Gilli, R., Paoletti, S., and Cesàro, A., 1994, NMR analysis of succinoglycans from different microbial sources: Partial assignment of their 1H and 13C NMR spectra and localization of the succinate and acetate groups, *Carbohydr. Res.* **265**:167–179.

Mellor, R. B., and Collinge, D. B., 1995, A simple model based on known plant defence reactions is sufficient to explain most aspects of nodulation, *J. Exp. Bot.* **46**:1–18.

Mort, A. J., and Bauer, W. D., 1982, Application of two new methods for cleavage of polysaccharides into specific oligosaccharide fragments, *J. Biol. Chem.* **257**:1870–1875.

Müller, P., Hynes, M. F., Kapp, D., Niehaus, K., and Pühler, A., 1988, Two classes of *Rhizobium meliloti* infection mutants differ in exopolysaccharide production and in coinoculation properties with nodulation mutants, *Mol. Gen. Genet.* **211**:17–26.

Müller, P., Keller, M., Weng, W. M., Quandt, J., Arnold, W., and Pühler, A., 1993, Genetic analysis of the *Rhizobium meliloti exoYFQ* operon: ExoY is homologous to sugar transferases and ExoQ represents a transmembrane protein, *Mol. Plant–Microbe Interact.* **6**:55–65.

Mulligan, J. T., and Long, S. R., 1985, Induction of Rhizobium meliloti nodC expression by plant exudate requires nodD, *Proc. Natl. Acad. Sci. USA* **82**:6609–6613.

Mulligan, J. T., and Long, S. R., 1989, A family of activator genes regulates expression of *Rhizobium meliloti* nodulation genes, *Genetics* **122**:7–18.

Mutaftschiev, S., Vasse, J., and Truchet, G., 1982, Exostructures of *Rhizobium meliloti*, *FEMS Microbiol. Lett.* **13**:171–175.

Napoli, C. A., and Hubbell, D. H., 1975, Ultrastructure of *Rhizobium*-induced infection threads in clover root hairs, *Appl. Microbiol.* **30**:1003–1009.

Newcomb, W., 1981, Nodule morphogenesis and differentiation, in *Biology of the Rhizobiaceae* (K. L. Giles and A. G. Atherly, eds.), pp. 245–298, Academic Press, New York.

Niehaus, K., Kapp, D., and Puehler, A., 1993, Plant defence and delayed infection of alfalfa pseudonodules induced by an exopolysaccharide (EPS I)-deficient *Rhizobium meliloti* mutant, *Planta* **190**:415–425.

Niehaus, K., Baier, R., Lorenzen, J., Meyer-Gattermann, P., Sieben, S., and Pühler, A., 1994, Plant defence in alfalfa pseudonodules induced by an exopolysaccharide (EPS I)-deficient symbiont *Rhizobium meliloti*, *Acta Horticult.* **381**:258–264.

Niehaus, K., Baier, R., Kohring, B., Flaschel, E., and Pühler, A., 1997a, Symbiotic suppression of the *Medicago sativa* plant defence system by *Rhizobium meliloti* oligosaccharides, in *Biological Fixation of Nitrogen for Ecology and Sustainable Agriculture* (A. Legocki, H. Bothe, and A. Pühler, eds.), pp. 111–114, Springer, Heidelberg.

Niehaus, K., Lorenzen, J., Lagares, A., and Pühler, A., 1997b, A *Rhizobium meliloti* lipopolysaccharide mutant effective in *Medicago sativa* but ineffective in *Medicago truncatula*, submitted.

Noel, K. D., VandenBosch, K. A., and Kulpaca, B., 1986, Mutations in *Rhizobium phaseoli* that lead to arrested development in infection threads, *J. Bacteriol.* **168**:1392–1401.

Noel, K. D., Duelli, D. M., Tao, H., and Brewin, N. J., 1996, Antigenic change in the lipopolysaccharide of *Rhizobium etli* CFN42 induced by exudates of *Phaseolus vulgaris*, *Mol. Plant–Microbe Interact.* **9**:180–186.

Østeras, M., Stanley, J., and Finan, T. M., 1995, Identification of *Rhizobium*-specific intergenic mosaic elements within essential two-component regulatory system of *Rhizobium* species, *J. Bacteriol.* **177**:5485–5494.

Ozga, D. A., Lara, J. C., and Leigh, J. A., 1994, The regulation of exopolysaccharide production is important at two levels of nodule development in *Rhizobium meliloti*, *Mol. Plant–Microbe Interact.* **7**:758–765.

Parke, D., Rivelli, M., and Ornston, L. N., 1985, Chemotaxis to aromatic and hydroaromatic acids: Comparison of *Bradyrhizobium japonicum* and *Rhizobium trifolii*, *J. Bacteriol.* **163**:417–422.

Parniske, M., Schmidt, P. E., Kosch, K., and Mueller, P., 1994, Plant defense response of host plants with determinate nodules induced by EPS-defective *exoB* mutants of *Bradyrhizobium japonicum*, *Mol. Plant–Microbe Interact.* **7**:631–638.

Perotto, S., Brewin, N. J., and Kannenberg, E. L., 1994, Cytological evidence for a host defense response that reduces cell and tissue invasion in pea nodules by lipopolysaccharide-defective mutants of *Rhizobium leguminosarum* strain 3841, *Mol. Plant–Microbe Interact.* **7**:99–112.

Petrovics, G., Putnoky, P., Reuhs, B. L., Kim, J., Thorp, T. A., Noel, K. D., Carlson, R. W., and Kondorosi, A., 1993, The presence of a novel type of surface polysaccharide in *Rhizobium meliloti* requires a new fatty acid synthase-like gene cluster involved in symbiotic nodule development, *Mol. Microbiol.* **8**:1083–1094.

Praveen, N., Webb, D. T., and Borthakur, D., 1996, *Leucaena leucocephala* nodules formed by a surface polysaccharide defective mutant of *Rhizobium* sp. strain TAL1145 are delayed in bacteroid development and nitrogen fixation, *Mol. Plant–Microbe Interact.* **9**:364–372.

Price, N. P., Jeyaretnam, B., Carlson, R. W., Kadrmas, J. L., Raetz, C. R., and Brozek, K. A., 1995, Lipid A biosynthesis in *Rhizobium leguminosarum*: Role of a 2-keto-3-deoxyoctulosonate-activated 4′ phosphatase, *Proc. Natl. Acad. Sci. USA* **92**:7352–7356.

Priefer, U. B., 1989, Genes involved in lipopolysaccharide production and symbiosis are clustered on the chromosome of *Rhizobium leguminosarum* biovar *viciae* VF39, *J. Bacteriol.* **171**:6161–6168.

Pull, S. P., Pueppke, S. G., Hymowitz, T., and Ord, J. H., 1978, Soybean lines lacking the 120,000-Dalton seed lectin, *Science* **200**:1277–1279.

Putnoky, P. E., Grosskopf, E., Ha, D. T. C., Kiss, G. B., and Kondorosi, A., 1988, *Rhizobium fix* genes mediate at least two communication steps in symbiotic nodule development, *J. Cell Biol.* **106**:597–607.

Puvanesarajah, V., Schell, F. M., Gerhold, D., and Stacey, G., 1987, Cell surface polysaccharides from *Bradyrhizobium japonicum* and a nonnodulating mutant, *J. Bacteriol.* **169**:137–141.

Quandt, J., Hillemann, A., Niehaus, K., Arnold, W., and Pühler, A., 1992, An osmorevertant of a *Rhizobium meliloti ndvB* deletion mutant forms infection threads but is defective in bacteroid development, *Mol. Plant–Microbe Interact.* **5**:420–427.

Reed, J. W., and Walker, G. C., 1991a, Acidic conditions permit effective nodulation of alfalfa by invasion-deficient *Rhizobium meliloti exoD* mutants, *Genes Dev.* **5**:2274–2287.

Reed, J. W., and Walker, G. C., 1991b, The *exoD* gene of *Rhizobium meliloti* encodes a novel function needed for alfalfa nodule invasion, *J. Bacteriol.* **173**:664–677.

Reed, J. W., Capage, M., and Walker, G. C., 1991a, *Rhizobium meliloti exoG* and *exoJ* mutations affect the *exoX–exoY* system for modulation of exopolysaccharide production, *J. Bacteriol.* **173**:3776–3788.

Reed, J. W., Glazebrook, J., and Walker, G. C., 1991b, The *exoR* gene of *Rhizobium meliloti* affects RNA levels of other *exo* genes but lacks homology to known transcriptional regulators, *J. Bacteriol.* **173**:3789–3794.

Reinhold, B. B., Chan, S. Y., Reuber, T. L., Marra, A., Walker, G. C., and Reinhold, V. N., 1994, Detailed structural characterization of succinoglycan, the major exopolysaccharide of *Rhizobium meliloti* Rm1021, *J. Bacteriol.* **176**:1997–2002.

Reuber, T. L., and Walker, G. C., 1993a, The acetyl substituent of succinoglycan is not necessary for alfalfa nodule invasion by *Rhizobium meliloti* Rm1021, *J. Bacteriol.* **175:**3653–3655.

Reuber, T. L., and Walker, G. C., 1993b, Biosynthesis of succinoglycan, a symbiotically important exopolysaccharide of *Rhizobium meliloti, Cell* **74:**269–280.

Reuber, T. L., Long, S., and Walker, G. C., 1991, Regulation of *Rhizobium meliloti exo* genes in free-living cells and in plants examined by using TnphoA fusions, *J. Bacteriol.* **173:**426–434.

Reuhs, B. L., Carlson, R. W., and Kim, J. S., 1993, *Rhizobium fredii* and *Rhizobium meliloti* produce 3-deoxy-D-manno-2-octulosonic acid-containing polysaccharides that are structurally analogous to group II K antigens (capsular polysaccharides) found in *Escherichia coli, J. Bacteriol.* **175:**3570–3580.

Reuhs, B. L., Kim, J. S., Badgett, A., and Carlson, R. W., 1994, Production of cell-associated polysaccharides of Rhizobium fredii USDA205 is modulated by apigenin and host root extract, *MPMI* **7:**240–247.

Reuhs, B. L., Williams, M. N., Kim, J. S., Carlson, R. W., and Cote, F., 1995, Suppression of the Fix− phenotype of *Rhizobium meliloti exoB* mutants by *lpsZ* is correlated to a modified expression of the K polysaccharide, *J. Bacteriol.* **177:**4289–4296.

Rietschel, E. T., Brade, H., Holst, O., Brade, L., Müller-Loennies, S., Mamat, U., Zähringer, U., Beckmann, F., Seydel, U., Brandenburg, K., Ulmer, A. J., Mattern, T., Heine, H., Schletter, J., Loppnow, H., Schönbeck, U., Flad, H.-D., Hauschildt, S., Schade, U. F., Padova, D., Kusumoto, S., and Schumann, R. R., 1996, Bacterial endotoxins: Chemical constitution, biological recognition, host response, and immunological detoxification, in *Pathology of Septic Shock* (E. T. Rietschel and H. Wagner, eds.), pp. 40–69, Springer, Berlin.

Robertsen, B. K., Aman, P., Darvill, A. G., McNeil, M., and Albersheim, P., 1981, Host–symbiont interactions V., The structe of acidic extracellular polysaccharides secreted by *Rhizobium leguminosarum* and *Rhizobium trifolii, Plant Physiol.* **67:**389–400.

Rolfe, B. G., Gresshoff, P. M., Shine, J., and Vincent, J. M., 1980, Interaction between a non-nodulating and an ineffective mutant of *Rhizobium trifolii* resulting in effective (nitrogen fixing) nodulation, *Appl. Environ. Microbiol.* **39:**449–452.

Rolfe, B. G., Carlson, R. W., Ridge, R. W., Dazzo, F. B., Mateos, P. F., and Pankhurst, C. E., 1996, Defective infection and nodulation of covers by exopolysaccharide mutants of *Rhizobium leguminosarum* bv. *trifolii, Austral. J. Physiol.* **23:**285–303.

Ross, P., Mayer, R., and Benziman, M., 1991, Cellulose biosynthesis and function in bacteria, *Microbiol. Rev.* **55:**35–58.

Roth, L. E., Jeon, K. W., and Stacey, G., 1988, Homology in endosymbiotic systems: The term "symbiosome," in *Molecular Genetics of Plant–Microbe Interactions* (R. Palacios and D. P. S. Verma, eds.), Minnesota. pp. 28–28, APS Press, St. Paul,

Russa, R., Urbanik Sypniewska, T., Shashkov, A. S., Kochanowski, H., and Mayer, H., 1995, The structure of the homopolymeric O-specific chain from the phenol soluble LPS of the *Rhizobium luti* type strain NSP2213, *Carbohydr. Polym.* **27:**299–303.

Russa, R., Bruneteau, M., Shashkov, A. S., Urbanik Sypniewska, T., and Mayer, H., 1996, Characterization of the lipopolysaccharides from *Rhizobium meliloti* strain 102F51 and its nonnodulating mutant WL113, *Arch. Microbiol.* **165:**26–33.

Salzer, P., Hebe, G., Reith, A., Zitterell-Haid, B., Stransky, H., Gaschler, K., and Hager, A., 1996, Rapid reactions of spruce cells to elicitors released from the ectomycorrhizal fungus *Hebeloma crustuliniforme,* and inactivation of these elicitors by extracellular spruce cell enzymes, *Planta* **198:**118–126.

Saxena, I. M., Kudlicka, K., Okuda, K., and Brown, R. M., 1994, Characterization of genes in the cellulose-synthesizing operon (*acs* operon) of *Acetobacter xylinum:* Implications for cellulose crystallization, *J. Bacteriol.* **176:**5735–5752.

Schaede, R., 1940, Die Knöllchen der adventiven Wasserwurzeln von *Neptunia oleraceae* und ihre Bakteriensymbiose, *Planta* **31:**1–21.

Schultze, M., and Kondorosi, A., 1995, What makes nodulation signals host-plant specific? *Trends Microbiol.* **3:**370–372.

Shiraishi, T., Yamada, T., Saitoh, K., Kato, T., Toyoda, K., Yoshioka, H., Kim, H. M., Ichinose, Y., Tahara, M., and Oku, H., 1994, Suppressors: Determinants of specificity produced by plant pathogens, *Plant Cell Physiol.* **35:**1107–1119.

Sindhu, S. S., Brewin, N. J., and Kannenberg, E. L., 1990, Immunochemical analysis of lipopolysaccharides from free-living and endosymbiotic forms of *Rhizobium leguminosarum, J. Bacteriol.* **172:**1804–1813.

Smith, G., Kijne, J., and Lugtenberg, B. J. J., 1987, Involvement of both cellulose fibrils and a Ca2+-dependent adhesin in the attachment of *Rhizobium leguminosarum* to pea root hair tips, *J. Bacteriol.* **169:**4294–4301.

Smith, G., Logman, T. J. J., Boerrigter, M. E. T. I., Kijne, J., and Lugtenberg, B. J. J., 1989, Purification and partial characterization of the *Rhizobium leguminosarum* biovar *viciae* Ca2+-dependent adhesin, which mediates the first step in attachment of cells of the family Rhizobiaceae to plant root hair tips, *J. Bacteriol.* **171:**4054–4062.

Smith, G., Swart, S., Lugtenberg, B. J., and Kijne, J. W., 1992, Molecular mechanisms of attachment of *Rhizobium* bacteria to plant roots, *Mol. Microbiol.* **20:**2897–2903.

Solheim, B., and Fjellheim, K. E., 1984, Rhizobial polysaccharide-degrading enzymes from roots of legumes, *Physiol. Plant* **62:**11–17.

Stacey, G., So, J. S., Roth, L. E., Lakshmi, S. B., and Carlson, R. W., 1991, A lipopolysaccharide mutant of *Bradyrhizobium japonicum* that uncouples plant from bacterial differentiation, *Mol. Plant–Microbe Interact.* **4:**332–340.

Tolmasky, M. E., Staneloni, R. J., and Leloir, L. F., 1982, Lipid-bound saccharides in *Rhizobium meliloti, J. Biol. Chem.* **257:**6751–5757.

Truchet, G., Barker, D. G., Camut, S., de Billy, F., Vasse, J., and Huguet, T., 1989, Alfalfa nodulation in the absence of *Rhizobium, Mol. Gen. Genet.* **219:**65–68.

Truchet, G., Roche, P., Lerouge, P., Vasse, J., Camut, S., De Billy, F., Promé, J. C., and Dénarié, J., 1991, Sulphated lipooligosaccharide signals of *Rhizobium meliloti* elicit root nodule organogenesis in alfalfa, *Nature* **351:**670–673.

Turgeon, B. G., and Bauer, W. D., 1985, Ultrastructure of infection-thread development during the infection of soybean by *Rhizobium japonicum, Planta* **174:**3403–3406.

Urzainqui, A., and Walker, G. C., 1992, Exogenous suppression of the symbiotic deficiencies of *Rhizobium meliloti exo* mutants, *J. Bacteriol.* **174:**3403–3406.

Uttaro, A. D., Cangelosi, G. A., Geremia, R. A., Nester, E. W., and Ugalde, R. A., 1990, Biochemical characterization of avirulent *exoC* mutants of *Agrobacterium tumefaciens, J. Bacteriol.* **172:**1640–1646.

VandenBosch, K. A., Bradley, D. J., Knox, J. P., Perotto, S., Butcher, G. W., and Brewin, N. J., 1989a, Common components of the infection thread matrix and the intercellular space identified by immunocytochemical analysis of pea nodules and uninfected roots, *EMBO J.* **8:**335–242.

VandenBosch, K. A., Brewin, N. J., and Kannenberg, E. L., 1989b, Developmental regulation of a *Rhizobium* cell surface antigen during growth of pea root nodules, *J. Bacteriol.* **171:**4537–4542.

van Eijsden, R., Diaz, C. L., de Pater, B. J., and Kijne, J. W., 1995, Sugar-binding activity of pea (*Pisum sativum*) lectin is essential for heterologous infection of transgenic white clover hairy roots by *Rhizobium leguminosarum* biovar *viciae, Plant Mol. Biol.* **29:**431–439.

van Spronsen, P. C., Bakhuizen, R., van Brussel, A. A., and Kijne, J. W., 1994, Cell wall degradation during infection thread formation by the root nodule bacterium *Rhizobium leguminosarum* is a two-step process, *Eur. J. Cell Biol.* **64:**88–94.

Varik, A., 1993, Biological roles of oligosaccharides: All of the theories are correct, *Glycobiology* **3**:97–130.

Vasse, J., DeBilly, F., Camut, S., and Truchet, J., 1990, Correlation between ultrastructural differentiation of bacteroids and nitrogen fixation in alfalfa nodules, *J. Bacteriol.* **172**:4295–4306.

Vesper, S. J., and Bauer, D. W., 1986, Role of pili (fimbriae) in attachment of *Bradyrhizobium japonicum* to soybean roots, *Appl. Environ. Microbiol.* **52**:134–141.

Vesper, S. L., Malik, N. S. A., and Bauer, W. D., 1987, Transposon mutants of *Bradyrhizobium japonicum* altered in attachment to host roots, *Appl. Environ. Microbiol.* **53**:1959–1961.

Vincent, J. M., 1980, Factors controlling the legume–*Rhizobium* symbiosis, in *Nitrogen Fixation*, Vol. 2 (W. H. Newton and W. H. Orme-Johnson, eds.), pp. 103–129, University Park Press, Baltimore, Maryland.

Wang, Y., and Hollingsworth, R. I., 1994, The structure of the *O*-antigenic chain of the lipopolysaccharide of *Rhizobium trifolii* 4s, *Carbohydr. Res.* **260**:305–317.

Whitfield, C., 1995, Biosynthesis of lipopolysaccharide *O*-antigens, *Trends Microbiol.* **3**:178–185.

Williams, M. N., Hollingsworth, R. I., Klein, S., and Signer, E. R., 1990, The sybiotic defect of *Rhizobium meliloti* exopolysaccharide mutants is suppressed by *lpsZ,⁺ a gene involved in lipopolysaccharide biosynthesis, J. Bacteriol.* **172**:2622–2632.

Wong, H. C., Fear, A. L., Calhoon, R. D., Eichinger, G. H., Mayer, R., Amikam, D., Benziman, M., Gelfand, D. H., Meade, J. H., Emerick, A. W., Bruner, R., Ben-Bassat, A., and Tal, R., 1990, Genetic organization of the cellulose synthase operon in *Acetobacter xylinum*, *Proc. Natl. Acad. Sci. USA* **87**:8130–8184.

Yang, C., 1992, Ultrastructure of infection mode of *Rhizobium meliloti* exopolysaccharide mutant in alfalfa, *Acta Microbiol. Sin.* **32**:191–197.

Yang, C., Signer, E. R., and Hirsch, A. M., 1992, Nodules initiated by *Rhizobium meliloti* exopolysaccharide mutants lack a discrete, persistent nodule meristem, *Plant Physiol.* **98**:143–151.

York, G. M., and Walker, G. C., 1997, The *Rhizobium meliloti exoK* gene and *prsD/prsE/exsH* genes are components of independent degradative pathways which contribute to production of low-molecular-weight succinoglycan, *Mol. Microbiol.* **25**:117–134.

Zevenhuizen, L. P. T. M., 1986, Selective synthesis of polysaccharides by *Rhizobium trifolii*, strain TA-1, *FEMS Microbiol. Lett.* **35**:43–47.

Zevenhuizen, L. P., and Faleschini, P., 1991, Effect of the concentration of sodium chloride in the medium on the relative proportions of poly- and oligo-saccharides excreted by *Rhizobium meliloti* strain YE-2SL, *Carbohydr. Res.* **209**:203–209.

Zevenhuizen, L. P. T. M., and van Neerven, A. R. W., 1983, Gel-forming capsular polysaccharide of *Rhizobium leguminosarum* and *Rhizobium trifolii*, *Carbohydr. Res.* **124**:166–171.

Zevenhuizen, L. P. T. M., Scholten-Koerselman, I., and Posthumus, M. A., 1980, Lipopolysaccharides of *Rhizobium*, *Arch. Microbiol.* **125**:1–8.

Zhan, H. J., and Leigh, J. A., 1990, Two genes that regulate exopolysaccharide production in *Rhizobium meliloti*, *J. Bacteriol.* **172**:5254–5259.

Zhan, H. J., Levery, S. B., Lee, C. C., and Leigh, J. A., 1989, A second exopolysaccharide of *Rhizobium meliloti* strain SU47 that can function in root nodule invasion, *Proc. Natl. Acad. Sci. USA* **86**:3055–3059.

Zhan, H. J., Lee, C. C., and Leigh, J. A., 1991, Induction of the second exopolysaccharide (EPSb) in *Rhizobium meliloti* SU47 by low phosphate concentrations, *J. Bacteriol.* **173**:7391–7394.

Zhang, Y., Hollingsworth, R. I., and Priefer, U. B., 1992, Characterization of structural defects in the lipopolysaccharides of symbiotically impaired *Rhizobium leguminosarum* biovar *viciae* VF-39 mutants, *Carbohydr. Res.* **231**:261–271.

*Chapter 4*

# The Symbiotic Interaction between *Azorhizobium caulinodans* and *Sesbania rostrata*

## Molecular Cross-Talk in a Beneficial Plant–Bacterium Interaction

Sofie Goormachtig, Peter Mergaert, Marc Van Montagu, and Marcelle Holsters

## 1. INTRODUCTION

### 1.1. General Context

Biological nitrogen fixation, the enzymatic reduction of atmospheric dinitrogen to ammonia, is carried out by a limited number of prokaryotic organisms. They all possess an oxygen-sensitive nitrogenase enzyme complex that functions at ambient temperature and pressure and requires high ATP consumption to drive electron transport to $N_2$.

The nitrogen-fixing organisms are divided into two major groups. The free-

**Sofie Goormachtig, Peter Mergaert, Marc Van Montagu, and Marcelle Holsters** Laboratorium voor Genetica, Department of Genetics, Flanders Interuniversity Institute for Biotechnology (VIB), Universiteit Gent, B-9000 Gent, Belgium.

*Subcellular Biochemistry, Volume 29: Plant–Microbe Interactions,* edited by Biswas and Das. Plenum Press, New York, 1998.

living nitrogen-fixers or diazotrophs directly benefit from fixation by assimilation of ammonia for growth. The symbiotic nitrogen-fixers pass the ammonia to a eukaryotic host and profit from nitrogen fixation in a more indirect, sophisticated way by the occupation of a particular ecological niche and/or by supporting the growth of the soil population through better feeding. In the latter group, the symbiosis between the rhizobia (a collective name for representatives of the Gram-negative genera *Azorhizobium, Bradyrhizobium, Rhizobium,* and *Sinorhizobium*) and leguminous host plants is of particular importance for agriculture, legumes being among the major crop plants. The rhizobia have established an elaborate endophytic interaction with plants (for a review, see Mylona *et al.*, 1995). They induce the formation of new organs (nodules) on the roots of a specific host. The bacteria invade the nodules and differentiate into intracellular bacteroids. Inside the nodules optimal conditions for nitrogenase functioning are created. The plants benefit in a very efficient way from the fixed nitrogen, which is supplied to them directly and according to their physiological needs.

Nitrogen-fixing nodules typically occur on roots; however, on some legumes, stem-located, aerial nodules are formed. This phenomenon has been described in members of the subfamilies *Papilionoideae* (*Aeschenomyne* sp., *Sesbania* sp., *Discolobium pulchellum*) and *Mimosoideae* (*Neptunia oleracea*) (Allen and Allen, 1981; Loureiro *et al.*, 1994; for a recent review, see Boivin *et al.*, 1997a). These legumes grow in waterlogged soils of tropical regions and they have dormant, adventitious root primordia on the stem. The stem-located root primordia are the sites of nodule initiation as they develop into nodules upon infection with an appropriate microbial partner. Stem nodulation, apart from its topology, is basically similar to root nodulation. However, the nodular vascular system is connected, via the vascular bundles of the adventitious root primordium, to the vascular system of the stem (James *et al.*, 1992).

A particularly well-studied case of stem nodulation occurs on *Sesbania rostrata* upon inoculation with its microsymbiont *Azorhizobium caulinodans* (Dreyfus and Dommergues, 1981). The interaction has been studied from applied and fundamental angles. A previous review has dealt with the earlier molecular genetic analysis of the system (de Bruijn, 1989). Here, we present an updated overview of the field. The emphasis will be on the particular features of this symbiosis: the special properties of the *A. caulinodans* symbiont and the many advantages presented by the predetermined nodulation sites of the host plant to study the initial stages of nodule development.

## 1.2. *Sesbania rostrata* and *Azorhizobium caulinodans*

*Sesbania rostrata* is an annual leguminous plant from the Sahel region of West Africa. As an adaptation to growth in waterlogged conditions during the rainy season, *S. rostrata* carries dormant adventitious root primordia all along its stem in abundant vertical rows. The primordia protrude through the stem cortex and epidermis, creating a circular fissure, where bacteria can invade and proliferate. These preformed primordia can develop into adventitious roots upon immersion in water, or they can develop into stem nodules upon inoculation with *A. caulinodans* (Duhoux and Dreyfus, 1982). *S. rostrata* grows very fast and stem-nodulated plants show a high rate of nitrogen fixation. These features make *S. rostrata* well fit as green manure in rice cultivation (see Section 5.1), whereas it has no nutritional value for people and cattle. Also, no breeding is being performed and the plant is generally not well characterized. The karyotype has been described as a diploid with $2n = 12$ chromosomes (Le Coq *et al.*, 1985) and the genome content has been calculated as 1187 Mbp per $1n$, which is about eightfold more than *Arabidopsis thaliana* (Arumuganathan and Earle, 1991). Like most tropical legumes, *S. rostrata* carries determinate nodules. The stem nodules are dark green due to photosynthesis pigments (See section 4.3), but also the root nodules become green when developing in the light. Light does not inhibit nodulation. Root nodules occur at specific places, that is, at the junction of lateral roots to the main root. On stems as well as on roots, bacterial invasion of the plant tissues occurs by intercellular infection (crack entry) and does not involve root hairs (Tsien *et al.*, 1983; Duhoux, 1984; Ndoye *et al.*, 1994).

*Azorhizobium caulinodans* is the most competitive symbiont of *S. rostrata* (see Section 4.2). It was isolated from stem nodules (Dreyfus and Dommergues, 1981) and originally called *Rhizobium* sp. strain ORS571. Two features set it apart from the other rhizobia: its taxonomic position and its dual capacity for free-living and symbiotic nitrogen fixation. *A. caulinodans* is exceptional among rhizobia because it can fix nitrogen in the free-living state and assimilate the ammonia for growth (Elmerich *et al.*, 1982), implying a regulatory mechanism to either assimilate the ammonia or donate it to the plant (see Section 2.2). The first taxonomic study of strain ORS571 (Jarvis *et al.*, 1986) showed that it belongs to the *Rhodopseudomonas palustris* rRNA branch of purple bacteria; *Bradyrhizobium japonicum* also belongs there, but ORS571 is quite distinct from both *Rhodopseudomonas* and *Bradyrhizobium*. Using different criteria (numerical analysis of many phenotypes, protein patterns, DNA–DNA and DNA–rRNA hybridization), the relationships of *S. rostrata* symbionts were further investigated by Dreyfus *et al.* (1988). The conclusion of that study was that ORS571 should be considered as a separate genus with *Xanthobacter* as closest relative. *Xanthobacter* spp. are diazotrophic bacteria that occur in diverse soil habitats and also have been found to be associated with rice (Reding *et al.*, 1991; Wiegel, 1992). The name *Azorhizobium caulinodans* was chosen, inspired by the stem (cauli)-nodulating capacity and by the diazotrophic properties of the strain (azorhizobium). The genus harbors only one other, as yet unnamed species (Rinaudo *et al.*, 1991). Recently, by comparison of 16S rRNA sequences, Rainey and Wiegel (1996) showed that *Xanthobacter flavus* is closely related to ORS571.

The question can be raised whether the two organisms should be classified in the same genus.

The host range of *A. caulinodans* for effective nodulation is very narrow. Although nodulation on roots of several *Sesbania* spp. has been reported, only on *S. rostrata* and *S. punctata* are nitrogen-fixing nodules formed (Boivin *et al.*, 1997a). *A. caulinodans* was also reported to induce $Fix^-$ nodules on *Phaseolus vulgaris* and *Leucaena leucocephala* (Waelkens *et al.*, 1995).

## 2. BACTERIAL SYMBIOTIC FUNCTIONS

### 2.1. Nodulation Genes and Nod Factors

#### 2.1.1. Signal Exchange between the Symbionts

The nodule development is a complex process governed by chemical communication between the two symbiotic partners (Dénarié *et al.*, 1992). The early signaling events share common features in all *Rhizobium*–plant interactions and can be described briefly as follows. Specific phenolic compounds (often flavonoids), exuded in the rhizosphere by legume roots, are recognized by the bacteria and induce the bacterial nodulation genes (originally called *nod* genes, but due to the large number of identified genes, the names *nol* and *noe* have been adopted for later discoveries); these genes encode proteins that synthesize and secrete Nod factors. Nod factors are lipo-chitooligosaccharides, consisting of an *N*-acetylglucosamine oligomer acylated at the nonreducing end and decorated with strain-specific substitutions; they are recognized by the host plant and trigger several plant responses related to nodulation. Rhizobia have a limited host range and can nodulate only specific legumes. At nodule initiation, the flavonoid–Nod factor signaling is the major host-range determinant (Mergaert *et al.*, 1997a).

#### 2.1.2. The Nodulation Genes of *Azorhizobium caulinodans*

In most *Rhizobium* and *Sinorhizobium* strains, *nod, nol,* and *noe* genes reside on large plasmids that often contain other symbiotic genes as well and that are therefore called symbiotic plasmids (pSym). In contrast, in *A. caulinodans, Bradyrhizobium,* and *Rhizobium loti,* the symbiotic genes including the *nod* genes are located on the chromosome. Nodulation genes are organized in different operons that form a regulon. The *nod* operons are often closely linked to each other and to other symbiotic functions (Dénarié *et al.*, 1992; Schultze *et al.*, 1994).

In *A. caulinodans* strain ORS571, three chromosomal *nod* loci were isolated, one regulatory and two flavonoid-inducible ones (Fig. 1). These loci are

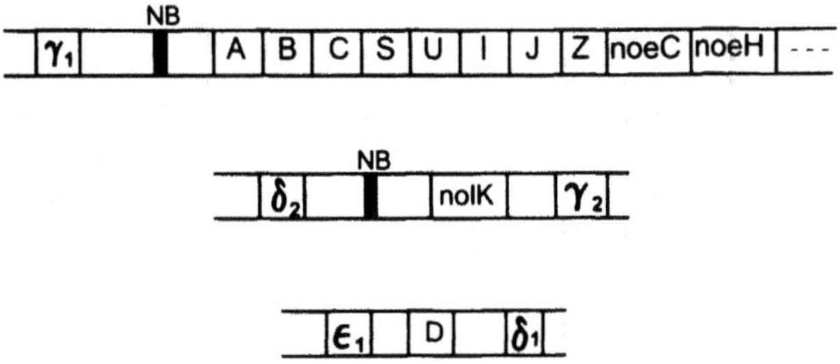


**FIGURE 1.** The nodulation loci of *Azorhizobium caulinodans*. The drawing is a schematic presentation of the three loci. The different genetic elements are not drawn proportionally to their real size. The *nod* genes are only indicated by the corresponding letter, while the *nol* and *noe* genes have their complete name. γ, δ, and ϵ are repeated elements flanking the nodulation loci. NB, *nod* box. See text for more details.

not closely linked to each other. One locus contains the constitutively expressed gene *nodD*, involved in the regulation of the two other *nod* loci (Goethals *et al.*, 1990). A first plant-inducible locus contains a large operon comprising the genes *nodABCSUIJZnoeCH* (Goethals *et al.*, 1989; Geelen *et al.*, 1993; Mergaert *et al.*, 1996). These genes are all transcribed from the same promoter upstream of *nodA*. Possibly there are more genes in this locus, since the sequence downstream of *noeH* is unknown. The second inducible locus contains a single gene, *nolK* (Goethals *et al.*, 1992a; Mergaert *et al.*, 1997b).

### 2.1.3. Regulation of Nodulation Genes

Nodulation genes have no known function in free-living bacteria and are symbiotically regulated (for a review, see Schultze *et al.*, 1994). The regulation of *nod* regulons in different rhizobia features three common elements: a *trans*-acting regulatory protein, member of the LysR family and encoded by the *nodD* gene; a *cis*-acting element, present in the *nod* gene promoters and termed the *nod* box, and finally a plant-derived flavonoid signal. NodD binds the *nod* box sequences and activates the transcription of the downstream-located genes in the presence of the flavonoid plant signals. Direct biochemical evidence for binding of flavonoids by NodD has never been demonstrated, but several observations are in line with this hypothesis. The NodD–flavonoid interaction contributes to host specificity, as different NodD proteins have different flavonoid specificities, restricting nodulation of a *Rhizobium* strain to those plants secreting inducers that are recognized by the NodD protein of that strain. In many rhizobia, this basic regulatory circuit is extended with additional regulatory aspects (see Schultze *et al.*, 1994, and references therein) such as multiple copies of the *nodD* gene-encoding proteins with different inducer specificities, autoregulation of the *nodD* genes, nitrogen regulation, the presence of a *nod* gene repressor, and two-component regulatory systems acting synergistically with NodD activation. Additionally, in indeterminate nodules, *nod* gene expression is repressed in the bacteroids by a mechanism that is not yet completely understood.

In *A. caulinodans*, only evidence for the most elementary form of *nod* gene regulation has been found. One *nodD* gene copy is present (Goethals *et al.*, 1990), it is constitutively expressed and shows no autoregulation (Geelen *et al.*, 1995a). It is absolutely required for the expression of the *nodABCSUIJZnoeCH* and *nolK* loci and for nodulation, as mutagenesis abolishes both. Transcription of the latter two loci is induced by NodD in the presence of the host plant-exuded flavanone liquiritigenin (Messens *et al.*, 1991) or the related flavanone naringenin (Goethals *et al.*, 1989). The *nod* box elements of the azorhizobial *nodABCSUIJZnoeCH* and *nolK* operons have a low degree of similarity to the conserved *nod* box elements of other rhizobia. From the comparison of the *nod* box sequences and from site-specific mutation analysis, a modular buildup of the *nod* box was proposed (Goethals *et al.*, 1992b). In this model, the *nod* box carries two (sometimes three) NodD-binding sites with the inverted repeat structure A-T-C-$N_9$-G-A-T. Promoters of other LysR-type-regulated genes may be organized in a similar way (Goethals *et al.*, 1992b; Schell, 1993). In contrast to the indeterminate nodule types, *nod* gene expression in *A. caulinodans* bacteroids is not repressed in the determinate nodules on *S. rostrata* (Van den Eede *et al.*, 1992).

### 2.1.4. Common Nodulation Genes

The *nodABC* genes of *R. meliloti* and *R. leguminosarum* were the first *nod* genes characterized. Later, it was found that these three genes are present in all rhizobia studied, including *A. caulinodans*. Therefore, *nodABC* are classified as "common" nodulation genes (Dénarié *et al.*, 1992; Schultze *et al.*, 1994, and references therein). They are absolutely essential for the nodulation ability, as inactivation of the genes results in bacteria that are unable to induce any plant response.

Two other genes, *nodIJ*, are also considered as common *nod* genes. Mutations in these genes have no drastic effect on nodulation and result only in a delay of nodulation in some strains (Dénarié *et al.*, 1992, and references therein).

In *A. caulinodans*, the common *nod* genes are located in one single operon, a situation that is found in most other rhizobia. However, in *A. caulinodans*, the *nodIJ* genes are separated from *nodABC* by the genes *nodSU*, whereas in other strains, *nodI* and *nodJ* are immediately downstream of *nodABC* [for an overview of the organization of nodulation genes in rhizobia, see Schultze *et al.*, (1994)]. Also in *B. japonicum*, the *nodABCSUIJ* genes are consecutive, but the upstream and downstream regions of the operon, composed of *nodYABCSUIJnolMNO*, are different (Nieuwkoop *et al.*, 1987; Göttfert *et al.*, 1990; Luka *et al.*, 1993). These differences indicate that complex genetic rearrangements occurred in rhizobia during the acquisition and inheritance of nodulation genes.

### 2.1.5. Strain-Specific Nodulation Genes

Other nodulation genes do not occur in all rhizobia, but are strain-specific. Some of these genes have been termed host-specific *nod* genes. These genes are necessary for or influence the nodulation of certain hosts; their inactivation alters (either restricts, extends, or switches) the host range of the mutant bacteria (Dénarié *et al.*, 1992). The *nodSUZ, nolK,* and *noeCH* genes of *A. caulinodans* can be considered as such genes. The *nodSU* genes are found in many tropical rhizobia, whereas *nodFE* genes are found in rhizobia from temperate legumes [for an overview of the occurrence of *nod* genes in different rhizobia, see Schultze *et al.* (1994)]. The *nodZ* gene is found in *B. japonicum* (Stacey *et al.*, 1994), *Rhizobium* strain NGR234 (Fellay *et al.*, 1995), and *A. caulinodans* (Mergaert *et al.*, 1996). A *nolK*-homologous gene is also found in *Rhizobium* sp. NGR234 (Fellay *et al.*, 1995). The genetic organization of *nodZ* and *nolK* in each of these bacteria is different. In *A. caulinodans, nodZ* is part of a large operon and transcribed from the *nodA* promoter. In *B. japonicum, nodZ* is under control of its own promoter, which does not contain a *nod* box, and expression of the gene is independent of *nodD*. In *Rhizobium* sp. NGR234, *nodZ* is preceded by a *nod* box immediately upstream of the gene and is part of an operon that also contains the *nodK*-homologous gene. Thus, again, these genes might have been subject to different rearrangements during evolution. The *noeCH* genes are thus far unique to *A. caulinodans*.

The influence of all these genes on the nodulation ability of *A. caulinodans* is difficult to assess on the basis of current data. Mutations (Tn*5* insertions) in *nodS, nodU, nodJ, nodZ,* and *noeC* all result in strains that have a more or less delayed and reduced nodulation phenotype on *S. rostrata* (Geelen *et al.*, 1993; W. D'Haeze, M. Fernández-López, and M. Holsters, unpublished results). Because these genes are located in an operon and Tn*5* insertions are known to be often polar, it is not possible to evaluate the role of each individual gene. In addition, the *nodZ* gene may be duplicated or another gene may have a similar function (Mergaert *et al.*, 1996). A mutation in *nolK* did not affect *S. rostrata* nodulation, but reduced nodulation of *Sesbania formosa* (Goethals *et al.*, 1992a). Double mutants with an insertion in *nolK* and *nodU* or in *nolK* and *nodZ* are much more severely affected and induce much fewer nodules on *S. rostrata* (W. D'Haeze and M. Holsters, unpublished results). Thus, there may be a synergistic effect of some *nod* functions. As mentioned above, *A. caulinodans* can also induce nodules on *L. leucocephala* and *P. vulgaris*. The *nodS* gene is needed for *P. vulgaris* nodulation, as an insertion in *nodS* abolishes nodulation, whereas a downstream-located mutation in *nodU* has no effect, a situation that is similar for *R. tropici* and *Rhizobium* sp. NGR234 (Waelkens *et al.*, 1995). Nodulation of *L. leucocephala* by *A. caulinodans* is abolished by both a *nodS* and a *nodU* mutation (Waelkens *et al.*, 1995), but the effect of downstream-located mutations is not known, and therefore no firm conclusions can be made.

### 2.1.6. Evolution of Nodulation Genes in *Azorhizobium caulinodans*

*Azorhizobium caulinodans* forms an interesting case for the study of the evolution of *nod* genes. First, azorhizobial *nod* genes are only distantly related to their counterparts in other rhizobia, indicating that they diverged a long time ago. Phylogenetic comparisons of *nodA* and *nodC* sequences demonstrated that the genes from rhizobia nodulating temperate legumes (e.g., *R. meliloti, R. leguminosarum* bvs. *viciae* and *trifolii*, and *R. galegae*) group together and the genes from rhizobia nodulating tropical legumes (e.g., *B. japonicum, B. elkanii, R. loti, R. tropici*, and *R. etli*) form a second cluster (Dobert *et al.*, 1994; Lindstrom *et al.*, 1995; Ueda *et al.*, 1995). The existence of two branches for the *nodB* sequences is less obvious. However, the *nodA, nodB*, and *nodC* genes of *A. caulinodans* are only distantly related to their counterparts in other rhizobia and *nodA* and *nodC* do not belong to either the tropical or the temperate cluster. Also, the genetic distance between *nodSU* of *A. caulinodans* and the *nodSU* genes of other rhizobia is much greater than the mutual genetic distance between *nodSU* genes of these rhizobia (Waelkens *et al.*, 1995).

Second, five of the six borders of the three *nod* loci are formed by genetic elements that are reiterated in the genome and that have homology to insertion elements (Goethals *et al.*, 1992a; Geelen *et al.*, 1995a). The border downstream of *noeH* has not yet been determined. Two γ elements, found upstream of *nodA* and downstream of *nolK* (Fig. 1), are related to IS*Rm3*, an insertion element found in *R. meliloti* (Goethals *et al.*, 1992a). Two δ elements, located respectively upstream of *nolK* and downstream of *nodD* (Fig. 1), are homologous to a repetitive element that is associated with *R. fredii nodABC* genes and to *Agrobacterium rhizogenes* Ri T-DNA sequences (Goethals *et al.*, 1992a). One ϵ element is found upstream of *nodD* (Fig. 1) and a second copy is present in the genome at an unknown location. This element has similarities to IS*51* from *Pseudomonas syringae* subspecies *savastanoi* and to *Agrobacterium tumefaciens* T-DNA sequences (Geelen *et al.*, 1995a). These repeated elements could be the relics of insertion elements that once played a role in the evolution of *A. caulinodans* nodulation genes.

A third observation concerns the GC content of the nodulation genes. The operons *nodABCSUIJZnoeCH, nodD*, and *nolK* have a GC content of approximately 55%, 47%, and 55%, respectively (Goethals *et al.*, 1992a; Geelen *et al.*, 1995a). Sequences outside these loci have a GC content of 67%, which is similar to the mean value of the genome (Dreyfus *et al.*, 1988), and the shift in GC content at the *nod* loci corresponds to the location of the repeated elements (Geelen *et al.*, 1995a).

These three observations suggest that *A. caulinodans* acquired its *nod* genes (and thus the ability to nodulate) by horizontal gene transfer and that it evolved from a diazotrophic (possibly a plant-associated *Xanthobacter* strain) to a symbiotic bacterium. Concerning the donor of the *nod* genes, hypotheses are more difficult to make. The *nod* genes present in *A. caulinodans* diverged a long time ago from the *nod* genes of other (known) rhizobia, but their low GC content that is different from the genomic GC content suggests that the *nod* genes were acquired relatively recently from (an) organism(s) with a GC content of 47–55% (or lower). As the genomes of rhizobia have a GC content of 59% or higher, the donor of the *A. caulinodans nod* genes may be an unknown organism. This may be interesting with respect to the question of how rhizobia acquired their nodulation genes and thereby the ability to interfere with plant development.

### 2.1.7. Nod Factor Signaling

Enzymes encoded by the nodulation genes are involved in the biosynthesis and secretion of Nod factor signals. The existence of secreted rhizobial factors causing root hair deformation was reported many years ago (Yao and Vincent, 1976; Ervin and Hubbell, 1985) and the production of such factors was found to be dependent on the presence and the flavonoid induction of nodulation genes (Zaat *et al.*, 1987; Faucher *et al.*, 1988, 1989). Such a factor was first purified and chemically characterized for *R. meliloti* (Lerouge *et al.*, 1990) and since then Nod factors from many rhizobia have been characterized. All these Nod factors belong to the same class of molecules and consist of a chitooligosaccharide (β-1, 4-linked *N*-acetylglucosamine residues; Fig. 2A) of which the nonreducing-end N-acetyl group is replaced by an N-acyl group (Fig. 2A, $R_1$). This lipo-chitooligosaccharide (LCO) basic structure, however, varies according to the *Rhizobium* strain and variations are possible in the chitooligosaccharide chain length (3–6 GlcNAc residues; Fig. 2A, $n = 1–3$), in the nature of the fatty acid, and in modifications that are added to the reducing and/or nonreducing end of the oligosaccharide (Fig. 2A, $R_{2/5}$) (for an overview, see Schultze *et al.*, 1994). Rhizobia make Nod factors with specific structures to meet the requirements for nodulation of their host plants and thus Nod factors are important determinants of the host range of rhizobia.

Azorhizobial Nod factors (NodARc) are mainly chitopentasaccharides, although a small fraction of tetrameric forms are produced (Fig. 2, $n = 3$ in the majority and $n = 2$ in a minority of the Nod factors). The N-acyl chain at the nonreducing end (Fig. 2, modification $R_1$) can be either palmitic (C16:0), vaccenic (C18:1), or stearic (C18:0) acid. The nonreducing end of the oligosac

A

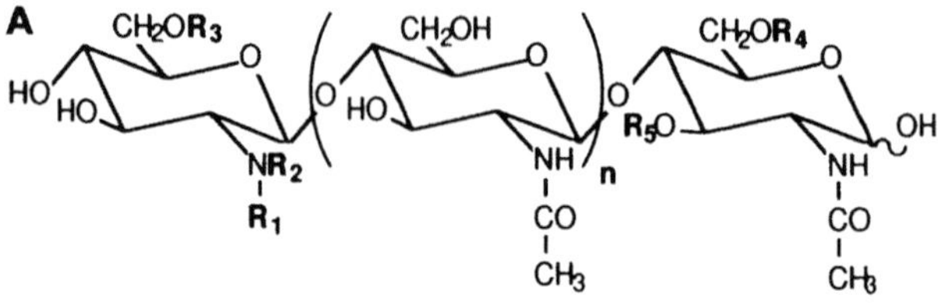


B $R_1$ = stearoyl ($C_{18:1}$) or vaccenoyl ($C_{18:0}$) or palmitoyl ($C_{16:0}$)
$R_2$ = methyl
$R_3$ = carbamoyl or H
$R_4$ = fucosyl or arabinosyl or H
$R_5$ = arabinosyl or H
n = 2 or 3

**FIGURE 2.** The chemical structure of Nod factors. (A) Structure of the lipo-chitooligosaccharide basis as it is found in all Nod factors. The value of *n* (1–3) determines the oligosaccharide length. $R_1$ is a fatty acid. $R_{2-5}$ are possible Nod factor modifications. (B) List of the *n* and R variables for the Nod factors of *A. caulinodans*.

charide is also substituted with an N-methyl group (Fig. 2, $R_2$) and a 6-O-carbamoyl group (Fig. 2, $R_3$). The reducing end of NodARc can have different forms. First, it can be unsubstituted (Fig. 2, $R_4$ and $R_5$ are hydrogen atoms). Second, it can be branched with a D-arabinosyl or L-fucosyl sugar on C6 (Fig. 2, $R_4$), but the former can also be on C3 (Fig. 2, $R_5$). Finally, it can be doubly substituted with the L-fucosyl group on C6 (Fig. 2, $R_4$) and the D-arabinosyl group on C3 (Fig. 2, $R_5$) (Mergaert *et al.*, 1993, 1997c). An identical collection of Nod factors is produced by strains from *Sinorhizobium saheli* and *Sinorhizobium teranga,* which are, just like *A. caulinodans,* symbionts of *S. rostrata* (Lorquin *et al.*, 1997). This Nod factor structure may thus be of crucial importance for nodulation of *S. rostrata*. However, we isolated several *A. caulinodans nod* gene mutants that produced Nod factors lacking one or several modifications and that still have the ability to nodulate *S. rostrata*. For example, a *nodU* mutant produces uncarbamoylated Nod factors (W. D'Haeze and M. Holsters, unpublished results), a *nodZ* or *noeC* mutant produces Nod factors lacking the arabinosyl branch, and Nod factors produced by a *nolK* mutant are not fucosylated (Mergaert *et al.*, 1996) (see below for a description of the enzymatic activity of the proteins encoded by these genes). Despite the aberrant Nod factor structures, all these strains still have the ability to nodulate *S. rostrata,* although for some the nodulation is delayed and the total number of nodules is reduced. In *R. meliloti,* a *nodH* mutation causes the absence of a sulfate modification that is present on the wild-type Nod factors and this results in a complete block of nodulation (Roche *et al.*, 1991). None of the modifications of azorhizobial Nod factors seems to be that important for *S. rostrata* nodulation. But individual Nod factor modifications may improve the efficiency of the Nod factor and mutants lacking a modification may be less efficient in nodulation. In the case of double mutants lacking several modifications, the efficiency of the Nod factors is probably so strongly reduced that the strains lose their nodulation ability almost completely (see above).

Purified Nod factors are able to mimic many aspects of the development of a genuine nodule induced by a *Rhizobium* strain (Mylona *et al.*, 1995; Dénarié *et al.*, 1996). On *S. rostrata* roots, at the basis of lateral roots, Nod factors induce formation and deformation of root hairs, initiation of nodule formation, and formation of nodule-like structures at nanomolar concentrations (Mergaert *et al.*, 1993). In batch cultures, *A. caulinodans* produces a complex mixture of Nod factors, as described above. Several of these different types of molecules were compared for their biological activity. Apparently, the type of fatty acid is of less importance, as no difference in activity between vaccenoylated or stearoylated Nod factors could be discerned (Mergaert *et al.*, 1993). Nod factors that are at the reducing terminus either unmodified, arabinosylated, fucosylated, or both arabinosylated and fucosylated are all active for root hair formation and deformation and initiation of nodule-like structures. But the molecules that have one or both branches are more potent and active at lower concentrations (Mergaert *et al.*, 1997c). Such assays using pure Nod factors lacking one or several modifications (the methyl and carbamoyl groups at the nonreducing end and the glycosyl branches at the reducing end), together with analysis of nodulation behavior and Nod factor structure of *nod* gene mutants, can be used to get a more detailed idea of the relative importance of the different modifications.

Three different roles were envisaged for the different parts of the Nod factor (Schultze and Kondorosi, 1995). First, it is generally accepted that the Nod factor signal is transduced to the plant cells via a highly specific receptor. Thus, part of the Nod factor structure may be involved in recognition by a receptor. The sulfate group of *R. meliloti* may have such a function. Second, Nod factors contain a chitooligosaccharide chain and may thus be substrates for chitinases. Nod factor modifications may protect the signals from degradation by chitinases or other Nod factor-degrading enzymes. Finally, specific acyl chains may influence the integration and diffusion of Nod factors in membranes and thereby target and concentrate Nod factors in the vicinity of receptors. Chitinase genes and a Nod factor-degrading activity are induced during the nodulation of *S. rostrata* (Goormachtig *et al.*, 1995; S. Goormachtig, unpublished results; see also below). Thus, the influence of the methyl, carbamoyl, arabinosyl, and fucosyl groups on the stability of Nod factors against degradation by these hydrolases will have to be examined.

### 2.1.8. Nod Factor Biosynthesis

The biosynthesis of Nod factors by the Nod proteins is described in detail in the chapter by Kamst *et al.* in this volume and another recent review on this topic has been published by Mergaert *et al.* (1997a). Therefore, we will only briefly describe here what is known in *A. caulinodans.* The biosynthetic pathway was unraveled using different approaches: the amino acid sequences of several Nod proteins showed similarities to proteins with known functions; for several of the Nod proteins, *in vitro* enzymatic activities were demonstrated; the structure of Nod factors produced by nodulation gene mutants was determined; and finally, combinations of nodulation genes were expressed in *Escherichia coli* and the Nod metabolites produced by these strains were characterized.

The synthesis starts with the NodC protein that polymerizes *N*-acetylglucosamine residues. NodC has *in vitro N*-acetyl glucosaminyl transferase activity and produces chitooligosaccharides with a polymerization degree up to 5, using UDP-*N*-acetylglucosamine as substrate (Geremia *et al.*, 1994). This activity was suggested by the homology of NodC of different rhizobia with yeast chitin synthases (Bulawa and Wasco, 1991; Atkinson and Long, 1992; Debellé *et al.*, 1992). An *E. coli* strain expressing *nodC* produced *in vivo* chitooligosaccharides and when *nodB* was expressed in addition, the chitooligosaccharides were deacetylated at the nonreducing end (Mergaert *et al.*, 1995). John *et al.* (1993) demonstrated that NodB of *R. meliloti* has *in vitro* chitin deacetylase activity removing the N-acetyl group of the nonreducing end of chitooligosaccharides. The NodS protein of *A. caulinodans* has *in vitro* methyltransferase activity and uses chitooligosaccharides deacetylated at the nonreducing end as substrate and *S*-adenosyl-L-methionine as methyl donor, in agreement with the presence of an *S*-adenosyl-L-methionine-binding domain in the protein (Geelen *et al.*, 1993, 1995b). An *E. coli* strain expressing *nodBCS* produced deacetylated and methylated chitooligosaccharides (Mergaert *et al.*, 1995). Thus, the nonreducing terminus of the oligosaccharides produced by NodC is deacetylated by NodB and subsequently methylated by NodS. LCOs were produced in *E. coli* when *nodABC*, *nodABCS*, or *nodABCSU* were expressed (Mergaert *et al.*, 1995). Thus, NodA is an acyltransferase and is acting after NodC, NodB, and NodS, forming a LCO structure. The acyltransferase activity was demonstrated *in vitro* for the NodA protein of *R. meliloti* (Atkinson *et al.*, 1994; Röhrig *et al.*, 1994).

The NodU protein transfers the carbamoyl group, NodZ transfers the fucosyl group, and NoeC, NoeH, or a protein encoded by a gene downstream of *noeH* transfers the arabinosyl group. These conclusions were suggested by structures of Nod factors produced by mutants (Mergaert *et al.*, 1996; W. D'Haeze and M. Holsters, unpublished results). Additional arguments were provided by the *E. coli* strain expressing *nodABCSU*, which produced carbamoylated Nod factors (Fernández-López *et al.*, 1996), and the NodU protein sequence that is

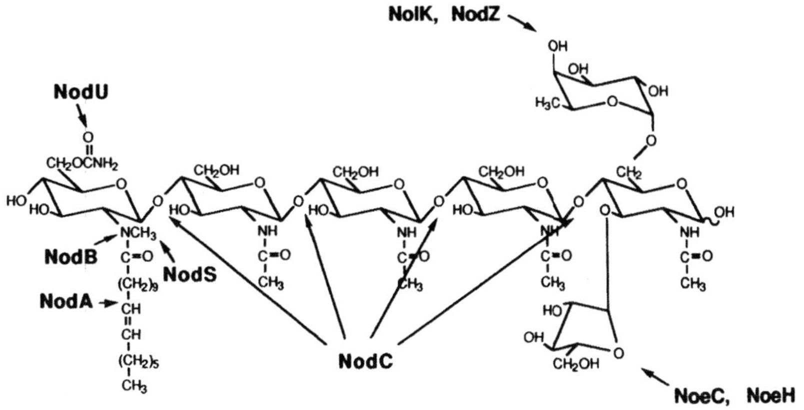


**FIGURE 3.** The function of Nod proteins in the biosynthesis of Nod factors of *A. caulinodans*. For details, see text.

similar to a carbamoyltransferase of *Nocardia lactamdurans*. For the NodZ protein, a fucosyltransferase activity was also demonstrated *in vitro* (Mergaert *et al.*, 1996). The exact position in the biosynthesis pathway of the carbamoyl, arabinosyl, and fucosyl transfer reactions still has to be determined experimentally. The NolK protein is involved in the synthesis of the GDP-fucose precursor (Mergaert *et al.*, 1996, 1997b). The arabinose precursor still has to be identified and its synthesis may also be encoded by *noeC, noeH,* and/or downstream genes. Figure 3 gives an overview of the Nod protein functions discussed above.

Nod factors are assembled at the cytoplasmic face of the inner membrane, once they are secreted to the extracellular medium. The NodIJ proteins are members of the family of ATP-binding cassette (ABC) transporters, suggesting a function of these proteins in Nod factor secretion (Vázquez *et al.*, 1993). Inactivation of NodJ in *A. caulinodans* resulted in a strain that is deficient in Nod factor secretion. Additionally, *E. coli* expressing *nodABCSU* produced LCOs that remained associated with the bacterial cells. However, when the *nodIJ* genes were also expressed, the LCOs became secreted. Therefore, it was concluded that the NodIJ proteins are part of a secretion complex dedicated to the transfer of Nod factors across the bacterial membranes to the exterior (Fernández-López *et al.*, 1996).

## 2.2. Nitrogen Fixation

### 2.2.1. Free-Living versus Symbiotic Nitrogen Fixation

The dual capacity of *A. caulinodans* for free-living and symbiotic nitrogen fixation is exceptional and interesting from a regulatory point of view. Only some photosynthetic bradyrhizobia, isolated from stem nodules of *Aeschenomyne* species (Alazard, 1990), and one *R. leguminosarum* bv. *trifolii* strain (Urban *et al.*, 1986) were shown to also grow on $N_2$, although at much lower $O_2$ concentration than *A. caulinodans*. In *A. caulinodans*, there is a coupling of nitrogen fixation and ammonium assimilation in culture and an uncoupling in bacteroids. *A. caulinodans* can fix nitrogen in culture while growing at fairly high temperature (37°C) and oxygen pressure (3%) (Elmerich *et al.*, 1982; Dreyfus *et al.*, 1983). Under these circumstances, 90% of the fixed nitrogen is retained inside the cells (Gebhardt *et al.*, 1984). This diazotrophic capacity was very helpful for the isolation of mutants affected in free-living fixation and assimilation. Such mutants were subsequently screened for symbiotic performance (Elmerich *et al.*, 1982; Donald *et al.*, 1985, 1986, 1988).

The nitrogenase enzyme complex was purified from diazotrophically grown cultures and the specific activity of the components—the Fe protein and the MoFe protein—was determined (Kush *et al.*, 1985). Evidence was obtained for a reversible inactivation, a switchoff by ammonia, of the Fe protein, perhaps by adenosine diphosphoribosylation, as described in *Rhodospirillum rubrum* and *Azospirillum* sp. (Zhang *et al.*, 1993). Such a mechanism was not found in other rhizobia.

For growth in medium with defined N and C sources, *A. caulinodans* requires the addition of the vitamins biotin, pantothenate, and nicotinic acid (Elmerich *et al.*, 1983). Nicotinate is needed as anabolic substrate for NAD synthesis; indeed, *A. caulinodans* is a NAD auxotroph (Kitts *et al.*, 1989). Nicotinate, however, can also, under conditions of N limitation, be degraded and used as a nitrogen source. For $N_2$-fixing growth in batch cultures, tenfold more nicotinate was required than for growth with ammonia, and nicotinate was metabolized at a high rate (Elmerich *et al.*, 1983). The available nicotinate seemed to limit growth during $N_2$ fixation. This interesting case of an essential growth factor that is broken down during $N_2$ fixation was subject to—often controversial—research and discussions. The first step in nicotinate catabolism was found to be the $O_2$-dependent hydroxylation of nicotinate to 6-OH nicotinate (Ludwig, 1986). Ludwig (1986) formulated the idea that oxidative catabolism of nicotinate would play a major role in maintaining a low intracellular $O_2$ concentration, necessary for free-living nitrogen fixation, and would as such be essential for $N_2$-fixing growth. As the nicotinate catabolism pathway was elucidated and mutants were isolated in the catabolism genes, complementation with different intermediate breakdown products clearly showed that the initial hydroxylation reaction is not responsible for the improved growth under nitrogen fixation; stimulation of $N_2$ fixation is not due to nicotinate hydroxylase-driven respiration, but to subsequent steps of nicotinate catabolism (Kitts *et al.*, 1989, 1992; Buckmiller *et al.*, 1991). Recently, it has been convincingly demonstrated that growth on $N_2$ is not dependent on nicotinate catabolism (Pronk *et al.*, 1995). However, supplementation of nicotinate to nitrogen-fixing batch cultures temporarily stimulates growth; this stimulation is now ascribed to an enhancement of intracellular pyridine-N pools (Kitts *et al.*, 1992).

In free-living nitrogen fixers such as *Klebsiella pneumoniae*, assimilation of ammonia occurs via the glutamine synthetase–glutamate synthase (GS–GOGAT) pathway and the expression of the genes coding for these enzymes is under control of the general nitrogen regulation system (*ntrBC*) (for a review, see Magasanik, 1982). In bacteroids of rhizobia, normally no assimilation takes place, ammonium is exported to the plant, and, in most cases, bacterial GS–GOGAT expression is switched off (for a review, see Miflin and Cullimore, 1984). *A. caulinodans* forms, not surprisingly because of its dual capacities, an exception among the rhizobia. Ammonium assimilation takes place via the GS–GOGAT pathway (Donald and Ludwig, 1984). Genes coding for GS and GOGAT were cloned (de Bruijn *et al.*, 1985; Hilgert *et al.*, 1987; Donald *et al.*, 1988). Their expression is not under the control of the general nitrogen regulation system *ntrBC* (Hilgert *et al.*, 1987). Mutations in the genes coding for GS and GOGAT yielded a Nif$^-$Nod$^+$Fix$^-$ phenotype (no nitrogen-fixing growth, no symbiotic $N_2$ fixation) (Donald and Ludwig, 1984; Hilgert *et al.*, 1987; Donald *et al.*, 1988). The Nif$^-$ phenotype of GS$^-$ and GOGAT$^-$ mutants was as expected and most probably due to feedback inhibition of nitrogenase activity (Kush *et al.*, 1985) and/or synthesis (via regulation of *nif* gene transcription) by accumulation of ammonia or glutamine. The Fix$^-$ phenotype was unexpected, as it is unlike the situation in most other rhizobia. The Fix$^-$ nodules, however, harbored only few bacteroids, probably indicating that $N_2$ fixation and assimilation play a role in azorhizobial proliferation during invasion. Other rhizobia are not dependent on nitrogen fixation for invasion. For example, some *S. teranga* and *S. saheli* strains (de Lajudie *et al.*, 1994) are symbionts of *S. rostrata* and can induce stem nodules, although they are not capable of free-living fixation and assimilation (Tomekpe *et al.*, 1996; Boivin *et al.*, 1997b). Therefore, these rhizobia must be supplemented by the plant or use stored nitrogen compounds. It seems then that *A. caulinodans* cannot use efficiently the nitrogen sources provided by *S. rostrata* or nitrogen storage products. *A. caulinodans* may thus be dependent on its own $N_2$ fixation and assimilation during invasion of the host tissues and therefore mutants defective in assimilation may be seriously disadvantaged and poorly infect and invade. This explanation fits with the observation of very similar Fix$^-$ phenotypes obtained with all *A. caulinodans* Nif$^-$ and Fix$^-$ mutants whose behavior was studied *in planta* [cf. *nifHDK*, *nifA*, *ntrC*, *nifB*, Ass$^-$ mutants; reviewed in de Bruijn (1989)]. All these mutants form, on roots as well as on stems, Fix$^-$ nodules that contain few bacteroids, a phenotype that is in general much more pronounced than that observed with similar Fix$^-$ mutants of *R. meliloti*, for example. This indicates that free-living $N_2$ fixation with assimilation during invasion is very important for *A. caulinodans*.

In relation to this discussion of nicotinate metabolism and free-living $N_2$ fixation, it seems plausible that during invasion of a putative host plant, the bacterial NAD auxotrophy, the obligate but limited supply of nicotinate by a nitrogen-starved host, and nicotinate catabolism coupled to nitrogen-fixing growth will contribute to the host-range limitation of *A. caulinodans* and could be a way of balancing invasive proliferation.

### 2.2.2. Nitrogen Fixation Genes

Like all nitrogen-fixing organisms, *A. caulinodans* harbors genes that code for the nitrogenase complex, its assembly, and functioning. These genes are homologous to well characterized *nif* genes of *K. pneumoniae* and *fix* genes of rhizobia and they are named after their homologs. They are essential for both free-living and symbiotic nitrogen fixation, with the possible exception of one operon (the *fixNOQP* operon, see below). In *A. caulinodans*, 15 polypeptides were found to be synthesized under nitrogen-fixing conditions (Denèfle *et al.*, 1987). Several *nif* and *fix* loci have been identified, either by mutation analysis or by probing of the *A. caulinodans* genome with cloned *nif* and *fix* genes from other organisms. For a recent overview of cloned *nif* and *fix* loci, see Kaminski *et al.* (1995). The first locus to be cloned from *A. caulinodans* contained the structural genes *nifHDK* encoding the components of the nitrogenase complex, i.e., the nitrogenase reductase and the two subunits of the nitrogenase (Elmerich *et al.*, 1982; Norel *et al.*, 1985). A second *nifH* copy (*nifH2*) is present elsewhere in the genome (Donald *et al.*, 1986). Nucleotide sequence, expression, and mutation analyses showed that the two *nifH* gene copies are closely related (only one amino acid change occurs in the NifH polypeptides) and that both genes are functional *ex planta* and *in planta* (Norel and Elmerich, 1987). Other genes were cloned that encode auxiliary functions for nitrogenase synthesis and function, such as enzymes that synthesize the cofactors of the nitrogenase, e.g., *nifE* (Denèfle *et al.*, 1987), electron transporters, e.g., *fixABCX* (Kaminski *et al.*, 1988; Arigoni *et al.*, 1991), and proteins involved in microaerobial respiration. Indeed, in aerobic diazotrophs and in the aerobic rhizobia, the large amounts of ATP that are required for nitrogenase functioning are generated by oxidative phosphorylation. To function under low free oxygen tensions, nitrogen-fixing organisms have special terminal cytochrome oxidases with high oxygen affinity. Oxidation of carbon sources yields reducing equivalents that donate electrons to $O_2$. Electron transport is coupled to the generation of a proton motive force that drives ATP synthesis. Especially in nodules, where extremely low free oxygen tensions prevail, bacteroids induce a new branch of the respiratory chain that supposedly terminates with a high-oxygen-affinity cytochrome oxidase (Preisig *et al.*, 1993). The respiratory chain of *A. caulinodans* is branched with at least four terminal oxidases, some of which function during aerobic, some during microaerobic growth (Kitts and Ludwig, 1994). Genes encoding several of them were identified (Kitts and Ludwig, 1994; Mandon *et al.*, 1994). Their role in symbiosis was studied by mutation analysis. The prevailing complexity may be related to the different ranges of oxygen tension that *A. caulinodans* can live in. There is, for example, a large difference in optimal oxygen concentrations during free-living (10 μM) and symbiotic (0.02 μM) fixation. The *fixNOQP* operon is induced under microaerobiosis. It encodes cytcbb$_3$ oxidase and was, therefore, renamed *cytNOQP* (Kaminski *et al.*, 1996). In *B. japonicum*, the oxidase is essential for symbiotic $N_2$ fixation and it is a candidate to be the bacteroid terminal oxidase (Preisig *et al.*, 1993). In *A. caulinodans*, mutations in *cytNOQP* had no effect on free-living fixation, but caused a 50% reduction of symbiotic fixation (Mandon *et al.*, 1994). This suggested the occurrence of still another oxidase in *A. caulinodans* bacteroids, perhaps a cytochrome bd complex (Kitts and Ludwig, 1994). This hypothesis has recently been confirmed (Kaminski *et al.*, 1996).

### 2.2.3. Regulation of *nif*/*fix* Gene Expression

A strict regulation of the expression of *nif* and *fix* genes is imposed by the high energy demands and oxygen sensitivity of the $N_2$ fixation process. In free-living $N_2$-fixing bacteria, oxygen tension and availability of fixed nitrogen are the key regulators. In symbiosis, it is mostly oxygen (Soupène *et al.*, 1995). In both free-living and symbiotic organisms, NifA is the master regulatory protein (Fig. 4). It is the transcriptional activator of the *nif* operons and works together with the $\sigma^{54}$ factor encoded by the *ntrA* gene (Merrick, 1992). Control of its expression and activity varies according to the organism. In *A. caulinodans* the regulation is fairly complex. The *nifA* gene was cloned and sequenced (Pawlowski *et al.*, 1987; Nees *et al.*, 1988; Ratet *et al.*, 1989). The *nifA* expression was found to be repressed by both rich nitrogen sources and oxygen (Ratet *et al.*, 1989). Nitrogen regulation of *nifA* expression is achieved by two two-component regulatory systems, NtrBC and NtrXY (Fig. 4) (Pawlowski *et al.*, 1987, 1991). The NtrB and NtrY proteins sense intracellular and extracellular nitrogen status, respectively. NtrC and NtrX are response regulators, transcriptional activators depending in this case on an as-yet-unidentified $\sigma^{54}$ factor, different from the *ntrA* product (Fig. 4) (Stigter *et al.*, 1993). Oxygen control of *nifA* expression is mediated by FixLJ (Fig. 4) (Kaminski and Elmerich, 1991), a two-component system, and the transcription factor FixK (Fig. 4) (Kaminski *et al.*, 1991). FixL is homologous to the *R. meliloti* FixL oxygen-sensing histidine kinase that activates the response regulator FixJ in response to low oxygen (Gilles-Gonzalez *et al.*, 1991). Activated FixJ switches on the expression of the *fixK* gene. In *A. caulinodans*, FixK in turn activates transcription of the *nifA* gene and some other operons that are activated under microaerobiosis, such as the *fixNOQP* operon (Fig. 4) (Mandon *et al.*, 1994). The complex organization of

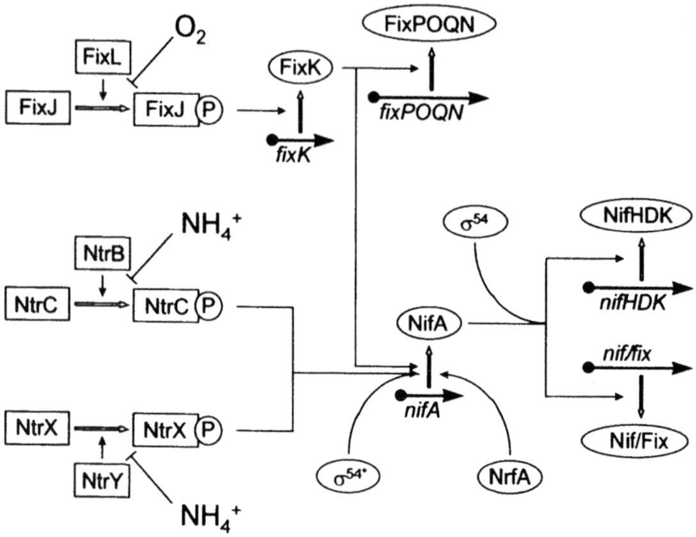


**FIGURE 4.** Regulation of *nif* and *fix* genes in *A. caulinodans*. The figure shows schematically the signal transduction cascade that senses the oxygen and ammonium levels and regulates accordingly the transcription of *nif* and *fix* genes. For more details, see text.

the *nifA* promoter reflects the many controlling factors that govern its activity (Loroch *et al.*, 1995). In addition to this sophisticated transcriptional control, a gene *nrfA* homologous to an RNA-binding *E. coli* protein was found to be essential for *nifA* expression, perhaps by exerting control at the translational level (Fig. 4) (Kaminski *et al.*, 1994). Finally, recently yet another regulatory gene has been identified that plays a role in regulation of nitrogen fixation and belongs to the LysR family of regulatory genes (Kaminski *et al.*, 1995).

## 2.3. Surface Polysaccharides Involved in Symbiosis

In the *Rhizobium*–legume interaction bacterial surface polysaccharides are known to be important for successful establishment of the symbiosis. Extracellular polysaccharides (EPS), lipopolysaccharides (LPS), capsular polysaccharides, and periplasmic oligo- or polysaccharides have all been implicated to a more or less important degree in the symbiosis. Defined functions, however, have not been ascribed. For an overview of this particular topic, see the chapter by Niehaus and Becker in this volume. A general picture that emerges is that of a greater importance of EPS in temperate interactions and of LPS in tropical symbioses.

In *A. caulinodans*, a random mutagenesis led to the identification of a mutant strain that is defective in nodulation because of a changed surface polysaccharide composition (Van den Eede *et al.*, 1987; Goethals *et al.*, 1994). This mutant, strain ORS571-X15, harbors a Tn*5* insertion in a gene that encodes an enzyme involved in the biosynthesis of the deoxysugar rhamnose. The surface polysaccharides of the mutant strain do not contain any rhamnose, in contrast to the wild type (W. D'Haeze and M. Holsters, unpublished results). The mutant strain produces normal amounts of Nod factors and upon inoculation of stems or roots of *S. rostrata* induces the formation of small bumps, reminiscent of Nod factor-induced structures. Surface-sterilized bumps did not contain bacteria; therefore the assumption is that the changed outer surface of the mutant bacteria interferes with invasion of the plant tissue.

## 2.4. *hem, hup,* and *egl* Genes

For the sake of completeness, we will briefly mention additional *A. caulinodans* genes that have been cloned and studied, but that are less directly or less obviously involved with the cross-talk between the symbiotic partners.

An azorhizobial *hemA* gene was characterized by Pawlowski *et al.* (1993). A mutant in this gene was severely defective in symbiosis, causing Fix$^-$ root nodules, and no stem nodules. The *hemA* gene encodes a δ-aminolevulinic acid synthase that carries out the first step in heme biosynthesis in rhizobia. The role of heme biosynthesis in symbiosis has recently been critically reviewed by O'Brian (1996), who unrooted the longstanding speculation that the heme prosthetic group of the plant protein leghemoglobin—the oxygen carrier in nodules—is synthesized by the bacterial symbiont. Instead, the symbiotic defects of *hem* mutants should be ascribed to the involvement of heme in cytochrome heme proteins that are expressed as part of the bacteroid electron transport system and in oxygen sensing by the FixL protein, a heme-containing histidine kinase.

Hydrogen is an obligatory by-product of nitrogenase action and, as such, is a cause of energy loss. Some rhizobia possess an uptake hydrogenase that recycles the hydrogen by oxidation and shuttles the electrons back to the electron transport system (for review, see Evans *et al.*, 1987; Arp, 1992). *A. caulinodans* has an uptake hydrogenase (de Vries *et al.*, 1984; Stam *et al.*, 1984) and hydrogenase-negative mutants were isolated (de Vries *et al.*, 1988). They were studied under free-living nitrogen-fixation conditions (Boogerd *et al.*, 1994), but their role in symbiosis remains to be investigated.

Finally, the putative role in symbiosis of an azorhizobial endoglucanase (encoded by the *egl* gene) was studied by reversed genetics (Geelen *et al.*, 1995c). At several stages during infection, invading bacteria have to cross plant cell walls and hydrolytic enzymes of bacterial origin might be implicated in this process. However, knocking out of the azorhizobial *egl* gene did not cause any effect on the establishment of root and stem nodules or on their functioning. The role of this gene needs to be further investigated. Recently, Finnie *et al.* (1997) described a glycanase from *R. leguminosarum* bv. *viciae* that may degradate high-molecular-weight EPS to a low-molecular-weight form. A mutant in the gene encoding the glycanase could be restored by introduction of the azorhizobial gene. Therefore, an alternative hypothesis could be that the azorhizobial endoglucanase is involved in the degradation of a bacterial polysaccharide rather than a plant cell wall polysaccharide.

## 3. THE PLANT PARTNER IN THE SYMBIOSIS

### 3.1. Stem Nodule Development

Depending on the host plant, two types of nodules occur. Indeterminate nodules are rod-shaped due to the presence of a persistent apical meristem. They generally appear on temperate legumes such as pea, alfalfa, and vetch. In longitudinal sections of mature indeterminate nodules, several zones can be observed, specified by typical morphological traits and gene expression (Newcomb, 1976; Vasse *et al.*, 1990; Franssen *et al.*, 1992). In distal–proximal direction, a meristem (zone I), a prefixation zone (II), an interzone (II–III), and a fixation zone (III) are recognized. In older nodules, a senescent zone (IV) appears proximal to the fixation zone. Determinate nodules have no persistent meristem and are round. They appear mostly on tropical legumes such as bean and soybean. Based on morphological studies, the cells of the central tissue were thought to pass synchronously through subsequent stages of development. However, recently it has been shown for bean and for *S. rostrata* nodules that, before the meristem activity ceases, different developmental zones occur simultaneously (Ndoye *et al.*, 1994; Taté *et al.*, 1994; Patriarca *et al.*, 1996; Goormachtig *et al.*, 1997).

Mature *S. rostrata* root and stem nodules are of the determinate type. Different microscopic techniques have been used to study their ontogeny (Tsien *et al.*, 1983; Duhoux, 1984; Ndoye *et al.*, 1994; Goormachtig *et al.*, 1997). A description of stem nodule development will be presented here (Fig. 5). Eventual differences with root nodules will be underlined. Results from *in situ* hybridizations on developing stem nodules using cell-cycle-phase-specific molecular markers [for a review on plant cell cycle, see Jacobs (1995) and Shaul *et al.* (1996a)] will be integrated in this description. Goormachtig *et al.* (1997) isolated a mitotic B1 cyclin clone (Sesro;*CycB1*;1), a cyclin-dependent kinase (CDK) homolog (*Cdc2-1Sr*), and a histone H4 gene homolog (*H4-1Sr*) from *S. rostrata* and used these clones as probes to follow meristem organization during stem nodule development. CDKs are the catalytic subunits of complexes with different cyclins that regulate progression through the cell cycle. Histone H4 and B1 cyclin gene expression are indicative for the DNA synthesis and the mitotic

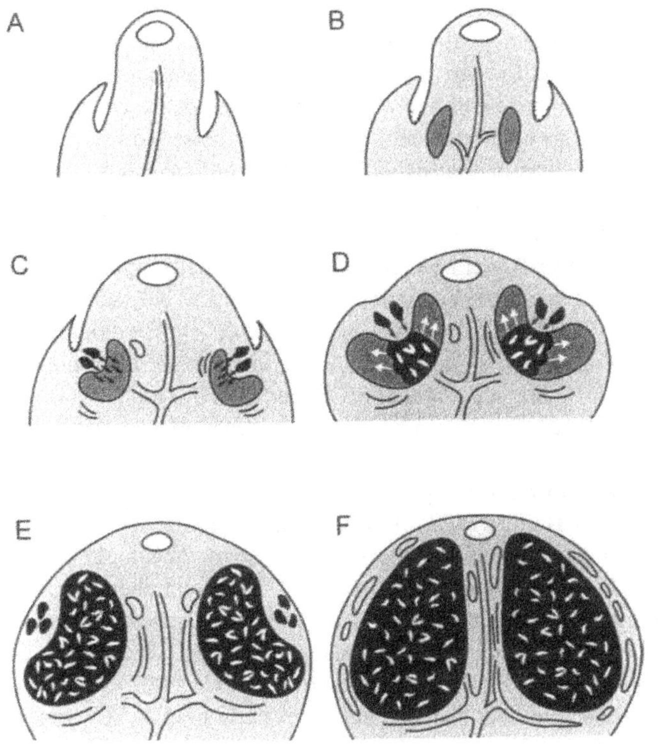


**FIGURE 5.** The development of *Sesbania rostrata* stem nodules. The drawing represents medial longitudinal sections of various developmental stages. (A) Uninfected root primordium; white spot represents dormant root meristem. (B) After inoculation, bacteria multiply in the fissure (small black spots) and globular nodule primordia are formed (two gray regions). (C) Infection pockets (black regions, two on each side) are formed from which the bacteria invade (black arrows) the nodule primordia, which now have an open-basket structure. Nodule vascular tissue is developing (parallel lines and open circles). (D) First cells differentiating into central tissue are located at the proximal side of the primordium, near the root stele. Black region represents infected cells wherein few uninfected cells (white cells) are present. At this stage, different zones are distinguished in the central tissue: an infected zone, an infection zone (gray regions with white arrows), and a meristem (distal gray part). (E) Young nodule; central tissue is completely differentiated. Growth happens via cell enlargement. (F) Mature nodule.

phase of the cell cycle, respectively (Hirt *et al.*, 1992; Nakayama and Iwabuchi 1993; Fobert *et al.*, 1994; Yang *et al.*, 1994; Kouchi *et al.*, 1995; Savouré *et al.*, 1995; Setiady *et al.*, 1995; Shaul *et al.*, 1996b). CDKs were found to be markers not only for cell divisions, but also for competence to divide (Martinez *et al.*, 1992; Hemerly *et al.*, 1993).

As mentioned earlier, *S. rostrata* stem nodules develop from predetermined dormant root primordia (Fig. 5A). A root primordium consists of a central vascular system connected to the vascular system of the stem and surrounded by cortical tissues. At the top, a dormant root meristem occurs (Fig. 5A). A root primordium can follow different developmental pathways depending on the growth conditions or external stimuli. When immersed in water, an adventitious root is formed due to the activation of the root meristem. When internodal explants are cultured on basal medium, without growth regulators, shoot buds originate from a newly formed meristem at the periphery of the dormant root meristem (Spencer-Barreto and Duhoux, 1994). Finally, upon bacterial infection, nodules appear due to the dedifferentiation of cortical cells at the base of the root primordium (Tsien *et al.*, 1983; Duhoux, 1984). During nodule formation the apical root meristem is preserved. Adventitious roots can still develop upon immersion of mature stem nodules (Duhoux and Dreyfus, 1982).

In uninduced root primordia (in the absence of azorhizobia), only a few exceptional cells of the apical root meristem expressed the mitotic cyclin homolog, indicating low cell-division activity. This was confirmed by a null-mitotic index (Spencer Barreto *et al.*, 1995). On the contrary, *H4-1Sr* transcripts were detected in many cells of the apical root meristem. Accumulation of histone *H4* transcripts has been reported as a typical S-phase event; therefore, the discrepancy between the abundance of expression of the two genes could be indicative of an arrest in late S phase or of endoreduplication. In either case a double DNA content is expected in the cells expressing *H4-1Sr*. However, flow-cytometric measurements revealed that the meristem cells have a 2C content (Spencer-Barreto *et al.*, 1995). Spencer-Barreto *et al.* (1995) proposed, based on several criteria (the low water content of the cells, the null-mitotic index, and the unimodal distribution of the DNA content), that the majority of the cells of the apical meristem are arrested in $G_0$–$G_1$. Therefore, in this particular case, the strong *H4-1Sr* expression in the dormant apical root meristem in the absence of DNA synthesis and division should probably be explained as an arrest in early S phase, a preparation of this tissue for a fast division response upon stimulation for adventitious root formation.

Due to the protrusion of the incipient adventitious root through the stem epidermis and cortex, a fissure is created wherein bacteria can proliferate (Fig. 5B; small black dots in fissure represent groups of actively dividing bacteria). Cellular damage may provide nutrients for bacterial growth during colonization of intercellular spaces (Duhoux, 1984). Root nodulation is also initiated by intercellular penetration of bacteria, in this case at the basis of lateral roots (Ndoye *et al.*, 1994). After intercellular infection, bacteria form densely populated infection pockets (Fig. 5B, black regions). Even before infection pockets are developed, cyclin and histone H4 gene transcripts can be detected in large vacuolated cells all over the cortex, indicating incipient dedifferentiation and division. About 2 days after infection, opposite the infection pockets, concentrations of small dividing cells constitute a nodule primordium. In the neighborhood, small groups of cells expressing the cell cycle genes prepare the first vascular elements of the nodule (Fig. 5B, C; parallel lines). From the infection pockets, intercellular infection threads (surrounded by cell wall material) grow inside toward the nodule primordial zones (Fig. 5C; black arrows indicate direction of penetration).

The nodule primordium then starts to surround an infection center. This peculiar stage was also observed for the root nodule development (Ndoye *et al.*, 1994). An open basket-like structure is formed that keeps growing at its distal end (oriented to the environment) with a half-dome-like peripheral meristem. This observation is supported by the distal-peripheral location of cells expressing the cyclin and histone gene homolog (Fig. 6A, B).

The first plant cells to be infected by bacteria are located at the proximal end of the nodule primordium, close to the vascular bundle of the root primordium (Fig. 5D; black region representing infected cells with white dots representing uninfected cells). At this stage, different developmental zones can be recognized in one longitudinal section: at the proximal end, an infected zone; next, an infection zone with infection threads; and at the distal end, a meristem. As nodulation progresses, the infection and infected zone move to more distally located regions, following the meristem (Fig. 5D; differentiation direction is indicated by white arrows). Finally, when division activity ceases, the last meristem-derived cells are invaded by bacteria and further growth of the central tissue occurs via cell enlargement (Fig. 5E).

Interestingly, from about 4 days after infection on, within the nodule periphery, a zone of small and flattened cells expressing both the Sesro;*CycB1;*1 and *H4-1Sr* genes is detected (Fig. 6A, B; arrows). This zone represents a peripheral meristem. Its function remains to be investigated; perhaps it is involved in periderm formation (Goormachtig *et al.*, 1997). In fully mature nodules (Fig. 5F), cell division and cell cycle gene expression are no longer detected.

The stem nodule development is basically always as described above, but can have somewhat different end products depending on the number of nodule primordia that arise per root primordium and on how fast they eventually coalesce. For example, a single nodule primordium can give rise to a stem nodule (Fig. 6C, D, G, H). This situation is similar to what usually happens in root nodulation at the base of a side root. On the stem, however, in most cases, several nodule primordia are formed per root primordium. These nodule primordia may either grow out as lobes separated by parenchyma or fuse early and form a ring structure around the central vascular bundle of the root primordium.

In conclusion, all observations indicate that development of root and stem nodules on *S. rostrata* shows many similarities to development of indeterminate nodules. Development is limited in time, though, and, when division stops, cell enlargement leads to mature nodules of the determinate type.

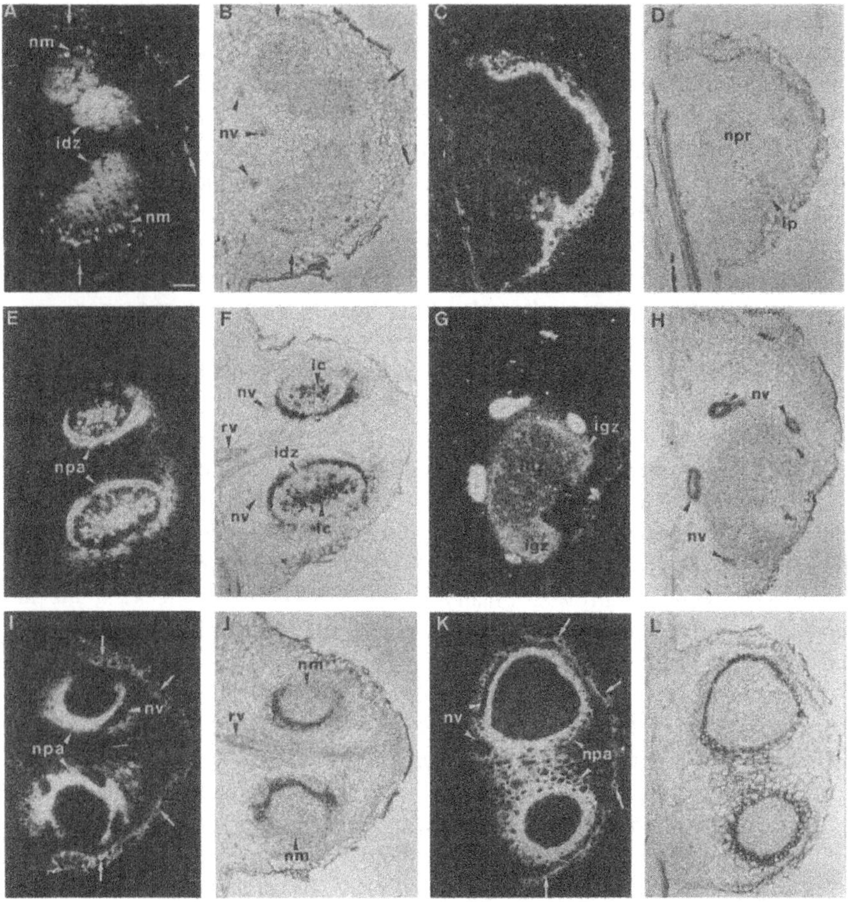


**FIGURE 6.** *In situ* hybridizations on sections of developing stem nodules. (A, C, E, G, I, K) Photographs taken with dark-field optics so that signals are seen as white spots; (B, D, F, H, J, L) photographs taken with bright-field optics, so that signals are seen as black spots. (A, B) *In situ* localization of *H4-1Sr*. Expression is seen in the meristem and in a peripheral cell layer (arrows). (C, D) *In situ* localization of *Srchi-24*. (E, F) *In situ* localization of *Srchi-13*. (G, H) *In situ* localization of *ENOD40*. (I–L) *In situ* localization of *SrENOD2*. Root primordia were infected for 3 days (I, J) and 6 days (A–H, K, L). Arrows point to the external cortical cell layer expressing *ENOD2*. Abbreviations: ic, infection center; idz, infected zone; igz, infection zone; ip, infection pocket; nm, nodule meristem; npa, nodule parenchyma; npr, nodule primordium; nv, nodule vascular tissue; rv, root vascular bundle. Bar = 100 μm.

## 3.2. *Sesbania rostrata* Genes Involved in Nodulation

### 3.2.1. Isolation of Nodule-Enhanced Genes

Nodulins are defined as proteins specifically found in nodules (van Kammen, 1984; Franssen *et al.*, 1992). These proteins and their corresponding genes are divided into two groups depending on the onset of gene expression (Gloudemans *et al.*, 1989). The late-nodulin (NOD) genes are expressed from the moment nitrogen fixation starts and the corresponding proteins are involved in the metabolic activities of the nodule. The most abundant proteins belonging to this class are the leghemoglobins. On the other hand, the early nodulins (ENOD) are produced before the onset of nitrogen fixation and they have a function during infection (root hair curling, infection thread formation) and nodule ontogeny (primordium formation, cell type differentiation). It becomes more and more clear, however, that these definitions have to get toned down, since several of the early nodulin genes (e.g., *ENOD40, ENOD2*) were found to be also expressed in other parts of the plant. Moreover, as will be discussed below (Section 3.2.7), expression of an *S. rostrata* leghemoglobin gene (a so-called late-nodulin gene) was detected during early stages of nodulation.

By comparing two-dimensional polyacrylamide gel electrophoresis patterns of *in vitro* translation products of poly(A)$^+$ RNA from uninfected *S. rostrata* roots and stems with poly(A)$^+$ RNA from infected roots and stems, at least 16 nodule-specific proteins have been detected and about 10 proteins appear to be nodule-enhanced (de Lajudie and Huguet, 1988).

Using the differential display technique [a polymerase chain reaction (PCR)-based technique to compare mRNA populations (Liang and Pardee, 1992)], gene expression during stem nodule development was studied (Goormachtig *et al.*, 1995). RNA samples were prepared from uninfected dormant root primordia and from root primordia infected with *A. caulinodans* for 2, 3, and 4 days (Goormachtig *et al.*, 1995). Thirty-five different primer combinations were used and 45 PCR bands that showed a clear enhancement or induction in the infected RNA samples were identified. Several of them were confirmed by RNA blot analysis to corresponded to induced genes (Goormachtig *et al.*, 1995). One third of the clones corresponded to low-abundance genes because no signal could be obtained using RNA blot analysis. One of these low-abundance *S. rostrata* clones was homologous to Gm#93, an early soybean nodulin gene (Kouchi and Hata, 1993; S. Goormachtig, unpublished results). At present, the very early stages (1–48 hr after inoculation) of stem nodule development are being investigated using the differential display approach (S. Lievens and M. Holsters, unpublished results). In the next subsection we will give an overview of the nodulin clones that were isolated and characterized from *S. rostrata*.

### 3.2.2. The *Srchi13* and *Srchi24* Genes Are Homologous to Class III Chitinase Genes

Two clones, identified by differential display, were shown to be homologous to acidic class III chitinase genes (Goormachtig *et al.*, 1995; S. Goormachtig *et al.*, in preparation). The two clones share 70% similarity. RNA blot analysis and reverse transcriptase PCR (RT-PCR) revealed that the two chitinase homolog have a different temporal expression pattern. The *Srchi24* gene is already strongly induced 1 day after infection, whereas *Srchi13* shows a slower induction and its expression peaks about 5 days after inoculation. Both genes are expressed in empty nodules formed after inoculation with an *A. caulinodans* LPS mutant (strain ORS571-X15) (Goethals *et al.*, 1994). Inoculation with a *nodA*$^-$ mutant (which does not produce Nod factors) does not induce the genes, suggesting that Nod factors trigger the expression in a direct or indirect way.

Nod factors have an *N*-acetylglucosamine backbone and were shown to be substrates for degradation by different chitinases (Roche *et al.*, 1991; Staehelin *et al.*, 1994a, b). If Srchi13 and Srchi24 proteins can degrade Nod factors, then Nod factors are inducing their own breakdown. Serving as signal molecules and active at very low concentrations, a strict regulation in time and space of Nod factors could be necessary for proper function. Using $^{14}C$-labeled Nod factors, it was shown that crude extracts of developing stem nodules (2–3 days post-inoculation) degrade Nod factors quickly (within 30 min), whereas crude extracts of uninfected root primordia hardly show any degrading activity even after overnight incubation (S. Goormachtig and M. Holsters, unpublished results). An induced Nod factor degradation activity was also detected for alfalfa (Staehelin *et al.*, 1995). Full-length cDNA clones of *Srchi13* and *Srchi24* will be expressed in *E. coli*. The enzymatic activity of the proteins will be studied and a correlation with the activity of the crude nodule extracts will be made.

*In situ* hybridizations on sections of developing stem nodules revealed a very different spatiotemporal expression pattern for both genes (S. Goormachtig *et al.*, in preparation). The *Srchi24* gene was weakly present in one of the outermost cell layers of the cortex of the root primordia and strongly induced upon inoculation (Fig. 6C, D). In mature nodules the signal disappeared. Why a chitinase is needed in this cell layer of a developing nodule is not known. From morphological studies, it is clear that infection happens via the fissure resulting from the protrusion of the root primordia (Duhoux, 1984). It is possible that the presence of a class III chitinase in the outer cell layers of the dome could function as a barrier to prevent entering of the bacteria. Indeed, class III chitinases have in most cases a lysozyme activity (for a review, see Collinge *et al.*, 1993). Alternatively, the chitinase could prevent action of the Nod factors produced by the bacteria attached to the surface of the root primordium. Both events would be part of a mechanism to control spatial organization of nodulation. The fissure itself is a microenvironment where the bacteria multiply actively (Duhoux, 1984), yielding sufficiently high concentrations of bacteria and Nod factors to overcome this chitinase activity and cause a plant response at this position. In alfalfa, chitinases were detected as a part of a hypersensitive response in root cells where infection threads were aborted and it was suggested that this phenomenon is one way for the plant to control infection and, thereby, nodulation (Vasse *et al.*, 1993). Taken together, these observations could indicate that chitinases can serve at different stages of nodulation as inspectors by controlling not only the number of nodules, but also the spatial organization of the nodules. Chitinases were described as being part of a general defense response since some of them were induced not only after pathogen attack, but also after application of different physical stresses (Collinge *et al.*, 1993). Also, during nodulation, chitinase expression in peripheral cell layers could protect the developing nodule against pathogen attack. Chitinases detected by Staehelin *et al.* (1992) in the cortex of mature soybean nodules could be involved in protection. However, in *S. rostrata* the chitinase expression disappeared in mature nodules.

*In situ* hybridizations using *Srchi13* sequences as a probe revealed a very different expression pattern (S. Goormachtig *et al.*, in preparation). No signal was detected in uninfected root primordia, whereas from 1 day after inoculation on, low expression was observed in the cortex of the root primordium. Once the nodule primordia were formed, strong expression was detected in the developing nodule parenchyma and in the uninfected cells of the central tissue (Fig. 6E, F). Also, very strong expression was observed around infection pockets and infection threads (Fig. 6E, F; infection center). Srchi13 could have a function in limiting the diffusion of Nod factors. Its presence in the uninfected cells could serve as a tool to prevent bacteria from entering these cells (i.e., by killing the bacteria via lysozyme activity). Again, also for Srchi13, a function as inspector can be deduced from the *in situ* pattern of expression. These hypotheses will now be tested. Some time ago it was proposed that pathogen attack and symbiosis have common molecular bases (Djordevic *et al.*, 1987). The spatial–temporal expression of chitinases during nodulation further supports this idea.

### 3.2.3. *Srextl* Is Homologous to Extensins

A cDNA clone, *Srext1*, isolated after differential display, has striking homology to genes coding for extensin-like proline-rich proteins (Goormachtig *et al.*, 1995). RNA blot analysis revealed that *Srextl* is expressed at low level in uninfected primordia. Between 2 and 4 hours after inoculation, expression is highly induced, stays on during development, and disappears in mature nodules. Infection with a *nodA*$^-$ mutant did not result in induction, whereas treatment with an LPS mutant of *A. caulinodans* caused transcript accumulation (S. Goormachtig and M. Holsters, unpublished results). This indicates that, as for *Srchi13* and *Srchi24*, Nod factors trigger the induction. *In situ* hybridization on developing nodules revealed that *Srextl* is first expressed in the nodule primordia and later in the nodule meristem. When the central tissue is completely differentiated, *Srextl* transcripts are detected in the nodule parenchyma (S. Goormachtig and M. Holsters, unpublished results).

Genes coding for proline-rich proteins were isolated from developing nodules of several legumes. The expression pattern of one of these, the *Medicago truncatula MtPRP4*, is very similar to that of *Srextl*. Besides expression in the primordia and meristem, expression was found in the nodule parenchyma, but, in contrast with *Srextl*, elevated levels of expression were also observed around the nodule vascular bundles (Wilson *et al.*, 1994).

Proline-rich proteins are localized in the cell walls and may confer specific functions or properties to the cells (for a review, see Showalter, 1993; Kieliszewski and Lamport, 1994). It is remarkable that one gene can be expressed in such different cell types (parenchymatous cells and meristematic cells). However, it cannot be ruled out that the *Srextl* antisense RNA probe recognizes different highly conserved transcripts. At present this question is under investigation.

### 3.2.4. *Srchrl* Is Homologous to Chalcone Reductases

*Srchrl* is homologous to genes coding for chalcone reductase enzymes (Goormachtig *et al.*, 1995). Chalcone reductase, in cooperation with chalcone synthase, catalyzes the formation of 2,4,4′-trihydroxychalcone or isoliquiritigenin and opens the 5-deoxyflavonoid pathway (Welle *et al.*, 1991). *Srchrl* transcripts were detected at low level in uninfected dormant root primordia. Three days after inoculation, the level of expression increased significantly in the developing nodules (Goormachtig *et al.*, 1995). From expression patterns obtained after infection with an *A. caulinodans nodA*$^-$ mutant and LPS mutant, it was concluded that also for *Srchrl*, Nod factors could be the (possibly indirect) inducers (S. Goormachtig and M. Holsters, unpublished results). *In situ* hybridizations revealed expression in the nodule parenchyma, in the uninfected cells of the central tissue, and in the pericycle of the nodule and root primordium vascular bundles (S. Goormachtig and M. Holsters, unpublished results).

Much speculation can be made about the function of this particular chalcone reductase. In pea nodules, chalcone synthase transcripts were detected mainly in the exterior part of the nodule meristem and in some neighboring cortical cell layers; a link with nodule morphogenesis was suggested (Yang *et al.*, 1992). Certain flavonoids were shown to act as auxin transport inhibitors and could be involved in changing hormone balances and influencing morphogenesis (Jacobs and Rubery, 1988). *Srchrl* transcripts were detected in the pericycle of vascular bundles, perhaps indicating that flavonoids produced in the pericycle could have a function as auxin transport inhibitors.

Another hypothesis is based on the observation that *Srchr1* expression in the nodule parenchyma and uninfected cells resembles partly the expression pattern of *Srchi13*, the chitinase gene homolog. Isoliquiritigenin and the flavanone liquiritigenin, produced in the first step of the 5-deoxyflavonoid pathway, are the main inducers of the *A. caulinodans nod* genes (Messens *et al.*, 1991). The elevated expression of *Srchr1* could bring about a higher production of these inducing compounds, a phenomenon that would be similar to the "Ini" effect (increased *nod* gene-inducing activity) observed in vetch (van Brussel *et al.*, 1990; Recourt *et al.*, 1991) upon nodulation with bacteria or treatment with Nod factors (Spaink *et al.*, 1991). In *S. rostrata* it was shown, using a *nodA*–β-glucuronidase (*gus*) fusion construct, that the bacterial *nod* genes are still expressed in bacteroids (Van den Eede *et al.*, 1992). Inducing flavanoids are probably required for this ongoing expression and they may be provided by the expression of biosynthetic genes in the nodule parenchyma. This would suggest that in the same cell layers of plant tissues, both an inducer of Nod factor production and a Nod factor degrador are made. Cooperation of both may be needed for a fine tuning of Nod factor action. Further experiments have to be done to test these hypotheses.

### 3.2.5. The Early Nodulin Gene *SrENOD40-1*

*ENOD40* is an early nodulin gene that is expressed in determinate as well as indeterminate nodules (Kouchi and Hata, 1993; Yang *et al.*, 1993; Asad *et al.*, 1994; Crespi *et al.*, 1994; Matvienko *et al.*, 1994; Vijn *et al.*, 1995; Papadopoulou *et al.*, 1996) and bears some interesting characteristics. *ENOD40* is first expressed in the root pericycle opposite protoxylem poles and in the incipient nodule primordia. In later stages of indeterminate nodule development, expression in the root pericycle disappears and *ENOD40* transcripts are mainly detected in the pericycle of the nodule vascular bundles and in the prefixation zone II, where cells are differentiating. Additional expression in infected and/or uninfected cells of the fixation zone III has also been reported (Matvienko *et al.*, 1994; Vijn *et al.*, 1995). In later stages of determinate nodule ontogeny, *ENOD40* transcripts are detected in the uninfected cells of the central tissue and in the cells of the boundary layer that separates the central tissue from the nodule parenchyma (Kouchi and Hata, 1993; Yang *et al.*, 1993; Papadopoulou *et al.*, 1996). For bean, it was reported that in older nodules *ENOD40* expression disappears in the pericycle of the connecting vascular tissue and finally also in the uninfected cells of the central region and of the boundary layer (Papadopoulou *et al.*, 1996). Accumulating evidence shows that *ENOD40* plays an important role in nodule development and in plant development in general. Introduction of an antisense *ENOD40* construct into *Medicago sativa* resulted in a loss of embryo development, whereas overexpressing of *ENOD40* led to the formation of teratomas (Crespi *et al.*, 1994). The *ENOD40* transcripts were first thought not to be translated because of the absence of long open-reading frames (ORF). However, recently it has convincingly been shown that both in legumes and tobacco a small peptide is synthesized encoded by a strongly conserved ORF in the 5′ part of the gene (van de Sande *et al.*, 1996). This peptide should be considered as a new plant growth regulator. Its expression in transgenic tobacco plants modulates the action of auxin (van de Sande *et al.*, 1996). These new data strengthen the idea of an involvement of *ENOD40* in nodule organogenesis by mediating phytohormone responses.

Two *S. rostrata ENOD40* homologs have been isolated from a developing stem nodule-specific cDNA library by hybridization with a soybean cDNA (Corich *et al.*, 1998). One full-length cDNA clone sequence was obtained (*SrENOD40-1*). In the 5′ part of the cDNA, a conserved small ORF is present that may code for a peptide of 12 amino acids. Semiquantitative RT-PCR showed the presence of *SrENOD40-1* transcripts in uninfected adventitious root primordia. Expression is significantly induced 8 hr after inoculation with *A. caulinodans* (Corich *et al.*, 1998). More details were obtained from *in situ* hybridizations. In uninfected root primordia, *ENOD40* transcripts were detected in parenchymatous cells around the central vascular bundle. One day after inoculation, the complete cortical region of the dormant root primordia showed expression. At 2 days, *ENOD40* is expressed in the globular nodule primordia and in the cells that form the connecting vascular bundles. With the onset of zonation (see Section 3.1), *ENOD40* transcripts were detected in differentiating cells (Fig. 6G, H; infection zone) (Corich *et al.*, 1998). Finally, in maturing nodules, when cell division ceases and nodule growth happens via cell enlargement, *SrENOD40-1* transcripts were detected in the uninfected cells of the central tissue and in the pericycle of the nodule vascular bundles. In 20-day-old nodules, expression in the central tissue diminished and remnants of expression were detected in the pericycle of the nodule vascular bundles.

By RNA blot and *in situ* hybridizations, *ENOD40* transcripts were also detected in stems, uninfected roots (lateral root primordia), and flowers of several legumes (Yang *et al.*, 1993; Asad *et al.*, 1994; Crespi *et al.*, 1994; Matvienko *et al.*, 1994; Papadopoulou *et al.*, 1996). Also in *S. rostrata*, expression was detected in other plant organs, not only in the adventitious root primordia, but also in procambial strands which form the shoot vascular system and around the vascular bundles of young leaves (Corich *et al.*, 1998).

### 3.2.6. The Early Nodulation Gene *SrENOD2*

*ENOD2* codes for a hydroxyproline-rich protein consisting of two repeated pentapeptide motifs and was first isolated from soybean (Franssen *et al.*, 1987, 1989). The proline-rich nature of the protein and the presence of a putative signal peptide suggest that the ENOD2 protein is located in the cell wall (Franssen *et al.*, 1989). *In situ* hybridization revealed that *ENOD2* is expressed in the nodule parenchyma of determinate as well as indeterminate nodules (van de Wiel *et al.*, 1990a; Allen *et al.*, 1991). A function of ENOD2 in establishing an oxygen barrier by determining the morphology of the parenchyma cells was proposed (Van de Wiel *et al.*, 1990a). However, by comparing oxygen changes and ENOD2 accumulation in wild-type alfalfa and in transgenic alfalfa plants with a lower ENOD2 level, no evidence could be provided for such a function (Hunt *et al.*, 1995). A successful rhizobial infection is not needed for *ENOD2* expression. The gene is expressed in empty nodules resulting from an *exo*$^-$ mutant (Dickstein *et al.*, 1988; Van de Wiel *et al.*, 1990b), in spontaneous alfalfa nodules (Truchet *et al.*, 1989), and in pseudonodules induced by auxin transport inhibitors (Hirsch *et al.*, 1989; Van de Wiel *et al.*, 1990b). This indicates that the function of ENOD2 is linked to nodule morphogenesis.

Two *ENOD2* cDNAs and one gene copy were isolated from *S. rostrata* (Strittmatter *et al.*, 1989; Dehio and de Bruijn 1992). It was shown that *ENOD2* transcripts accumulate in *S. rostrata* roots after application of cytokinins in a time- and concentration-dependent manner (Dehio and de Bruijn, 1992). Other hormones (such as auxins, abscisic acid, gibberellic acid) and physical stresses failed to induce this response (Dehio and de Bruijn, 1992). Evidence was obtained that the cytokinin effect is a posttranscriptional event and that cytokinins cause *SrENOD2* transcript accumulation and stabilization (Silver *et al.*, 1996) by a mechanism that requires translation and involves protein kinases and phosphatases. A basal *SrENOD2* expression level was detected in untreated *S. rostrata* roots (Silver *et al.*, 1996). *In situ* hybridizations revealed that *ENOD2*, like *Srchi24*, is expressed in the outermost cortical cell layer of the dormant root primordium and that this expression stays on until about 7 days after inoculation (when the nodule parenchyma is completely differentiated) (Fig. 6I–L; arrows). Expression in this exterior cortical cell layer could reflect a function for ENOD2 in constructing an oxygen barrier in the root primordium since aerial roots are exposed to higher concentrations of oxygen in comparison to roots in the soil and a low oxygen concentration may be needed for root development. However, a function in preventing desiccation of the root primordium could also be envisaged. It would be very interesting to study *SrENOD2* expression in aerial roots of other plants. The *SrENOD2* gene was used as a molecular marker to study the development of the nodule parenchyma by *in situ* hybridizations (Goormachtig *et al.*, 1998). Symbiosis-specific transcripts were first detected in outmost cell layers of the proximal part of the nodule primordium, near the root primordium stele (Fig. 6I, J). During further development, *SrENOD2* expression progressed gradually in a distal direction and finally the nodule parenchyma surrounded the complete central tissue (Fig. 6K, L). This is similar to the *ENOD2* expression patterns described during nodule development of pea, soybean, and alfalfa (Van de Wiel *et al.*, 1990a; Allen *et al.*, 1991).

Strittmatter *et al.* (1989) isolated two *ENOD2* cDNAs that showed a higher homology in the 3′ noncoding region than in the coding region. These authors suggested that the 3′ noncoding region could have a regulatory function. Studies of *gus* expression using the *ENOD2* promoter and various parts of the 5′- and 3′-untranslated regions (UTR) showed that the 3′-UTR is important to direct *ENOD2* expression in the nodule parenchyma (Chen and de Bruijn, 1995). This is somehow contradictory with the results obtained for the soybean *ENOD2(B)* promoter because it was demonstrated that this promoter is sufficient to direct *gus* expression in the parenchyma of both determinate and indeterminate nodules (Lauridsen *et al.*, 1993).

### 3.2.7. *Sesbania rostrata* Leghemoglobins

Leghemoglobins belong to the most abundant proteins in a functional nodule. They provide $O_2$ for bacteroid respiration in the infected cells where free $O_2$ tension must be very low (<15 nM) to allow nitrogenase activity (Appleby, 1984). Leghemoglobins and their corresponding genes were isolated from various legumes (de Bruijn and Schell, 1992) and some of the promoters were extensively investigated (de Bruijn *et al.*, 1990; Sánchez *et al.*, 1991). In *S. rostrata* stem and root nodules, seven forms of leghemoglobins were identified using anion-exchange chromatography and isoelectric focusing techniques. One form (VI) was shown to be more abundant in stem nodules (Bogusz *et al.*, 1987). Using chimeric promoter–*gus* fusions introduced into *Lotus corniculatus* via *Agrobacterium rhizogenes* transformation, the *Srglb3* promoter was analyzed in detail (Szabados *et al.*, 1990; de Bruijn *et al.*, 1994; Szczyglowski *et al.*, 1994). A 78-bp nodule-infected cell expression (NICE) element was delimited that was sufficient for targeting *Srglb3* expression in the infected cells (Szczyglowski *et al.*, 1994). Two other elements were identified that positively enhance gene expression (Szabados *et al.*, 1990). The NICE element contains several sequence motifs that are conserved among various leghemoglobin promoters. Mutagenesis of these motifs showed that they are necessary for gene expression (Szczyglowski *et al.*, 1994). Upstream of the NICE element, an AT-rich element (ATRE-BS2*) was identified that was shown to interact specifically with a DNA-binding factor isolated from nodule extracts (Metz *et al.*, 1988; de Bruijn *et al.*, 1988; Jacobsen *et al.*, 1990). Based on sequence conservation other ATRE elements were found throughout the *Srglb3* promoter (de Bruijn, 1989; Welters *et al.*, 1993). These elements are not essential for nodule-specific expression, but probably serve as general weak enhancers (Szabados *et al.*, 1990; Welters *et al.*, 1993). Finally, a GC-rich enhancer element (BBS1) was identified that interacts specifically with a bacterially derived binding protein named AcBBF1 (Welters *et al.*, 1993). The gene corresponding to this binding factor was isolated and the deduced amino acid sequence revealed a well-defined helix–turn–helix motif at the C-terminal end, confirming its DNA-binding properties (de Bruijn *et al.*, 1994). Promoter–*gus* fusions containing a mutated BBS1 element resulted in a 30% reduction of GUS activity, indicating that also the BBS1–AcBBF1 interaction plays a role in optimizing leghemoglobin gene expression in the nodule (Welters *et al.*, 1993). Also for the NICE element, a bacterial-binding factor was found that is distinct from AcBBF1 (de Bruijn *et al.*, 1994). However, the *in vivo* significance and the detailed binding site remain to be determined (de Bruijn *et al.*, 1994). These promoter analyses show that the leghemoglobin promoter is organized in a complex fashion. It had been thought that leghemoglobin genes are only induced when nitrogen fixation starts and this idea was mainly deduced from RNA blot analysis. However, using differential display, a PCR band was identified that corresponded to the *Srlb6* gene (Goormachtig *et al.*, 1995) and that was present already 2 days after inoculation. Why and where this leghemoglobin gene is expressed during the early stages are unknown. *In situ* hybridization and semiquantitative PCR will be performed to get more insight into these intriguing questions.

## 4. ECOLOGICAL ASPECTS OF STEM NODULATION

### 4.1. Inhibition by Nitrate

Nitrate in soil or plant growth medium has inhibitory effects on all stages of nodule development and functioning (Streeter, 1988). Dreyfus and Dommergues (1980) and Ratet *et al.* (1988) reported that stem nodulation and nitrogen fixation in stem nodules were not affected by relatively high nitrate concentrations supplied to the roots of *S. rostrata*, while root nodulation and fixation by root nodules were inhibited. However, a recent study by Kwon and Beevers (1993) questioned these results. It was found that, although the number of stem nodules was not affected, nitrate strongly inhibited growth of nodules and nitrogen fixation activity. Equally, addition of nitrate to plants, which had developed stem nodules in the absence of nitrate, led to inhibition of further nodule growth and nitrogen fixation, indicating that there is a systemic response to nitrate also in *S. rostrata* stem nodulation.

### 4.2. Suppression of Stem Nodulation by Root Nodulation

In nature, *A. caulinodans* is the most competitive symbiont of stem nodules of *S. rostrata*. In an ecological study of bacterial symbionts present in stem and root nodules of *S. rostrata* from different regions in Senegal, Robertson *et al.* (1995) observed that 90% of *S. rostrata* stem nodules harbored *A. caulinodans*, whereas only 39–48% of root nodules contained *A. caulinodans*, and other nodules harbored mostly representatives of the genus *Sinorhizobium*. Sinorhizobia are seldom found in stem nodules in nature. Part of the reason is probably the more abundant presence of *A. caulinodans* on the leaf and the stem surfaces of the host plant (Adebayo *et al.*, 1989; Robertson *et al.*, 1995). Another aspect that explains the greater stem specificity of *A. caulinodans* follows from a study by Tomekpe *et al.* (1996) on the effect of root nodulation on subsequent stem nodulation. These authors compared the stem nodulation behavior of *A. caulinodans* and of two different *Sinorhizobium* symbionts of *S. rostrata* under controlled conditions. They found that under experimental conditions in which roots are not inoculated, all strains nodulated stems very well. However, the *Sinorhizobium* strains were much more sensitive to suppression of stem nodulation by previous root inoculation. The introduction of the regulatory *nodD* gene of *A. caulinodans* diminished this sensitivity, probably by enhanced *nod* gene expression and subsequent Nod factor production. Indeed, *A. caulinodans* and *Sinorhizobium* strains that nodulate *S. rostrata* were reported to produce the same population of Nod factors (Lorquin *et al.*, 1997). However, the Nod factor production is not well induced in the *Sinorhizobium* strains. A more efficient production of mitogenic Nod factors at the stem-located infection sites may make the bacteria more fit to overcome systemic suppression.

### 4.3. Oxygen and Stem Nodulation

Stem nodules are exposed to the air and thus surrounded by higher oxygen concentrations than root nodules in a soil environment. Additionally, the cortex of stem nodules is green due to the presence of chloroplasts. In a recent study, James *et al.* (1996) demonstrated the presence of photosystem II components, responsible for oxygen generation, in the middle and inner cortex of *S. rostrata* stem nodules. Photosynthesis is probably most active in middle cortical cells, which contain intact chloroplasts. These authors also provided evidence for the occurrence of a protective oxygen diffusion barrier that is more strongly developed in stem than in root nodules. Two protective devices contribute to an enhanced oxygen diffusion barrier in stem nodules and the slowing down of oxygen diffusion to the bacteroid-containing central tissue. One is the enhanced presence of glycoproteins in the intercellular spaces of the cortex of stem nodules as demonstrated by MAC265 antibodies that recognize a (hydroxy)proline-rich protein in soybean and bean. This enhanced glycoprotein deposition and hence occlusion of intercellular spaces creates a water-glycoprotein-filled diffusion barrier. Another adaptation consists of close packing of cortical cells to reduce the size of intercellular spaces. In root nodules, intercellular spaces were rarely occluded to the same degree (James *et al.*, 1996).

## 5. AGRICULTURAL APPLICATIONS

### 5.1. *Sesbania rostrata* as Green Manure

If a legume is an economically important crop plant, the benefit of symbiotic nitrogen fixation is most direct. Legumes that are not consumed may be used as green manure for subsequent crop yield improvement. *S. rostrata* falls in the latter category. From an applied viewpoint, its major interests are the extremely fast growth rate, its high nitrogen-fixing potential when nodulated, and the fact that it grows happily in waterlogged conditions that are compatible with lowland rice. Although nodulation of *S. rostrata* is subject to nitrate inhibition and suppression of nodulation, in nature it can be nodulated extremely profusely (several hundreds of nodules per stem; Dreyfus and Dommergues, 1981) and it has probably the highest known nitrogen fixation rate. These features make *S. rostrata* very well fit as a green manure for lowland rice and, more generally, for improving soil fertility and sustainability. Several reports of small- and large-scale field experiments have demonstrated this potential. For an overview, we refer to Ladha *et al.* (1992).

## 5.2. Association of *Azorhizobium caulinodans* with Rice and Other Crops

Extension of symbiotic nitrogen fixation to nonlegume cereal plants such as rice is a longstanding and challenging goal for biotechnology. Several reports describe the formation of nodule-like structures on roots of cereals by rhizobia in the presence or absence of phytohormones or cell-wall-degrading enzymes (reviewed by Cocking *et al.*, 1994; de Bruijn *et al.*, 1995). In several of these studies, *A. caulinodans* strain ORS571 was used and reported to inhabit such nodule-like structures or to enter and infect by crack entry the root cortex where the lateral root emerged.

*Azorhizobium caulinodans* is a good candidate for extension of symbiotic nitrogen fixation to nonlegume crops for several reasons. It can fix nitrogen in the free-living state under relatively high oxygen tension (Elmerich *et al.*, 1982). Thus, differentiation of the bacteria to bacteroids, a *conditio sine qua non* for nitrogen fixation for most rhizobia, and the formation of sophisticated oxygen barrier, such as, for example, provided by legume nodules, are probably not required to obtain nitrogen fixation *in planta*. Crack entry of lateral roots is a less specialized way of entry than root hair invasion and therefore probably the easiest mode of infection of roots. This is the type of infection that is employed by *A. caulinodans* for nodulation of *S. rostrata*. Gough *et al.* (1997) demonstrated that *A. caulinodans* tagged with a *lacZ* reporter gene is able to form intercellular infection pockets in the cortical cell layers of roots of *Arabidopsis thaliana* and wheat. Compared to *R. meliloti* and *Azospirillum brasiliense*, *A. caulinodans* was better performing. Although flavonoids stimulated the infection, it was independent of the nodulation genes. In another recent report, *A. caulinodans* and *B. elkanii* strains have been used for rice root inoculations (Ladha *et al.*, 1996). These authors found that these bacteria were able to invade emerging lateral roots and change their growth to "short thick lateral roots"

(STLRs). Again, this phenotype was not dependent on the nodulation genes. Instead, a mutant of *B. elkanii* deficient in indole-3-acetic acid production was no longer able to form STLRs. A final argument making *A. caulinodans* attractive for this type of experimentation is provided by Ladha *et al.* (1989), who found that *A. caulinodans* can survive very well in soil and rhizosphere of rice under *S. rostrata*-rice rotation and that rice seedlings inoculated with *A. caulinodans* showed high nitrogenase activity, suggesting an active diazotrophic interaction between rice and *A. caulinodans*. The latter may be a reflection of the very close relatedness of *A. caulinodans* with *Xanthobacter flavus* strains that were isolated from rice rhizosphere (Rainey and Wiegel, 1996).

## 6. PROSPECTS FOR FUTURE RESEARCH

The *A. caulinodans–S. rostrata* interaction is becoming a very well studied model with knowledge accumulating on different features. For the future we can expect interesting progress in the understanding of the regulation of the dual capacity for free-living and symbiotic fixation. What is the mechanism or what are the signals that determine the switch in bacterial metabolism inside the plant cells?

From the viewpoint of understanding the molecular mechanism of nodule induction and development, we want once more to emphasize the enormous advantage of the stem-located multiple predetermined infection sites for the molecular analysis of the plant genes for nodule development. The reproducible and simultaneous response upon triggering the nodulation sites allows the collection of material with great precision and increase of resolution. These features were exploited in the differential display approach for the identification of new plant genes and will equally be an advantage for experiments directed at the study of Nod factor signal perception and transduction.

From a totally different focus, *A. caulinodans* is interesting for the study of evolutionary aspects of the development of symbiotic functions. Questions can be asked about the origin and way of acquisition of symbiotic loci in this genus.

Finally, there is a growing interest in $N_2$-fixing organisms that could establish an endophytic, beneficial relation with such important crops as rice and wheat. *A. caulinodans* can be included as a good candidate because of its legume nodulation capacity, crack entry way of plant invasion, free-living nitrogen fixation, and rice association.

To end, we express our hope that a breakthrough will come in methods to transform *S. rostrata*, which have been unsuccessful until now. This will be most interesting for applied as well as for fundamental research.

ACKNOWLEDGMENTS. We thank our past collaborators and present colleagues for their many contributions. We thank Martine De Cock, Christiane Germonprez,

Karel Spruyt, and Rebecca Verbanck for layout, art work, and photography. This work has been supported by grants from the Belgian Programme on Interuniversity Poles of Attraction (Prime Minister's Office, Science Policy Programming, No. 38) and in part by the European Communities' BIOTECH Programme, as part of the Project of Technological Priority 1993–1996, and the Human Capital and Mobility Programme (CHRX-CT94-0656). S.G. was a Research Assistant and P.M. and M.H. are a Post-doctoral fellow and a Research Director of the Fund for Scientific Research (Flanders), respectively.

## 7. REFERENCES


Adebayo, A., Watanabe, I., and Ladha, J. K., 1989, Epiphytic occurrence of *Azorhizobium caulinodans* and other rhizobia on host and non-host legumes, *Appl. Environ. Microbiol.* **55:**2407–2409.

Alazard, D., 1990, Nitrogen fixation in pure culture by rhizobia isolated from stem nodules of tropical *Aeschynomene* species, *FEMS Microbiol. Lett.* **68:**177–182.

Allen, O. N., and Allen, E. K., 1981, *The Leguminosae: A Source Book of Characteristics, Uses, and Nodulation,* University of Wisconsin Press, Madison, Wisconsin.

Allen, T., Raja, S., and Dunn, K., 1991, Cells expressing ENOD2 show differential spatial organization during the development of alfalfa root nodules, *Mol. Plant–Microbe Interact.* **4:**139–146.

Appleby, C. A., 1984, Leghemoglobin and *Rhizobium* respiration, *Annu. Rev. Plant Physiol.* **35:**443–478.

Arigoni, F., Kaminski, P. A., Hennecke, H., and Elmerich, C., 1991, Nucleotide sequence of the *fixABC* region of *Azorhizobium caulinodans* ORS571: Similarity of the *fixB* product with eukaryotic flavoproteins, characterization of *fixX*, and identification of *nifW*, *Mol. Gen. Genet.* **225:**514–520.

Arp, D. J., 1992, Hydrogen cycling in symbiotic bacteria, in *Biological Nitrogen Fixation* (G. Stacey, R. H. Burris, and H. J. Evans, eds.), pp. 432–460, Chapman & Hall, New York, NY.

Arumuganathan, K., and Earle, E. D., 1991, Nuclear DNA content of some important plant species, *Plant Mol. Biol. Rep.* **9:**208–218.

Asad, S., Fang, Y., Wycoff, K. L., and Hirsch, A. M., 1994, Isolation and characterization of cDNA and genomic clones of MsENOD40; transcripts are detected in meristematic cells of alfalfa, *Protoplasma* **183:**10–23.

Atkinson, E. M., and Long, S. R., 1992, Homology of *Rhizobium meliloti* NodC to polysaccharide polymerizing enzymes, *Mol. Plant–Microbe Interact.* **5:**439–442.

Atkinson, E. M., Placic, M. M., Hindsgaul, O., and Long, S. R., 1994, Biosynthesis of *Rhizobium meliloti* lipooligosaccharide Nod factors: NodA is required for an *N*-acyltransferase activity, *Proc. Natl. Acad. Sci. USA* **91:**8418–8422.

Bogusz, D., Kortt, A. A., and Appleby, C. A., 1987, *Sesbania rostrata* root and stem nodule leghemoglobins: Purification and relationship among the seven major components, *Arch. Biochem. Biophys.* **254:**263–271.

Boivin, C., Ndoye, I., Molouba, F., de Lajudie, P., Dupuy, N., and Dreyfus, B., 1997a, Stem nodulation in legumes: Diversity, mechanisms and unusual characters, *Crit. Rev. Plant Sci.* **16:**1–30.

Boivin, C., Ndoye, I., Lortet, G., Ndiaye, A., de Lajudie, P., and Dreyfus, B., 1997b, The *Sesbania* root symbionts *Sinorhizobium saheli* and *S. teranga* bv. sesbaniae can form stem nodules on *Sesbania rostrata*, although they are less adapted to stem nodulation than *Azorhizobium caulinodans*, *App. Environ. Microbiol.* **63:**1040–1047.

Boogerd, F. C., Ferdinandy-van Vlerken, M. M. A., Mawadza, C., Pronk, A. F., Stouthamer, A. H., and van Verseld, H. W., 1994, Nitrogen fixation and hydrogen metabolism in relation to the dissolved oxygen tension in chemostat cultures of the wild type and a hydrogenase-negative mutant of *Azorhizobium caulinodans*, *Appl. Environ. Microbiol.* **60:**1859–1866.

Buckmiller, L. M., Lapointe, J. P., and Ludwig, R. A., 1991, Cloning of *Azorhizobium caulinodans* nicotinate catabolism genes and characterization of their importance in $N_2$ fixation, *J. Bacteriol.* **173:**2017–2025.

Bulawa, C. E., and Wasco, W., 1991, Chitin and nodulation, *Nature* **353:**710.

Chen, R., and De Bruijn, F. J., 1995, Regulation and function of the *Sesbania rostrata* early nodulin gene *ENOD2*, in *Nitrogen Fixation: Fundamentals and Applications* (I. A. Tikhonovich, N. A. Provorov, V. I. Romanov, and W. E. Newton, eds.), p. 502, Kluwer, Dordrecht, The Netherlands.

Cocking, E. C., Webster, G., Batchelor, C. A., and Davey, M. R., 1994, Nodulation of non-legume crops. A new look, *Agro-Food-Industry Hi-Tech* **5**(1):21–24.

Collinge, D. B., Kragh, K. M., Mikkelsen, J. D., Nielsen, K. K., Rasmussen, U., and Vad, K., 1993, Plant chitinases, *Plant J.* **3:**31–40.

Corich, V., Goormachtig, S., Lievens, S., Van Montagu, M., and Holsters, M., 1998, Patterns of *ENOD40* gene expression in stem-borne nodules of *Sesbania rostrata*, *Plant Mol. Biol.*, in press.

Crespi, M., Jurkevitch, E., Poiret, M., d'Aubenton-Carafa, Y., Petrovics, G., Kondorosi, E., and Kondorosi, A., 1994, *enod40*, a gene expressed during nodule organogenesis, codes for a non-translatable RNA involved in plant growth, *EMBO J.* **13:**5099–5112.

Debellé, F., Rosenberg, C., and Dénarié, J., 1992, The *Rhizobium*, *Bradyrhizobium*, and *Azorhizobium* NodC proteins are homologous to yeast chitin synthases, *Mol. Plant–Microbe Interact.* **5:**443–446.

De Bruijn, F. J., 1989, The unusual symbiosis between the diazotrophic stem-nodulating bacterium *Azorhizobium caulinodans* ORS571 and its host, the tropical legume *Sesbania rostrata*, in *Plant–Microbe Interactions: Molecular and Genetic Perspectives*, Vol. 3 (T. Kosuge and E. W. Nester, eds.), pp. 457–504, McGraw-Hill, New York, N.Y.

De Bruijn, F. J., and Schell, J., 1992, Regulation of plant genes specifically induced in developing and mature nitrogen-fixing nodules: *cis*-acting elements and *trans*-acting factors, in *Control of Plant Gene Expression* (D. P. S. Verma, ed.), pp. 241–258, CRC Press, Boca Raton, Florida.

De Bruijn, F. J., Rossbach, S., and Schell, J., 1985, Cloning and characterization of glutamine synthetase genes from *Rhizobium meliloti* 1021, *R. sesbania* ORS571 and *Agrobacterium tumefaciens* C58, in *Nitrogen Fixation Research Progress* (H. J. Evans, P. J. Bottomley, and W. E. Newton, eds), p. 218, Martinus Nijhoff, Den Haag, The Netherlands.

De Bruijn, F. J., Jensen, E. O., Metz, B. A., Hoffman, H. J., Welters, P., Marcker, K. A., and Schell, J., 1988, Interaction of *trans*-acting factors with distinct DNA elements in the promoter regions of the *Glycine max* and *Sesbania rostrata* leghemoglobin genes, in *Molecular Genetics of Plant–Microbe Interactions* (R. Palacios and D. P. S. Verma, eds.), pp. 333–339, American Phytopathological Society Press, St. Paul, Minnesota.

De Bruijn, F. J., Chen, R., Fujimoto, S. Y., Pinaev, A., Silver, D., and Szczyglowski, K., 1994, Regulation of nodule gene expression, *Plant Soil* **161:**59–68.

De Bruijn, F. J., Chen, R., Dehio, C., Goel, A., Reiser, S., and Szczglowski, K., 1995, Hormones and nodule formation: Cytokinin induction of the *Sesbania rostrata* early nodulin gene *Enod2*, in *Nodulation and Nitrogen Fixation in Rice: Potential and Prospects* (G. S. Khush and J. Bennett, eds.), pp. 33–38, International Rice Research Institute, Manila, The Philippines.

Dehio, C., and De Bruijn, F. J., 1992, The early nodulin gene *SrEnod2* from *Sesbania rostrata* is inducible by cytokinin, *Plant J.* **2:**117–128.

De Lajudie, P., and Huguet, T., 1988, Plant gene expression during effective and ineffective nodule development of the tropical stem-nodulated legume *Sesbania rostrata*, *Plant Mol. Biol.* **10:**537–548.

De Lajudie, P., Willems, A., Pot, B., Dewettinck, D., Maestrojuan, G., Neyra, M., Collins, M. D., Dreyfus, B., Kersters, K., and Gillis, M., 1994, Polyphasic taxonomy of rhizobia: Emendation of the genus *Sinorhizobium* and description of *Sinorhizobium meliloti* comb. nov., *Sinorhizobium saheli* sp. nov., and *Sinorhizobium teranga* sp. nov. *Int. J. Syst. Bacteriol.* **44:**715–733.

Dénarié, J., Debellé, F., and Rosenberg, C., 1992, Signaling and host range variation in nodulation, *Annu. Rev. Microbiol.* **46:**497–531.

Dénarié, J., Debellé, F., and Promé, J.-C., 1996, Rhizobium lipo-chitooligosaccharide nodulation factors: Signaling molecules mediating recognition and morphogenesis, *Annu. Rev. Biochem.* **65:**503–535.

Denèfle, P., Kush, A., Norel, F., Paquelin, A., and Elmerich, C., 1987, Biochemical and genetic analysis of the *nifHDKE* region of *Rhizobium* ORS571, *Mol. Gen. Genet.* **207:**280–287.

De Vries, W., Stam, H., and Stouthamer, A. H., 1984, Hydrogen oxidation and nitrogen fixation in rhizobia, with special attention focused on strain ORS571, *Ant. Leeuwenhoek* **50:**505–524.

Dickstein, R., Bisseling, T., Reinhold, V. N., and Ausubel, F. M., 1988, Expression of nodule-specific genes in alfalfa root nodules blocked at an early stage of development, *Genes Dev.* **2:**677–687.

Djordjevic, M. A., Gabriel, D. W., and Rolfe, B. G., 1987, *Rhizobium*—The refined parasite of legumes, *Annu. Rev. Phytopathol.* **25:**145–168.

Dobert, R. C., Breil, B. T., and Triplett, E. W., 1994, DNA sequence of the common nodulation genes of *Bradyrhizobium elkanii* and their phylogenetic relationship to those of other nodulating bacteria, *Mol. Plant–Microbe Interact.* **7:**564–572.

Donald, R. G. K., and Ludwig, R. A., 1984, *Rhizobium* sp. strain ORS571 ammonium assimilation and nitrogen fixation, *J. Bacteriol.* **158:**1144–1151.

Donald, R. G. K., Raymond, C. K., and Ludwig, R. A., 1985, Vector insertion mutagenesis of *Rhizobium* sp. strain ORS571: Direct cloning of mutagenized DNA sequences, *J. Bacteriol.* **162:**317–323.

Donald, R. G. K., Nees, D. W., Raymond, C. K., Loroch, A. I., and Ludwig, R. A., 1986, Characterization of three genomic loci encoding *Rhizobium* sp. strain ORS571 $N_2$ fixation genes, *J. Bacteriol.* **165:**72–81.

Donald, R. G. K., Lapointe, J., and Ludwig, R. A., 1988, Characterization of the *Azorhizobium sesbaniae* ORS571 genomic locus encoding NADPH-glutamate synthase, *J. Bacteriol.* **170:**1197–1204.

Dreyfus, B., and Dommergues, Y., 1980, Non-inhibition de la fixation d'azote atmosphérique par l'azote combiné chez une légumineuse à nodules caulinaires, *Sesbania rostrata*, *C. R. Acad. Sci. Paris D* **291:**767–770.

Dreyfus, B., and Dommergues, Y. R., 1981, Nitrogen-fixing nodules induced by *Rhizobium* on the stem of the tropical legume *Sesbania rostrata*, *FEMS Microbiol. Lett.* **10:**313–317.

Dreyfus, B. L., Elmerich, C., and Dommergues, Y. R., 1983, Free-living *Rhizobium* strain able to grow on $N_2$ as the sole nitrogen source, *Appl. Environ. Microbiol.* **45:**711–713.

Dreyfus, B., Garcia, J. L., and Gillis, M., 1988, Characterization of *Azorhizobium caulinodans* gen. nov., sp. nov., a stem-nodulating nitrogen-fixing bacterium isolated from *Sesbania rostrata*, *Int. J. Syst. Bacteriol.* **38:**89–98.

Duhoux, E., 1984, Ontogénèse des nodules caulinaires du *Sesbania rostrata* (légumineuses), *Can. J. Bot.* **62:**982–994.

Duhoux, E., and Dreyfus, B. L., 1982, Nature des sites d'infection par le *Rhizobium* de la tige de la légumineuse *Sesbania rostrata* Brem, *C. R. Hebd. Séances Acad. Sci. Paris* **294:**407–411.

Elmerich, C., Dreyfus, B. L., Reysset, G., and Aubert, J.-P., 1982, Genetic analysis of nitrogen fixation in a tropical fast-growing *Rhizobium*, *EMBO J.* **1:**499–503.

Elmerich, C., Dreyfus, B., and Aubert, J.-P., 1983, Nicotinic acid requirement and degradation by *Sesbania rhizobium* strain ORS571, *FEMS Microbiol. Lett.* **19**:281–284.

Ervin, S. E., and Hubbell, D. H., 1985, Root hair deformations associated with fractionated extracts from *Rhizobium trifolii*, *Appl. Environ. Microbiol.* **49**:61–68.

Evans, H. J., Harker, A. R., Papen, H., Russell, S. A., Hanus, F. J., and Zuber, M., 1987, Physiology, biochemistry, and genetics of the uptake hydrogenase in rhizobia, *Ann. Rev. Microbiol.* **41**:335–361.

Faucher, C., Maillet, F., Vasse, J., Rosenberg, C., van Brussel, A. A. N., Truchet, G., and Dénarié, J., 1988, *Rhizobium meliloti* host range *nodH* gene determines production of an alfalfa-specific extracellular signal, *J. Bacteriol.* **170**:5489–5499.

Faucher, C., Camut, S., Dénarié, J., and Truchet, G., 1989, The *nodH* and *nodQ* host range genes of *Rhizobium meliloti* behave as avirulence genes in *R. leguminosarum* bv. *viciae* and determine changes in the production of plant-specific extracellular signals, *Mol. Plant–Microbe Interact.* **2**:291–300.

Fellay, R., Perret, X., Viprey, V., Broughton, W. J., and Brenner, S., 1995, Organization of host-inducible transcripts on the symbiotic plasmid of *Rhizobium* sp. NRG234, *Mol. Microbiol.* **16**:657–667.

Fernández-López, M., D'Haeze, W., Mergaert, P., Verplancke, C., Promé, J.-C., Van Montagu, M., and Holsters, M., 1996, Role of *nodI* and *nodJ* in lipo-chitooligosaccharide secretion in *Azorhizobium caulinodans* and *Escherichia coli*, *Mol. Microbiol.* **20**:993–1000.

Finnie, C., Hartley, N. M., Findlay, K. C., and Downie, J. A., 1997, The *Rhizobium leguminosarum prsDE* genes are required for secretion of several proteins, some of which influence nodulation, symbiotic nitrogen fixation, and exopolysaccharide modification, *Mol. Microbiol.* **25**:135–146.

Fobert, P. R., Coen, E. S., Murphy, G. J. P., and Doonan, J. H., 1994, Patterns of cell division revealed by transcriptional regulation of genes during the cell cycle in plants, *EMBO J.* **13**:616–624.

Franssen, H. J., Nap, J.-P., Gloudemans, T., Stiekema, W., van Dam, H., Govers, F., Louwerse, J., van Kammen, A., and Bisseling, T., 1987, Characterization of cDNA for nodulin-75 of soybean: A gene product involved in early stages of root nodule development, *Proc. Natl. Acad. Sci. USA* **84**:4495–4499.

Franssen, H. J., Thompson, D. V., Idler, K., Kormelink, R., van Kammen, A., and Bisseling, T., 1989, Nucleotide sequence of two soybean ENOD2 early nodulin genes encoding Ngm-75, *Plant Mol. Biol.* **14**:103–106.

Franssen, H. J., Vijn, I., Yang, W. C., and Bisseling, T., 1992, Developmental aspects of the Rhizobium–legume symbiosis, *Plant Mol. Biol.* **19**:89–107.

Gebhardt, C., Turner, G. L., Gibson, A. H., Dreyfus, B. L., and Bergersen, F. J., 1984, Nitrogen-fixing growth in continuous culture of a strain of *Rhizobium* sp. isolated from stem nodules on *Sesbania rostrata*, *J. Gen. Microbiol.* **130**:843–848.

Geelen, D., Mergaert, P., Geremia, R. A., Goormachtig, S., Van Montagu, M., and Holsters, M., 1993, Identification of *nodSUIJ* genes in Nod locus *1* of *Azorhizobium caulinodans:* Evidence that *nodS* encodes a methyltransferase involved in Nod factor modification, *Mol. Microbiol.* **9**:145–154.

Geelen, D., Goethals, K., Van Montagu, M., and Holsters, M., 1995a, The *nodD* locus from *Azorhizobium caulinodans* is flanked by two repetitive elements, *Gene* **164**:107–111.

Geelen, D., Leyman, B., Mergaert, P., Klarskov, K., Van Montagu, M., Geremia, R., and Holsters, M., 1995b, NodS is a *S*-adenosyl-L-methionine-dependent methyltransferase that methylates chitin fragments deacetylated at the non-reducing end, *Mol. Microbiol.* **17**:387–397.

Geelen, D., Van Montagu, M., and Holsters, M., 1995c, Cloning of an *Azorhizobium caulinodans* endoglucanase gene and analysis of its role in symbiosis, *Appl. Environ. Microbiol.* **61**:3304–3310.

Geremia, R. A., Mergaert, P., Geelen, D., Van Montagu, M., and Holsters, M., 1994, The NodC protein of *Azorhizobium caulinodans* is a *N*-acetylglucosaminyl transferase, *Proc. Natl. Acad. Sci. USA* **91**:2669–2673.

Gilles-Gonzalez, M. A., Ditta, G. S., and Helinski, D. R., 1991, A haemoprotein with kinase activity encoded by the oxygen sensor of *Rhizobium meliloti*, *Nature* **350**:170–172.

Gloudemans, T., Bhuvaneswari, T. V., Moerman, M., van Brussel, T., van Kammen, A., and Bisseling, T., 1989, Involvement of *Rhizobium leguminosarum* nodulation genes in gene expression in pea root hairs, *Plant Mol. Biol.* **12**:157–167.

Goethals, K., Gao, M., Tomekpe, K., Van Montagu, M., and Holsters, M., 1989, Common *nodABC* genes in *Nod* locus 1 of *Azorhizobium caulinodans:* Nucleotide sequence and plant-inducible expression, *Mol. Gen. Genet.* **219**:289–298.

Goethals, K., Van den Eede, G., Van Montagu, M., and Holsters, M., 1990, Identification and characterization of a functional *nodD* gene in *Azorhizobium caulinodans* strain ORS571, *J. Bacteriol.* **172**:2658–2666.

Goethals, K., Mergaert, P., Gao, M., Geelen, D., Van Montagu, M., and Holsters, M., 1992a, Identification of a new inducible nodulation gene in *Azorhizobium caulinodans*, *Mol. Plant–Microbe Interact.* **5**:405–411.

Goethals, K., Van Montagu, M., and Holsters, M., 1992b, Conserved motifs in a divergent *nod* box of *Azorhizobium caulinodans* ORS571 reveal a common structure in promoters regulated by LysR-type proteins, *Proc. Natl. Acad. Sci. USA* **89**:1646–1650.

Goethals, K., Leyman, B., Van den Eede, G., Van Montagu, M., and Holsters, M., 1994, An *Azorhizobium caulinodans* ORS571 locus involved in lipopolysaccharide production and nodule formation on *Sesbania rostrata* stems and roots, *J. Bacteriol.* **176**:92–99.

Goormachtig, S., Valerio-Lepiniec, M., Szczglowski, K., Van Montagu, M., Holsters, M., and de Bruijn, F. J., 1995, Use of differential display to identify novel *Sesbania rostrata* genes enhanced by *Azorhizobium caulinodans* infection, *Mol. Plant–Microbe Interact.* **8**:816–824.

Goormachtig, S., Alves-Ferreira, M., Van Montagu, M., Engler, G., and Holsters, M., 1997, Expression of cell cycle genes during *Sesbania rostrata* stem nodule development, *Mol. Plant–Microbe Interact.* **10**:316–325.

Goormachtig, S., Van Montagu, M., and Holsters, M., 1998, The early nodulin gene *ENOD2* shows different expression patterns during *Sesbania rostrata* stem nodule development, *Mol. Plant-Microbe Interact.* **11**:in press.

Göttfert, M., Hitz, S., and Hennecke, H., 1990, Identification of *nodS* and *nodU*, two inducible genes inserted between the *Bradyrhizobium japonicum nodYABC* and *nodIJ* genes, *Mol. Plant–Microbe Interact.* **3**:308–316.

Gough, C., Galera, C., Vasse, J., Webster, G., Cocking, E. C., and Dénarié, J., 1997, Specific flavonoids promote intercellular root colonization of *Arabidopsis thaliana* by *Azorhizobium caulinodans* ORS571, *Mol. Plant–Microbe Interact.* **10**:560–570.

Hemerly, A. S., Ferreira, P. C. G., de Almeida Engler, J., Van Montagu, M., Engler, G., and Inzé, D., 1993, *cdc2a* expression in *Arabidopsis thaliana* is linked with competence for cell division, *Plant Cell* **5**:1711–1723.

Hilgert, U., Schell, J., and de Bruijn, F. J., 1987, Isolation and characterization of Tn*5*-induced NADPH-glutamate synthase (GOGAT$^-$) mutants of *Azorhizobium sesbaniae* ORS571 and cloning of the corresponding *glt* locus, *Mol. Gen. Genet.* **210**:195–202.

Hirsch, A. M., Bhuvaneswari, T. V., Torrey, J. G., and Bisseling, T., 1989, Early nodulin genes are induced in alfalfa root outgrowths elicited by auxin transport inhibitors, *Proc. Natl. Acad. Sci. USA* **86**:1244–1248.

Hirt, H., Mink, M., Pfosser, M., Bögre, L., Györgyey, J., Jonak, C., Gartner, A., Dudits, D., and Heberle-Bors, E., 1992, Alfalfa cyclins: Differential expression during the cell cycle and in plant organs, *Plant Cell* **4**:1531–1538.

Hunt, S., Wycoff, K. L., Layzell, D. B., and Hirsch, A. M., 1995, Is the early nodulin ENOD2 associated with regulation of nodule permeability to oxygen? in *Nitrogen Fixation: Fundamentals and Applications* (I. A. Tikhonovich, N. A. Provorov, V. I. Romanov, and W. E. Newton, eds.), p. 504, Kluwer, Dordrecht, The Netherlands.

Jacobs, M., and Rubery, P. H., 1988, Naturally occurring auxin transport regulators, *Science* **241**:346–349.

Jacobs, T. W., 1995, Cell cycle control, *Annu. Rev. Plant Physiol. Plant Mol. Biol.* **46**:317–339.

Jacobsen, K., Laursen, N. B., Jensen, E. Ø., Marcker, A., Poulsen, C., and Marcker, K. A., 1990, HMG I-like proteins from leaf and nodule nuclei interact with different AT motifs in soybean nodulin promoters, *Plant Cell* **2**:85–94.

James, E. K., Sprent, J. I., Sutherland, J. M., McInroy, S. G., and Minchin, F. R., 1992, The structure of nitrogen fixing root nodules on the aquatic mimosoid legume *Neptunia plena*, *Ann. Bot.* **69**:173–180.

James, E. K., Iannetta, P. P. M., Nixon, P. J., Whiston, A. J., Peat, L., Crawford, R. M. M., Sprent, J. I., and Brewin, N. J., 1996, Photosystem II and oxygen regulation in *Sesbania rostrata* stem nodules, *Plant Cell Environ.* **19**:895–910.

Jarvis, B. D. W., Gillis, M., and De Ley, J., 1986, Intra- and intergeneric similarities between the ribosomal ribonucleic acid cistrons of *Rhizobium* and *Bradyrhizobium* species and some related bacteria, *Int. J. Syst. Bacteriol.* **36**:129–138.

John, M., Röhrig, H., Schmidt, J., Wieneke, Y., and Schell, J., 1993, *Rhizobium* NodB protein involved in nodulation signal synthesis is a chitooligosaccharide deacetylase, *Proc. Natl. Acad. Sci. USA* **90**:625–629.

Kaminski, P. A., and Elmerich, C., 1991, Involvement of *fixLJ* in the regulation of nitrogen fixation in *Azorhizobium caulinodans*, *Mol. Microbiol.* **5**:665–673.

Kaminski, P. A., Norel, F., Desnoues, N., Kush, A., Salzano, G., and Elmerich, C., 1988, Characterization of the *fixABC* region of *Azorhizobium caulinodans* ORS571 and identification of a new nitrogen fixation gene, *Mol. Gen. Genet.* **214**:496–502.

Kaminski, P. A., Mandon, K., Arigoni, F., Desnoues, N., and Elmerich, C., 1991, Regulation of nitrogen fixation in *Azorhizobium caulinodans*: Identification of a *fixK*-like gene, a positive regulator of *nifA*, *Mol. Microbiol.* **5**:1983–1991.

Kaminski, P. A., Desnoues, N., and Elmerich, C., 1994, The expression of *nifA* in *Azorhizobium caulinodans* requires a gene product homologous to *Escherichia coli* HF-I, an RNA-binding protein involved in the replication of phage Qβ RNA, *Proc. Natl. Acad. Sci. USA* **91**:4663–4667.

Kaminski, P. A., Michel-Reydellet, N., Desnoues, N., and Elmerich, C. 1995, Regulation of free-living and symbiotic nitrogen fixation in *Azorhizobium caulinodans*, in *Nitrogen Fixation: Fundamentals and Applications* (I. A. Tikhonovich, N. A. Provorov, V. I. Romanov, and W. E. Newton, eds.), pp. 183–187, Kluwer, Dordrecht, The Netherlands.

Kaminski, P. A., Kitts, C. L., Zimmerman, Z., and Ludwig, R. A., 1996, *Azorhizobium caulinodans* uses both cytochrome *bd* (quinol) and cytochrome *cbb3* (cytochrome *c*) terminal oxidases for symbiotic $N_2$ fixation, *J. Bacteriol.* **178**:5989–5994.

Kieliszewski, M. J., and Lamport, D. T. A., 1994, Extensin: Repetitive motifs, functional sites, post-translational codes, and phylogeny, *Plant J.* **5**:157–172.

Kitts, C. L., and Ludwig, R. A., 1994, *Azorhizobium caulinodans* respires with at least four terminal oxidases, *J. Bacteriol.* **176**:886–895.

Kitts, C. L., Schaechter, L. E., Rabin, R. S., and Ludwig, R. A., 1989, Identification of cyclic intermediates in *Azorhizobium caulinodans* nicotinate catabolism, *J. Bacteriol.* **171**:3406–3411.

Kitts, C. L., Lapointe, J. P., Lam, V. T., and Ludwig, R. A., 1992, Elucidation of the complete *Azorhizobium* nicotinate catabolism pathway, *J. Bacteriol.* **174**:7791–7797.

Kouchi, H., and Hata, S., 1993, Isolation and characterization of novel nodulin cDNAs representing genes expressed at early stages of soybean nodule development, *Mol. Gen. Genet.* **238:**106–119.

Kouchi, H., Sekine, M., and Hata, S., 1995, Distinct classes of mitotic cyclins are differentially expressed in the soybean shoot apex during the cell cycle, *Plant Cell* **7:**1143–1155.

Kush, A., Elmerich, C., and Aubert, J.-P., 1985, Nitrogenase of *Sesbania Rhizobium* strain ORS571: Purification, properties and "switch-off" by ammonia, *J. Gen. Microbiol.* **131:**1765–1777.

Kwon, D.-K., and Beevers, H., 1993, Adverse effects of nitrate on stem nodules of *Sesbania rostrata* (Brem), *New Phytol.* **125:**345–350.

Ladha, J. K., Garcia, M., Miyan, S., Padre, A. T., and Watanabe, I., 1989, Survival of *Azorhizobium caulinodans* in the soil and rhizosphere of wetland rice under *Sesbania rostrata*-rice rotation, *Appl. Environ. Microbiol.* **55:**454–460.

Ladha, J. K., Pareek, R. P., and Becker, M., 1992, Stem nodulating legume-*Rhizobium* symbiosis and its agronomic use in lowland rice, in *Advances in Soil Science*, Vol. 20 (B. A. Stewart, ed.), pp. 147–192, Springer-Verlag, NY.

Ladha, J. K., So, R., Angeles, O. R., Hernandez, R., Ramos, M. C., and Reddy, P. M., 1996, Rhizobial invasion and induction of phenotypic changes in rice roots are independent of Nod factors, Abstract presented at the 8th International Congress on Molecular Plant–Microbe Interactions, Knoxville, Tennessee, July 14–19, 1996 (#L-11).

Lauridsen, P., Franssen, H., Stougaard, J., Bisseling, T., and Marcker, K. A., 1993, Conserved regulation of the soybean early nodulin *ENOD2* gene promoter in determinate and indeterminate transgenic root nodules, *Plant J.* **3:**483–492.

Le Coq, C., Guervin, C., and Duhoux, E., 1985, Caryologie du *Sesbania rostrata* Brem. I. Etude chromosomique, *Rev. Cytol. Biol. Veg. Bot.* **8:**245–254.

Lerouge, P., Roche, P., Faucher, C., Maillet, F., Truchet, G., Promé, J. C., and Dénarié, J., 1990, Symbiotic host-specificity of *Rhizobium meliloti* is determined by a sulphated and acetylated glucosamine oligosaccharide signal, *Nature* **344:**781–784.

Liang, P., and Pardee, A. B., 1992, Differential display of eukaryotic messenger RNA by means of the polymerase chain reaction, *Science* **257:**967–971.

Lindstrom, K., Paulin, L., Roos, C., and Suominen, L., 1995, Nodulation genes of *Rhizobium galegae*, in *Nitrogen Fixation: Fundamentals and Applications* (I. A. Tikhonovich, N. A. Provorov, V. I. Romanov, and W. E. Newton, eds.), pp. 365–370, Kluwer, Dordrecht, The Netherlands.

Loroch, A. I., Nguyen, B. G., and Ludwig, R. A., 1995, Interactive regulation of *Azorhizobium nifA* transcription via overlapping promoters, *J. Bacteriol.* **177:**7210–7221.

Lorquin, J., Lortet, G., Ferro, M., Mear, N., Promé, J.-C., and Boivin, C., 1997, *Sinorhizobium teranga* bv. acaciae ORS1073 and *Rhizobium* sp. strain ORS1001, two distantly related *Acacia*-nodulating strains, produce similar Nod factors that are O carbamoylated, N methylated, and mainly sulfated, *J. Bacteriol.* **179:**3079–3083.

Loureiro, M. F., de Faria, S. M., James, E. K., Pott, A., and Franco, A. A., 1994, Nitrogen fixing stem nodules of the legume, *Discolobium pulchellum* Benth, *New Phytol.* **128:**283–295.

Ludwig, R. A., 1986, *Rhizobium* sp. strain ORS571 grows synergistically on $N_2$ and nicotinate as N sources, *J. Bacteriol.* **165:**304–307.

Luka, S., Sanjuan, J., Carlson, R. W., and Stacey, G., 1993, *nolMNO* genes of *Bradyrhizobium japonicum* are co-transcribed with *nodYABCSUIJ*, and *nolO* is involved in the synthesis of the lipo-oligosaccharide nodulation signals, *J. Biol. Chem.* **268:**27053–27059.

Magasanik, B., 1982, Genetic control of nitrogen assimilation in bacteria, *Annu. Rev. Genet.* **16:**135–168.

Mandon, K., Kaminski, P. A., and Elmerich, C., 1994, Functional analysis of the *fixNOQP* region of *Azorhizobium caulinodans*, *J. Bacteriol.* **176:**2560–2568.

Martinez, M. C., Jørgensen, J.-E., Lawton, M. A., Lamb, C. J., and Doerner, P. W., 1992, Spatial pattern of *cdc2* expression in relation to meristem activity and cell proliferation during plant development, *Proc. Natl. Acad. Sci. USA* **89:**7360–7364.

Matvienko, M., Van De Sande, K., Yang, W.-C., van Kammen, A., Bisseling, T., and Franssen, H., 1994, Comparison of soybean and pea ENOD40 cDNA clones representing genes expressed during both early and late stages of nodule development, *Plant Mol. Biol.* **26:**487–493.

Mergaert, P., Van Montagu, M., Promé, J.-C., and Holsters, M., 1993, Three unusual modifications, a D-arabinosyl, an *N*-methyl, and a carbamoyl group, are present on the Nod factors of *Azorhizobium caulinodans* strain ORS571, *Proc. Natl. Acad. Sci. USA* **90:**1551–1555.

Mergaert, P., D'Haeze, W., Geelen, D., Promé, D., Van Montagu, M., Geremia, R., Promé, J.-C., and Holsters, M., 1995, Biosynthesis of *Azorhizobium caulinodans* Nod factors. Study of the activity of the NodABCS proteins by expression of the genes in *Escherichia coli, J. Biol. Chem.* **270:**29217–29223.

Mergaert, P., D'Haeze, W., Fernández-López, M., Geelen, D., Goethals, K., Promé, J.-C., Van Montagu, M., and Holsters, M., 1996, Fucosylation and arabinosylation of Nod factors in *Azorhizobium caulinodans:* Involvement of *nolK, nodZ* as well as *noeC* and/or downstream genes, *Mol. Microbiol.* **21:**409–419.

Mergaert, P., Van Montagu, M., and Holsters, M., 1997a, Control of rhizobial host range by Nod factors, *Mol. Microbiol.* **25:**811–817.

Mergaert, P., Van Montagu, M., and Holsters, M., 1997b, The nodulation gene *nolK* of *Azorhizobium caulinodans* is involved in the formation of GDP-fucose from GDP-mannose, *FEBS Lett.* **409:**312–316.

Mergaert, P., Ferro, M., D'Haeze, W., Van Montagu, M., Holsters, M., and Promé, J.-C., 1997c, Nod factors of *Azorhizobium caulinodans* strain ORS571 can be glycosylated with an arabinosyl group, a fucosyl group, or both, *Mol. Plant–Microbe Interact.* **10:**683–687.

Merrick, M. J., 1992, Regulation of nitrogen fixation genes in free-living and symbiotic bacteria, in *Biological Nitrogen Fixation* (G. Stacey, R. H., Burris, and H. J. Evans, eds.), pp. 835–876, Chapman & Hall, New York, NY.

Messens, E., Geelen, D., Van Montagu, M., and Holsters, M., 1991, 7,4′-Dihydroxyflavanone is the major *Azorhizobium nod* gene-inducing factor present in *Sesbania rostrata* seedling exudate, *Mol. Plant–Microbe Interact.* **4:**262–267.

Metz, B. A., Welters, P., Hoffmann, H. J., Jensen, E. Ø., Schell, J., and de Bruijn, F. J., 1988, Primary structure and promoter analysis of leghemoglobin genes of the stem-nodulated tropical legume *Sesbania rostrata:* Conserved coding sequences, *cis*-elements and *trans*-acting factors. *Mol. Gen. Genet.* **214:**181–191.

Miflin, B. J., and Cullimore, J. V., 1984, Nitrogen assimilation: A joint endeavour, in *Genes Involved in Microbe–Plant Interactions* (D. P. S. Verma, and T. Hohn, eds.), pp. 129–178, Springer, Vienna, Austria.

Mylona, P., Pawlowski, K., and Bisseling, T., 1995, Symbiotic nitrogen fixation, *Plant Cell* **7:**869–885.

Nakayama, T., and Iwabuchi, M., 1993, Regulation of wheat histone gene expression, *Crit. Rev. Plant Sci.* **12:**97–110.

Ndoye, I., de Billy, F., Vasse, J., Dreyfus, B., and Truchet, G., 1994, Root nodulation of *Sesbania rostrata, J. Bacteriol.* **176:**1060–1068.

Nees, D. W., Stein, P. A., and Ludwig, R. A., 1988, The *Azorhizobium caulinodans nifA* gene: Identification of upstream-activating sequences including a new element, the 'anaerobox,' *Nucleic Acids Res.* **16:**9839–9853.

Newcomb, W., 1976, A correlated light and electron microscopic study of symbiotic growth and differentiation in *Pisum sativum* root nodules, *Can J. Bot.* **54:**2163–2186.

Nieuwkoop, A. J., Banfalvi, Z., Deshmane, N., Gerhold, D., Schell, M. G., Sirotkin, K. M., and Stacey, G., 1987, A locus encoding host range is linked to the common nodulation genes of *Bradyrhizobium japonicum, J. Bacteriol.* **169:**2631–2638.

Norel, F., and Elmerich, C., 1987, Nucleotide sequence and functional analysis of the two *nifH* copies of *Rhizobium* ORS571, *J. Gen. Microbiol.* **133:**1563–1576.

Norel, F., Desnoues, N., and Elmerich, C., 1985, Characterization of DNA sequences homologous to *Klebsiella pneumoniae nifH, D, K* and *E* in the tropical *Rhizobium* ORS571, *Mol. Gen. Genet.* **199:**352–356.

O'Brian, M. R., 1996, Heme synthesis in the Rhizobium–legume symbiosis: A palette for bacterial and eukaryotic pigments, *J. Bacteriol.* **178:**2471–2478.

Papadopoulou, K., Roussis, A., and Katinakis, P., 1996, *Phaseolus* ENOD40 is involved in symbiotic and non-symbiotic organogenetic processes: Expression during nodule and lateral root development, *Plant Mol. Biol.* **30:**403–417.

Patriarca, E. J., Taté, R., Fedorova, E., Riccio, A., Defez, R., and Iaccarino, M., 1996, Down-regulation of the *Rhizobium ntr* system in the determinate nodule of *Phaseolus vulgaris* identifies a specific developmental zone, *Mol. Plant–Microbe Interact.* **9:**243–251.

Pawlowski, K., Ratet, P., Schell, J., and de Bruijn, F. J., 1987, Cloning and characterization of *nifA* and *ntrC* genes of the stem nodulating bacterium ORS571, the nitrogen fixing symbiont of *Sesbania rostrata:* Regulation of nitrogen fixation (*nif*) genes in the free living versus symbiotic state, *Mol. Gen. Genet.* **206:**207–219.

Pawlowski, K., Klosse, U., and de Bruijn, F. J., 1991, Characterization of a novel *Azorhizobium caulinodans* ORS571 two-component regulatory system, NtrY/NtrX, involved in nitrogen fixation and metabolism, *Mol. Gen. Genet.* **231:**124–138.

Pawlowski, K., Gough, S. P., Kannangara, C. G., and de Bruijn, F. J., 1993, Characterization of a 5-aminolevulinic acid synthase mutant of *Azorhizobium caulinodans* ORS571, *Mol. Plant–Microbe Interact.* **6:**35–44.

Preisig, O., Anthamatten, D., and Hennecke, H., 1993, Genes for a microaerobically induced oxidase complex in *Bradyrhizobium japonicum* are essential for a nitrogen-fixing endosymbiosis, *Proc. Natl. Acad. Sci. USA* **90:**3309–3313.

Pronk, A. F., Stouthamer, A. H., Van Verseveld, H. W., and Boogerd, F. C., 1995, Nicotinate catabolism is dispensable and nicotinate anabolism is crucial in *Azorhizobium caulinodans* growing in batch culture and chemostat culture on $N_2$ as the N source, *J. Bacteriol.* **177:**75–81.

Rainey, F. A., and Wiegel, J., 1996, 16S ribosomal DNA sequence analysis confirms the close relationship between the genera *Xanthobacter, Azorhizobium,* and *Aquabacter* and reveals a lack of phylogenetic coherence among *Xanthobacter* species, *Int. J. Syst. Bacteriol.* **46:**607–610.

Ratet, P., Pawlowski, K., Meyer A., Schell, J., and de Bruijn, F. J., 1988, Regulation of nitrogen fixation (*nif*) genes of *Azorhizobium caulinodans* ORS571 in culture and *in planta*, *J. Plant Physiol.* **132:**405–411.

Ratet, P., Pawlowski, K., Schell, J., and de Bruijn, F. J., 1989, The *Azorhizobium caulinodans* nitrogen-fixation regulatory gene, *nifA*, is controlled by the cellular nitrogen and oxygen status, *Mol. Microbiol.* **3:**825–838.

Recourt, K., Schripsema, J., Kijne, J. W., van Brussel, A. A. N., and Lugtenberg, B. J. J., 1991, Inoculation of *Vicia sativa* subsp. *nigra* roots with *Rhizobium leguminosarum* biovar *viciae* results in release of *nod* gene activating flavanones and chalcones, *Plant Mol. Biol.* **16:**841–852.

Reding, H. K., Hartel, P. G., and Wiegel, J., 1991, Effect of *Xanthobacter*, isolated and characterized from rice roots, on growth of wetland rice, *Plant Soil* **138:**221–229.

Rinaudo, G., Orenga, S., Fernandez, M. P., Meugnier, H., and Bardin, R., 1991, DNA homologies among members of the genus *Azorhizobium* and other stem- and root-nodulating bacteria isolated from the tropical legume *Sesbania rostrata*, *Int. J. Syst. Bacteriol.* **41:**114–120.

Robertson, B. K., Dreyfus, B., and Alexander, M., 1995, Ecology of stem-nodulating *Rhizobium* and *Azorhizobium* in four vegetation zones of Senegal, *Microb. Ecol.* **29:**71–81.

Roche, P., Debellé, F., Maillet, F., Lerouge, P., Faucher, C., Truchet, G., Dénarié, J., and Promé, J.-C., 1991, Molecular basis of symbiotic host specificity in *Rhizobium meliloti: nodH* and *nodPQ* genes encode the sulfation of lipo-oligosaccharide signals, *Cell* **67:**1131–1143.

Röhrig, H., Schmidt, J., Wieneke, U., Kondorosi, E., Barlier, I., Schell, J., and John, M., 1994, Biosynthesis of lipooligosaccharide nodulation factors: *Rhizobium* NodA protein is involved in *N*-acylation of the chitooligosaccharide backbone, *Proc. Natl. Acad. Sci. USA* **91**:3122–3126.

Sánchez, F., Padilla, J. E., Pérez, H., and Lara, M., 1991, Control of nodulin genes in root-nodule development and metabolism, *Annu. Rev. Plant Physiol. Plant Mol. Biol.* **42**:507–528.

Savouré, A., Fehér, A., Kaló, P., Petrovics, G., Csanádi, G., Szécsi, J., Kiss, G., Brown, S., Kondorosi, A., and Kondorosi, E., 1995, Isolation of a full-length mitotic cyclin cDNA clone *CycIIIMs* from *Medicago sativa:* Chromosomal mapping and expression, *Plant Mol. Biol.* **27**:1059–1070.

Schell, M. A., 1993, Molecular biology of the LysR family of transcriptional regulators, *Annu. Rev. Microbiol.* **47**:597–626.

Schultze, M., and Kondorosi, Á., 1995, What makes nodulation signals host-plant specific? *Trends Microbiol.* **3**:370–372.

Schultze, M., Kondorosi, E., Ratet, P., Buiré, M., and Kondorosi, A., 1994, Cell and molecular biology of *Rhizobium*–plant interactions, *Int. Rev. Cytol.* **156**:1–75.

Setiady, Y. Y., Sekine, M., Hariguchi, N., Yamamoto, T., Kouchi, H., and Shinmyo, A., 1995, Tobacco mitotic cyclins: Cloning, characterization, gene expression and functional assay, *Plant J.* **8**:949–957.

Shaul, O., Van Montagu, M., and Inzé, D., 1996a, Regulation of cell division in *Arabidopsis*, *Crit. Rev. Plant Sci.* **15**:97–112.

Shaul, O., Mironov, V., Burssens, S., Van Montagu, M., and Inzé, D., 1996b, Two *Arabidopsis* cyclin promoters mediate distinctive transcriptional oscillation in synchronized tobacco BY-2 cells, *Proc. Natl. Acad. Sci. USA* **93**:4868–4872.

Showalter, A. M., 1993, Structure and function of plant cell wall proteins, *Plant Cell* **5**:9–23.

Silver, D. L., Pinaev, A., Chen, R., and de Bruijn, F. J., 1996, Posttranscriptional regulation of the *Sesbania rostrata* early nodulin gene *SrEnod2* by cytokinin, *Plant Physiol.* **112**:559–567.

Soupène, E., Foussard, M., Boistard, P., Truchet, G., and Batut, J., 1995, Oxygen as a key developmental regulator of *Rhizobium meliloti* $N_2$-fixation gene expression with the alfalfa root nodule, *Proc. Natl. Acad. Sci. USA* **92**:3759–3763.

Spaink, H. P., Sheeley, D. M., van Brussel, A. A. N., Glushka, J., York, W. S., Tak, T., Geiger, O., Kennedy, E. P., Reinhold, V. N., and Lugtenberg, B. J. J., 1991, A novel highly unsaturated fatty acid moiety of lipo-oligosaccharide signals determines host specificity of *Rhizobium*, *Nature* **354**:125–130.

Spencer-Barreto, M. M., and Duhoux, E., 1994, Root-to-shoot primordium conversion on *Sesbania rostrata* Brem. stem explants, *J. Exp. Bot.* **45**:1851–1857.

Spencer-Barreto, M. M., Cottignies, A., Chamel, A., and Duhoux, E., 1995, Caractéristiques physiologiques des ébauches racinaires adventives de la tige de *Sesbania rostrata* Brem (*Leguminosae*), *Acta bot. Gallica* **142**:763–771.

Stacey, G., Luka, S., Sanjuan, J., Banfalvi, Z., Nieuwkoop, A. J., Chun, J. Y., Forsberg, L. S., and Carlson, R., 1994, *nodZ*, a unique host-specific nodulation gene, is involved in the fucosylation of the lipooligosaccharide nodulation signal of *Bradyrhizobium japonicum*, *J. Bacteriol.* **176**:620–633.

Staehelin, C., Müller, J., Mellor, R. B., Wiemken, A., and Boller, T., 1992, Chitinase and peroxidase in effective ($fix^+$) and ineffective ($fix^-$) soybean nodules, *Planta* **187**:295–300.

Staehelin, C., Schultze, M., Kondorosi, E., Mellor, R. B., Boller, T., and Kondorosi, A., 1994a, Structural modifications in *Rhizobium meliloti* Nod factors influence their stability against hydrolysis by root chitinases, *Plant J.* **5**:319–330.

Staehelin, C., Granado, J., Müller, J., Wiemken, A., Mellor, R. B., Felix, G., Regenass, M., Broughton, W. J., and Boller, T., 1994b, Perception of *Rhizobium* nodulation factors by tomato cells and inactivation by root chitinases, *Proc. Natl. Acad. Sci. USA* **91**:2196–2200.

Staehelin, C., Schultze, M., Kondorosi, É., and Kondorosi, Á., 1995, Lipo-chitooligosaccharide nodulation signals from *Rhizobium meliloti* induce their rapid degradation by the host plant alfalfa, *Plant Physiol.* **108:**1607–1614.

Stam, H., van Verseveld, H. W., de Vries, W., and Stouthamer, A. H., 1984, Hydrogen oxidation and efficiency of nitrogen fixation in succinate-limited chemostat cultures of *Rhizobium* ORS571, *Arch. Microbiol.* **139:**53–60.

Stigter, J., Schneider, M., and de Bruijn, F. J., 1993, *Azorhizobium caulinodans* nitrogen fixation (*nif*/*fix*) gene regulation: Mutagenesis of the *nifA* −24/−12 promoter element, characterization of a *ntrA*(*rpoN*) gene, and derivation of a model, *Mol Plant–Microbe Interact.* **6:**238–252.

Streeter, J., 1988, Inhibition of legume nodule formation and $N_2$ fixation by nitrate, *CRC Crit. Rev. Plant Sci.* **7:**1–23.

Strittmatter, G., Chia, T.-F., Trinh, T. H., Katagiri, F., Kuhlemeier, C., and Chua, N.-H., 1989, Characterization of nodule-specific cDNA clones from *Sesbania rostrata* and expression of the corresponding genes during the initial stages of stem nodules and root nodules formation, *Mol. Plant–Microbe Interact.* **2:**122–127.

Szabados, L., Ratet, P., Grunenberg, B., and de Bruijn, F. J., 1990, Functional analysis of the *Sesbania rostrata* leghemoglobin *glb*3 gene 5′-upstream region in transgenic *Lotus corniculatus* and *Nicotiana tabacum* plants, *Plant Cell* **2:**973–986.

Szczglowski, K., Szabados, L., Fujimoto, S. Y., Silver, D., and de Bruijn, F. J., 1994, Site-specific mutagenesis of the nodule-infected cell expression (NICE) element and the AT-rich element ATRE-BS2* of the *Sesbania rostrata* leghemoglobin *glb*3 promoter, *Plant Cell* **6:**317–332.

Taté, R., Patriarca, E. J., Riccio, A., Defez, R., and Iaccarino, M., 1994, Development of *Phaseolus vulgaris* root nodules, *Mol. Plant–Microbe Interact.* **7:**582–589.

Tomekpe, K., Dreyfus, B., and Holsters, M., 1996, Root nodulation of *Sesbania rostrata* suppresses stem nodulation by *Sinorhizobium teranga* but not *Azorhizobium caulinodans*, *Can. J. Microbiol.* **42:**187–190.

Truchet, G., Barker, D. G., Camut, S., de Billy, F., Vasse, J., and Huguet, T., 1989, Alfalfa nodulation in the absence of *Rhizobium*, *Mol. Gen. Genet.* **219:**65–68.

Tsien, H. C., Dreyfus, B. L., and Schmidt, E. L., 1983, Initial stages in the morphogenesis of nitrogen-fixing stem nodules of *Sesbania rostrata*, *J. Bacteriol.* **156:**888–897.

Ueda, T., Suga, Y., Yahiro, N., and Matsuguchi, T., 1995, Phylogeny of Sym plasmids of rhizobia by PCR-based sequencing of a *nodC* segment, *J. Bacteriol.* **177:**468–472.

Urban, J. E., Davis, L. C., and Brown, S. J., 1986, *Rhizobium trifolii* 0403 is capable of growth in absence of combined nitrogen, *Appl. Environ. Microbiol.* **52:**1060–1067.

van Brussel, A. A. N., Recourt, K., Pees, E., Spaink, H. P., Tak, T., Wijffelman, C. A., Kijne, J. W., and Lugtenberg, B. J. J., 1990, A biovar-specific signal of *Rhizobium leguminosarum* bv. viciae induces increased nodulation gene-inducing activity in root exudate of *Vicia sativa* subsp. *nigra*, *J. Bacteriol.* **172:**5394–5401.

Van den Eede, G., Dreyfus, B., Goethals, K., Van Montagu, M., and Holsters, M., 1987, Identification and cloning of nodulation genes from the stem-nodulating bacterium ORS571, *Mol. Gen. Genet.* **206:**291–299.

Van den Eede, G., Deblaere, R., Goethals, K., Van Montagu, M., and Holsters, M., 1992, Broad host range and promoter selection vectors for bacteria that interact with plants, *Mol. Plant–Microbe Interact.* **5:**228–234.

van de Sande, K., Pawlowski, K., Czaja, I., Wieneke, U., Schell, J., Schmidt, J., Walden, R., Matvienko, M., Wellink, J., van Kammen, A., Franssen, H., and Bisseling, T., 1996, Modification of phytohormone response by a peptide encoded by *ENOD40* of legumes and a nonlegume, *Science* **273:**370–373.

Van de Wiel, C., Scheres, B., Franssen, H., van Lierop, M.-J., van Lammeren, A., van Kammen, A., and Bisseling, T., 1990a, The early nodulin transcript ENOD2 is located in the nodule parenchyma (inner cortex) of pea and soybean root nodules, *EMBO J.* **9:**1–7.

Van de Wiel, C., Norris, J. H., Bochenek, B., Dickstein, R., Bisseling, T., and Hirsch, A. M.,

1990b, Nodulin gene expression and ENOD2 localization in effective, nitrogen-fixing and ineffective, bacteria-free nodules of alfalfa, *Plant Cell* **2:**1009–1017.

van Kammen, A., 1984, Suggested nomenclature for plant genes involved in nodulation and symbiosis, *Plant Mol. Biol. Rep.* **2:**43–45.

Vasse, J., de Billy, F., Camut, S., and Truchet, G., 1990, Correlation between ultrastructural differentiation of bacteroids and nitrogen fixation in alfalfa nodules, *J. Bacteriol.* **172:**4295–4306.

Vasse, J., de Billy, F., and Truchet, G., 1993, Abortion of infection during the *Rhizobium meliloti*–alfalfa symbiotic interaction is accompanied by a hypersensitive reaction, *Plant J.* **4:**555–566.

Vázquez, M., Santana, O., and Quinto, C., 1993, The NodI and NodJ proteins from *Rhizobium* and *Bradyrhizobium* strains are similar to capsular polysaccharide secretion proteins from Gram-negative bacteria, *Mol. Microbiol.* **8:**369–377.

Vijn, I., Yang, W.-C., Pallisgård, N., Jensen, E. Ø., van Kammen, A., and Bisseling, T., 1995, *VsENOD5, VsENOD12* and *VsENOD40* expressing during *Rhizobium*-induced nodule formation on *Vicia sativa* roots, *Plant Mol. Biol.* **28:**1111–1119.

Waelkens, F., Voets, T., Vlassak, K., Vanderleyden, J., and van Rhijn, P., 1995, The *nodS* gene of *Rhizobium tropici* strain CIAT899 is necessary for nodulation on *Phaseolus vulgaris* and on *Leucaena leucocephala, Mol. Plant–Microbe Interact.* **8:**147–154.

Welle, R., Schröder, G., Schiltz, E., Grisebach, H., and Schröder, J., 1991, Induced plant responses to pathogen attack. Analysis and heterologous expression of the key enzyme in the biosynthesis of phytoalexins in soybean (*Glycine max* L. Merr. cv. Harosoy 63), *Eur. J. Biochem.* **196:**423–430.

Welters, P., Metz, B., Felix, G., Palme, K., Szczglowski, K., and de Bruijn, F. J., 1993, Interaction of a rhizobial DNA-binding protein with the promoter region of a plant leghemoglobin gene, *Plant Physiol.* **102:**1095–1107.

Wiegel, J., 1992, The genus *Xanthobacter*, in *The Prokaryotes. A Handbook on the Biology of Bacteria: Ecophysiology, Isolation, Identification, Applications*, Vol. III, 2nd ed. (A. Balows, H. G. Trüper, M. Dworkin, W. Harder, and K.-H. Schleifer, eds.), pp. 2365–2383, Springer-Verlag, Berlin, Germany.

Wilson, R. C., Long, F., Maruoka, E. M., and Cooper, J. B., 1994, A new proline-rich early nodulin from *Medicago truncatula* is highly expressed in nodule meristematic cells, *Plant Cell* **6:**1265–1275.

Yang, W.-C., Canter Cremers, H. C. J., Hogendijk, P., Katinakis, P., Wijffelman, C. A., Franssen, H., van Kammen, A., and Bisseling, T., 1992, *In-situ* localization of chalcone synthase mRNA in pea root nodule development, *Plant J.* **2:**143–151.

Yang, W.-C., Katinakis, P., Hendriks, P., Smolders, A., de Vries, F., Spee, J., van Kammen, A., Bisseling, T., and Franssen, H., 1993, Characterization of *GmENOD40*, a gene showing novel patterns of cell-specific expression during soybean nodule development, *Plant J.* **3:**573–585.

Yang, W.-C., de Blank, C., Meskiene, I., Hirt, H., Bakker, J., van Kammen, A., Franssen, H., and Bisseling, T., 1994, *Rhizobium* Nod factors reactivate the cell cycle during infection and nodule primordium formation, but the cycle is only completed in primordium formation, *Plant Cell* **6:**1415–1426.

Yao, P. Y., and Vincent, J. M., 1976, Factors responsible for the curling and branching of clover root hairs by *Rhizobium, Plant Soil* **45:**1–16.

Zaat, S. A. J., van Brussel, A. A. N., Tak, T., Pees, E., and Lugtenberg, B. J. J., 1987, Flavonoids induce *Rhizobium leguminosarum* to produce *nodDABC* gene-related factors that cause thick, short roots and root hair responses on common vetch, *J. Bacteriol.* **169:**3388–3391.

Zhang, Y., Burris, R. H., Ludden, P. W., and Roberts, G. P., 1993, Posttranslational regulation of nitrogenase activity by anaerobiosis and ammonium in *Azospirillum brasilense, J. Bacteriol.* **175:**6781–6788.

*Chapter 5*

# Interaction between *Frankia* and Actinorhizal Plants

Changhui Guan, Katharina Pawlowski, and Ton Bisseling

## 1. NITROGEN-FIXING PLANT–BACTERIAL SYMBIOSES

A significant part of the nitrogen in living organisms is derived from atmospheric dinitrogen which gets incorporated into organic compounds by biological or chemical nitrogen fixation. Biological nitrogen fixation is an energy-consuming process performed by the enzyme nitrogenase, which is irreversibly denatured by oxygen. Nitrogenase is formed only by prokaryotes, which in some cases fix nitrogen in symbiosis with higher plants. Two groups of bacteria can enter symbioses with higher plants which lead to the formation of special organs, the root nodules, where the bacteria fix nitrogen while being hosted inside plant cells. The product of nitrogen fixation, ammonium, is exported to the plant, while the plant in turn provides its symbiont with energy sources. *Azorhizobium, Bradyrhizobium,* and *Rhizobium* enter symbioses with leguminous plants (with one exception, *Parasponia* from the *Ulmaceae* family; Trinick, 1979), and actinomycetes of the genus *Frankia* induce nodules on the roots of a diverse group of dicotyledonous plants, mostly woody shrubs, which are collectively referred to

**Changhui Guan, Katharina Pawlowski, and Ton Bisseling** Department of Molecular Biology, Agricultural University, 6703 HA Wageningen, The Netherlands. *Present address for C. G.:* Max-Planck-Institut für Züchtungsforschung, D-50829 Köln, Germany. *Present address for K. P.:* Albrecht-von-Haller-Institut für Pflanzenwissenschaften, Universität Göttingen, 37073 Göttingen, Germany.
*Subcellular Biochemistry, Volume 29: Plant–Microbe Interactions,* edited by Biswas and Das. Plenum Press, New York, 1998.

as actinorhizal plants (Benson and Silvester, 1993). While rhizobia are unicellular, *Frankia* usually grows in hyphal form, but can also form sporangia and, under certain conditions, nitrogen-fixing vesicles (Benson and Silvester, 1993). Vesicles are formed at the ends of hyphae or on short side branches either in the free-living state under nitrogen starvation and aerobic conditions, or in symbiosis. Within the vesicles, nitrogenase is protected from oxygen (Parsons *et al.*, 1987) by the vesicle envelope, which consists of multiple layers of hopanoids, bacterial steroid lipids (Berry *et al.*, 1993). In contrast to *Frankia*, rhizobia can use $N_2$ as nitrogen source only in symbiosis (with one exception, *Azorhizobium caulinodans* ORS571; reviewed by de Bruijn, 1989).

In the next two sections different aspects of actinorhizal symbioses will be discussed. Because of the phylogenetic relationship between legumes and actinorhizal plants (Soltis *et al.*, 1995), and since more data are available on the first system, data on legume nodules will be included in the discussion of the interaction between *Frankia* and actinorhizal plants for comparative purposes.

## 2. STRUCTURE OF NITROGEN-FIXING ROOT NODULES

The presence of nodular structures (actinorhizae) on the roots of actinorhizal plants was first reported for alder in 1829 (Meyen, 1829). Since then, actinorhizae have been found among over 25 genera of dicotyledonous plants belonging to eight different families (Benson and Silvester, 1993). Actinorhizal root nodules represent coralloid structures composed of several modified lateral roots (lobes). Actinorhizal nodule lobes contain a central vascular system, and infected cells in the expanded cortex. Since the nodule lobe meristem develops only in one direction, root caps are lacking. In some symbioses, the nodule lobe meristem can redifferentiate to form an agraviotropically growing root, the so-called "nodule root," which grows out of the tip of the lobe. Nodule roots show no root hairs and no well-developed root cap, and do not contain cells which can become infected by *Frankia* (Torrey and Callaham, 1978). Thus, at a certain stage the meristem can be reprogrammed. Nodule roots contain large air spaces and serve as oxygen diffusion pathways to aerate the nodule tissue. In other actinorhizal symbioses, nodules are aerated via lenticels (reviewed by Silvester *et al.*, 1990). Due to the presence of the apical meristem, actinorhizal nodule lobes have an indeterminate growth pattern, and their tissues are of graded age (Fig. 1). Starting from the apical meristem, four different zones have been defined in a mature nodule lobe (Ribeiro *et al.*, 1995). The meristematic zone (zone 1) consists of small dividing cells. The infection zone (zone 2) contains enlarging cortical cells. Some of them are infected by *Frankia* and in turn become larger than uninfected cells while becoming filled with branching hyphae (Lalonde, 1980; Schwintzer *et al.*, 1982; reviewed by Berry and Sunell, 1990). The fixation zone (zone 3) is characterized by formation of vesicles by the

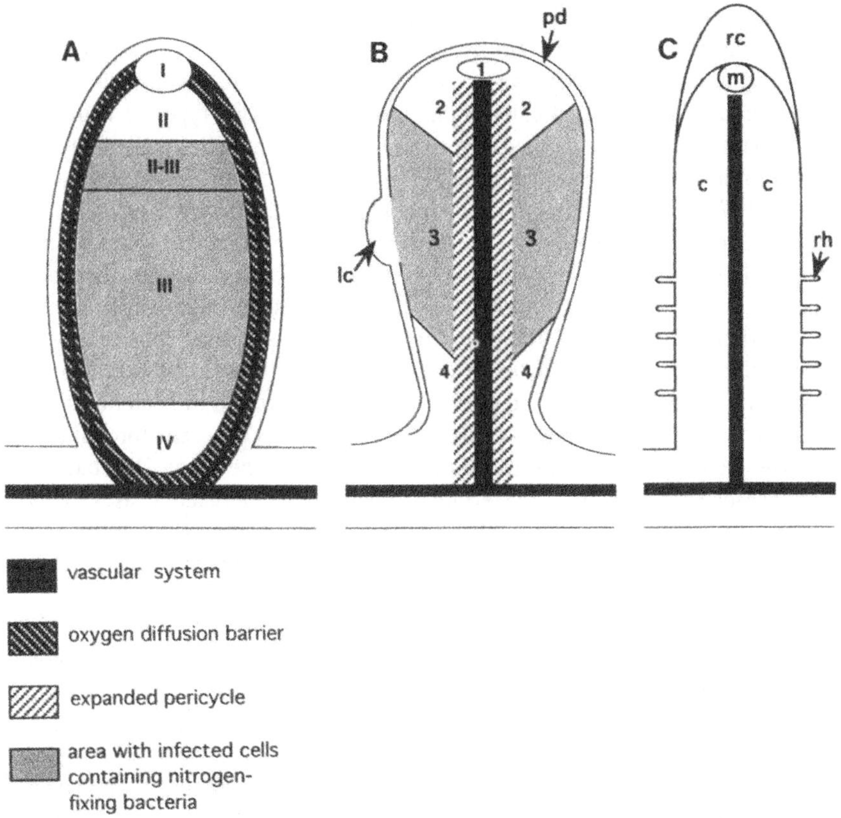


**FIGURE 1.** Comparison of an indeterminate legume nodule and an actinorhizal nodule lobe. (A) Legume nodules contain a peripheral vascular system and infected cells in the central tissue. An oxygen diffusion barrier is present in the nodule parenchyma surrounding the nodule vascular system. Since the nodule parenchyma is interrupted by the meristem (zone I), an oxygen gradient is formed that extends from the distal to the proximal end of the nodule. In the prefixation zone, the oxygen concentration is low enough to allow the induction of nitrogen-fixation (*nif*) gene expression (Soupène *et al.*, 1995). II, Prefixation zone; II–III, interzone; III, fixation zone; IV, senescence zone. (B) Actinorhizal nodule lobes contain a central vascular system and are surrounded by a periderm (pd), which in case of *Alnus* is interrupted by lenticels (lc). Due to the activity of the distal nodule meristem (1), a developmental gradient is formed in the cortex. In the infection zone (2), cortical cells enlarge and some of them are infected and gradually filled by *Frankia* hyphae. In the fixation zone (3), *Frankia* vesicles have formed and nitrogen-fixation (*nif*) genes are expressed, resulting in nitrogen fixation. In the senescence zone (4), plant and bacterial material is degraded in the infected cells. In *Alnus* nodule lobes, the vascular system contains a multilayered pericycle. (C) For comparison, a lateral root is shown. c, Cortex; rc, root cap; rh, root hair; m, meristem.

bacteria where bacterial nitrogen-fixation (*nif*) genes are transcribed and nitrogenase is formed (Huss-Danell and Bergman, 1990; Pawlowski *et al.*, 1995). In the senescence zone (zone 4), the infected cells degenerate, and degradation of the host cytoplasm and *Frankia* hyphae occurs (Schwintzer *et al.*, 1982).

Leguminous root nodules represent stemlike structures with peripheral vascular bundles and infected cells in the central tissue. Two types of leguminous root nodules have been defined, the determinate and the indeterminate nodules. Indeterminate nodules are characterized by a persistent distal meristem (Newcomb, 1976) like actinorhizal nodule lobes. Due to its activity, a developmental gradient from the meristem to the proximal senescence zone is present in the central tissue of the nodule, by which the central tissue can be divided into specific zones (Vasse *et al.*, 1990) (Fig. 1). The meristem (zone I) is followed by the prefixation zone (zone II), where infection of the cortical cells takes place. Inside the infected cells, rhizobia differentiate into their symbiotic form, the bacteroids. In the so-called interzone II–III, the expression of bacterial nitrogen fixation (*nif*) genes is induced (Yang *et al.*, 1991) and proceeds throughout the nitrogen fixation zone (zone III). In the senescence zone (zone IV), bacteroids are degraded by the plant. In determinate nodules, the meristem ceases to divide at an early stage of development (Newcomb *et al.*, 1979). As a result, all the cells of the central tissue are at a similar stage of development at any given time point.

The interaction between host plants and microsymbionts starts with signal exchange and recognition of the symbiotic partners. From this step until nodule formation and functioning, many genes from both partners participate in the process. Some of them are nodule-specific plant genes, called "nodulin genes" (van Kammen, 1984) in *Rhizobium*–legume symbioses and "actinorhizin genes" in actinorhizal symbioses (Tremblay *et al.*, 1986). Nodulins are grouped into early and late nodulins based on the time course of their expressions (Nap and Bisseling, 1990). Previously the term "actinorhizin gene" was used for actinorhizal nodule-specific genes (Tremblay *et al.*, 1986) and applied to plant as well as bacterial genes. In contrast, in legume–*Rhizobium* symbioses, the term "nodulin gene" only referred to nodule-specific plant genes (van Kammen, 1984). However, several plant genes that were assumed to be nodule-specific were later found to be expressed at low levels in nonsymbiotic tissues by more sensitive detection methods. These genes are still being called nodulins. We will use the term "actinorhizal nodulin gene" to refer to plant genes expressed at elevated levels in actinorhizal nodules compared to lateral roots. The terms "early actinorhizal nodulin gene" and "late actinorhizal nodulin gene" will be used as well. The most convenient distinction between early and late actinorhizal nodulin genes is obtained by determining gene expression patterns along the developmental gradient of nodule cortical cells using *in situ* hybridizations. Actinorhizal nodulin genes expressed at highest levels in the meristematic zone

(zone 1) and/or in the infection zone (zone 2), i.e., before the onset of nitrogen fixation, will be termed early actinorhizal nodulin genes. Those expressed at highest levels in the fixation zone (zone 3) and/or senescence zone (zone 4) will be termed late actinorhizal nodulin genes.

Molecular biological studies on the plant aspects of legume and actinorhizal symbioses date back to the 1980s when oxygen transport proteins, i.e., leghemoglobins in legume nodules and hemoglobins in actinorhizal nodules, were investigated (Appleby, 1984; Roberts *et al.*, 1985; Fleming *et al.*, 1987). In general, extensive studies have been performed on legume nodules (for reviews, see Appleby, 1984; Sanchez *et al.*, 1991; Mylona *et al.*, 1995), while studies on the molecular aspects of actinorhizal nodule formation were hampered by the lack of methods to isolate high-quality DNA and RNA, since most actinorhizal plants represent trees or woody shrubs. The development of techniques for isolating high-molecular-weight DNA and purifying RNA from actinorhizal plants paved the way to initiate studies on actinorhizal nodule development.

## 3. NODULE FORMATION AND FUNCTIONING

The induction of nodule formation and the establishment of the symbiosis have been studied in greater detail in *Rhizobium*/legume than in actinorhizal symbioses. Therefore, the data available on rhizobial symbioses will be summarized first for each process.

### 3.1. Nodule Induction

In most cases, infection of legume plants by rhizobia starts by root hair deformation (Bauer, 1981). While rhizobia attach to the root hair, the root hair deforms as a response to rhizobial Nod factors (Lerouge *et al.*, 1990; Heidstra *et al.*, 1994), and the bacteria get trapped in the root hair curl. By hydrolyzing the local root hair cell wall (Callaham and Torrey, 1981), the bacteria enter the epidermal cell in an infection thread, which proceeds to grow through the cortical cells. Simultaneously, cortical cells are mitotically activated, giving rise to the nodule primordium (Dart, 1977). Infection threads grow toward the primordium and then the bacteria are released into the cytoplasm of the host cells, surrounded by a plant-derived peribacteroid membrane (PBM). The nodule primordium thereupon develops into a mature nodule, while the bacteria differentiate into their endosymbiotic form, the bacteroids (Newcomb, 1976). Bacteroids can form nitrogenase and fix nitrogen. In a few cases, however, infection follows a so-called crack-entry mechanism during which bacteria enter the plant roots through gaps in the epidermis (reviewed by Sprent, 1989). While in some plants infection threads are formed after infection by crack entry (James *et al.*, 1992), in others, infection proceeds intercellularly until bacteria are taken up into cells of the nodule primordium surrounded by a PBM (Chandler, 1978; Chandler *et al.*, 1982).

*Frankia* can infect its host plants in two different ways, root hair infection or intercellular penetration. Which of these ways is realized depends on the host plant species (Miller and Baker, 1985; Racette and Torrey, 1989). During root hair infection, *Frankia* hyphae are trapped in the deformed root hair. The trapped *Frankia* hypha branches and initiates the digestion of the primary cell wall of the root hair, and the host plant in turn deposits cell wall material around the site of digestion (Callaham *et al.*, 1979; Berry *et al.*, 1983). Continued digestion by the bacteria and triggered cell wall building by the host plant leads to the formation of a tubular ingrowth which is termed encapsulation (Lalonde and Knowles, 1975; Newcomb and Wood, 1987). This structure is analogous to the infection thread wall in *Rhizobium*–legume symbioses. Mitotic activity in the root cortex and cell expansion result in the formation of the prenodule, whose cells are infected by encapsulated *Frankia* hyphae. Subsequently, one or several nodule lobe primordia are initiated in the root pericycle (reviewed by Berry and Sunell, 1990). *Frankia* hyphae grow from the root cortex into the developing nodule lobe primordium through cortical cells. Upon entering the primordium, once again, *Frankia* hyphae branch and invade numerous cells of the young nodule lobe cortex, being surrounded by encapsulating material derived from the host cell wall as well as by the invaginating host plasma membrane. During intercellular infection, *Frankia* hyphae penetrate the root epidermis between epidermal cells and colonize the root cortex intercellularly (Miller and Baker, 1985; Racette and Torrey, 1989; Liu and Berry, 1991). No prenodule is formed, but the colonization of the cortex directly leads to the induction of a nodule primordium in the root pericycle. *Frankia* hyphae infect cells of the young nodule lobe cortex. When infected cells have been filled with hyphae, *Frankia* begins to differentiate vesicles where nitrogenase is formed and nitrogen fixation starts (Huss-Danell and Bergman, 1990).

### 3.2. Bacterial Nod Factors

For legumes, nodule induction both via root hair infection and crack entry depend on *Rhizobium*-derived lipo-chitooligosaccharides, the so-called Nod factors (Lerouge *et al.*, 1990; Goethals *et al.*, 1990; reviewed by Dénarié and Cullimore, 1993), which play a key role in the induction of the initial stages of nodulation. In general, Nod factors consist of an oligomer of three to five β-1,4-linked *N*-acetylglucosamines bearing a fatty acid on the nonreducing terminal sugar residue. They can have various other substitutions on both terminal sugar residues. Nod factors have been shown to be involved in the host specificity of rhizobia, and the structures and specificities of several such molecules have been determined (reviewed by Van Rhijn and Vanderleyden, 1995). The bacterial genes involved in Nod factor synthesis are termed *nod* (nodulation) genes. These genes are not expressed in free-living bacteria, with the exception of *nodD*, which is expressed constitutively. Its product NodD presumably is able to bind to specific flavonoids secreted by the roots of the host plant (Goethals *et al.*, 1992). Upon activation by flavonoids, NodD acts as a transcriptional activator of the other *nod* genes which encode enzymes involved in the synthesis of Nod factors (reviewed by Van Rhijn and Vanderleyden, 1995).

Due to the striking similarities between the initial infection steps of *Rhizobium*–legume symbioses and actinorhizal symbioses, similar signaling mechanisms between the microsymbionts and the hosts are expected. However, investigations to detect any DNA sequences homologous to the *nod* genes in the *Frankia* genome have failed (Simonet *et al.*, 1990). This may be due to different codon usage in both types of bacteria, as might be enforced by their different G + C content (An *et al.*, 1993), or could mean that the *Frankia nod* genes are not conserved at the DNA level, while enzymes with analogous functions may yet exist in *Frankia*. However, no functional complementation of *Rhizobium nod* genes by *Frankia* DNA has been observed (Reddy *et al.*, 1992). In recent studies, it was found that independent of the presence of its host plant, *Frankia* strain ArI3 secretes a factor eliciting root hair deformation on *Alnus* into the growth medium (Van Ghelue *et al.*, 1997). Culture filtrate of *Frankia* strain ArI3 only led to deformation of the root hairs of *Alnus glutinosa* and not of non-host-plant species tested, indicating a certain degree of specificity. However, the chemical nature of the root hair deformation factor that is being purified from this *Frankia* strain is not known (Van Ghelue *et al.*, 1997).

## 3.3. Symbiotic Nitrogen Fixation

### 3.3.1. Symbiosis-Specific Differentiation

In *Rhizobium*–legume interactions, the intracellular bacteria differentiate into their symbiosis-specific form, the bacteroids, after endocytotic uptake from the infection thread into nodule primordium cells (see above). Both plant (Häser *et al.*, 1992) and bacterial (Glazebrook *et al.*, 1993) mutants have been identified that are specifically defective in bacteroid differentiation, and bacterial mutants specifically defective in the release of bacteria from the infection thread are known (De Maagd *et al.*, 1989). Thus, the processes of bacteria internalization and bacteroid differentiation may require some signal exchange between the symbiotic partners. Bacterial *nod* genes are expressed in the distal part of the prefixation zone II (Schlaman *et al.*, 1991), indicating that Nod factors may play a role in signal exchange within the nodule. However, since bacterial release and bacteroid development can be impaired in bacterial strains with functional *nod* genes, other bacterial and/or plant signals must also play a role in these steps of development.

In all actinorhizal genera except *Casuarina* and *Allocasuarina,* the onset of nitrogen fixation is associated with the appearance of *Frankia* vesicles in the infected host cortical cells (reviewed by Berry and Sunell, 1990). Within the nodule tissue, the extent of vesicle formation and the shape, septation, and spatial organization of the vesicles are controlled by the host plant. Symbiotic vesicles can be spherical, club-shaped, elliptical, or filamentous (reviewed by Baker and Mullin, 1992). Also in (*Allo-*)*Casuarina* nodules, some differentiation of *Frankia* is likely to occur, since bacterial *nif* gene expression is only induced after the infected cells have been completely filled with *Frankia* hyphae (Gherbi *et al.,* 1994). Thus, like *Rhizobium,* symbiotic *Frankia* shows some host-plant-specific differentiation, indicating that, as in legumes, plant factors are involved in this process.

### 3.3.2. The Interface between the Symbionts

Root nodules provide a proper environment to allow efficient nitrogen fixation by the microsymbiont and regulated nutrient exchange between both symbionts. The nutrient exchange is regulated by plant-derived membranes that in all cases surround the "intracellular" microsymbiont. In legume nodules, bacteroids are enclosed in peribacteroid membranes (PBMs). They form the interface between the symbiotic partners across which signals and metabolites are exchanged and prevent a defense response by the plant against the "intracellular" bacteria (reviewed by Verma, 1992).

The PBM of legume nodules has a phospholipid (Perotto *et al.*, 1995) and protein composition that is different from that of the plasma membrane (Verma, 1992) and that (presumably) endows it with specialized functions. It has been suggested that the PBM has adopted some properties of the tonoplast membranes (Verma, 1992), even that the symbiosome (the PBM with enclosed bacteroids) has properties of the lytic compartment that is continuously being neutralized by ammonia exported by the bacteroids (Kannenberg and Brewin, 1989). This is supported by the fact that soybean nodulin-26, an integral PBM protein, is targeted to the vacuolar membrane when expressed in tobacco (Zhang and Roberts, 1995). Because the PBM constitutes the interface between bacteroids and host plants, it plays an important role in controlling the exchange of metabolites. These metabolites include ammonium, the product of nitrogen fixation, which is exported by the bacteroids (Schubert, 1986), and carbon sources (Werner, 1992) and probably also assimilated ammonium (De Bruijn *et al.*, 1989), which are supplied by the host to the bacteroids.

In actinorhizal nodules, the interface between the microsymbiont and the host plant is the invaginated plasma membrane formed around the growing hyphae. Within this membrane, *Frankia* is still surrounded by the pectinaceous cell wall-like encapsulating material. Since the invaginated plasma membrane has to control the exchange of nutrient between *Frankia* and host plant, it can be assumed to differ from the rest of the plasma membrane in protein content. However, nodule-specific plant proteins that are incorporated into the invaginated plasma membrane have not yet been identified.

## 4. MOLECULAR BIOLOGY AND PHYSIOLOGY OF ACTINORHIZAL NODULES

The *Alnus–Frankia* symbiosis was taken by a few groups (Mullin *et al.*, 1993; Pawlowski *et al.*, 1994b) as a system to study molecular aspects on actinorhiza development and functioning because of the availability of basic knowledge about *Alnus* nodule physiology (reviewed by Huss-Danell, 1990) and the development of *Alnus* cell and tissue culture systems (Séguin and Lalonde, 1990). Furthermore, several *Frankia* strains that form actinorhizae on *Alnus* have been cultured (Benson and Silvester, 1993). Other groups have concentrated on the *Casuarina–Frankia* symbiosis (Jacobsen-Lyon *et al.*, 1995; Gherbi *et al.*, 1997).

Before suitable RNA isolation methods were available, actinorhizal gene expression was studied at the protein level. For example, polypeptide patterns of uninfected roots and nodules of *Alnus* were compared (Séguin and Lalonde, 1993). By this approach, five nodule-specific polypeptides could be detected, but it was unclear whether they were encoded by the host plant or by the microsymbiont. Improvements in poly(A) RNA isolation from *Alnus* made it possible to construct nodule cDNA libraries (Mullin *et al.*, 1993; Pawlowski *et al.*, 1994a). By differential screening of nodule cDNA libraries, several cDNA clones representing actinorhizal nodulin genes have been isolated. The plant origin of these cDNAs has been confirmed by DNA gel blot hybridization, also showing that all of them are encoded by small gene families. The *A. glutinosa* actinorhizal nodulin genes identified this way are listed in Table I.

### 4.1. Early Actinorhizal Nodulins from *A. glutinosa*

Two early actinorhizal nodulin gene families, namely *ag12* and *ag164/agNT84*, have been identified in *A. glutinosa* nodules. The *ag12* gene is highly expressed in nodules, and at very low levels in shoot tips (Ribeiro *et al.*, 1995). In nodules, the expression is localized mainly in infected cells of zone 2, where bacterial *nif* genes are not yet expressed, and at much lower levels in infected cells of zone 3. The *ag12* gene encodes a subtilisin-like serine protease. The presence of a putative signal peptide at the N-terminus indicates that Ag12 is

**Table I**
**Actinorhizal Nodulin cDNA Clones Isolated from *A. glutinosa***

| Clone name | Where expressed | Expression site in the nodule | Characteristics | References |
|---|---|---|---|---|
| pAg3-1 | Every organ | Infected cells of zones 2 and 3, pericycle | Sucrose synthase | Van Ghelue *et al.*, 1996 |
| pAg6 | Every organ | Infected cells of zones 2 and 3, pericycle | Homolog of yeast *thi4*; involved in thiazole synthesis | Ribeiro *et al.*, 1996 |
| pAg11 | Every organ | Infected cells of zone 3; pericycle | Glutamine synthetase | Guan *et al.*, 1996 |
| pAg12 | Nodules and shoot tip | Infected cells of zone 2, weaker in zone 3 | Subtilisin-like protease | Ribeiro *et al.*, 1995 |
| pAg13 | Every organ | Infected cells at the beginning of zone 4; pericycle | Proline- and glutamate-rich protein | Guan *et al.*, 1997 |
| pAg118 | Nodules and roots | Infected cells of zone 3 | Acetylornithine transaminase | Guan *et al.*, 1996 |
| pAg135 | Nodules, flowers, and shoot tips | N.D. | Unknown | Pawlowski *et al.*, 1994a |
| pAg144 | Every organ | Infected cells of zones 2 and 3; pericycle | Enolase | Van Ghelue *et al.*, 1996 |
| pAg164, pAgNt84 | Nodules | Infected cells of zone 2 | Glycine- and histidine-rich protein | Pawlowski *et al.*, 1997 |
| pAg172 | Nodules | N.D. | Transmembrane transporter | Pawlowski *et al.*, 1994b |
| pAgNg203-118 | Nodules and developing fruits | Infected cells at the beginning of zone 4 | Cysteine protease (AgNOD-CP1) | Goetting-Minesky and Mullin, 1994; Pawlowski, Goetting-Minesky, and Mullin, unpublished results |

probably active in the extracellular space. Based on its expression patterns, Ag12 has been postulated to be involved in processing proteins that are part of the cell wall-like matrix material surrounding the bacteria, or in processing other proteins with undetermined function. Since the corresponding gene is expressed in early stages of nodule development, this protease is most likely not involved in processes related to senescence.

The *ag164* and *agNT84* genes have also been found to be expressed in the infected cells of zone 2, but unlike *ag12*, no expression could be detected in infected cells of zone 3 (Pawlowski *et al.*, 1997). Like Ag12, the Ag164 and AgNt84 polypeptides contain N-terminal signal peptides, indicating that they are taken up into the endomembrane system. The putative mature proteins are small (6.12 and 7.54 kD, respectively) and rich in alternating glycine and histidine residues. Such sequence motifs have been found to be involved in metal ion binding by yeast proteins (Conklin *et al.*, 1992), and AgNt84 could be shown to bind metal ions (Pawlowski *et al.*, 1997). Putative extracellular metal-binding nodulins have also been identified in legume nodules. For example, in pea, nodulins ENOD3/14 and NOD6 have been proposed to bind metal ions (Kardailsky *et al.*, 1993). The members of the nodulin-A family in soybean (Fortin *et al.*, 1985; Sandal *et al.*, 1987) and *Phaseolus vulgaris* (Npv30; Campos *et al.*, 1995) also have been suggested to bind metal ions. Nitrogenase function requires molybdenum and iron, while both rhizobia and *Frankia* need cobalt for survival (Hewitt and Bond, 1966; Dixon and Wheeler, 1983). Thus, the host has to provide metal ions for the microsymbiont. Ag164/AgNT84 and the legume nodulins mentioned above might be involved in the transport of such ions. Alternatively, they might represent structural proteins.

## 4.2. Nodule Metabolism

Several late actinorhizal nodulin genes were found to encode enzymes involved in nodule metabolism. This is not surprising, since the host plant has to create a microenvironment in the nodule to support bacterial nitrogen fixation. The host supplies carbohydrates for plant and bacterial respiration in nitrogen-fixing nodules (Akkermans *et al.*, 1981; Vance and Heichel, 1991) and carbon backbones for the assimilation of ammonium, the product of bacterial nitrogen fixation (Schubert, 1986). Physiological studies have provided insight into carbon and nitrogen metabolic pathways in actinorhizal nodules (reviewed by Huss-Danell, 1990). Several cDNAs encoding enzymes involved in C and N metabolism have been cloned from legumes (for review, see Sanchez *et al.*, 1991) and actinorhizal plants (Table I). In legumes, developmental and metabolic control are combined in the regulation of genes involved in nodule nitrogen metabolism (reviewed by Hirel *et al.*, 1993). Some genes are expressed at high levels in nodules compared to other plant organs, but with the exception of soybean

(Roche *et al.*, 1993), no strictly nodule-specific GS genes exist. This aspect has not been studied in *A. glutinosa*, because all *A. glutinosa* cDNAs isolated are encoded by small multigene families and analysis has not distinguished among different family members.

### 4.2.1. C-Metabolism-Related Gene Expression

The main transport form of carbohydrate from the shoot to the nodule in plant is sucrose (Pate, 1962), which can be degraded by sucrose synthase (Morell and Copeland, 1985). This enzyme is present at high levels in both legume and actinorhizal nodules (Thummler and Verma, 1987; Van Ghelue *et al.*, 1996). In the case of *Rhizobium*-induced nodules, bacteroids express a dicarboxylic acid uptake system. Isolated bacteroids take up dicarboxylic acids, and mutants in this uptake are symbiotically ineffective (Ronson *et al.*, 1987; reviewed by Werner, 1992), indicating that dicarboxylic acids are likely to be the carbon source supplied by the plant to the intracellular bacteria. It has been suggested that nodulin nodulin-26 transports dicarboxylic acids to the bacteroids (Ouyang *et al.*, 1991). However, its low substrate specificity *in vitro* indicates that it is more likely to form a pore responsible for the uptake of ions or small metabolites in general (Weaver *et al.*, 1994). The form of carbon that is supplied to symbiotic *Frankia* in actinorhizal nodules is not clear, yet a malate/aspartate shuttle between host and microsymbiont has been suggested (Akkermans *et al.*, 1981). Thus, as in legume nodules, *Frankia* may be fed $C_4$-dicarboxylic acids in actinorhizal nodules. In this case, in actinorhizal nodules sucrose would have to be metabolized to phosphoenolpyruvate (PEP) to provide the carbon source for the microsymbiont. Subsequent carbon dioxide fixation by PEP carboxylase would lead to the formation of dicarboxylic acids which can be reduced to malate. This hypothesis is supported by the fact that high activities of PEP carboxylase have been found in actinorhizal (McClure *et al.*, 1983) as well as legume nodules (King *et al.*, 1986).

Nodule-enhanced sucrose synthase and enolase cDNAs have been cloned in *A. glutinosa*. The expression patterns of both genes in the nodule have been analyzed by *in situ* hybridization (Van Ghelue *et al.*, 1996). The transcripts of both genes are found in the infected cells of zone 2 and zone 3, with highest expression levels at the beginning of zone 3, and in the pericycle of the nodule vascular bundle. The gene *ag6(thi1)* is another gene related to C metabolism found to be expressed at high levels in *Alnus* nodules. The encoded protein (Ag6) is involved in the biosynthesis of thiazole (Ribeiro *et al.*, 1996), which is precursor of thiamine, a cofactor in glycolysis (Praekelt and Meacock, 1992; Praekelt *et al.*, 1994). The *ag6(thi1)* transcripts can be detected in all organs of *Alnus* tested. In agreement with the function of thiazole in glycolysis, the expression pattern of *ag6(thi1)* in nodules is the same as that of sucrose synthase and enolase

(Ribeiro *et al.*, 1996). The most striking difference in sucrose synthase expression between legume and actinorhizal nodules is that sucrose synthase expression could not be detected in the starch-containing uninfected cortical cells of *A. glutinosa* nodules. Whether this is a problem of detection limits, or whether in these cells invertase is metabolizing sucrose for starch biosynthesis, as has been suggested for the uninfected cells of soybean nodules (Morell and Copeland, 1984), remains to be examined.

### 4.2.2. N-Metabolism-Related Gene Expression

The form in which nitrogen is exported from nodules to the root xylem depends on the plant. In general, temperate legumes, which form indeterminate nodules, export amides, whereas tropical legumes, which form determinate nodules, export ureides. Actinorhizal plants mostly export amides, with the exception of *Alnus* sp. and *Casuarina equisetifolia*, which are citrulline exporters (Schubert, 1986; Sellstedt and Atkins, 1991). In all cases, ammonium is exported by the microsymbiont as the first product of nitrogen fixation and is assimilated in the cytoplasm of nodule cells via the glutamine synthetase (GS)/glutamate synthase (GOGAT) pathway (Schubert, 1986). Subsequently, glutamate is metabolized into nitrogen transport forms.

In legume nodules, the products of several late nodulin genes play a role in this metabolism. In ureide-producing determinate legume nodules, the assimilation of ammonium by GS and the biosynthesis of ureides are spatially separated to some extent. Ammonium assimilation seems to take place in both infected and uninfected cells of the central tissue of soybean nodules, since GS is expressed in both cell types (Miao *et al.*, 1991). The biosynthesis of ureides, however, seems to take place in the uninfected cells, since uricase (nodulin-35), one of the key enzymes in purine oxidation, which catalyzes the oxidation of uric acid to allantoin, and allantoinase, which catalyzes the next step in purine oxidation, have been localized in these cells only (Hanks *et al.*, 1981; Nguyen *et al.*, 1985). The uninfected cells of determinate nodules also seem to be involved in the transport of fixed nitrogen to the vascular system (Selker, 1988). In indeterminate legume nodules, however, uninfected cells of the central tissue seem only to have storage functions (Bauchrowitz *et al.*, 1995).

From *A. glutinosa*, two cDNA clones were isolated that have been found to encode products related to nitrogen metabolism. pAg11 encodes a glutamine synthetase (GS) and pAg118 encodes an acetylornithine transaminase (AOTA) which is involved in the biosynthesis of citrulline (Guan *et al.*, 1996). GS transcripts could be found in the infected cells of zone 3, in agreement with the localization of GS in this cell type by immunocytochemical studies (Hirel *et al.*, 1982), and in the pericycle of the vascular system. AOTA mRNA was found at highest levels in the infected cells of zone 3. These data suggest that in *A.*

*glutinosa* nodules, citrulline synthesis mainly takes place in the nitrogen-fixing infected cortical cells. None of actinorhizal nodulin genes has been found to be expressed in the uninfected cortical cells of *A. glutinosa* nodules.

### 4.2.3. Gene Expression in the Pericycle of the Nodule Vascular Bundle

One of the characteristics of *Alnus* nodule lobes is a multilayered pericycle consisting of small cells with dense cytoplasm that surrounds the central vascular system (Burgess and Peterson, 1987a,b). The pericycle forms an interface between xylem/phloem and nodule cortical cells and has been suggested to have a function in nutrient transport (Burgess and Peterson, 1987a,b). So far, the actinorhizal nodulin genes which are expressed in the pericycle of the nodule vascular bundle in *Alnus* are *ag11* (GS), *ag3-1* (sucrose synthase), *ag144* (enolase), *ag6l(thil),* and *ag13* (Table I). The expression of GS, sucrose synthase, enolase, and *ag6(thil),* in combination with the high amount of mitochondria found in pericycle cells, indicates high glycolytic activities, as would be required for transport processes (Burgess and Peterson, 1987a,b). GS has been implicated to function in reassimilation of ammonium for nitrogen transport in other systems (Edwards *et al.*, 1990). Ag13 is also present in senescent infected cells (see below), so its function in the pericycle is unclear.

### 4.2.4. Oxygen Control

The enzyme nitrogenase is highly oxygen-sensitive (reviewed by Burris, 1991). On the other hand, the large amount of energy required for nitrogen fixation has to be generated by oxidative processes; thus, there is a high demand for oxygen in nodules. Different strategies are used in different symbiotic interactions to cope with this paradox.

In legume nodules, a low oxygen tension in the central part of the nodule is achieved by a combination of a high metabolic activity of the microsymbiont and an oxygen diffusion barrier in the periphery of the nodule, that is, in the nodule parenchyma (Witty *et al.*, 1986) (Fig. 1A). Oxygen is supposed to diffuse via the intercellular spaces. The nodule parenchyma contains very few and small intercellular spaces; and it has been suggested that the expression of nodulin genes encoding putative extracellular proteins such as ENOD2 in this tissue might contribute to the formation of the oxygen barrier (van de Wiel *et al.*, 1990). In the infected cells of the central part of the nodule, high amounts of the oxygen carrier protein leghemoglobin facilitate oxygen diffusion to the sites of respiration. In this way, the microsymbiont is provided with sufficient oxygen to generate energy within a low overall oxygen concentration (Appleby, 1984). In contrast to rhizobia, *Frankia* can provide its own system of oxygen protection for nitrogenase by forming vesicles (Benson and Silvester, 1993). However, vesicle formation during symbiosis does not take place in all actinorhizal symbioses and does not always seem to provide full oxygen protection of nitrogenase (Tjepkema, 1983; Kleemann *et al.*, 1994). In *Casuarina* symbioses, where no vesicles are formed, an oxygen diffusion barrier is established around infected cells by lignification of the cell walls (Berg and McDowell, 1987; Zeng *et al.*, 1989). In addition, high amounts of the oxygen transport protein hemoglobin, the equivalent of leghemoglobin, are present in the infected cells of *Casuarina* nodules (Fleming *et al.*, 1987; Tjepkema and Asa, 1987; Jacobsen-Lyon *et al.*, 1995). Also in *Myrica* nodules, an oxygen diffusion barrier is present (Zeng and Tjepkema, 1994) and high amounts of hemoglobin have been found, although *Frankia* does form vesicles in this symbiosis (Tjepkema and Asa, 1987).

The oxygen diffusion barriers around the infected cells of *Casuarina* and *Myrica* differ from that found in legume nodules in one important aspect. They allow optimal nitrogen fixation under atmospheric oxygen concentration, but in other oxygen conditions, nitrogen fixation by *Casuarina* and *Myrica* nodules is suboptimal (reviewed by Silvester *et al.*, 1990). However, in nodules of *Coriaria*, which do not contain hemoglobin, a variable oxygen diffusion barrier reminiscent of legume nodules seems to exist (Silvester and Harris, 1989). In summary, in different actinorhizal symbioses, diverse ways have been taken to solve the oxygen dilemma.

### 4.2.5. The Senescence Zone: Degradation of Endosymbiont and Plant Cytoplasm

Senescence-related processes have been analyzed in several plant organs, but in general are far from understood. During legume nodule senescence, hemoglobin and bacteroids are degraded by infected cell-specific proteases (Pladys *et al.*, 1991), but it is not yet known how bacteroids become accessible to proteases. In fact, a protease inhibitor has been found specifically in the peribacteroid space of senescent infected cells (Manen *et al.*, 1991). In some mutant legume nodules, premature senescence of bacteroids occurs (Egli *et al.*, 1991; Pladys and Vance, 1993), indicating that the timing of endosymbiont degradation is plant-controlled.

Two *A. glutinosa* actinorhizal nodulin genes have been found to be expressed in infected cortical cells of the senescence zone 4, where nitrogen fixation has been switched off and host cytoplasm as well as *Frankia* bacteria are subjected to degradation. The *ag13* gene encodes a glutamate- and proline-rich polypeptide which contains a putative signal peptide, suggesting an extracellular location (Guan *et al.*, 1997). A homolog of *ag13* is expressed during fruit development in kiwifruit (Ledger and Gardner, 1994). Since senescence and fruit development show strong similarities (Vercher and Carbonell, 1991; Smart, 1994), Ag13 is probably involved in plant cell senescence rather than in endosymbiont degradation. There is evidence for an increase of calcium in the cell wall during senescence (Iwahori *et al.*, 1994), and Ag13 shows some sequence homology with calcium-binding proteins. In either case, a function in senescence does not explain the expression of *ag13* in the pericycle of the nodule vascular bundle (Guan *et al.*, 1997). Another actinorhizal nodulin gene that might be involved in senescence encodes a cysteine protease, AgNOD-CP1 (Goetting-Minesky and Mullin, 1994), and is expressed in infected cells of zone 4 (P. Goetting-Minesky, K. Pawlowski, and B. C. Mullin, unpublished observations). Cysteine protease activities have been found correlated with senescence of plant organs in several systems (Pladys and Vance, 1993; Hensel *et al.*, 1993; Valpuesta *et al.*, 1995; Drake *et al.*, 1996). The occurrence of proteolysis during senescence is supposed to play a role in recovery of nitrogenous compounds from the senescing tissues (Pfeiffer *et al.*, 1983a,b; Pladys *et al.*, 1991; Pladys and Vance, 1993). Altogether, our knowledge on nodule senescence is scanty, and more data are needed to understand the mechanism by which the degradation of the microsymbiont is induced.

## 5. PLANT MODEL SYSTEMS FACILITATE STUDIES ON ACTINORHIZAL NODULES

As mentioned before, molecular and genetic analyses of actinorhizal systems are difficult, because most actinorhizal plants are woody plant species with long generation times and high contents of polyphenols. Thus, whenever possible, model systems have to be used to analyze the function of actinorhizal nodulins.

*Arabidopsis thaliana* has turned out to be a model plant system useful also for the study of actinorhizal symbioses. The accumulated information from the random cDNA sequencing project on *Arabidopsis* (Höfte *et al.*, 1993; Newman *et al.*, 1994; Cooke *et al.*, 1996) has made it possible to obtain *Arabidopsis* homologs of the *A. glutinosa* cDNAs. Several cDNAs isolated from *A. glutinosa* nodules have turned out to have homologs among the random cDNA sequences from *A. thaliana* (Table II). Based on the comparison of the *A. glutinosa* and the *Arabidopsis* sequences, conserved amino acid sequences could be identified. This would allow the cloning of homologs from any other system by polymerase chain reaction using degenerate primers. The development and morphology of *Arabidopsis* has been well characterized, which makes it a convenient system to study gene expression. *Arabidopsis* also represents a good system for transformation for gene function studies. Actinorhizal nodulin cDNA homologs of *Arabidopsis* can be used for overexpression or antisense expression studies, in order to analyze the function of the corresponding genes. Furthermore, availability of quick mapping procedures and an integrated genetic map for *Arabidopsis* (Lister and Dean, 1993; Hauge *et al.*, 1993) can lead to the identification of the

**Table II**
**Actinorhizal Nodulin Homologs in *Arabidopsis***

| *A. glutinosa* gene | *Arabidopsis* homolog | References |
|---|---|---|
| *ag3-1* | *atsus1* | Martin *et al.*, 1993; Van Ghelue *et al.*, 1996 |
| *ag6(thi1)* | *ara6/tz* | Ribeiro *et al.*, 1996; Machado *et al.*, 1996 |
| *ag11* | *atgsr1* | Peterman and Goodman, 1991; Guan *et al.*, 1996 |
| *ag12* | *ara12* | Ribeiro *et al.*, 1995 |
| *ag13* | *ara21371* | Guan *et al.*, 1997 |
| *ag118* | T43861[a], Z34670[b] | Guan *et al.*, 1996 |
| *ag135* | *ms2* | Pawlowski *et al.*, 1994a; Aarts *et al.*, 1993 |
| *ag144* | *atenge* | Van der Straeten *et al.*, 1991; Van Ghelue *et al.*, 1996b |
| *ag172* | *atchl-1* | Tsay *et al.*, 1993; Pawlowski *et al.*, 1994b |
| *agNg203-118* | *sag12* | Goetting-Minesky and Mullin, 1994; Gan and Amasino, 1995 |

[a]Accession number in GenBank DNA sequence databases (National Center of Biotechnology Information, National Library of Medicine, and National Institutes of Health, Bethesda, Maryland).
[b]Accession number in EMBL DNA sequence databases (European Molecular Biology Laboratory, European Bioinformatics Institute, Cambridge, U.K.).

function of the *Arabidopsis* clones. When the map of the *Arabidopsis* genome has become more saturated with mutant loci in the near future, it will be easy to assign mutant phenotypes to cDNAs. Thus, the comparative analysis of homologous cDNAs from the nonactinorhizal plant *Arabidopsis* can contribute to the understanding of the function of actinorhiza-specific cDNAs.

If actinorhizal nodulin cDNAs show homology with yeast genes, functional analysis of the yeast homolog is possible (Oliver, 1996). In the optimal case, direct complementation of yeast mutants can be used to assign a function to an actinorhizal nodulin. RFLP mapping in *Arabidopsis* and yeast complementation have been applied successfully for the characterization of *ag6(thi1)* (Ribeiro *et al.*, 1996; see above).

## 6. CONCLUDING REMARKS

Actinorhizal symbioses represent a more versatile system than symbioses between rhizobia and legumes, as can be deduced from the variability of the arrangement of infected cells and of nodule aeration/oxygen protection mechanisms (reviewed by Silvester *et al.*, 1990). This is probably due to the fact that the host plants nodulated by *Frankia* comprise a diverse group of plant species from eight different families (Benson and Silvester, 1993). Due to this variability, the knowledge obtained from the studies on actinorhizal symbioses would be very valuable in understanding the diverse ways by which plants can create a proper environment for nitrogen-fixing symbionts.

## 7. REFERENCES


Aarts, M. G. M., Dirkse, W. B., Stiekema, W. J., and Pereira, A., 1993, Transportation tagging of a male sterility gene in *Arabidopsis, Nature* **363:**715–717.

Akkermans, A. D. L., Huss-Danell, K., and Roelofsen, W., 1981, Enzymes of the tricarboxylic acid cycle and the malate-aspartate shuttle in the $N_2$-fixing endophyte of *Alnus glutinosa, Physiol. Plant.* **53:**289–294.

An, C. S., Wills, J. W., Riggsby, W. S., and Mullin, B. C., 1993, Deoxyribonucleic acid base composition of 12 *Frankia* isolates, *Can. J. Bot.* **61:**2859–2862.

Appleby, C. A., 1984, Leghemoglobin and *Rhizobium* respiration, *Annu. Rev. Plant Physiol.* **35:**443–478.

Baker, D. D., and Mullin, B. C., 1992, Actinorhizal symbiosis, in *Biological Nitrogen Fixation* (G. Stacey, R. H. Burris, and H. J. Evans, eds.), pp. 259–292, Chapman and Hall, New York.

Bauchrowitz, M. A., Barker, D. G., and Truchet, G., 1995, Lectin genes are expressed throughout root nodule development and during nitrogen-fixation in the *Rhizobium–Medicago* symbiosis, *Plant J.* **9:**31–43.

Bauer, W. D., 1981, Infection of legumes by rhizobia, *Annu. Rev. Plant Physiol.* **32:**407–449.

Benson, D. R., and Silvester, W. B., 1993, Biology of *Frankia* strains, actinomycete symbionts of actinorhizal plants, *Microbiol. Rev.* **57:**293–319.

Berg, R. H., and McDowell, L., 1987, Cytochemistry of the wall of infected cells in *Casuarina actinorhizae, Can. J. Bot.* **66:**2038–2047.

Berry, A. M., and Sunell, L. A., 1990, The infection process and nodule development, in *The Biology of Frankia and Actinorhizal Plants* (C. R. Schwintzer and J. D. Tjepkema, eds.), pp. 61–81, Academic Press, San Diego.

Berry, A. M., Torrey, J. G., and McCully, M. E., 1983, The fine structure of the root hair wall and surface mucilage in the actinorhizal host *Alnus rubra*, in *Plant Molecular Biology* (R. Goldberg, ed.), pp. 319–327, Liss, New York.

Berry, A. M., Harriot, O. T., Moreau, R. A., Osman, S. F., Benson, D. R., and Jones, A. D., 1993, Hopanoid lipids compose the *Frankia* vesicle envelope, presumptive barrier of oxygen diffusion to nitrogenase, *Proc. Natl. Acad. Sci. USA* **90:**6091–6094.

Burgess, D., and Peterson, R. L., 1987a, Development of *Alnus japonica* root nodules after inoculation with *Frankia* strain HFPArI3, *Can. J. Bot.* **65:**1647–1657.

Burgess, D., and Peterson, R. L., 1987b, Effect of nutrient conditions on root nodule development in *Alnus japonica, Can. J. Bot.* **65:**1658–1670.

Burris, R. H., 1991, Nitrogenases, *J. Biol. Chem.* **266:**9339–9342.

Callaham, D., and Torrey, J. G., 1981, The structural basis of infection of root hairs of *Trifolium repens* by *Rhizobium, Can. J. Bot.* **59:**1647–1664.

Callaham, D., Newcomb, W., Torrey, J. G., and Peterson, R. L., 1979, Root hair infection in actinomycete-induced root nodule initiation in *Casuarina, Myrica,* and *Comptonia, Bot. Gaz. (Suppl.)* **140:**S1–S9.

Campos, F., Carsolio, C., Kuin, H., Bisseling, T., Rocha-Sosa, M., and Sánchez, F., 1995, Characterization and gene expression of nodulin Npv30 from common bean, *Plant Physiol.* **109:**363–370.

Chandler, M. R., 1978, Some observations on infection of *Arachis hypogaea* L. by *Rhizobium, J. Exp. Bot.* **29:**749–755.

Chandler, M. R., Date, R. A., and Roughley, R. J., 1982, Infection and root nodule development in *Stylosanthes* species by *Rhizobium, J. Exp. Bot.* **33:**47–57.

Conklin, D. S., McMaster, J. A., Culbertson, M. R., and Kung, C., 1992, *COT1*, a gene involved in cobalt accumulation in *Saccharomyces cerevisiae, Mol. Cell Biol.* **12:**3678–3688.

Cooke, R., Raybak, M., Laudié, M., Grellet, F., Delseny, M., Morris, P.-C., Guerrier, D., Giraudat, J., Quigley, F., Clabault, G., Li, Y.-F., Mache, R., Krivitzky, M., Jean-Jacques Gy, I., Kreis, M., Lecharny, A., Parmentier, Y., Marbach, J., Fleck, J., Clément, B., Philipps, G., Hervé, C., Bardet, C. Tremousaygue, D., Lescure, B., Lacomme, C., Roby, D., Jourjon, M.-F., Chabrier, P., Charpenteau, J.-L., Desprez, T., Amselem, J., Chiapello, H., and Höfte, H., 1996, Further progress towards a catalogue of all *Arabisopsis* genes: Analysis of a set of 5000 non-redundant ESTs, *Plant J.* **9**:101–124.

Dart, P. J., 1977, Infection and development of leguminous nodules, in *A Treatise on Dinitrogen Fixation*, Vol. 3 (R. W. F. Hardy and W. S. Silver, eds.), pp. 367–472, Wiley, New York.

De Bruijn, F. J., 1989, The unusual symbiosis between the diayotrophic stem-nodulating bacterium *Azorhizobium caulinodans* ORS571 and its host, the tropical legume *Sesbania rostrata:* Plant and bacterial aspects, in *Plant–Microbe Interactions. Molecular Genetics and Perspective*, Vol. 3 (T. Kosuge and E. W. Nester, eds.), pp. 457–504, McGraw-Hill, New York.

De Bruijn, F. J., Rossbach, S., Schneider, M., Messmer, S., Szeto, W. W., Ausubel, F. M., and Schell, J., 1989, *Rhizobium meliloti* 1021 has three differentially regulated loci involved in glutamine biosynthesis, none of which is essential for symbiotic nitrogen fixation, *J. Bacteriol.* **171**:1673–1682.

De Maagd, R. A., Rao, A. S., Mulders, I. H. M., Goosen-De Roo, L., van Loosdrecht, M. C. M., Wijfelman, C. A., and Lugtenberg, B. J. J., 1989, Isolation and characterization of mutants of *Rhizobium leguminosarum* bv. *viciae* 248 with altered lipopolysaccharides: Possible role of surface charge or hydrophobicity in bacterial release from the infection thread, *J. Baceriol.* **171**:1143–1150.

Dénarié, J., and Cullimore, J., 1993, Lipo-oligosaccharide nodulation factors: A new class of signalling molecules mediating recognition and morphogenesis, *Cell* **74**:951–954.

Dixon, R. O. D., and Wheeler, C. T., 1983, Biochemical, physiological and environmental aspects of symbiotic nitrogen fixation, in *Biological Nitrogen Fixation in Forest Ecosystems: Foundations and Application* (J. C. Gordon and C. T. Wheeler, eds.), pp. 107–171, Nijhoff/Junk, The Hague.

Drake, R., John, I., Farrell, A., Cooper, W., Schuch, W., and Grierson, D., 1996, Isolation and analysis of cDNAs encoding tomato cysteine proteases expressed during leaf senescence, *Plant Mol. Biol.* **30**:755–767.

Edwards, J. E., Walker, E. L., and Coruzzi, G. M., 1990, Cell-specific expression in transgenic plants reveals nonoverlapping roles for chloroplast and cytosolic glutamine synthetase, *Proc. Natl. Acad. Sci. USA* **87**:3249–3263.

Egli, M. A., Larson, R. J., Hruschka, W. R., and Vance, C. P., 1991, Synthesis of nodulins and nodule-enhanced polypeptides by plant gene-controlled ineffective alfalfa nodules, *J. Exp. Bot.* **42**:969–977.

Fleming, A. I., Wittenberg, J. B., Wittenberg, B. A., Dudman, W. F., and Appleby, C. A., 1987, The purification, characterization and ligand-binding kinetics of hemoglobins from root nodules of the non-leguminous *Casuarina glauca–Frankia* symbiosis, *Biochim. Bioiphys. Acta* **911**:209–220.

Fortin, M. G., Zechelowska, M., and Verma, D. P. S., 1985, Specific targeting of membrane nodulins to the bacteroid-enclosing compartment in soybean nodules, *EMBO J.* **4**:3041–3046.

Gan, S., and Amasino, R. M., 1995, Inhibition of leaf senescence by autoregulated production of cytokinin, *Science* **270**:1986–1988.

Gherbi, H., Duhoux, E., Franche, C., Pawlowski, K., Berry, A., and Bogusz, D., 1997, Cloning of a full-length symbiotic hemoglobin cDNA and *in situ* localization of the corresponding mRNA in *Casuarina glauca* root nodule, *Physiol. Plant.* **99**:608–616.

Glazebrook, J., Ichige, A., and Walker, G. C., 1993, A *Rhizobium meliloti* homolog of the *Escherichia coli* peptide-antibiotic transport protein SbmA is essential for bacteroid development, *Genes Dev.* **7**:1485–1497.

Goethals, K., Van den Eede, G., Van Montagu, M., and Holsters, M., 1990, Identification and characterization of a functional *nodD* gene in *Azorhizobium caulinodans* ORS571, *J. Bacteriol.* **172:**2658–2666.

Goethals, K., Van Montagu, M., and Holsters, M., 1992, Conserved motifs in a divergent *nod* box of *Azorhizobium caulinodans* ORS571 reveals a common structure in promoters regulated by LysR-type proteins, *Proc. Natl. Acad. Sci. USA* **89:**1646–1650.

Goetting-Minesky, M. P., and Mullin, B. C., 1994, Differential gene expression in an actinorhizal symbiosis: Evidence for a nodule-specific cysteine proteinase, *Proc. Natl. Acad. Sci. USA* **91:**9891–9895.

Guan, C., Ribeiro, A., Akkermans, A. D. L., Jing, Y., van Kammen, A., Bisseling, T., and Pawlowski, K., 1996, Nitrogen metabolism in actinorhizal nodules of *Alnus glutinosa:* Expression of glutamine synthetase and acetylornithine transaminase, *Plant Mol. Biol.* **32:**1177–1184.

Guan, C., Akkermans, A. D. L., Bisseling, T., and Pawlowski, K., 1997, *agl3* is expressed in *Alnus glutinosa* nodules in infected cells during endosymbiont degradation and in the nodule pericycle, *Physiol. Plant.* **99:**601–607.

Hanks, J. F., Tolbert, N. E., and Schubert, K. R., 1981, Localization of enzymes of ureide biosynthesis in peroxisomes and microsomes of nodules, *Plant Physiol.* **68:**65–69.

Häser, A., Robinson, D. L., Duc, G., and Vance, C. P., 1992, A mutation in *Vicia faba* results in ineffective nodules with impaired bacteroid differentiation and reduced synthesis of late nodulins, *J. Exp. Bot.* **43:**1397–1407.

Hauge, B. M., Hanley, S. M., Cartinhour, S., Cherry, J. M., Goodman, H. M., Koornneef, M., Stam, P., Chang, C., Kempin, S., Medrano, L., and Meyrowitz, E. M., 1993, An integrated genetic/RFLP map of the *Arabidopsis thaliana* genome, *Plant J.* **3:**745–754.

Heidstra, R., Geurts, R., Franssen, H., Spaink, H. P., van Kammen, A., and Bisseling, T., 1994, Root hair deformation activity of nodulation factors and their fate on *Vicia sativa*, *Plant Physiol.* **105:**787–797.

Hensel, L. L., Grbic, V., Baumgarten, D. A., and Bleecker, A. B., 1993, Developmental and age-related processes that influence the longevity and senescence of photosynthetic tissues in *Arabidopsis*, *Plant Cell* **5:**553–564.

Hewitt, E. J., and Bond, G., 1966, The cobalt requirement of non-legume root nodule plants, *J. Exp. Bot.* **52:**480–491.

Hirel, B., Perrot-Rechenmann, C., Maudinas, B., and Gadal, P., 1982, Glutamine synthetase in alder (*Alnus glutinosa*) root nodules. Purification, properties and cytoimmunochemical localization, *Physiol. Plant.* **55:**197–203.

Hirel, B., Miao, G. H., and Verma, D. P. S., 1993, Metabolic and developmental control of glutamine synthetase genes in legumes and non-legume plants, in *Control of Plant Gene Expression* (D. P. S. Verma, ed.), pp. 443–458, CRC Press, Boca Raton, Florida.

Höfte, H., Desprez, T., Amselem, J., Chiapello, H., Caboche, M., Moisan, A., Jourjon, M. F., Charpenteau, J. L., Berthomieu, P., Guerrier, D., Giraudat, J., Quigley, F., Thomas, F., Yu, D. Y., Mache, R., Raynal, M., Cooke, R., Grellet, F., Delseny, M., Partmentier, Y., Marcillac, G., Gigot, C., Fleck, J., Philipps, G., Axelos, M., Bardet, C., Tremousaygue, D., and Lescure, B., 1993, An inventory of 1152 expressed sequence tags obtained by partial sequencing of cDNAs from *Arabidopsis thaliana*, *Plant J.* **4:**1051–1061.

Huss-Danell, K., 1990, The physiology of actinorhizal nodules, in *The Biology of Frankia and Actinorhizal Plants* (C. R. Schwintzer and J. D. Tjepkema, eds.), pp. 129–156, Academic Press, San Diego, California.

Huss-Danell, K., and Bergman, B., 1990, Nitrogenase in *Frankia* from root nodules of *Alnus incana* (L.) Moench: Immunolocalization of the Fe- and MoFe-proteins during vesicle differentiation, *New Phytol.* **116:**443–455.

Iwahori, S., Kair, R., and Nishi, M., 1994, Subcellular localization of calcium ions within cells of the abscission layer of citrus leaf explants during abscission, *J. Jpn. Soc. Hort. Sci.* **63**:45–50.

Jacobsen-Lyon, K., Jensen, E. Ø., Jørgensen, J.-E., Marcker, K. A., Peacock, W. J., and Dennis, E. S., 1995, Symbiotic and nonsymbiotic hemoglobin genes of *Casuarina glauca*, *Plant Cell* **7**:213–223.

James, E. K., Sprent, J. I., Sutherland, J. M., McInroy, S. G., and Minchin, F. R., 1992, The structure of nitrogen fixing root nodules on the aquatic mimosoid legume *Neptunia plena*, *Ann. Bot.* **69**:173–180.

Kannenberg, E. L., and Brewin, N. J., 1989, Expression of a cell surface antigen from *Rhizobium leguminosarum* 3841 is regulated by oxygen and pH, *J. Bacteriol.* **171**:4543–4548.

Kardailsky, I., Yang, W.-C., Zalensky, A., van Kammen, A., and Bisseling, T., 1993, The pea late nodulin gene *PsNOD6* is homologous to the early nodulin genes *ENOD3/14* and is expressed after the leghemoglobin genes, *Plant Mol. Biol.* **23**:1029–1037.

King, B. J., Layzell, D. B., and Canvin, D. T., 1986, The role of dark carbon dioxide fixation in root nodules of soybean, *Plant Physiol.* **81**:200–205.

Kleemann, G., Alskog, G., Berry, A. M., and Huss-Danell, K., 1994, Lipid composition and nitrogenase activity of symbiotic *Frankia* (*Alnus incana*) in response to different oxygen concentrations, *Protoplasma* **183**:107–115.

Lalonde, M., 1980, Techniques and observations of the nitrogen fixing *Alnus* root nodule symbiosis, in *Recent Advances in Biological Nitrogen Fixation* (N. S. Subba Rao, ed.), pp. 421–434, Oxford and IBH, New Delhi.

Lalonde, M., and Knowles, R., 1975, Ultrastructure, composition and biogenesis of the encapsulation material surrounding the endophyte in *Alnus crispa* var. *mollis* root nodules, *Can. J. Bot.* **53**:1951–1971.

Ledger, S. E., and Gardner, R. C., 1994, Cloning and characterization of five cDNAs for genes differentially expressed during fruit development of kiwifruit (*Actinidia deliciosa* var. *deliciosa*), *Plant Mol. Biol.* **25**:877–886.

Lerouge, P., Roche, P., Faucher, C., Maillet, F., Truchet, G., Promé, J.-C., and Dénarié, J., 1990, Symbiotic host-specificity of *Rhizobium meliloti* is determined by a sulphated and acylated glucosamine oligosaccharide signal, *Nature* **344**:781–784.

Lister, C., and Dean, C., 1993, Recombinant inbred lines for mapping RFLP and phenotypic markers in *Arabidopsis thaliana*, *Plant J.* **4**:745–750.

Liu, Q., and Berry, A. M., 1991, The infection process and nodule initiation in the *Frankia–Ceanothus* root nodule symbiosis: A structural and histochemical study, *Protoplasma* **163**:82–92.

Machado, C. R., Costa de Oliveira, R. L., Boiteux, S., Praekelt, U. M., Meacock, P. A., and Menck, C. F. M., 1996, *Thi1*, a thiamine biosynthetic gene in *Arabidopsis thaliana*, complements bacterial defects in DNA repair, *Plant Mol. Biol.* **31**:585–593.

Manen, J. F., Simon, P., van Slooten, J.-C., Østerås, M., Frutiger, S., and Hughes, G. J., 1991, A nodulin specifically expressed in senescent nodules of winged bean is a protease inhibitor, *Plant Cell* **3**:259–270.

Martin, T., Frommer, W. B., Salanoubat, M., and Willmitzer, L., 1993, Expression of an *Arabidopsis* sucrose synthase gene indicates a role in metabolization of sucrose both during phloem loading and in sink organs, *Plant J.* **4**:367–377.

McClure, P. R., Coker, G. T., and Schubert, K. R., 1983, Carbon dioxide fixation in roots and nodules of *Alnus glutinosa*. I. Role of phosphoenolpyruvate carboxylase and carbamyl phosphate synthetase in dark $CO_2$ fixation, citrulline synthesis, and $N_2$ fixation, *Plant Physiol.* **71**:652–657.

Meyen, J., 1829, Über das Hervorwachsen parasitischer Gebilde aus den Wurzeln anderer Pflanzen, *Flora* **12:**49–64.

Miao, G.-H., Hirel, B., Marsolier, M. C., Ridge, R. W., and Verma, D. P. S., 1991, Ammonia-regulated expression of a soybean gene encoding cytosolic glutamine synthetase in transgenic *Lotus corniculatus*, *Plant Cell* **3:**11–22.

Miller, I. M., and Baker, D. D., 1985, The initiation, development and structure of root nodules in *Elaeagnus angustifolia* L. (Elaeagnaceae), *Protoplasma* **128:**107–119.

Morell, M., and Copeland, L., 1984, Enzymes of sucrose breakdown in soybean nodules. Alkaline invertase, *Plant Physiol.* **74:**1030–1034.

Morell, M., and Copeland, L., 1985, Sucrose synthase of soybean nodules, *Plant Physiol.* **78:**149–154.

Mullin, B. C., Goetting-Minesky, P., and Twigg, P., 1993, Differential gene expression in the development of actinorhizal root nodules, in: *New Horizons in Nitrogen Fixation* (R. Palacios, J. Mora, and W. E. Newton, eds.), pp. 309–314, Kluwer, Dordrecht, The Netherlands.

Mylona, P., Pawlowski, K., and Bisseling, T., 1995, Symbiotic nitrogen fixation, *Plant Cell* **7:**869–885.

Nap, J. P., and Bisseling, T., 1990, Developmental biology of a plant–prokaryote symbiosis: The legume root nodule, *Science* **250:**948–954.

Newcomb, W., 1976, A correlated light and electron microscopic study of symbiotic growth and differentiation in *Pisum sativum* rootnodules, *Can. J. Bot.* **54:**2163–2186.

Newcomb, W., and Wood, S. M., 1987, Morphogenesis and fine structure of *Frankia* (Actinomycetales): The microsymbiont of nitrogen-fixing actinorhizal root nodules, *Int. Rev. Cytol.* **109:**1–88.

Newcomb, W., Sippel, D., and Peterson, R. L., 1979, The early morphogenesis of *Glycine max* and *Pisum sativum* root nodules, *Can. J. Bot.* **57:**2603–2616.

Newman, T., de Bruijn, F. J., Green, P., Keegstra, K., Kende, H., McIntosh, L., Ohlrogge, J., Raikhel, N., Somerville, S., Thomashow, M., Retzel, E., and Somerville, C., 1994, Genes galore: A summary of methods for accessing results from large-scale partial sequencing of anonymous *Arabidopsis* cDNA clones, *Plant Physiol.* **106:**1241–1255.

Nguyen, T., Zelechowska, M., Forster, V., Bergmann, H., and Verma, D. P. S., 1985, Primary structure of the soybean nodulin-35 gene encoding uricase II localized in the peroxisomes of uninfected cells of nodules, *Proc. Natl. Acad. Sci. USA* **82:**5040–5044.

Oliver, S. G., 1996, From DNA sequence to biological function, *Nature* **379:**597–600.

Ouyang, L.-J., Whelan, J., Weaver, C. D., Roberts, D. M., and Day, D. A., 1991, Protein phosphorylation stimulates the rate of malate uptake across the peribacteroid membrane of soybean nodules, *FEBS Lett.* **293:**188–190.

Parsons, R., Silvester, W. B., Harris, S., Gruijters, W. T. M., and Bullivant, S., 1987, *Frankia* vesicles provide inducible and absolute oxygen protection for nitrogenase, *Plant Physiol.* **83:**728–731.

Pate, J. S., 1962, Root exudation studies on the exchange of $^{14}C$-labeled organic substances between the roots and shoot of the nodulated legumes, *Plant Soil* **17:**333–356.

Pawlowski, K., Guan, C., Ribeiro, A., van Kammen, A., Akkermans, A. D. L., and Bisseling, T., 1994a, Genes involved in *Alnus glutinosa* nodule development, in *Proceedings of the 1st European Nitrogen Fixation Conference* (G. B. Kiss and G. Endre, eds.), pp. 220–224. Officina Press, Szeged, Hungary.

Pawlowski, K., Ribeiro, A., Guan, C., van Kammen, A., Akkermans, A. D. L., and Bisseling, T., 1994b, Differential gene expression in root nodules of *Alnus glutinosa*, in *Nitrogen Fixation with Non-Legumes* (N. A. Hegazi, M. Fayez, and M. Monib, eds.), pp. 185–190, The American University in Cairo Press, Cairo.

Pawlowski, K., Akkermans, A. D. L., van Kammen, A., and Bisseling, T., 1995, Expression of *Frankia nif* genes in nodules of *Alnus glutinosa*, *Plant Soil* **170:**371–376.

Pawlowski, K., Twigg, P., Dobritsa, S., Guan, C., and Mullin, B. C., 1997, A nodule-specific gene family from *Alnus glutinosa* encodes glycine- and histidine-rich proteins expressed in the early stages of actinorhizal nodule development, *Mol. Plant–Microbe Interact.* **10:**656–664.

Perotto, S., Donovan, N., Drobak, B. K., and Brewin, N. J., 1995, Differential expression of a glycosyl inositol phospholipid antigen on the peribacteroid membrane during pea nodule development, *Mol. Plant–Microbe Interact.* **8:**560–568.

Peterman, T. K., and Goodman, H. M., 1991, The glutamine synthetase gene family of *Arabidopsis thaliana:* Light-regulation and differential expression in leaves, roots and seeds, *Mol. Gen. Genet.* **230:**145–154.

Pfeiffer, N. E., Malik, N. S. A., and Wagner, F. W., 1983a, Reversible dark-induced senescence of soybean root nodules, *Plant Physiol.* **71:**393–399.

Pfeiffer, N. E., Malik, N. S. A., and Wagner, F. W., 1983b, Proteolytic activity in soybean root nodules, *Plant Physiol.* **71:**797–802.

Pladys, D., and Vance, C. P., 1993, Proteolysis during development and senescence of effective and plant gene-controlled ineffective alfalfa nodules, *Plant Physiol.* **103:**379–384.

Pladys, D., Dimitrijevic, L., and Rigaud, J., 1991, Localization of a protease in protoplast preparations in infected cells of French bean nodules, *Plant Physiol.* **97:**1174–1180.

Praekelt, U. M., and Meacock, P. A., 1992, *MOL1*, a *Saccharomyces cerevisiae* gene that is highly expressed in early stationary phase during growth on molasses, *Yeast* **8:**699–710.

Praekelt, U. M., Byrne, K. L., and Meacock, P. A., 1994, Regulation of *THI4* (*MOL1*), a thiamine-biosynthetic gene of *Saccharomyces cerevisiae*, *Yeast* **10:**481–490.

Racette, S., and Torrey, J. G., 1989, Root nodule initiation in *Gymnostoma* (Casuarinaceae) and *Shepherdia* (Elaeagnaceae) induced by *Frankia* strain HFPGpI1, *Can. J. Bot.* **67:**2873–2879.

Reddy, A., Bochenek, B., and Hirsch, A. M., 1992, A new *Rhizobium meliloti* symbiotic mutant isolated after introducing *Frankia* DNA sequence into a *nodA*::Tn*5* strain, *Mol. Plant–Microbe Interact.* **5:**62–71.

Ribeiro, A., Akkermans, A. D. L., van Kammen, A., Bisseling, T., and Pawlowski, K., 1995, A nodule-specific gene encoding a subtilisin-like protease is expressed in early stages of actinorhizal nodule development, *Plant Cell* **7:**785–794.

Ribeiro, A., Praekelt, U., Akkermans, A. D. L., Meacock, P., van Kammen, A., Bisseling, T., and Pawlowski, K., 1996, Identification of *agthi1*, encoding an enzyme involved in biosynthesis of the thiamine precursor thiazole, in actinorhizal nodules of *Alnus glutinosa*, *Plant J.* **10:**361–368.

Roberts, M. P., Jafar, S., and Mullin, B. C., 1985, Leghemoglobin-like sequences in the DNA of four actinorhizal plants, *Plant Mol. Biol.* **5:**333–337.

Roche, D., Temple, S. J., and Sengupta-Gopalan, C., 1993, Two classes of differentially regulated glutamine synthetase genes are expressed in the soybean nodule: A nodule-specific class and a constitutively expressed class, *Plant Mol. Biol.* **22:**971–983.

Ronson, C. W., Astwood, P. M., Nixon, B. T., and Ausubel, F. M., 1987, Deduced products of C4-dicarboxylate transport regulatory genes of *Rhizobium leguminosarum* are homologous to nitrogen regulatory gene products, *Nucleic Acids Res.* **15:**7921–7935.

Sanchez, F., Padilla, J. E., Perez, H., and Lara, M., 1991, Control of nodulin genes in root-nodule development and metabolism, *Annu. Rev. Plant Physiol. Plant Mol. Biol.* **42:**507–528.

Sandal, N. N., Bojsen, K., and Marcker, K. A., 1987, A small family of nodule specific genes from soybean, *Nucleic Acids Res.* **15:**1507–1519.

Schlaman, H. R. W., Horvath, B., Vijgenboom, E., Okker, R. J. H., and Lugtenberg, B. J. J., 1991,

Evidence for a new negative regulation mechanism involved in the suppression of nodulation gene expression of *Rhizobium leguminosarum* bv. *viciae*, *J. Bacteriol.* **173:**4277–4287.

Schubert, K. R., 1986, Products of biological nitrogen fixation in higher plants: Synthesis, transport, and metabolism, *Annu. Rev. Plant Physiol.* **37:**539–574.

Schwintzer, C. R., Berry, A. M., and Disney, L. D., 1982, Seasonal patterns of root nodule growth, endophyte morphology, nitrogenase activity and shoot development in *Myrica gale*, *Can. J. Bot.* **60:**746–757.

Séguin, A., and Lalonde, M., 1990, Micropropagation, tissue culture, and genetic transformation of actinorhizal plants and *Betula*, in *The Biology of Frankia and Actinorhizal Plants* (C. R. Schwintzer and J. D. Tjepkema, eds.), pp. 215–238, Academic Press, San Diego, California.

Séguin, A., and Lalonde, M., 1993, Modification of polypeptide patterns during nodule development in the *Frankia–Alnus* symbiosis, *Symbiosis* **15:**135–149.

Selker, J. M. L., 1988, Three-dimensional organization of uninfected tissue in soybean root nodules and its relation to cell specialization in the central region, *Protoplasma* **147:**178–190.

Sellstedt, A., and Atkins, C. A., 1991, Composition of amino compounds transported in xylem of *Casuarina* sp., *J. Exp. Bot.* **42:**1493–1497.

Silvester, W. B., and Harris, S. L., 1989, Nodule structure and nitrogenase activity of *Coriaria arborea* in response to varying $pO_2$, *Plant Soil* **118:**97–109.

Silvester, W. B., Harris, S. L., and Tjepkema, J. D., 1990, Oxygen regulation and hemoglobin, in *The Biology of Frankia and Actinorhizal Plants* (C. R. Schwintzer and J. D. Tjepkema, eds.), pp. 157–176, Academic Press, San Diego, California.

Simonet, P., Normand, P., Hirsch, A. M., and Akkermans, A. D. L., 1990, The genetics of the *Frankia*–actinorhizal symbiosis, in *Molecular Biology of Symbiotic Nitrogen Fixation* (P. M. Gresshoff, ed.), pp. 77–109, CRC Press, Boca Raton, Florida.

Smart, C. M., 1994, Gene expression during leaf senescence, *New Phytol.* **126:**419–448.

Soltis, D. E., Soltis, P. S., Morgan, D. R., Swensen, S. M., Mullin, B. C., Dowd, J. M., and Martin, P. G., 1995, Chloroplast gene sequence data suggest a single origin of the predisposition for symbiotic nitrogen fixation in angiosperms, *Proc. Natl. Acad. Sci. USA* **92:**2647–2651.

Soupène, E., Foussard, M., Boistard, P., Truchet, G., and Batut, J., 1995, Oxygen as a key developmental regulator of *Rhizobium meliloti* $N_2$-fixation gene expression within the alfalfa root nodule, *Proc. Natl. Acad. Sci. USA* **92:**3759–3763.

Sprent, J. I., 1989, Which steps are essential for the formation of functional legume nodules? *New Phytol.* **111:**129–153.

Thummler, F., and Verma, D. P. S., 1987, Nodulin-100 of soybean is the subunit of sucrose synthase regulated by the availability of free heme in nodules, *J. Biol. Chem.* **262:**14730–14736.

Tjepkema, J. D., 1983, Oxygen concentration within the nitrogen-fixing root nodules of *Myrica gale* L., *Am. J. Bot.* **70:**59–63.

Tjepkema, J. D., and Asa, D. J., 1987, Total and CO-reactive heme content of actinorhizal nodules and the roots of some non-nodulated plants, *Plant Soil* **100:**225–236.

Torrey, J. G., and Callaham, D., 1978, Determinate development of nodule roots in actinomycete-induced root nodules of *Myrica gale*, *Can. J. Bot.* **56:**1357–1364.

Tremblay, F. M., Périnet, P., and Lalonde, M., 1986, Tissue culture of *Alnus* spp. with regard to symbioses, in *Biotechnology in Agriculture and Forestry* (Y. P. S. Bajaj, ed.), pp. 87–100, Springer-Verlag, Berllin.

Trinick, M. J., 1979, Structure of nitrogen-fixing nodules formed by *Rhizobium* on roots of *Parasponia andersonii*, *Appl. Environ. Microbiol.* **55:**2046–2055.

Tsay, Y. F., Schroeder, J. I., Feldmann, K. A., and Crawford, N. M., 1993, The herbicide sensitivity gene *CHL1* of *Arabidopsis* encodes a nitrate-inducible nitrate transporter, *Cell* **72:**705–713.

Valpuesta, V., Lange, N. E., Guerrero, C., and Reid, M. S., 1995, Up-regulation of a cysteine protease accompanies the ethylene-insensitive senescence of daylily (*Hemerocallis*) flowers, *Plant Mol. Biol.* **28:**575–582.

Vance, C. P., and Heichel, G. H., 1991, Carbon in $N_2$-fixation: Limitation or exquisite adaption, *Annu. Rev. Plant Physiol. Plant Mol. Biol.* **42:**373–392.

Van der Straeten, D., Rodrigues-Pousada, R. A., Goodman, H. M., and Van Montagu, M., 1991, Plant enolase: Gene structure, expression, and evolution, *Plant Cell* **3:**719–735.

Van de Wiel, C., Scheres, B., Franssen, H., van Lierop, M. J., van Lammeren, A., van Kammen, A., and Bisseling, T., 1990, The early nodulin transcript *ENOD2* is located in the nodule parenchyma (inner cortex) of pea and soybean root nodules, *EMBO J.* **9:**1–7.

Van Ghelue, M., Ribeiro, A., Solheim, B., Akkermans, A. D. L., Bisseling, T., and Pawlowski, K., 1996, Sucrose synthase and enolase expression in actinorhizal nodules of *Alnus glutinosa:* Comparison with legume nodules, *Mol. Gen. Genet.* **250:**437–446.

Van Ghelue, M., Løvaas, E., Ringø, E., and Solheim, B., 1997, Early interactions between *Alnus glutinosa* (L.) Gaertn. and *Frankia* strain ArI3, *Physiol. Plant.* **99:**579–587.

Van Kammen, A., 1984, Suggested nomenclature for plant genes involved in nodulation and symbiosis, *Plant Mol. Biol. Rep.* **2:**43–45.

Van Rhijn, P., and Vanderleyden, J., 1995, The *Rhizobium*-plant symbiosis, *Microbiol. Rev.* **59:**124–142.

Vasse, J., de Billy, F., Camut, S., and Truchet, G., 1990, Correlation between ultrastructural differentiation of bacteroids and nitrogen fixation in alfalfa nodules, *J. Bacteriol.* **172:**4295–4306.

Vercher, Y., and Carbonell, J., 1991, Changes in the structure of ovary tissues and in the ultrastructure of mesocarp cells during ovary senescence or fruit development induced by plant growth substances in *Pisum sativum*, *Physiol. Plant* **81:**518–526.

Verma, D. P. S., 1992, Signals in root nodule organogenesis and endocytosis of *Rhizobium*, *Plant Cell* **4:**373–382.

Weaver, C. D., Shower, N. H., Louis, C. F., and Roberts, D. M., 1994, Nodulin-26, a nodule specific symbiosome membrane protein from soybean, is an ion channel, *J. Biol. Chem.* **269:**17858–17862.

Werner, D., 1992, Physiology of nitrogen-fixing legume nodules: Compartments and functions, in *Biological Nitrogen Fixation* (G. Stacey, R. H. Burris, and H. J. Evans, eds.), pp. 399–431, Chapman & Hall, New York.

Witty, J. F., Minchin, F. R., Skøt, L., and Sheely, J. E., 1986, Nitrogen fixation and oxygen in root nodules, *Oxford Surv. Plant Cell Biol.* **3:**275–315.

Yang, W.-C., Horvath, B., Hontelez, J., van Kammen, A., and Bisseling, T., 1991, *In situ* localization of *Rhizobium* mRNAs in pea root nodules: *nifA* and *nifH* localization, *Mol. Plant–Microbe Interact.* **4:**464–468.

Zeng, S., and Tjepkema, J. D., 1994, The wall of the infected cell may be the major diffusion barrier in nodules of *Myrica gale* L., *Soil Biol. Biochem.* **5:**633–639.

Zeng, S., Tjepkema, J. D., and Berg, R. H., 1989, Gas diffusion pathway in nodules of *Casuarina cunninghamiana*, *Plant Soil* **118:**119–123.

Zhang, Y., and Roberts, D. M., 1995, Expression of soybean nodulin 26 in transgenic tobacco. Targeting to the vacuolar membrane and effects on floral and seed development, *Mol. Biol. Cell* **6:**109–117.

*Chapter 6*

# Strategies to Engineer Plants Resistant to Bacterial and Fungal Diseases

Günter Strittmatter, Koen Goethals, and Marc Van Montagu

## 1. INTRODUCTION

Fungal and bacterial diseases are major factors limiting crop production worldwide. Control of damage caused by microbial pathogens is mainly based on the application of agrochemicals, breeding for resistant varieties, and various crop husbandry strategies, such as crop rotation. The tremendous investment in the development and use of agrochemicals has to be understood as a clear indication that plant breeding has not been able to keep pace with the rapid evolution of pathogenicity in microorganisms and to match the efficiency and flexibility of phytopharmacy. For farmers, however, planting the "disease control principle" with the seed would represent a potentially attractive alternative, as such a process could reduce the costs and the level of technology required to grow a certain crop. In addition, the environment would be less burdened with potentially harmful compounds. Alltogether, these considerations stimulate plant breeders to look for new approaches of breeding disease resistance.

---

**Günter Strittmatter** Plant Genetic Systems, B-9000 Gent, Belgium; *present address:* KWS Kleinwanzlebener Saatzucht AG, Institut für Pflanzenzüchtung, D-37555 Einbeck, Germany. **Koen Goethals and Marc Van Montagu** Laboratorium voor Genetica, Department of Genetics, Flanders Interuniversity Institute for Biotechnology (VIB), Universiteit Gent, B-9000 Gent, Belgium.

*Subcellular Biochemistry, Volume 29: Plant–Microbe Interactions,* edited by Biswas and Das. Plenum Press, New York, 1998.

This need for novel breeding strategies comes together with a continuous improvement of molecular genetic technologies and consequently also with a more detailed understanding of plant–pathogen interactions at the molecular level. In general, molecular genetic tools allow the comparatively rapid identification, isolation, and controlled recombination *in vitro* of genes and gene segments, and it renders possible the transfer of novel combinations far beyond the limits of sexual crosses and somatic hybridization (Kung, 1993). Thereby, these technologies have the capacity to improve the possibilities and efficiency of producing genetic variability and selecting desirable traits.

The first race-specific resistance genes have recently been molecularly cloned, which are genetic determinants of complete resistance to distinct races of a pathogen species, and in most cases are associated with the establishment of the hypersensitive response (HR) in the host plant (for reviews see Dangl, 1995; Dong, 1995; Staskawicz *et al.*, 1995). The isolation of these genes now allows the detailed functional analysis of their gene products, which very likely represent central components of the signal transduction chain that connects the recognition of a potential pathogenic organism with the activation of defense reactions. In parallel, host genes that are required for the function of race-specific resistance gene products have been identified (Freialdenhoven *et al.*, 1994; Hammond-Kosack *et al.*, 1994b), and an increasing number of rapid changes of host cell metabolism upon pathogen attack are being described (Hammond-Kosack *et al.*, 1996; May *et al.*, 1996; Levine *et al.*, 1994; Koch *et al.*, 1994; Martini *et al.*, 1993; Bradley *et al.*, 1992; for review see Kombrink and Somssich, 1995). Detailed knowledge of the signal transduction processes in resistant plants and of the primary defense reactions will certainly point to new directions for the modification (improvement) of pathogen resistance in otherwise susceptible host plants by genetic engineering.

To our knowledge, no reports have been published on the generation of disease resistance by modifying the complex processes of recognition between hosts and pathogens or of signal transduction chains in the host. Likewise, no examples are known where more than one of the coordinate steps of multienzyme pathways for the synthesis of protective compounds, such as phytoalexins or lignin, have been engineered. Nevertheless, substantial improvements of resistance have been achieved, at least at the level of greenhouse or growth-chamber testing, by expressing single proteins that are either toxic to pathogens, interfere with the reproductive cycle of pathogens, or inhibit the toxicity of compounds produced by pathogens. In this chapter we will summarize these approaches and also give attention to concepts whose feasibility has not been proven, but which have the potential, in our view, to play a major role in future plant breeding. For further reference see the review articles by Michelmore (1995), Shah *et al.* (1995), and Lamb *et al.* (1992).

## 2. RESISTANCE TO BACTERIAL PATHOGENS

The development of technologies to engineer plants for enhanced resistance toward bacterial pathogens has lagged considerably behind that for insect, viral, and fungal pathogens. Two major approaches to enhance crop resistance to bacterial infection have been employed: the engineering of plants for the production of antibacterial enzymes and peptides that kill or retard the pathogen, and the engineering of resistance to phytotoxic compounds produced by the pathogen. The growing understanding of the diverse molecular mechanisms underlying the reactions of plants to bacterial infection holds promise for future progress in the domain. Current acquisition of details that govern the recognition of bacterial pathogens and their capacity to colonize the host plant will lead to modifications that restrain these processes in genetically engineered crops.

### 2.1. Antibacterial Enzymes and Peptides

Lysozymes are defined as hydrolases with specific activity toward the major bacterial cell wall component, peptidoglycan (murein). As such they function as bacteriolytic enzymes that degrade the reinforcing cell wall of both Gram-negative and Gram-positive bacteria. Although increased lysozyme activity has been detected in many different plant species as a result of pathogen infection, it has been found that the purified activity was mostly connected with the presence of a higher chitinase activity (Düring, 1993, and references therein). In fact, plant lysozymes are classified as vacuolar, basic chitinases with an additional but low lysozyme activity (Majeau *et al.*, 1990; Trudel *et al.*, 1989; Boller, 1987). It was further suggested that strict targeting of these enzymes to the vacuole in combination with their expression profile delays their activity toward intercellular growth of pathogenic bacteria (Düring *et al.*, 1993). The intercellular pathogens would thus accumulate to massive cell densities before triggering adequate defense responses and the resulting weak lysozyme activity would be insufficient to cope with the ensuing infection.

To enhance the general bacteriolytic activity of plants, foreign lysozyme genes with higher specific activity have been transferred to tobacco. Transformation with a hen egg-white lysozyme cDNA encoding also the native signal peptide resulted in the constitutive intercellular accumulation of the enzyme. This was correlated with the ability of plant extracts to inhibit growth of several plant pathogenic bacteria (Trudel *et al.*, 1992, 1995). However, no other phenotypic characteristics of the transgenic tobacco plants in the greenhouse or on the field were reported. In another approach, the bacteriophage T4 lysozyme gene was engineered by fusing its coding sequence to the signal peptide sequence of the barley α-amylase gene. Introduction of this construct into potato resulted in low-level expression, but efficient secretion of the protein to the intercellular space in transgenic plants. This was accompanied by a marked increase of their resistance to the soft rot pathogen *Erwinia carotovora* ssp. *atroseptica* under laboratory as well as greenhouse conditions (Düring, 1993; Düring *et al.*, 1993). Optimization of the expression, secretion, and activity of lysozyme genes in transgenic plants will possibly lead to a further enhancement of resistance.

Other lysozyme genes could be tested for their capacity to mediate resistance in transgenic plants. Tobacco plants produce extracellular proteins with strong lysozyme activity that exhibit a potent bacteriolytic potential (Stintzi *et al.*, 1993a). Possibly, constitutive or rapid and strong pathogen-responsive expression of the corresponding genes in transgenic plants would result in a reduced infection potential of bacteria, allowing a more successful response by the endogenous defense program.

Several organisms produce small peptides that exert antimicrobial activity. This represents an ancient defense strategy that is adopted by diverse animals, plants, and also microbes themselves (Broekaert *et al.*, 1995; Rao, 1995; Bohlman, 1994; Ganz and Lehrer, 1994; Hoffmann and Hétru, 1992; Bevins and Zasloff, 1990; Boman and Hultmark, 1987). For some peptides, antimicrobial activities to a whole range of fungi and both Gram-positive and Gram-negative bacteria have been described. Others are more specific in their activity against fungi, whereas still others exert inhibitory effects on fungi, bacteria, plants, and animal cells. In plants, the production of such antimicrobial peptides can be constitutive or organ-specific, but in several cases is enhanced via different signaling pathways upon infection with particular pathogens (Broekaert *et al.*, 1997).

In one study, tobacco plants harboring the barley α-thionin (α-HTH) gene under control of the CaMV 35S promoter revealed a markedly reduced infection by *Pseudomonas syringae* pvs. *tabaci* and *syringae* when leaves were infiltrated with bacteria (Carmona *et al.*, 1993). However, in an independent study using the same approach, transgenic tobacco plants were not resistant to infections with *P. syringae* pv. *tabaci* or *P. solanacearum*. Furthermore, α-HTH expression in tomato did not substantially increase resistance to the pathogens *P. solanacearum*, *Clavibacter michiganensis* ssp. *michiganensis* or *Xanthomonas campestris* pv. *vesicatoria*, although some of these strains were highly sensitive to the toxin *in vitro* (Florack *et al.*, 1994; Florack and Stiekema, 1994). The latter authors therefore question the applicability of this approach to increasing resistance of crops under field conditions. The difference in results might be attributed to the use of different strains or to the intracellular accumulation of the thionins in the vacuoles of the transgenic plants (Florack *et al.*, 1994), in combination with variation in the inoculation techniques. Targeting of the peptide to the intercellular spaces of plants where pathogenic bacteria proliferate before ensuing further pathogenesis might result in a noticeable effect on propagation of pathogenic bacteria. As thionins can show diverse inhibiting activities toward mammalian cells *in vitro* (Florack and Stiekema, 1994), care should be taken to evaluate their possible negative side effects upon overexpression in crops.

Plant defensins constitute another class of small, mostly antifungal peptides, some of which also show toxicity toward *P. solanacearum* and *C. michiganensis* (Moreno *et al.*, 1994). As plant defensins cause no detrimental effects on mammalian cells *in vitro,* they represent an interesting group of peptides for engineering resistance in heterologous hosts. However, the effect of expressing defensins in transgenic plants on pathogen attack has only been tested for fungi; results from these investigations will be described below.

Lipid transfer proteins (LTPs) from plants have been proven to exhibit a strong inhibitory effect on bacteria and fungi without affecting human cell lines *in vitro* (Cammue *et al.*, 1995). In response to pathogen attack, expression of some LTP genes is enhanced, whereas that of others is repressed, suggesting the involvement of these proteins in the establishment of plant–pathogen interactions. In compatible interactions between virulent races of *P. syringae* pv. *japonica* or *X. campestris* pv. *translucens* and susceptible cultivars of barley, the expression of three LTP genes was repressed (Garcia-Olmedo *et al.*, 1995). Therefore, inhibition of LTP production possibly results in sensitivity to pathogen attack. In tobacco, constitutive expression of an LTP from barley, which has been shown to be effective in inhibiting the propagation of *C. michiganensis* ssp. *sepedonicus* and *P. solanacearum* (Molina *et al.*, 1993), resulted in a drastic reduction of disease symptoms upon inoculation of leaves with *P. syringae* (Garcia-Olmedo *et al.*, 1995).

Still other types of small antimicrobial peptides are being isolated and characterized from plants, further extending the potential for their use in genetic engineering programs. In one case the heterologous expression of an *Amaranth* hevein-type peptide in tomato proved to be successful in reducing infection by the fungus *Trichoderma hamatum* (Lee and Raikhel, 1995). Although this peptide also revealed growth inhibition of Gram-positive bacteria *in vitro* (Broekaert *et al.*, 1992), no reports on the protection of transgenic plants against bacterial infection have been published.

A strategy that has been applied with variable success is the introduction of animal peptide genes into plants. The cecropins, produced by a diverse group of insects, build a large family of antimicrobial peptides that are lytic for both Gram-negative and Gram-positive bacteria, but do not affect eukaryotic cells (Hultmark *et al.*, 1982; Jaynes *et al.*, 1987). Transformation of tobacco with a gene encoding a peptide, Shiva-1, that is 46% identical with cecropin B under control of a constitutively active or a wound-responsive promoter rendered the plants more resistant to bacterial wilt caused by *P. solanacearum*. In these experiments, synthesis of the Shiva-1 peptide could indeed be detected in the transgenic plants, but, interestingly, in parallel experiments with transgenic plants carrying a bona fide cecropin B type gene (encoding SB-37, with 96% identity to cecropin B), no peptide accumulation could be found and the plants displayed no increase in resistance (Jaynes *et al.*, 1993). In accordance with these results, only extremely low levels of peptide were obtained by other groups using cecropin B-type transgenes in tobacco and consequently no effect on resistance was observed (Hightower *et al.*, 1994; Florack *et al.*, 1995). Although these authors cannot exclude the faulty translation of the cecropin mRNA, there was compelling evidence for proteolytic degradation of the cecropin B peptides in the plants, preventing their accumulation to an active concentration. Apparently, differences in the primary sequence of the Shiva-1 and cercopin B peptides determine their stability in plant cells. Assessment of the prerequisites for stable peptide production in transgenic plants is required in order to generate crops with higher protein concentrations.

A whole array of additional antimicrobial peptides await testing of their activity in transgenic plants. The apidaecins comprise an interesting group of peptides found in the hemolymph of bees (Casteels *et al.*, 1994). They have inhibitory effects on a whole range of plant-associated bacteria without affecting eukaryotic cells. Another example is provided by the magainins, amphibian peptides that are likewise restricted in activity toward microorganisms (Bevins and Zasloff, 1990). Still other peptides with varying activities toward eukaryotes have been described (Rao, 1995). Their potential as a means to increase plant resistance to bacterial pathogens justifies further investigations. The expression pattern, targeting, spectrum, and degree of toxicity and stability could be modulated by selecting different promoters, by introducing or manipulating specific targeting sequences, and by modifying their amino acid sequences.

A drawback of some of the approaches described above is their unspecified action toward bacteria in general. Although potentially useful with respect to pathogen resistance, the production of general bacteriocides by the plant could have severe effects on the colonization by beneficial bacteria that contribute to the plant's fitness in the soil. Furthermore, one can envision similar effects on secondary engineering attempts using *Agrobacterium*-mediated transformation. This should be considered when evaluating particular constructs for their use in crop improvement.

## 2.2. Detoxification of Bacterial Toxins or Adaptation of Their Targets

During infection of plants, several bacterial pathogens produce nonspecific toxins that are involved in the development of disease symptoms and that support the infection (Mitchell, 1984). In principle, there are several different ways for plants to cope with the toxic effects: inactivation of the toxin, target adaptation, or sequestering processes. In fact, the first plant resistance gene to be cloned, the maize *Hm1* gene, which confers resistance to race 1 isolates of the fungus *Cochliobolus carbonum,* turned out to encode an NADPH-dependent HC-toxin reductase which inactivates the HC-toxin produced by virulent isolates of the fungus (Johal and Briggs, 1992; Meeley *et al.*, 1992).

Engineering plants for resistance to toxins produced by microorganisms represents an elegant way to prevent the damage caused by infection. A good source for resistance genes is the toxin-producing organism themselves, as they have to avoid the toxic effects on their own metabolism. This approach was applied for engineering resistance to the *P. syringae* pv. *phaseolicola* toxin, phaseolotoxin (de la Fuente-Martinez *et al.*, 1992; de la Fuente-Martinez and Herrera-Estrella, 1993). Phaseolotoxin is responsible for chlorosis on plants through the irreversible inhibition of the chloroplast enzyme ornithine carbamoyltransferase (OCTase) which results in a block of arginine biosynthesis. Resistance of the bacterium to its own toxin is in part attributed to the presence of an OCTase isozyme that is insensitive to the toxin (Mosqueda *et al.*, 1990; Staskawicz *et al.*, 1980). Transformation of tobacco with the gene for this isozyme, engineered for targeting to the chloroplasts, rendered the plants insensitive to chlorosis induced by purified toxin. Transgenic plants showed an HR-like reaction, but no disease symptoms, after infection with *P. syringae* pv. *phaseolicola*, whereas nontransgenic plants were very vulnerable to this pathogen (for which beans are the native host), in some cases even suffering from systemic spread of the disease (de la Fuente-Martinez *et al.*, 1992).

In a second strategy, the gene for a tabtoxin acetylating enzyme of the tobacco pathogen *P. syringae* pv. *tabaci* was introduced into tobacco (Anzai *et al.*, 1989). Tabtoxin produced by the pathogen during infection provokes chlorosis by the inhibition of glutamine synthetase, resulting in the accumulation of toxic levels of ammonia. Transgenic tobacco expressing the acetylating enzyme inactivated the toxin and no longer showed chlorosis when treated with purified toxin or when inoculated with the pathogen.

These two examples exemplify the power of employing particular traits of microorganisms to expand the resistance of plants to pathogens. Isolation of microbial genes that confer resistance to phytotoxins by easy and fast screening methods (de la Fuente-Martinez and Herrera-Estrella, 1993) could lay the foundation for novel engineering strategies that protect plants from pathogen attack. Furthermore, as this type of resistance is specified toward a particular pathogenic character, no major effects on the remnant epiphytic microflora are to be expected.

## 2.3. Oxidative Burst

The transient accumulation of active oxygen species (AOS), such as $H_2O_2$, the superoxide anion radical ($O_2^-$) and the hydroxyl radical ($OH^\bullet$), is one of the most rapid responses in plants after pathogen attack (Baker and Orlandi, 1995). This process is referred to as oxidative burst (Lamb *et al.*, 1989; Mehdy, 1994). Besides its possibly toxic effect on pathogenic organisms, $H_2O_2$ has also been shown to be involved in several defense responses of the host plant. $H_2O_2$

produced during the oxidative burst induces the biosynthesis of phytoalexins and the cross-linking of hydroxyproline-rich cell wall glycoproteins, as demonstrated for plant cell suspension culture systems (Brisson *et al.*, 1994; Degoussée *et al.*, 1994; Davis *et al.*, 1993; Bradley *et al.*, 1992; Apostol *et al.*, 1989). In several pathosystems, incompatible host–pathogen interactions are characterized by a biphasic increase of AOS concentration: the first peak is followed by a rapid second burst of AOS; compatible interactions show only one peak of AOS accumulation (Baker *et al.*, 1991; Glazener *et al.*, 1991; Chai and Doke, 1987). The $H_2O_2$ bursts produced during incompatible interactions have been implicated, on the one hand, as a trigger of hypersensitive cell death of challenged cells and, on the other hand, as a diffusible signal molecule for the activation in adjacent cells of genes encoding cell protective enzymes, such as glutathione *S*-transferase (Levine *et al.*, 1994). Furthermore, in conjunction with salicylic acid (SA), $H_2O_2$ may also be involved in the induction of defense responses associated with systemic acquired resistance (SAR) (Klessig and Malamy, 1994; Chen *et al.*, 1993; Chai and Doke, 1987), although $H_2O_2$ is probably not the actual systemic signal and is probably not involved in SA-mediated systemic activation of pathogenesis-related (PR)-gene expression (Chamnongpol *et al.*, 1996; Bi *et al.*, 1995; Neuenschwander *et al.*, 1995). Repression of catalase synthesis to enhance $H_2O_2$ production and consequent PR-gene expression has proven an unsuccessful approach to enhancing systemic pathogen resistance in tobacco (Chamnongpol *et al.*, 1996).

The enzyme glucose oxidase from *Aspergillus niger* catalyzes the oxidation of β-D-glucose by molecular oxygen, yielding gluconic acid and $H_2O_2$. In transgenic potato plants which contained the glucose oxidase gene under control of the cauliflower mosaic virus (CaMV) 35S promoter, the $H_2O_2$ concentration increased by about two- to threefold, in both leaf and tuber tissues (Wu *et al.*, 1995). These elevated levels of $H_2O_2$ were correlated with a substantially reduced growth of *Erwinia carotovora* ssp. *carotovora*, the causal agent of soft rot, on tubers of transgenic plants when compared to tubers from nontransgenic control plants. Surprisingly, the constitutively increased concentration of $H_2O_2$ did not lead to any obvious morphological defects in the transgenic plants. The elucidation of the precise mechanism of $H_2O_2$-mediated resistance awaits further investigation. Apart from the direct antimicrobial effect of $H_2O_2$, a general activation of defense reactions by the elevated $H_2O_2$ concentration in the host plant has to be taken into consideration.

## 3. RESISTANCE TO FUNGAL PATHOGENS

### 3.1. Antifungal Proteins

Several distinct features of fungal cell structure and metabolism are potential targets for fungitoxic or fungistatic compounds. The constitutive or pathogen-inducible expression of genes encoding proteins with such an activity in transgenic plants represents an obvious strategy to protect plants against pathogenic fungi.

Cell walls of many fungi contain chitin and β-1,3-glucans as major structural polysaccharides (Wessels and Sietsma, 1981). Very likely, enzymatic degradation of these components by endochitinases and β-1,3-endoglucanases, both belonging to the group of pathogenesis-related proteins (Cutt and Klessig, 1992), is part of the natural defense mechanisms employed by plants upon infection with pathogenic fungi. Five classes (I–V) of endochitinases and three major classes (I–III) of β-1,3-endoglucanases have been identified in tobacco (Melchers *et al.*, 1994; Collinge *et al.*, 1993; Stintzi *et al.*, 1993b; Meins *et al.*, 1992; Mikkelsen *et al.*, 1992; Ward *et al.*, 1991a). Class I hydrolases are located in vacuoles and exhibit a strong inhibitory effect on fungal growth *in vitro* (Cornelissen and Melchers, 1993; Sela-Buurlage *et al.*, 1993; Leah *et al.*, 1991; Mauch *et al.*, 1991). Class II hydrolases, which are very similar to class I proteins at the level of their primary structure, but are targeted to the extracellular space, have no comparable effect on fungal growth *in vitro* (Cornelissen and Melchers, 1993; Sela-Buurlage *et al.*, 1993). The inhibitory effect of class I hydrolases is substantially increased when mixtures of different enzymes are applied to the *in vitro* assay, indicating their synergistic action; in combination with a class I hydrolase, also class II chitinase (but not class II glucanase) exerts antifungal activity (Sela-Buurlage *et al.*, 1993). The specific activity of class I hydrolases is not changed when proteins are modified for extracellular targeting in transgenic plants.

The expression of chitinases, like various other pathogenesis-related proteins, is developmentally and organ-specifically regulated in noninfected plants (Memelink *et al.*, 1990; Neale *et al.*, 1990; Samac *et al.*, 1990; Lotan *et al.*, 1989). In response to pathogen attack, the concentrations of chitinase mRNA and protein increase (Cutt and Klessig, 1992; Bol *et al.*, 1990). The outcome of the interaction between a host plant and a pathogenic fungus is thought to depend largely on the timing and intensity of defense response activation. Therefore, in one of the first approaches to engineering disease resistance the bean endochitinase CH5B gene, encoding a class I protein, was fused to the CaMV 35S promoter, conferring high constitutive expression in a broad spectrum of plant cells (Broglie *et al.*, 1991). When compared to control plants, transgenic tobacco plants harboring the chimeric construct in a homozygous state showed a two- to fourfold increase in chitinase activity in the roots and a 23- to 44-fold increase in the leaves before fungal infection. The phytopathological performance of transgenic plants was tested after infection with *Rhizoctonia solani*, an endemic, chitinous, soilborne fungus that infects numerous plant species. Seedling mortality and the loss of root fresh weight were substantially reduced in transgenic plants as compared to nontransgenic control plants. This effect was positively correlated with the amount of bean chitinase expressed in the individual transgenic lines. An improvement of quantitative resistance to *Rhizoctonia solani* was also observed in canola plants transformed with the 35S-chitinase construct. However, no impact on the survival was found when transgenic tobacco was infected with the nonchitinous fungal pathogen *Pythium aphanidermatum*. Most likely, the inhibitory effect of chitinase on fungal growth is based on the hydrolysis of newly formed chitin leading to the destruction of growing hyphal tips, an assumption which is supported by investigations on the effect of chitinase on the powdery mildew, *Erysiphe graminis* (Toyoda *et al.*, 1991).

The protective capacity of constitutive chitinase expression against *Rhizoctonia solani* was further proven in transgenic tobacco plants with chitinases from *Nicotiana tabacum* (Vierheilig *et al.*, 1993) or the bacterium *Serratia marcescens* (Logemann *et al.*, 1993) and in rice with a rice chitinase (Lin *et al.*, 1995). In *Brassica napus*, the constitutive synthesis of a chimeric endochitinase encoded by a fusion between a tomato and a tobacco chitinase gene provided field tolerance to *Cylindrosporium concentricum*, *Sclerotinia sclerotiorum*, and *Phoma lingam*, each causing economically important diseases (Grison *et al.*, 1996). In these studies, protection against fungal disease development was demonstrated in field trials with the T3 generation of transformants, indicating the stability of the phenotype as well as its potential relevance for application in agricultural practice. The specific chitinase activities in transgenic plants were two- to fivefold higher than in nontransgenic control plants, supporting the assumption that also a moderate increase of this enzyme activity is sufficient to reduce the susceptibility of *Brassica napus* to these pathogens. The best protection level was obtained against *Cylindrosporium*, whereas the effect on the development of blackleg disease caused by *Phoma lingam* was restricted to one of six transgenic lines.

To make use of combinatorial effects, chitinase genes have been combined with glucanase genes. Constitutive coexpression of the rice RCH10 basic chitinase gene and the alfalfa AGLU1 acidic glucanase gene in tobacco has been achieved by crossing transgenic parental lines harboring fusions of the individual genes with the CaMV 35S enhancer or a CaMV 35S double promoter (Zhu *et al.*, 1994). Protection against the fungal pathogen *Cercospora nicotianae* has been substantially greater in transgenic lines containing both genes than in lines transformed with either one of the two genes. Results even implied a synergistic interaction between the transgenes such that a combination of chitinase and glucanase genes is more effective than double doses of either transgene individually. A further increase of the protection level has been obtained in plants homozygous for both chitinase and glucanase transgenes. In all these experiments disease development was determined by measuring lesion areas on leaves. A similar synergistic effect has been observed in tobacco plants constitutively coexpressing a class II chitinase and a class II β-1,3-glucanase from barley, which both accumulate in the intercellular space of leaves (Jach *et al.*, 1995). In these experiments, the two genes were cotransformed on a single T-DNA. Evaluation of disease development caused by *Rhizoctonia solani* revealed significantly enhanced protection against fungal attack in plants expressing both barley genes when compared to isogenic lines expressing a single barley transgene to a similar level. The reduction of the disease index in plants harboring the two genes exceeded the reduction expected from an addition of the effects contributed by each individual gene.

Besides genes encoding glucanases or chitinases, other PR genes, such as a PR1a gene or an osmotin gene from tobacco, have been proven to mediate enhanced resistance to fungal disease when constitutively expressed in tobacco or potato (Liu *et al.*, 1994; Alexander *et al.*, 1993). Protection of transgenic tobacco plants against various fungal pathogens has been reported as a result of high-level expression of proteins associated with systemic acquired resistance (SAR) (Lawton *et al.*, 1993). The majority of these SAR proteins belong to the group of PR proteins. Typically, their mRNA accumulates not only in pathogen-infected (primary) leaves, but also in noninfected upper (secondary) leaves (Ward *et al.*, 1991b).

Ribosome-inactivating proteins (RIPs) from plants have been used as another tool to interfere specifically with fungal metabolism. RIPs inhibit translation in target cells by N-glycosidic cleavage of 28S rRNA (Stirpe *et al.*, 1992; Endo *et al.*, 1988). Most importantly, RIPs do not disturb the function of "self" ribosomes, but exhibit varying degrees of inhibition toward ribosomes of distantly related species, including fungal ribosomes (Stirpe and Hughes, 1989; Roberts and Selitrennikoff, 1986). The enzymes are among the most potent natural toxins, the best known of which is ricin. They are divided into two groups: single-chain type 1 proteins, and type 2 proteins consisting of two chains one of which bears a galactose-specific lectin domain that can bind to cell surfaces. Similar to PR proteins, purified type 1 RIP from barley inhibits fungal growth *in vitro*, and this effect is synergistically enhanced by class I chitinases and β-1,3-glucanases (Leah *et al.*, 1991). On the basis of these observations, Logemann *et al.* (1992) used the expression of a barley type 1 RIP gene under control of potato *wun1* promoter which confers wound- and pathogen-responsive transcription in the epidermis of leaves, stems, and roots from tobacco (Siebertz *et al.*, 1989). A synergistic effect of ribosome-inactivating protein and a class II β-1,3-glucanase or a class II chitinase has been demonstrated for transgenic tobacco plants when assaying resistance to *Rhizoctonia solani* (Jach *et al.*, 1995).

Finally, a group of small, cysteine-rich proteins designated plant defensins have been tested for their protective capacity in transgenic plants. These proteins were originally isolated from *Raphanus sativus* seeds, and have been proven to exhibit antifungal activity *in vitro* (Broekaert *et al.*, 1992; Terras *et al.*, 1992). They are released during seed germination after disruption of the seed coat, and the amount of protein in the microenvironment of the seed is sufficient to inhibit fungal growth. Plant defensin transcripts are barely detectable in healthy uninfected leaves, but accumulate systemically at high levels after localized fungal infection. One of the plant defensin genes, Rs-AFP2, has been fused to the CaMV 35S promoter and transferred into tobacco. The integration of this chimeric construct into the tobacco genome conferred enhanced resistance to the foliar pathogen *Alternaria longipes* (Terras *et al.*, 1995). The reduction of average lesion area after fungal infection was dependent on the amount of plant defensin produced in the transgenic lines.

### 3.2. Phytoalexins

For several pathosystems a positive correlation has been described between the resistance of host plants to certain pathogens and the concentration of phytoalexins accumulating in host plants in response to pathogen attack (Nicholson and Hammerschmid, 1982, and references therein). Investigations on the pea/*Nectria haematococca* pathosystem provided genetic evidence for a causal relationship between the synthesis of phytoalexins and resistance: the *pda* gene in the fungus enables the pathogen to overcome the resistance of the host plant; *pda* encodes an enzyme, pisatin demethylase, which detoxifies the phytoalexin pisatin from pea (Schäfer *et al.*, 1989; vanEtten *et al.*, 1989; Weltring *et al.*, 1988). However, results arguing against a causal relationship between the accumulation of phytoalexins and resistance have also been found (Glazebrook and Ausubel, 1994). Phytoalexin-deficient mutants of *Arabidopsis thaliana* were not affected in their resistance to avirulent strains of *Pseudomonas syringae*. The finding that two of the mutants displayed enhanced sensitivity to isogenic virulent *Pseudomonas syringae* pathogens suggests that the phytoalexin may play a role in limiting the growth of virulent bacteria.

Stilbenes are typical phytoalexins of *Vitis vinifera* and *Picea sitchensis* involved in the defense of these plant species against fungal attack (Woodward and Pearce, 1988; Stein and Blaich, 1985). The accumulation of stilbenes and the activity of genes for key enzymes of their biosynthesis is induced by various biotic and abiotic signals (Melchior and Kindl, 1991; Vornam *et al.*, 1988; Langcake and Pryce, 1977). The precursors for the biosynthesis of stilbenes are present in most plant species, including tobacco, but they lack the enzyme stilbene synthase. Transfer of a stilbene synthase gene from grapevine to tobacco led to the accumulation of the stilbene resveratrol in tansgenic plants in response to treatment with fungal elicitor or infection with the broad-range fungal pathogen *Botrytis cinerea* (Hain *et al.*, 1993). In two of three transgenic tobacco lines, the percentage of diseased leaf area was reduced after inoculation with *Botrytis cinerea* when compared to nontransgenic control plants. The level of resistance was directly correlated with the rapidity of stilbene synthase gene activation and the concentration of resveratrol accumulating in transgenic leaves. Furthermore, the improved resistance phenotype was stably inherited by progeny of primary transformants. In potato plants, the synthesis of resveratrol also caused a substantial reduction of leaf or tuber damage after infection with the late blight fungus *Phytophthora infestans* (Dr. Rüdiger Hain, Bayer AG, Leverkusen, Germany, personal communication). According to these results, the synthesis of phytoalexins plays a crucial role in the efficient defense of the host plant, at least in some pathosystems.

### 3.3. Simulation of Hypersensitive Cell Death in Susceptible Hosts

True pathosystems are characterized by cultivar/race-specific interactions: resistance of a certain host cultivar to a particular physiological race of a pathogen species is determined by the presence of resistance alleles in the host and avirulence alleles in the pathogen. This type of genetic interdependence between host cultivars and pathogen races controlling the outcome of their interaction was first postulated in Flor's "gene-for-gene hypothesis" (Flor, 1955, 1971). Often, race-specific resistance is associated with rapid and localized cell death at infection sites in the host plant (hypersensitive response, HR) limiting the spread and propagation of the pathogen. Recently, various race-specific resistance genes have been molecularly cloned. Sequence analysis of these genes has indicated striking structural similarities between the different gene products (Dangl, 1995; Staskawicz *et al.*, 1995): all share leucine reach repeats (LRRs), which might be responsible for protein–protein interactions, most of them contain a conserved nucleotide-binding site (NBS), and some of them exhibit homology to parts of the *Drosophila* Toll protein and the mammalian interleukin-1 receptor; the structural features of at least one resistance gene product, Cf9, suggest an extracellular localization of the majority of the protein and membrane anchoring, typical characteristics of a receptor for extracellular ligands from a pathogen. Functional analysis of the resistance gene products and other components of the signal transduction chain which connects the perception of a signal from the pathogen with the activation of defense responses in the attacked host tissue will provide the basis for novel engineering strategies. These might apply the induction of hypersensitive cell death in otherwise susceptible host cultivars by manipulating components of the signal transduction chain.

Introgression of race-specific resistance genes was frequently applied in classical breeding programs to generate novel varieties with resistance to fungal diseases. However, the nature of this type of resistance implies that protection against only a limited number of pathogen races (those containing the avirulence genes corresponding to the newly introgressed resistance genes) is provided. In addition, large-scale growth of varieties containing race-specific resistance genes in monoculture led to the rapid selection and epidemic spread of new virulent races. Most likely, the same problems will be faced if transgenic plants carrying such race-specific resistance genes from other cultivars are grown in the field.

On the other hand, transformation of plants will allow us to combine different resistance genes and provide a longer lasting broad-spectrum resistance more rapidly than classical breeding tools.

A strategy for genetic engineering hypersensitive cell death in a wide range of host–pathogen interactions has been proposed by de Wit (1992). In this model, the generation of hypersensitive cell death is based on the strictly controlled coexpression of an avirulence allele from a pathogen and the corresponding resistance allele from a host plant. In the transgenic plants, the resistance allele should be transcribed under control of a constitutively active promoter or its cognate promoter, whereas the expression of the avirulence allele has to be regulated by a promoter with very specific features. To mediate broad-spectrum resistance, it should be responsive to a wide range of pathogens; however, transcription must not be activated in any noninfected tissue or in response to environmental stimuli other than pathogen attack. One regulatory region which at least comes close to fulfilling these requirements is a promoter fragment from the potato defence gene *gst1* (formerly called *prp1-1;* Strittmatter *et al.*, 1996; Martini *et al.*, 1993). This promoter fragment mediates rapid and local transcriptional activation in response to various pathogenic organisms; it is not active in noninfected tissues, except root tips. This promoter could, for example, be fused to the *avr9* gene from the tomato pathogen *Cladosporium fulvum,* and the chimeric construct could then be transferred into a genome harboring the corresponding *Cf9* allele from tomato. Previous investigations have already proven that apoplastic accumulation of the *avr9* gene product in *Cf9*-containing tomato lines is lethal to the plant cells (Hammond-Kosack *et al.*, 1994a). In these experiments, *Cf0* tomato lines harboring a constitutively expressed *avr9* gene were crossed with *Cf9* tomato plants. The occurrence of cell death was developmentally regulated, in that necrosis was not detected until 10 days after sowing.

An alternative strategy for artificial generation of rapid and localized death of host cells at infection sites has been developed by Strittmatter *et al.* (1995). This approach, which has been tested for the interaction between potato leaves and *Phytophthora infestans,* relies on the use of two chimeric genes. On the one hand, a *gst1* (*prp1-1*) promoter fragment mediating rapid and local transcriptional activation selectively in response to pathogen attack is used to drive the expression of the bacterial ribonuclease barnase. On the other hand, transgenic plants constitutively express barstar, a specific inhibitor of barnase (Hartley, 1989). The barnase gene, under control of a tapetum-specific promoter, and the barstar gene have previously been used for the generation of nuclear-encoded male sterility and for the restoration of fertility, respectively (Mariani *et al.*, 1992, 1990). Rapid and high-level synthesis of barnase in the vicinity of infection sites should initiate necrosis of host cells during early stages of compatible interactions and therefore restrict the growth and propagation of biotrophic pathogenic fungi also in this type of interaction, analogous to the naturally occurring hypersensitive cell death in incompatible interactions. The efficiency of the two-component system relies on the balance between the constitutive barstar expression and the pathogenesis-responsive barnase expression. Strittmatter *et al.* (1995) postulate that the level of barstar is exceeded by the level of barnase only in the close vicinity of infection sites, thereby leading to cell death specifically in infected host tissue. On detached leaves from transgenic potato lines harboring the two-component system, sporulation of the late-blight fungus *Phytophthora infestans* was considerably reduced. Treatment of leaves from transgenic lines with ethylene, a chemical inducer of *prp1-1* promoter activity, led to rapid tissue destruction, indirectly indicating the inducibility and action of barnase. The reduction of fungal sporulation cosegregated with the transgenes in an F1 population, providing further evidence for the correlation between the presence of the barnase gene and the inhibition of fungal reproduction. Presently, the feasibility of the concept is being tested in field trials. In these experiments an answer will be sought to the question of whether the artificial cell death can be restricted to defined areas of the transgenic plants under massive infection pressure or whether complete plants are comitting suicide under these conditions. In both cases, an inhibition of the epidemic spread of late-blight disease on the field can be expected with the transgenic potato lines. Special emphasis will also be put on the question of whether the suicide construct provokes pleiotropic effects under natural conditions.

### 3.4. Oxidative Burst

As described in Section 2, an elevation of the $H_2O_2$ concentration was achieved by transformation of potato plants with a glucose oxidase gene from *Aspergillus niger* under control of the CaMV 35S promoter (Wu *et al.*, 1995). This effect was correlated not only with resistance to soft rot bacteria in tubers, but also with an inhibition of disease development after inoculation of leaves with the late-blight fungus *Phytophthora infestans;* in these experiments the rate of lesion growth on leaves after inoculation with fungal spores was evaluated. These results indicate that a modification of $H_2O_2$ formation in transgenic plants has the potential to mediate resistance to a very broad spectrum of pathogens.

## 4. CLOSING REMARKS

In this review, several examples were given of successful attempts to engineer resistance to bacterial and/or fungal pathogens in transgenic plants. However, various reports were also cited describing unsuccessful approaches. The reasons for this can be diverse, ranging from an inability to express the transgenes at adequate levels to the synthesis of a gene product that is unstable or that is targeted to the wrong cellular compartments.

Producing transgenic crops which exhibit the expected phenotype relies on a variety of parameters, most of which are not fully understood or are still difficult to manipulate (de Block, 1993). Transformation using *Agrobacterium tumefaciens* can now be achieved for a broad spectrum of crops, but some recalcitrant species await further optimization of the procedure. In most cases the identification of a well-performing line needs the screening of 30–to 50 or even more independent transformants where the transgene has integrated at different places in the genome. Effects of insert position on the expression of a transgene are still difficult to manage and its control awaits further research on the genome organization of plants (Finnegan and McElroy, 1994). This should refine procedures that so far strongly depend on empirical research.

Once a gene is introduced into the plant genome and is transcribed at an acceptable level, still other factors will influence its expression and effect on the phenotype of the transgenic plant. The processing and stability of its mRNA and its possible impact on the expression of endogenous genes as well as other transgenes should be considered and optimized (Meyer, 1995). As exemplified by the approach using antibacterial peptides, negative results with model plants do not necessarily mean that the adopted methodology is not effective in principle. Efficient translation of the gene products and their concomitant processing, targeting, correct folding, and stability should be taken into account and adjusted if necessary. This is especially the case for prokaryotic genes and their encoded proteins that have evolved to be active in a completely different context, e.g., with an optimized codon usage and requiring specific conditions for correct protein folding and stability. Rules to optimize these parameters for plant cells are still largely unknown and will rely on ongoing basic research.

Finally, future strategies for engineering pathogen resistance in plants will apply results from molecular-ecological studies on the manifold ways in which nature controls the complex interactions between host plants and pathogenic microorganisms. In this way, the natural diversity could be used as a limitless source of inspiration.

## 5. REFERENCES

Alexander, D., Goodman, R. M., Gut-Rella, M., Glascock, C., Weymann, K., Friedrich, L., Maddox, D., Ahl-Goy, P., Luntz, T., Ward, E., and Ryals, J., 1993, Increased tolerance to two oomycete pathogens in transgenic tobacco expressing pathogenesis-related protein 1a, *Proc. Natl. Acad. Sci. USA* **90:**7327–7331.

Anzai, H., Yoneyama, K., and Yamaguchi, I., 1989, Transgenic tobacco resistant to a bacterial disease by the detoxification of a pathogenic toxin, *Mol. Gen. Genet.* **219:**492–494.

Apostol, I., Heinstein, P. F., and Low, P. S., 1989, Rapid stimulation of an oxidative burst during elicitation of cultured plant cells, *Plant Physiol.* **90:**109–116.

Baker, C. J., and Orlandi, E. W., 1995, Active oxygen in plant/pathogen interactions, *Annu. Rev. Phytopathol.* **33:**299–321.

Baker, C. J., O'Neill, N. R., Keppler, L. D., and Orlandi, E. W., 1991, Early responses during plant–bacteria interactions in tobacco cell suspensions, *Phytopathology* **81:**1504–1507.

Bevins, C. L., and Zasloff, M., 1990, Peptides from frog skin, *Annu. Rev. Biochem.* **59:**395–414.

Bi, Y. M., Kenton, P., Mur, L., Darby, R., and Draper, J., 1995, Hydrogen peroxide does not function downstream of salicylic acid in the induction of PR protein expression, *Plant J.* **8:**235–245.

Bohlmann, H., 1994, The role of thionins in plant protection, *Crit. Rev. Plant Sci.* **13:**1–16.

Bol, J. F., Linthorst, H. J. M., and Cornelissen, B. J. C., 1990, Plant pathogenesis-related proteins induced by virus infection, *Annu. Rev. Phytopathol.* **28:**113–138.

Boller, T., 1987, Hydrolytic enzymes in plant disease resistance, in *Plant–Microbe Interactions: Molecular and Genetic Perspectives*, Vol. 2 (T. Kosuge and E. W. Nester, eds.), pp. 385–414, Macmillan, New York.

Boman, H. G., and Hultmark, D., 1987, Cell-free immunity in insects, *Annu. Rev. Microbiol.* **41:**103–126.

Bradley, D. J., Kjellbom, P., and Lamb, C. J., 1992, Elicitor- and wound-induced oxidative cross-linking of a proline-rich plant cell wall protein: A novel, rapid defense response, *Cell* **70:**21–30.

Brisson, L. F., Tenhaken, R., and Lamb, C. J., 1994, Function of oxidative cross-linking of cell wall structural proteins in plant disease resistance, *Plant Cell* **6:**1703–1712.

Broekaert, W. F., Mariën, W., Terras, F. R. G., de Bolle, M. F. C., Proost, P., van Damme, J., Dillen, L., Claeys, M., Rees, S. B., Vanderleyden, J., and Cammue, B. P. A., 1992, Antimicrobial peptides from *Amaranthus caudatus* seeds with sequence homology to the cysteine/glycine-rich domain of chitin-binding proteins, *Biochemistry* **31:**4308–4314.

Broekaert, W. F., Terras, F. R. G., Cammue, B. P. A., and Osborn, R. W., 1995, Plant defensins: Novel antimicrobial peptides as components of the host defense system, *Plant Physiol.* **108:**1353–1358.

Broekaert, W. F., Cammue, B. P. A., de Bolle, M. F. C., Thevissen, K., de Samblanx, G. W., and Osborn, R. W., 1997, Antimicrobial peptides from plants, *Crit. Rev. Plant Sci.* **16:**297–323.

Broglie, K., Chet, I., Holliday, M., Cressman, R., Biddle, P., Knowlton, S., Mauvais, C. J., and Broglie, R., 1991, Transgenic plants with enhanced resistance to the fungal pathogen *Rhizoctonia solani*, *Science* **254:**1194–1197.

Cammue, B. P. A., Thevissen, K., Hendriks, M., Eggermont, K., Goderis, I. J., Proost, P., van Damme, J., Osborn, R. W., Geurbette, F., Kader, J. C., and Broekaert, W. F., 1995, A potent antimicrobial protein from onion (*Allium cepa* L.) seeds showing sequence homology to plant lipid transfer proteins, *Plant Physiol.* **109:**445–455.

Carmona, M. J., Molina, A., Fernandez, J. A., Lopez-Fando, J. J., and Garcia-Olmedo, F., 1993, Expression of the α-thionin gene from barley in tobacco confers enhanced resistance to bacterial pathogens, *Plant J.* **3:**457–462.

Casteels, P., Romagnolo, J., Castle, M., Casteels-Josson, K., Erdjument-Bromage, H., and Tempst, P., 1994, Biodiversity of apidaecin-type peptide antibiotics, *J. Biol. Chem.* **269:**26107–26115.

Chai, H. B., and Doke, N., 1987, Activation of the potential of potatoleaf tissue to react hypersensitively to *Phytophthora infestans* by cytospore germination fluid and the enhancement of this potential by calcium ions, *Physiol. Mol. Plant Pathol.* **30:**27–37.

Chamnongpol, S., Willekens, H., Langebartels, C., van Montagu, M., Inzé, D., and van Camp, W., 1996, Transgenic tobacco with a reduced catalase activity develops necrotic lesions and induces pathogenesis-related expression under high light, *Plant J.* **10:**491–503.

Chen, Z., Silva, H., and Klessig, D. F., 1993, Active oxygen species in the induction of plant systemic acquired resistance by salicylic acid, *Science* **162:**1883–1886.

Collinge, D. B., Kragh, K. M., Mikkelsen, J. D., Nielsen, K. K., Rasmussen, U., and Vad, K., 1993, Plant chitinases, *Plant J.* **3:**31–40.

Cornelissen, B. J. C., and Melchers, L. S., 1993, Strategies for control of fungal diseases with transgenic plants, *Plant Physiol.* **101**:709–712.

Cutt, J. R., and Klessig, D. F., 1992, Pathogenesis-related proteins, in *Genes Involved in Plant Defense* (T. Boller and F. Meins, eds.), pp. 209–243, Springer-Verlag, Berlin.

Dangl, J. L., 1995, Pièce de résistance: Novel classes of plant disease resistance genes, *Cell* **80**:363–366.

Davis, D., Merida, J., Legendre, L., and Low, P. S., 1993, Independent elicitation of the oxidative burst and phytoalexin formation in cultural plant cells, *Phytochemistry* **32**:607–611.

De Block, M., 1993, The cell biology of plant transformation: Current state, problems, prospects and the implications for the plant breeding, *Euphytica* **71**:1–14.

Degoussée, N., Triantaphylidès, C., and Montillet, J.-L., 1994, Involvement of oxidative processes in the signaling mechanisms leading to the activation of glyceollin synthesis in soybean (*Glycine max*), *Plant Physiol.* **104**:945–952.

De la Fuente-Martinez, J. M., Mosqueda-Cano, G., Alvares-Morales, A., and Herrera-Estrella, L., 1992, Expression of a bacterial phaseolotoxin-resistant ornithyl transcarbamylase in transgenic tobacco confers resistance to *Pseudomonas syringae* pv. *phaseolicola*, *Biotechnology* **10**:905–909.

De la Fuente-Martinez, J. M., and Herrera-Estrella, L., 1993, Strategies to design transgenic plants resistant to toxins produced by pathogens, *AgBiotech News Inf.* **5**:295N–299N.

De Wit, P. J. G. M., 1992, Molecular characterization of gene-for-gene systems in plant–fungus interactions and the application of avirulence genes in control of plant pathogens, *Annu. Rev. Phytopathol.* **30**:391–418.

Dong, X., 1995, Finding the missing pieces in the puzzle of plant disease resistance, *Proc. Natl. Acad. Sci. USA* **92**:7137–7139.

Düring, K., 1993, Can lysozymes mediate antibacterial resistance in plants, *Plant Mol. Biol.* **23**:209–214.

Düring, K., 1994, Differential patterns of bacteriolytic activities in potato in comparison of bacteriophage T4 and hen egg-white lysozymes, *J. Phytopathol.* **141**:159–164.

Düring, K., Porsch, P., Fladung, M., and Lörz, H., 1993, Transgenic potato plants resistant to the phytopathogenic bacterium *Erwinia carotovora*, *Plant J.* **3**:587–598.

Endo, Y., Tsurugi, K., and Ebert, R. F., 1988. The mechanism of action of barley toxin: A type 1 ribosome-inactivating protein with RNA N-glycosidase activity, *Biochem. Biophys. Acta* **954**:224–226.

Finnegan, J., and McElroy, D., 1994, Transgene inactivation: Plants fight back! *Biotechnology* **12**:883–888.

Flor, H. H., 1955, Host–parasite interaction in flax rust—Its genetics and other implications, *Phytopathology* **45**:680–685.

Flor, H. H., 1971, Current status of the gene-for-gene concept, *Annu. Rev. Phytopathol.* **9**:275–298.

Florack, D. E. A., and Stiekema, W. J., 1994, Thionins: Properties, possible biological roles and mechanisms of action, *Plant Mol. Biol.* **26**:25–37.

Florack, D. E. A., Dirkse, W. G., Visser, B., Heidekamp, F., and Stiekema, W. J., 1994, Expression of biologically active hordothionins in tobacco: Effects of pre- and pro-sequences at the amino and carboxyl termini of the hordothionin precursor on mature protein expression and sorting, *Plant Mol. Biol.* **24**:83–96.

Florack, D., Allefs, S., Bollen, R., Bosch, D., Visser, B., and Stiekema, W., 1995, Expression of giant silkmoth cecropin B genes in tobacco, *Transgenic Res.* **4**:132–141.

Freialdenhoven, A., Scherag, B., Hollrichter, B., Collinge, D. B., Thordal-Christensen, H., and Schulze-Lefert, P., 1994, *Nar-1* and *Nar-2*, two loci required for $Mla_{12}$-specified race-specific resistance to powdery mildew in barley, *Plant Cell* **6**:983–994.

Ganz, T., and Lehrer, R. I., 1994, Defensins, *Immunology* **6**:584–589.

Garcia-Olmedo, F., Molina, A., Segura, A., and Moreno, M., 1995, The defensive role of nonspecific lipid-transfer proteins in plants, *Trends Microbiol.* **3**:72–74.

Glazebrook, J., and Ausubel, F. M., 1994, Isolation of phytoalexin-deficient mutants of *Arabidopsis thaliana* and characterization of their interactions with bacterial pathogens, *Proc. Natl. Acad. Sci. USA* **91**:8955–8959.

Glazener, J. A., Orlandi, E. W., Harmon, G. L., and Baker, C. J., 1991, An improved method for monitoring active oxygen in bacteria-treated suspension cells using luminol-dependent chemiluminescence, *Physiol. Mol. Plant Pathol.* **39**:123–133.

Grison, R., Grezes-Besset, B., Schneider, M., Lucante, N., Olsen, L., Leguay, J.-J., and Toppan, A., 1996, Field tolerance to fungal pathogens of *Brassica napus* constitutively expressing a chimeric chitinase gene, *Nature Biotechnol.* **14**:643–646.

Hain, R., Reif, H.-J., Krause, E., Langebartels, R., Kindl, H., Vornam, B., Wiese, W., Schmelzer, E., Schreier, P. H., Stöcker, R. H., and Stenzel, K., 1993, Disease resistance results from foreign phytoalexin expression in a novel plant, *Nature* **361**:153–156.

Hammond-Kosack, K. E., Harrison, K., and Jones, J. D. G., 1994a, Developmentally regulated cell death on expression of the fungal avirulence gene *Avr9* in tomato seedlings carrying the disease-resistance gene *Cf9*, *Proc. Natl. Acad. Sci. USA* **91**:10445–10449.

Hammond-Kosack, K. E., Jones, D. A., and Jones, J. D. G., 1994b, Identification of two genes required in tomato for full *Cf-9*-dependent resistance to *Cladosporium fulvum*, *Plant Cell* **6**:361–374.

Hammond-Kosack, K. E., Silverman, P., Raskin, I., and Jones, J. D. G., 1996, Race-specific elicitors of *Cladosporium fulvum* induce changes in cell morphology and the synthesis of ethylene and salicylic acid in tomato plants carrying the corresponding *Cf* disease resistance gene, *Plant Physiol.* **110**:1381–1394.

Hartley, R. W., 1989, Barnase and barstar: Two small proteins to fold and fit together, *Trends Biochem. Sci.* **14**:450–454.

Hightower, R., Baden, C., Penzes, E., and Dunsmuir, P., 1994, The expression of cecropin peptide in transgenic tobacco does not confer resistance to *Pseudomonas syringae* pv. *tabaci*, *Plant Cell Rep.* **13**:295–299.

Hoffmann, J. A., and Hétru, C., 1992, Insect defensins: Inducible antibacterial peptides, *Immunology Today* **13**:411–415.

Hultmark, D., Engström, A., Bennich, H., Kapur, R., and Boman, H. G., 1982, Insect immunity: Isolation and structure of cecropin D and four minor antibacterial components from *Cecropia* pupae, *Eur. J. Biochem.* **127**:207–217.

Jach, G., Görnhardt, B., Mundy, J., Logemann, J., Pinsdorf, E., Leah, R., Schell, J., and Maas, C., 1995, Enhanced quantitative resistance against fungal disease by combinatorial expression of different barley antifungal proteins in transgenic tobacco, *Plant J.* **8**:97–109.

Jaynes, J. M., Xanthopoulos, K. G., Destéfano-Beltran, L., and Dodds, J. H., 1987, Increasing bacterial disease resistance utilising antibacterial genes from insects, *BioEssays* **6**:263–270.

Jaynes, J. M., Nagpala, P., Destéfano-Beltran, L., Huang, J. H., Kim, J. H., Denny, T., and Cetiner, S., 1993, Expression of a cecropin B lytic peptide analog in transgenic tobacco confers enhanced resistance to bacterial wilt caused by *Pseudomonas solanacearum*, *Plant Sci.* **89**:43–53.

Johal, G. S., and Briggs, S. P., 1992, Reductase activity encoded by the *Mh1* resistance gene in maize, *Science* **258**:985–987.

Klessig, D. F., and Malamy, J., 1994, The salicylic acid signal in plants, *Plant Mol. Biol.* **26**:1439–1458.

Koch, C., Noga, G., and Strittmatter, G., 1994, Photosynthetic electron transport is differentially affected during early stages of cultivar/race-specific interactions between potato and *Phytophthora infestans*, *Planta* **193**:551–557.

Kombrink, E., and Somssich, I. E., 1995, Defense responses of plants to pathogens, *Adv. Bot. Res.* **21**:1–34.

Kung, S., 1993, Introduction: From hybrid plants to transgenic plants, in *Transgenic Plants,* Vol. 1 (S. Kung and R. Wu, eds.), pp. 1–12, Academic Press, San Diego, California.

Lamb, C. J., Lawton, M. A., Dron, M., and Dixon, R. A., 1989, Signals and transduction mechanisms for activation of plant defense against microbial attack, *Cell* **56**:215–224.

Lamb, C. J., Ryals, J. A., Ward, E. R., and Dixon, R. A., 1992, Emerging strategies for enhancing crop resistance to microbial pathogens, *Biotechnology* **10**:1436–1445.

Langcake, P., and Pryce, R. J., 1977, The production of resveratrol and the viniferins by grapevines in response to ultraviolet irradiation, *Phytochemistry* **16**:1193–1196.

Lawton, K., Uknes, S., Friedrich, L., Gaffney, T., Alexander, D., Goodman, R., Metraux, J. P., Kessmann, H., Ahl-Goy, P., Gut-Rella, M., Ward, E., and Ryals, J., 1993, The molecular biology of systemic acquired resistance, in *Mechanisms of Plant Defense Responses* (B. Fritig and M. Legrand, eds.), pp. 422–432, Kluwer, Dordrecht, The Netherlands.

Leah, R., Tommerup, H., Svendsen, I., and Mundy, J., 1991, Biochemical and molecular characterization of three barley seed proteins with antifungal properties, *J. Biol. Chem.* **266**:1564–1573.

Lee, H. I., and Raikhel, N. V., 1995, Prohevein is poorly processed but shows enhanced resistance to a chitin-binding fungus in transgenic tomato plants, *Braz. J. Med. Biol. Res.* **28**:743–750.

Levine, A., Tenhaken, R., Dixon, R., and Lamb, C. J., 1994, $H_2O_2$ from the oxidative burst orchestrates the plant hypersensitive disease resistance response, *Cell* **79**:583–593.

Lin, W., Anarutha, C. S., Datta, K., Potrykus, I., Muthukrishnan, S., and Datta, S. K., 1995, Genetic engineering of rice for resistance to sheath blight, *Biotechnology* **13**:686–691.

Liu, D., Raghothama, K. G., Hasegawa, P. M., and Bressan, R. A., 1994, Osmotin overexpression in potato delays development of disease symptoms, *Proc. Natl. Acad. Sci. USA* **91**:1888–1892.

Logemann, J., Jach, G., Tommerup, H., Mundy, J., and Schell, J., 1992, Expression of a barley ribosome-inactivating protein leads to increased fungal protection in transgenic tobacco plants, *Biotechnology* **10**:305–308.

Logemann, J., Jach, G., Logemann, S., Leah, R., Wolf, G., Mundy, J., Oppenheim, A., Chet, I., and Schell, J., 1993, Expression of a ribosome inhibiting protein (RIP) or a bacterial chitinase leads to fungal resistance in transgenic plants, in *Mechanisms of Plant Defense Responses* (B. Fritig and M. Legrand, eds.), pp. 446–448, Kluwer, Dordrecht, The Netherlands.

Lotan, T., Ori, N., and Fluhr, R., 1989, Pathogenesis-related proteins are developmentally regulated in tobacco flowers, *Plant Cell* **1**:881–887.

Majeau, N., Trudel, J., and Asselin, A., 1990, Diversity of cucumber chitinase isoforms and characterisation of one seed basic chitinase with lysozyme activity, *Plant Sci.* **68**:9–16.

Mariani, C., de Beukeleer, M., Truettner, J., Leemans, J., and Goldberg, R. B., 1990, Induction of male sterility in plants by a chimaeric ribonuclease gene, *Nature* **347**:737–741.

Mariani, C., Gossele, V., de Beukeleer, M., de Block, M., Goldberg, R. B., de Greef, W., and Leemans, J., 1992, A chimaeric ribonuclease-inhibitor gene restores fertility to male sterile plants, *Nature* **357**:384–387.

Martini, N., Egen, M., Rüntz, I., and Strittmatter, G., 1993, Promoter sequences of a potato pathogenesis-related gene mediate transcriptional activation selectively upon fungal infection, *Mol. Gen. Genet.* **236**:179–186.

Mauch, F., Mauch-Mani, B., and Boller, T., 1988, Antifungal hydrolases in pea tissue, *Plant Physiol.* **88**:936–942.

May, M. J., Hammond-Kosack, K. E., and Jones, J. D. G., 1996, Involvement of reactive oxygen species, glutathione metabolism, and lipid peroxidation in the *Cf*-dependent defense response of tomato cotyledons induced by race-specific elicitors of *Cladosporium fulvum, Plant Physiol.* **110**:1367–1379.

Meeley, R. B., Johal, G. S., Briggs, S. P., and Walton, J. D., 1992, A biochemical phenotype for a disease resistance gene of maize, *Plant Cell* **4:**71–77.

Mehdy, M. C., 1994, Active oxygen species in plant defense against pathogens, *Plant Physiol.* **105:**467–472.

Meins, F., Neuhaus, J.-M., Sperisen, C., and Ryals, J., 1992, The primary structure of plant pathogenesis-related glucanohydrolases and their genes, in *Genes Involved in Plant Defense* (T. Boller and F. Meins, eds.), pp. 245–282, Springer-Verlag, Berlin.

Melchers, L. S., Apotheker-de Groot, M., van der Knaap, J. A., Ponstein, A. S., Sela-Buurlage, M. A., Bol, J. F., Cornelissen, B. J. C., van den Elzen, P. J. M., and Linthorst, H. J. M., 1994, A new class of tobacco chitinases homologous to bacterial exo-chitinases displays antifungal activity, *Plant J.* **5:**469–480.

Melchior, F., and Kindl, H., 1991, *Arch. Biochem. Biophys.* **288:**552–557.

Memelink, J., Linthorst, H. J. M., Schilperoort, R. A., and Hoge, H. C., 1990, Tobacco genes encoding acidic and basic pathogenesis-related proteins display different expression patterns, *Plant Mol. Biol.* **14:**119–126.

Meyer, P., 1995, *Gene Silencing in Higher Plants and Related Phenomena in Other Eukaryotes*, Springer-Verlag, Berlin.

Michelmore, R., 1995, Molecular approaches to manipulation of disease resistance genes, *Annu. Rev. Phytopathol.* **15:**393–427.

Mikkelsen, J. D., Berglund, L., Nielsen, K. K., Christiansen, H., and Bojsen, K., 1992, Structure of endochitinase genes from sugar beets, in *Advances in Chitin and Chitosan* (C. J. Brine, P. A. Sandford, and J. P. Zikakis, eds.), pp. 344–353, Elsevier Science Publishers, London.

Mitchell, R. E., 1984, The relevance of non-host specific toxins in the expression of virulence by pathogens, *Annu Rev. Phytopathol.* **22:**215–245.

Molina, A., Segura, A., and Garcia-Olmedo, F., 1993, Lipid transfer proteins (nsLTPs) from barley and maize leaves are potent inhibitors of bacterial and fungal plant pathogens, *FEBS* **316:**119–122.

Moreno, M., Segura, A., and Garcia-Olmedo, F., 1994, Pseudothionin-St1, a potato peptide active against potato pathogen, *Eur. J. Biochem.* **223:**135–139.

Mosqueda, G., van den Broek, G., Saucedo, O., Bailey, A. M., Alvarez-Morales, A., and Herrera-Estrella, L., 1990, Isolation and characterisation of the gene from *Pseudomonas syringae* pv. *phaseolicola* encoding the phaseolotoxin-insensitive ornithine carbamoyltransferase, *Mol. Gen. Genet.* **222:**461–466.

Neale, A. D., Wahleithner, J. A., Lund, M., Bonnett, H. T., Kelly, A., Meeks-Wagner, D. R., Peacock, W. J., and Dennis, E. S., 1990, Chitinase, β-1,3-glucanase, osmotin and extensin are expressed in tobacco explants during flower formation, *Plant Cell* **2:**673–684.

Neuenschwander, U., Vernooij, B., Friedrich, L., Uknes, S., Kessmann, H., and Ryals, J., 1995, Is hydrogen peroxide a second messenger of salicylic acid in systemic acquired resistance? *Plant J.* **8:**227–233.

Nicholson, R. L., and Hammerschmid, R., 1982, Phenolic compounds and their role in disease resistance, *Annu. Rev. Phytopathol.* **30:**369–389.

Rao, A. G., 1995, Antimicrobial peptides, *Mol. Plant–Microbe Interact.* **8:**6–13.

Roberts, W. K., and Selitrennikoff, C. P., 1986, Isolation and partial characterization of two antifungal proteins from barley, *Biochim. Biophys. Acta* **880:**161–170.

Samac, D. A., Hironoka, C. M., Yallay, P. E., and Shah, D. M., 1990, Isolation and characterization of the genes encoding basic and acidic chitinase in *Arabidopsis thaliana*, *Plant Physiol.* **93:**907–914.

Schäfer, W., Straney, D., Ciuffetti, L., VanEtten, H. D., and Yoder, O. C., 1989, One enzyme makes a fungal pathogen, but not a saprophyte, virulent on a new host plant, *Science* **246:**247–249.

Sela-Buurlage, M. B., Ponstein, A. S., Bres-Vloemans, S. A., Melchers, L. S., van den Elzen, P. J. M., and Cornelissen, B. J. C., 1993, Only specific tobacco (*Nicotiana tabacum*) chitinases and β-1,3-glucanases exhibit antifungal activity, *Plant Physiol.* **101**:857–863.

Shah, D. M., Rommens, C. M. T., and Beachy, R. N., 1995, Resistance to diseases and insects in transgenic plants: Progress and applications to agriculture, *TIBTECH* **13**:362–368.

Siebertz, B., Logemann, J., Willmitzer, L., and Schell, J., 1989, *Cis*-analysis of the wound-inducible promoter *wun1* in transgenic tobacco plants and histochemical localization of its expression, *Plant Cell* **1**:961–968.

Staskawicz, B. J., Panopoulos, N. J., and Hoogenraad, N. J., 1980, Phytotoxin-insensitive ornithine carbamoyltransferase of *Pseudomonas syringae* pv. *phaseolicola*. Basis for immunity to phaseolotoxin, *J. Bacteriol.* **142**:720–723.

Staskawicz, B. J., Ausubel, F. M., Baker, B. J., Ellis, J. G., and Jones, J. D. G., 1995, Molecular genetics of plant disease resistance, *Science* **268**:661–667.

Stein, U., and Blaich, R., 1985, Untersuchungen über Stilbenproduktion und Botrytisanfälligkeit bei Vitis-Arten, *Vitis* **24**:75–87.

Stintzi, A., Geoffroy, D., Bersuder, D., Fritig, B., and Legrand, M., 1993a, cDNA cloning and expression studies of tobacco class III chitinases-lysozymes, in *Developments in Plant Pathology*, Vol. 2 (B. Fritig and M. Legrand, eds.), pp. 312–315, Kluwer, Dordrecht, The Netherlands.

Stintzi, A., Heitz, T., Prasad, V., Wiedemann-Merdinoglu, S., Kauffmann, S., Geoffroy, P., Legrand, M., and Fritig, B., 1993b, Plant 'pathogenesis-related' proteins and their role in defense against pathogens, *Biochimie* **75**:687–706.

Stirpe, F., and Hughes, R. C., 1989, Specificity of ribosome inactivating proteins with RNA N-glycosidase activity, *Biochem. J.* **262**:1001–1002.

Stirpe, F., Barbieri, L., Batelli, M. G., Soria, M., and Lappi, D. A., 1992, Ribosome-inactivating proteins from plants: Present status and future prospects, *Biotechnology* **10**:405–412.

Strittmatter, G., Janssens, J., Opsomer, C., and Botterman, J., 1995, Inhibition of fungal disease development in plants by engineering controlled cell death, *Biotechnology* **13**:1085–1089.

Strittmatter, G., Gheysen, G., Gianinazzi-Pearson, V., Hahn, K., Niebel, A., Rohde, W., and Tacke, E., 1996, Infections with various types of organisms stimulate transcription from a short promoter fragment of the potato *gst1* gene, *Mol. Plant–Microbe Interact.* **9**:68–73.

Terras, F. R. G., Schoofs, H. M. E., de Bolle, M. F. C., van Leuven, F., Rees, S. B., Vanderleyden, J., Cammue, B. P. A., and Broekaert, W. F., 1992, Analysis of two novel classes of antifungal proteins from radish (*Raphanus sativus* L.) seeds, *J. Biol. Chem.* **267**:15301–15309.

Terras, F. R. G., Eggermont, K., Kovaleva, V., Raikhel, N. V., Osborn, R. W., Kester, A., Rees, S. B., Torrekens, S., van Leuven, F., Vanderleyden, J., Cammue, B. P. A., and Broekaert, W. F., 1995, Small cysteine-rich antifungal proteins from radish: Their role in host defense, *Plant Cell* **7**:573–588.

Toyoda, H., Matsuda, Y., Yamaga, T., Ikeda, S., Morita, M., Tamai, T., and Ouchi, S., 1991, Suppression of the powdery mildew pathogen by chitinase microinjected into barley coleoptile epidermal cells, *Plant Cell Rep.* **10**:217–220.

Trudel, J., Audy, P., and Asselin, A., 1989, Electrophoretic forms of chitinase activity in Xanthi-nc tobacco, healthy and infected with tobacco mosaic virus, *Mol. Plant–Microbe Interact.* **2**:315–324.

Trudel, J., Potvin, C., and Asselin, A., 1992, Expression of active hen egg-white lysozyme in transgenic tobacco, *Plant Sci.* **87**:55–67.

Trudel, J., Potvin, C., and Asselin, A., 1995, Secreted hen lysozyme in transgenic tobacco: Recovery of bound enzyme and *in vitro* growth inhibition of plant pathogens, *Plant Sci.* **106**:55–62.

VanEtten, H. D., Matthews, D. E., and Matthews, P. S., 1989, Phytoalexin detoxification: Importance for pathogenicity and practical implications, *Annu. Rev. Phytopathol.* **27**:143–164.

Vierheilig, H., Alt, M., Neuhaus, J.-M., Boller, T., and Wiemken, A., 1993, Colonization of transgenic *N. sylvestris* plants, expressing different forms of *N. tabacum* chitinase, by the root pathogen *Rhizotonia solani* and by the mycorrhizal symbiont *Glomus mosseae*, *Mol. Plant–Microbe Interact.* **6:**261–264.

Vornam, B., Schön, H., and Kindl, H., 1988, Control of gene expression during induction of cultured peanut cells: mRNA levels, protein synthesis and enzyme activity of stilbene synthase, *Plant Mol. Biol.* **10:**235–243.

Ward, E. R., Payne, G. B., Moyer, M. B., Williams, S. C., Dincher, S. S., Sharkey, K. C., Beck, J. H., Taylor, H. T., Ahl-Goy, P., Meins, F., and Ryals, J. A., 1991a, Differential regulation of β-1,3-glucanase messenger RNAs in response to pathogen infection, *Plant Physiol.* **96:**390–397.

Ward, E. R., Uknes, S. J., Williams, S. C., Dincher, S. S., Wiederhold, D. L., Alexander, D. C., Ahl-Goy, P., Metraux, J. P., and Ryals, J. A., 1991b, *Plant Cell* **3:**1085–1094.

Weltring, K.-M., Turgeon, B. G., Yoder, O. C., and VanEtten, H. D., 1988, Isolation of a phytoalexin-detoxification gene from the plant pathogenic fungus *Nectria haematococca* by detecting its expression in *Aspergillus nidulans*, *Gene* **68:**335–344.

Wessels, J. G. H., and Sietsma, J. H., 1981, Fungal cell walls: A survey, in *Encyclopedia of Plant Physiology*, New Series, Vol. 13B (W. Tanner and F. A. Loewus, eds.), pp. 352–394, Springer-Verlag, Berlin.

Woodward, S., and Pearce, R. B., 1988, The role of stilbenes in resistance of Sitka spruce (*Picea sitchensis* (Bong.) Carr.) to entry of fungal pathogens, *Physiol. Mol. Plant Pathol.* **33:**127–149.

Wu, G., Shortt, B. J., Lawrence, E. B., Levine, E. B., Fitzsimmons, K. C., and Shah, D. M., 1995, Disease resistance conferred by expression of a gene encoding $H_2O_2$-generating glucose oxidase in transgenic potato plants, *Plant Cell* **7:**1357–1368.

Zhu, Q., Maher, E. A., Masoud, S., Dixon, R. A., and Lamb, C. J., 1994, Enhanced protection against fungal attack by constitutive co-expression of chitinase and glucanase genes in transgenic tobacco, *Biotechnology* **12:**807–812.

*Chapter 7*

# Plant–Fungal Interactions and Plant Disease

Wolfgang Knogge

## 1. INTRODUCTION

Fungi are eukaryotic, carbon-heterotrophic microorganisms most of which are living saprophytically by degrading dead organic material. However, a small minority of them have acquired the capability to recruit nutrients from living plants, thereby usually causing disease in the host. However, plants are far from being the unprotected prey of any potential pathogen. Instead, they successfully ward off most offenders through the utilization of a large array of passive and active defense mechanisms. Hence, fungal pathogenicity and high virulence, the extreme form of which has become apparent in several devastating epidemics of crop plants, require the interplay of a number of factors and processes on both the fungal and the plant side. Over the years, many review articles have covered the manifold plant resistance reactions (Alexander *et al.*, 1994; Bowles, 1990; Kombrink and Somssich, 1995; Lamb, 1994; Tzeng and DeVray, 1993), so that this topic will be dealt with only briefly in the present chapter. Instead, molecular aspects of fungal pathogenicity and virulence will be illustrated using a number of selected examples. In addition, recent results on the mechanisms of genetically defined resistance of plants will be described that may allow the development of a general concept of the plant defense system.

---

**Wolfgang Knogge** Department of Biochemistry, Max-Planck-Institut für Züchtungsforschung, D-50829 Cologne, Germany.

*Subcellular Biochemistry, Volume 29: Plant–Microbe Interactions*, edited by Biswas and Das. Plenum Press, New York, 1998.

## 2. FUNGAL INVASION OF PLANTS

In order to implement the host state in plants, fungal microorganisms have evolved strategies to attach to plant surfaces, invade the plant tissue, optimize growth in the plant, and propagate. Bacteria and viruses, as well as some opportunistic fungal parasites, but also some serious fungal plant pathogens such as the rusts, invade plants through natural openings or wounds. However, most pathogenic fungi possess mechanisms to actively traverse the plant's outer structural barriers such as the cuticle and the epidermal cell wall in the case of leaf pathogens.

### 2.1. Attachment and Appressorium Formation

After landing on a plant leaf, an airborne spore of a fungal pathogen germinates, attaches firmly to the hydrophobic leaf surface, and usually produces a specialized infection structure, an appressorium, at the end of the germ tube. Not much is known about the attachment process and the chemical nature of the adhesives produced to enable a tight aggregation (for review see Mendgen and Deising, 1993). However, for the rice blast fungus, *Magnaporthe grisea*, it was recently suggested that a specific type of fungal protein, a hydrophobin, is involved in surface recognition and formation of appressoria as well as of condia (Beckerman and Ebbole, 1996; Talbot *et al.*, 1996). Hydrophobins are small, secreted hydrophobic proteins with a characteristic cysteine pattern, C-CC-C-C-CC-C (Wessels, 1994), that have been found in aerial hyphae and several developmental structures of fungi. In addition, for two hydrophobins, cerato-ulmin from the Dutch elm disease pathogen, *Ophiostoma novo-ulmi* (Bowden *et al.*, 1994), and cryparin from the chestnut blight fungus, *Cryphonectria parasitica* (Zhang *et al.*, 1994), a role as phytotoxins has been proposed, but not confirmed (Bowden *et al.*, 1996; Carpenter *et al.*, 1992).

During appressorium formation, disease symptom development, and conidiation, *M. grisea* expresses a pathogenicity gene, *MPG1* (Talbot *et al.*, 1993). This gene encodes a hydrophobin and is necessary for infection-related development of the fungus on rice leaves and for full pathogenicity toward susceptible cultivars. Conidia from gene disruption (Mpg1$^-$) mutants of *M. grisea* and from a wild-type isolate were allowed to germinate on Teflon membranes. Detachment studies using different solvents either uncapable or capable of dissolving hydrophobins *in vitro* revealed that appressoria from Mpg1$^-$ strains could be more easily removed from the membranes than those of the isogenic wild type (Talbot *et al.*, 1996), indicating that the *MPG1* gene product together with other compounds is involved in the attachment of fungal structures to the leaf surface. In addition, however, the MPG1 protein may act as a morphogenetic signal for appressorium morphogenesis (Beckerman and Ebbole, 1996). Evidence comes from the observation that Mpg1⁻ strains display a reduced frequency of appressorium development. This phenotype, but not the easier detachment from membranes, could be remediated by cAMP, indicating that this compound acts downstream of the *MPG1* function. In addition, it was shown that the *cpkA* gene encoding the cAMP-dependent protein kinase catalytic subunit is required for appressorium formation and pathogenesis of *M. grisea* (Mitchell and Dean, 1995). Therefore, cAMP appears to be part of a signaling pathway that is initiated by fungal surface attachment and that causes the development of appressoria (Lee and Dean, 1993).

The surface of plants carries a complex mixture of hydrophobic materials collectively called wax. These compounds may also serve as signals during early stages of a plant–fungus interaction (Kolattukudy *et al.*, 1995). Germination and appressorium formation of spores of *Colletotrichum gloeosporioides* are selectively induced by surface wax of its host, avocado, but not by other waxes (Podila *et al.*, 1993). Conversely, avocado wax did not induce these developmental processes in other *Colletotrichum* species. Among the different wax fractions, the fatty alcohols were found to show the highest activity. Synthetic aliphatic *n*-fatty alcohols with 24 or more carbons also induced germination and appressorium formation in *C. gloeosporioides*. Such alcohols are present in many plant waxes and may be released by the activity of fungal hydrolytic enzymes. The inactivity of the fatty alcohol fraction of waxes from other species suggests the presence of inhibitors of fungal development of unknown structure, as was indicated by the inhibition of the inducing activity of avocado wax by the addition of other plant waxes (Kolattukudy *et al.*, 1995).

## 2.2. Penetration

Following attachment and appressorium formation, the next step in pathogenesis is the invasion of plant tissues. For this purpose, fungal phytopathogens have evolved two different mechanisms, enzymatic and mechanical penetration. During germination and penetration, fungi generally secrete a mixture of hydrolytic enzymes including cutinases, cellulases, pectinases, and proteases. These types of enzymes are also required by saprophytes and are therefore unlikely to represent the tools specifically evolved by fungi to implement pathogenesis. It is likely, however, that their structures and biosynthetic regulation are adapted to the specific needs on particular host plants.

The plant cuticle needs to be pierced before other pathogenetic mechanisms can become effective. Therefore, enzymatic degradation of cutin, the structural polymer of the plant cuticle, has been postulated to be crucial for fungal pathogenicity, and cutinase to be a key player in the penetration process. Several lines of evidence indicated the pivotal role of cutinase in the process that leads to the entrance of the fungal infection peg (reviewed by Kolattukudy, 1985). Inhibition of the enzyme using different chemical inhibitors or cutinase-specific antibodies was shown to prevent infection by *Nectria haematococca* (*Fusarium solani* f.sp. *pisi*) of pea stem segments with intact cuticle, but not of those with mechanically breached cuticle. Furthermore a wound pathogen, *Mycosphaerella* sp., naturally lacking cutinase, was able to penetrate only after transformation with the cutinase gene, *CUT1*, from *N. haematococca* (Dickman *et al.*, 1989). However, cutinase-deficient mutants of *N. haematococca* (Stahl and Schäfer, 1992) and of *M. grisea* (Sweigard *et al.*, 1992), generated by transformation-mediated gene disruption showed the same infectivity on their respective hosts as wild-type strains. While for *M. grisea* this result can be explained by a different penetration mechanism (see below), in case of *N. haematococca* the assay conditions appear to be important. Since this fungus can also penetrate through the stomata, a drastically reduced infectivity of a cutinase-deficient mutant was only detectable when low inoculum titers were used (Rogers *et al.*, 1994). Consequently, the capability to degrade cutin indeed appears to be important for the infectivity of *N. haematococca*. However, the relative role of cutinase for penetration may vary in different fungi and other mechanisms may compensate for the lack of this enzyme. In addition, cutinase may be involved in prepenetration processes by altering the adhesive properties of the cuticle, thus facilitating fungal attachment to plant surfaces (Nicholson and Epstein, 1991), and by releasing hydroxy fatty acids that trigger the expression of the cutinase gene in germinating fungi (Woloshuk and Kolattukudy, 1986).

Alternatively or in addition to hydrolytic enzymes, some fungi have developed a mechanism to mechanically penetrate the cuticle of host plants. After firm attachment of the fungal appressorium to the plant surface the porosity of the appressorium wall is drastically reduced by melanin incorporation, allowing the establishment of a high turgor pressure (>8 Mpa; Howard *et al.*, 1991). This pressure is focused effectively to a small area at the base of the appressorium which is kept free of wall material and melanin. From this penetration pore, an infection peg develops and pierces through the cuticle and cell wall, possibly assisted by hydrolytic enzymes (reviewed by Mendgen and Deising, 1993). For *M. grisea*, it could be demonstrated that melanin-deficient mutants were unable to infect intact plants, but some mutants retained pathogenicity on leaves with abraded epidermis (Kubo and Furusawa, 1991). Furthermore, melanized appressoria of *M. grisea* were capable of developing penetration pegs through plastic membranes (Howard *et al.*, 1991). In addition, pathogenicity of a melanin-nonproducing albino mutant of the cucumber pathogen, *Colletotrichum lagenarium*, could be restored by transformation with a melanin biosynthetic gene (Kubo *et al.*, 1991). These results suggest that for mechanically penetrating fungi, melanin is an essential factor that may not require additional enzyme activities.

Other fungal species such as the rusts have not acquired a direct penetration mechanism and bypass the plant cuticle and outer cell walls by entering through the stomata. Instead of a penetration device, these fungi developed a not yet fully understood mechanism that utilizes chemical and topographical features of the host surface to precisely locate the stomata and to differentiate infection structures (Hoch *et al.*, 1987). Using artificial substrates, it was demonstrated that uredospore germlings of *Uromyces appendiculatus*, the causal agent of bean rust, recognize ridges of 0.4–0.7 μm height that are similar to the erected lips of the stomatal guard cells. In response, all bean rust races (Allen *et al.*, 1991a) as well as many other rust species (Allen *et al.*, 1991b) formed appressoria that were morphologically and functionally similar to those formed over stomata (Hoch *et al.*, 1987). Therefore, a sharp elevation change by about $^1/_{10}$ of the hyphal diameter appears to be the signal for cell differentiation, whereas the width of the ridge did not have an impact (Hoch *et al.*, 1987). The mechanism of this thigmosensing is still unknown, but changes in the arrangement of the cell's cytoskeleton appear to be involved (Hoch *et al.*, 1987; Tucker *et al.*, 1986).

Recently, in patch-clamp studies using protoplasts from germ tubes of *U. appendiculatus* a mechanosensitive ion channel was identified that passes a variety of cations, including $Ca^{2+}$ (Zhou *et al.*, 1991). It is therefore discussed that this channel may transduce the membrane stress induced by the leaf topography into an ion influx that may trigger the differentiation process. Furthermore, perturbation of germ tubes using glass micropipettes revealed that the most responsive region is within the apical 10 μm of the hyphae. This suggests that the thigmoreceptor(s) are localized at the tip of the growing germ tubes. It is this region to which appressorium formation is expected to be restricted (Corrêa and Hoch, 1995).

These examples demonstrate that fungal penetration is a process of varying complexity in different fungi that is likely to be controlled by a combination of several factors. In addition to fungal compounds, these factors also include physical plant surface features as well as chemical activators or inhibitors of fungal spore germination and germ tube penetration.

## 3. FUNGAL PATHOGENICITY AND VIRULENCE STRATEGIES

### 3.1. Fungal Toxins in Pathogenesis

Following penetration, the next important mechanism to enable or facilitate further fungal development in the plant is in many cases the secretion of toxins or plant hormone-like compounds to manipulate the plant tissue. Interference with a plant's physiology can be very rude, killing plant cells to gain access to the plant's nutrient reservoir, or more subtle, redirecting the cell machinery in the pathogen's favor (Keen, 1986). A frequently followed strategy to affect the host plant appears to be the production of phytotoxins. These compounds often impair very basic cellular functions that occur widely among plants. Host sensitivity to a particular toxin requires the presence of a specific toxin target. Alternatively, sensitivity may be caused by the absence of an efficient detoxification mechanism in the plant.

Although phytotoxins have been identified in a broad spectrum of pathogens, in most cases their actual contribution to pathogenesis remains unclear. Remarkable exceptions are the host-selective toxins, for which genetic and biochemical studies revealed their function as the determinants of host specificity; resistance or susceptibility to the toxin-producing fungus always correlates with insensitivity or sensitivity to the toxin. In contrast, host-nonselective toxins are active on host and nonhost species. They may be very important, but not sufficient for fungal pathogenesis on particular hosts. Alternatively, they may have been crucial in earlier stages of fungal evolution toward pathogenicity, but their role may be lessened in most plants by detoxification or other still unknown mechanisms.

### 3.1.1. Host-Nonselective Toxins

The mode of action of only a small portion of fungal phytotoxins has been disclosed. From studies with several toxins, however, the plant plasma membrane-localized $H^+$-ATPase emerged as a preferred target. This enzyme is a master regulator of the plant physiology (Serrano, 1989). By mediating ATP-dependent $H^+$ extrusion, an inwardly directed electrochemical proton gradient is generated at the plasma membrane that is required for a number of "uphill" ion and solute transport processes against concentration gradients. In this way, the $H^+$-ATPase is also involved in various turgor-related processes and in the direct regulation of the intracellular and extracellular pH (Briskin and Hanson, 1992; Michelet and Boutry, 1995).

The $H^+$-ATPase is activated by the carbotricyclic terpene glycoside fusicoccin, the major phytotoxic metabolite of the peach and almond pathogen *Fusicoccum amygdali* (Marrè, 1979; Marrè and Ballarin-Denti, 1985). As a consequence, the uptake of $K^+$ (and other cations), $Cl^-$, and water by the stomatal guard cells is increased, causing the disease-typical wilting of leaves by irreversibly opening the stomata (Ballio, 1991). The plasma membrane-localized fusicoccin receptor is a member of the 14-3-3 superfamily of eukaryotic regulatory proteins (Korthout and De Boer, 1994; Oecking *et al.*, 1994) that is present in all higher plants (C. Meyer *et al.*, 1993). In animals, these proteins have been found to be involved in signaling processes through protein kinase-related events and protein–protein interaction (Aitken *et al.*, 1992; Burbelo and Hall, 1995; Ferl, 1996). In plants, a protein kinase was recently characterized as putative effector of the fusicoccin receptor (Van der Hoeven *et al.*, 1996). This suggests that fusicoccin affects a signal transduction pathway in plants that leads to $H^+$-ATPase stimulation through alterations in the enzyme's phosphorylation state (cf. also Vera-Estrella *et al.*, 1994). The toxicity of fusicoccin in all higher plants (Aducci *et al.*, 1995) and the ubiquitous distribution of the fusicoccin receptor are in contradiction to the limited number of host species of *F. amygdali*. Despite the general potential of fusicoccin, additional, as-yet-unknown factors must therefore be responsible for restricting fungal development to only few host species.

Different from fusicoccin, sensitivity to tentoxin, the nonselective toxin from the broad-host-range pathogen *Alternaria alternata*, varies among plant species. Some families, such as the *Brassicaceae*, are insensitive, while others, such as the *Apiaceae* and *Asteraceae*, are sensitive to the toxin. In contrast, the toxin may be species-specific in some families, such as the *Solanaceae*, *Fabaceae*, and *Poaceae*, that comprise both sensitive and insensitive species (Durbin and Uchytil, 1977). Tentoxin is a cyclic tetrapeptide, cyclo(*N*-methyl-L-alanyl-L-leucyl-*N*-methyl-*trans*-dehydrophenylalanylglycyl (Meyer *et al.*, 1975), that causes seedling chlorosis and the arrest of growth in sensitive plants. Results from numerous studies on tentoxin activity indicate a very complex mode of action, including a number of conflicting reports (for review see Klotz, 1988). Recently, however, another fundamental plant process, the energy transfer in the light-driven photophosphorylation in chloroplasts, was suggested to be affected by tentoxin. The specific toxin target appears to be the $\alpha\beta$ subunit complex of the chloroplast coupling factor 1. Sequence comparison of the $\beta$-subunit-encoding *atpB* gene from six closely related *Nicotiana* species revealed a glutamate residue at position 83 to correlate with tentoxin insensitivity and an aspartate residue with sensitivity (Avni *et al.*, 1992). In addition, the presence of the $\alpha$ subunit is required for tentoxin binding (Hu *et al.*, 1993). However, it has not yet been unequivocally demonstrated that sensitivity of a plant species to tentoxin is indeed responsible for its susceptibility to the pathogen.

### 3.1.2. Host-Selective Toxins

The highest level of specificity is seen with the host-selective toxins. In contrast to the host-nonselective toxins, sensitivity to this group of toxins is restricted to particular cultivars of a plant species. Host-selective toxins were mainly found in two fungal genera, *Cochliobolus* (*Helminthosporium*) and *Alternaria*. *Cochliobolus* spp. have caused a number of devastating epidemics of cereal crop plants. Consequently, a great deal of research interest was devoted to unraveling the molecular mechanisms underlying the development of these diseases (for review see Walton and Panaccione, 1993). Best studied are the interactions of *C. carbonum* and *C. heterostrophus* (formerly *Helminthosporium may-*

*dis*) and their common host species maize, which will therefore serve to explain the two basic specificity-determining mechanisms, specific detoxification and presence of a specific toxin target.

A cultivar-specific detoxification mechanism is responsible for the resistance of maize to *C. carbonum*. Five fungal races can be distinguished based on their host specificity and lesion morphology. Races 1–3 have been described based on the lesion type they produce on maize (Welz and Leonard, 1993). Race 0 is nearly avirulent on maize (Welz and Leonard, 1988), whereas race 4 causes unusually large lesions on some maize lines (Dodd and Hooker, 1990). Race 1, the causal agent of Northern corn leaf spot of maize, produces the host-selective toxin HC-toxin, a cyclic tetrapeptide (D-prolyl-L-alanyl-D-alanyl-L-Aeo; Aeo = 2-amino-9,10-epoxy-8-oxodecanoic acid). It induces large lesions on maize genotypes homozygous for the recessive *hm1* gene, which conditions sensitivity to the toxin (Walton, 1993).

The toxin target is a ubiquitous enzyme, histone deacetylase. By inhibiting this enzyme the toxin appears to interfere with reversible histone acetylation and thus with the control of fundamental cellular processes such as formation of chromatin structure, cell cycle progression, and gene expression (Brosch *et al.*, 1995). It remains to be demonstrated whether the inhibition of defense gene expression in response to fungal attack is also included in the toxin effects. If this is the case, HC-toxin may also be regarded as a specific suppressor of the host defense response (Walton *et al.*, 1994a; see also next paragraph).

Early genetic studies indicated a single Mendelian locus, *TOX2*, to control the production of HC-toxin in race 1 of *C. carbonum* (Nelson and Ullstrup, 1964; Scheffer *et al.*, 1967). Molecular analysis revealed, however, that genetic cosegregation studies with plant pathogenic fungi may not always be sufficient to interpret the situation at the molecular level. It turned out that the *TOX2* locus of *C. carbonum*, race 1, is in fact a complex locus containing multiple copies of multiple genes that only segregated as a single locus in crosses between race 1 and race 2 because the latter has no equivalent (Walton *et al.*, 1994b). The locus contains two copies of a 22-kb region of DNA that is completely missing from toxin-nonproducing strains. This region includes the *HTS1* gene encoding HC-toxin synthetase, a multifunctional cyclic peptide synthetase (Scott-Craig *et al.*, 1992; Panaccione *et al.*, 1992). *HTS1* homologs have also been detected in other filamentous fungi producing cyclic peptides (Nikolskaya *et al.*, 1995). Among these is a related fungal species, *C. victoriae*, that synthesizes the host-selective toxin victorin (TOX3), a cyclic pentapeptide that causes susceptibility of oats carrying the *Vb* gene (Scheffer and Livingston, 1984).

Disruption of both *HTS1* copies in race 1 of *C. carbonum* led to a loss of toxin formation and to nonpathogenicity of the mutants. In contrast, mutants in which only one copy was disrupted retained both the ability to produce the toxin and pathogenicity (Panaccione *et al.*, 1992). However, transfer of the complete

22-kb DNA region into a wild-type Tox2⁻ strain failed to generate HC-toxin-producing transformants, indicating that *HST1* is necessary, but not sufficient for HC-toxin synthesis (Walton *et al.*, 1994b). Subsequently, three additional genes, *TOXA*, *TOXC*, and *TOXD*, with a possible role in HC-toxin production have been found (Ahn and Walton, 1996; Panaccione *et al.*, 1992; Walton *et al.*, 1994b). *TOXA* encodes a putative HC-toxin efflux pump and is present in two copies located immediately 5′ of each *HTS1* copy. *TOXC* is present in three copies in strain SB111 of *C. carbonum*, race 1. Gene disruption experiments indicate that the gene product is required for HC-toxin formation, probably for the synthesis of Aeo. *TOXD* is also present in three copies in strain SB111. This gene, however, does not yet have any demonstrated role in HC-toxin biosynthesis. All copies of the three genes as well as both copies of the *HTS1* gene are located on the same 3.5-Mb chromosome in strain SB111, thus expanding the *TOX2* locus to >540 kb (Ahn and Walton, 1996).

Resistance of maize to race 1 of *C. carbonum* and insensitivity to HC-toxin are based on a single dominant gene, *Hm1*, encoding HC-toxin reductase (HCTR), which became the first plant resistance gene to be cloned (Johal and Briggs, 1992; Meeley *et al.*, 1992). HCTR inactivates HC-toxin by NADPH-dependent reduction of the 8-keto group of the functionally essential amino acid Aeo (Kim *et al.*, 1987; Meeley and Walton, 1991, 1993). Interestingly, HCTR activity was also found in other monocots such as wheat, barley, oat, and sorghum, but not in dicots (Meeley and Walton, 1993), indicating that the enzyme may also contribute to host species specificity.

The absence of a specific target protein is the cause of insensitivity of maize to a different host-selective toxin, T-toxin, and resistance to race T of *C. heterostrophus*. Cytoplasmic male sterility (*cms*), known in three types (T, C, and S), is maternally inherited and leads to pollen abortion. To avoid mechanical emasculation, *cms-T* cytoplasm was exploited in hybrid corn production until in 1969 and 1970 a new severe epidemic occurred called Southern corn leaf blight. The causal agent of this disease was race T of *C. heterostrophus*, which was found to produce a family of toxic linear long-chain polyketides collectively called T-toxin (Kono and Daly, 1979; Kono *et al.*, 1981). Interestingly, very similar phytotoxins (Danko *et al.*, 1984) targeting the same plant protein are produced by *Mycosphaerella zeae-maydis* (*Phyllosticta maydis*), a fungus causing yellow corn blight on *cms-T* maize (Levings *et al.*, 1995).

T-toxin production is controlled by the *TOX1* locus in *C. heterostrophus*, race T. The Tox1⁻ race 0 is also pathogenic to maize, but without exerting the extreme virulence of race T. The genomes of both races differ by about 1 Mb of Tox1⁺-unique DNA. In addition, a chromosomal rearrangement has occurred in race T. Like *TOX2* in *C. carbonum*, *TOX1* turned out not to be a single Mendelian locus. Instead, results from a combination of classical and molecular genetic experiments suggest it to be at least two loci. Races T and 0 differ by a reciprocal chromosomal translocation, and the translocation breakpoint was found to be linked to *TOX1* (Tzeng *et al.*, 1992). Both *TOX1* loci appear to be located at or near this breakpoint on the two chromosomes involved in the translocation. Both loci are required for T-toxin biosynthesis and for high virulence on *cms-T* maize. One of the loci was shown to encode a polyketide synthase that is necessary for T-toxin biosynthesis, while the function of the other locus is unknown (Turgeon *et al.*, 1995).

The T-toxin target is a 13-kDa integral protein of the maize inner mitochondrial membrane (Dewey *et al.*, 1987; Braun *et al.*, 1990) that is not essential for normal mitochondrial activity. Strong correlative evidence supports, however, a causal relationship between its expression and cytoplasmic male sterility (Rhoads *et al.*, 1995). This protein is encoded by the T-*urf13* gene that originated by a series of recombinational events involving other mitochondrial genes and that is only present in *cms-T* mitochondria (Rhoads *et al.*, 1995). Expression of the URF13 protein in *E. coli* and several eukaryotic organisms caused their sensitivity to T-toxin (Levings *et al.*, 1995). Interaction of T-toxin with URF13 leads to the formation of a hydrophilic pore consisting of at least six transmembrane α-helices across the inner mitochondrial membrane. A rapid depermeabilization of the inner mitochondrial membrane occurs leading to leakage of $NAD^+$ and other essential solutes (Matthews *et al.*, 1979). As a consequence, mitochondrial bioenergetics is disrupted and the affected cells die (Levings and Siedow, 1992; Siedow *et al.*, 1995). This suggests that URF13 acts as a ligand-gated, pore-forming receptor for T-toxin (Rhoads *et al.*, 1995).

In summary, the interactions of HC-toxin and T-toxin with their respective hosts exemplify the two alternative mechanisms that determine host specificity. A monogenically inherited active defense factor, HCTR, is responsible for HC-toxin inactivation resulting in incompatibiliy as the specific interaction. In contrast, the specific interaction of a unique fungal molecule, T-toxin, and a unique plant target, URF13, gives rise to compatibility as the specific interaction.

## 3.2. Fungal Suppressors of the Plant Defense Response

### 3.2.1. Triggering of the Active Plant Defense Response

Invasion of plant tissues by most potential pathogens is stopped by the plant active defense machinery. A multitude of reactions has been described that are activated upon pathogen attack. Among these reactions is the initiation of ion fluxes across the plasma membrane, the generation of active oxygen species (oxidative burst), the formation of compounds to strengthen the plant cell wall, the induction of rapid local cell death, the so-called hypersensitive response (HR), as well as the biosynthesis of fungitoxic compounds, the phytoalexins (for reviews see Alexander *et al.*, 1994; Bowles, 1990; Kombrink and Somssich,

1995; Lamb, 1994; Tzeng and DeVray, 1993). In addition, many defense genes are not only activated locally at the site of infection, but also systemically in distant non-infected plant parts, a phenomenon called systemic acquired resistance (Ryals *et al.*, 1994).

Prerequisite for the activation of the plant defense response is an efficient surveillance system that allows the rapid recognition of the threat. It has long been known that crude fungal wall preparations trigger defense reactions in plant cells. Hence, plants have evolved the capacity to perceive signals that are either exposed on fungal surfaces or secreted by invading pathogens. In addition, plants are capable of recognizing fractions of their own cell walls such as oligogalacturonates that are released through the activity of extracellular fungal enzymes. Activators of plant defense reactions are termed exogenous elicitors if they are synthesized by pathogens, whereas endogenous elicitors originate from the plant. Unraveling the molecular basis of the plant recognition system has been a research goal for many years, but only recently were the first molecules identified that are involved in perception and transduction of defense-related signals.

A major limitation in studying the mechanisms underlying plant cell responses to pathogenic microorganisms has been the lack of defined and fully characterized elicitors. However, several purified elicitors such as specific chitin- and β-glucan-derived oligosaccharides and oligopeptide elicitors from various fungi have meanwhile been analyzed extensively (for reviews see Côté and Hahn, 1994; Ebel and Cosio, 1994; Ebel *et al.*, 1995; Ebel and Scheel, 1994; Knogge, 1997) and plant cell membrane-localized, defense-associated binding proteins and receptors, respectively, were identified (Cheong and Hahn, 1991; Cosio *et al.*, 1992; Basse *et al.*, 1992; Nürnberger *et al.*, 1995). Binding of an elicitor to a receptor is thought to initiate one or several signaling pathways that lead to the activation of plant defense reactions and to plant resistance (Lamb *et al.*, 1989; Lamb, 1994).

### 3.2.2. Evidence for the Role of Fungal Suppressors in Pathogenesis

Successful fungal pathogens need to avoid recognition or to eliminate, overcome, or escape the host defense response. Avoidance of recognition may be accomplished by either eliminating or disguising putative elicitors through structural modification. This may be difficult, however, if elicitors represent important basic structural components, for instance, of the hyphal wall. An alternative may therefore be the production of suppressors that inhibit or sufficiently delay plant defense reactions (Bushnell and Rowell, 1981; Heath, 1981). These molecules may interfere either directly with elicitor-receptor binding, with various steps in the signaling process, with gene activation, or even with the activity of defense-related compounds from the plant.

The existence of suppressors was postulated based on the observation that plant tissue can be conditioned toward general susceptibility (Heath, 1982). The successful infection by virulent fungal races frequently renders the tissue more susceptible to normally avirulent fungi. This has been interpreted such that the activity of fungal molecules suppresses the plant defense response (Ouchi *et al.*, 1979). Alternatively, a metabolic state of susceptibility may be induced in plant tissues during establishment of a compatible interaction that precludes or delays the triggering of the active plant defense (Heath, 1982). Thus, nonoccurrence or delay of defense reactions may be the consequence rather then the cause of a successful colonization. Although highly intriguing, the biological relevance of suppressors remains unresolved.

Suppressors of plant defense reactions have been isolated from only a few fungal pathogens (for review see Beissmann and Kogel, 1992), the first one being detected in *Phytophthora infestans* (Doke *et al.*, 1979). After inoculation with a virulent race of *P. infestans*, potato tuber discs showed a suppression of tissue browning and phytoalexin accumulation upon inoculation with an avirulent fungal race. In contrast, inoculation of the tissue with an avirulent race did not cause protection from a subsequent infection with a virulent race. These observations suggested the presence of HR suppressors in the virulent race (Garas *et al.*, 1979). Such compounds were subsequently isolated from mycelia and zoospore suspensions (Doke *et al.*, 1979) and partially characterized revealing $\beta$-1 $\rightarrow$ 3 and $\beta$-1 $\rightarrow$ 6 linked neutral and anionic glucans of 17–23 glucose units, the anionic character being due to phosphorylation. These glucans suppressed tissue browning, phytoalexin biosynthesis, and the protoplast HR (Doke and Tomiyama, 1980) possibly by interfering with elicitor binding (Doke *et al.*, 1979). This suggested their involvement in suppressing the potato defense response during infection (Doke *et al.*, 1980). Furthermore, the higher suppressor activity of the glucans from a virulent race as compared to those from an avirulent race was interpreted as evidence for their function as determinants of race-cultivar specificity. This is difficult to conceive, however, in light of genetic data arguing strongly that avirulence to single potato resistance genes is controlled by single dominant genes in the fungus (Shaw, 1991). Hence, the outcome of a fungus–plant interaction would be decided by a factor produced by an avirulent fungal race. The actual relevance of the glucan compounds during the interaction of *P. infestans* and potato remains to be demonstrated.

Suppressor-determined race-cultivar specificity was also discussed for the interaction of soybean and *Phytophthora sojae*. An extracellular fungal glycoprotein, an invertase, inhibited phytoalexin accumulation in wounded soybean cotyledons elicited by a cell wall glucan (Ziegler and Pontzen, 1982). It could be demonstrated that the glycomoiety containing mainly mannose residues was responsible for this effect. The enzyme analyzed from three fungal races possessed different carbohydrate structures (Ziegler and Albersheim, 1977) and the inhibition appeared to be race-specific. However, also in this interaction proof for the function of the suppressor *in vivo* and a genetic analysis of suppressor activity are missing.

For *P. sojae*, an additional suppressor has been described, the storage polysaccharide mycolaminaran. This intracellular β-1,6-branched β-1,3-glucan was highly purified from fungal cytosol and shown to suppress elicitor-induced phytoalexin production in soybean cotyledons (Yoshikawa and Sugimoto, 1993). Binding studies revealed that mycolaminaran inhibited the binding of the glucanase-released elicitor from fungal cell walls to soybean microsomal membranes by competing for the elicitor-binding site. However, the biological relevance of mycolaminaran as a suppressor compound *in vivo* needs to be substantiated, because earlier results had also indicated a relatively weak elicitor activity of this polysaccharide (Keen *et al.*, 1983). In addition, it is not clear whether this glucan actually occurs extracellularly during fungal growth in the plant. The competing effect may therefore be fortuitous and simply due to a structural similarity to the β-glucan elicitor, allowing binding to the same site without triggering defense responses.

The *Pgt* elicitor from the wheat stem rust fungus, *Puccinia graminis* f.sp. *tritici* (Kogel *et al.*, 1988), was detected not only in germ tube walls, but also in apoplastic fluids from infected susceptible wheat leaves (Beissmann and Kogel, 1992), without eliciting resistance symptoms such as an HR. The purified elicitor was active, however, in resistant as well as in susceptible noninfected leaves, suggesting the presence of pathogen-derived suppressor molecules in susceptible leaves. Preliminary data indicated that the HR was prevented by apoplastic fluids from infected leaves as well as by uredospore germination fluids (Beissmann and Kogel, 1992). From the latter source, a 56-kDa protein termed ESP56 was purified (K. H. Kogel, personal communication) that inhibited the *Pgt* elicitor-inducible β-1,3-glucanase accumulation in wheat leaves. Upon injection into leaves prior to inoculation, ESP56 enhanced the susceptibility of a genotypically resistant wheat cultivar to the fungal pathogen. Furthermore, upon coinjection into leaves the protein strongly suppressed the *Pgt* elicitor-inducible phenylalanine ammonia-lyase (PAL) activity, whereas the activity of two other elicitors, chitooligosaccharides and DEAE dextran, was not affected. First studies on the mode of action of ESP56 revealed that *Pgt* inhibited elicitor binding to wheat plasma membranes in a noncompetitive manner. In addition, the effective concentration for suppression of elicitor-induced defense responses was found to be the same as for the inhibition of elicitor binding. However, the suppressor appears to counteract only the activity of the nonspecific *Pgt* elicitor, since preliminary results indicate that defense reactions induced in incompatible interactions in a cultivar-specific manner are not inhibited (K. H. Kogel, personal communication). This suggests that the general elicitor–suppressor interaction is superimposed by specific elicitors determining race-cultivar specificity (Heath, 1981).

Best studied is the suppressor from the pea pathogen *Mycosphaerella pinodes*.

Pycnospore germination fluids of strain OMP-1 contain both a high-molecular-mass polysaccharide elicitor (~70 kDa) and a low-molecular-mass suppressor fraction (<5 kDa). From the suppressor-active fraction two compounds were purified and termed suppescins A and B (Shiraishi *et al.*, 1992). Both are mucin-type glycopeptides with related structures, (A) α-Gal*N*Ac-*O*-Ser-Ser-Gly and (B) β-Gal(1 → 4)-α-Gal*N*Ac-*O*-Ser-Ser-Gly-Asp-Glu-Thr, indicating their common biosynthetic origin. Most studies were performed using a partially purified fraction containing both supprescins (Shiraishi *et al.*, 1978). Treatment of pea leaves with the suppressors increased the infection frequency of several nonpathogenic fungi (Shiraishi *et al.*, 1978, 1991b). Furthermore, the suppressor specificity coincided with the host range of *M. pinodes*. The fungus is capable of infecting different leguminous species to varying degrees. When species that are highly susceptible to the pathogen were treated with the supprescins, they became highly susceptible to the nonpathogen *Alternaria alternata*. Conversely, supprescin treatment of species that are less susceptible to *M. pinodes* resulted in low *A. alternata* infection levels. This indicates that the supprescins may be the determinants of host species specificity for *M. pinodes* (Oku *et al.*, 1980; Shiraishi *et al.*, 1991a, b).

Elicitors from *M. pinodes* and two other *Mycosphaerella* species induced the accumulation of the phytoalexin pisatin in pea leaves. In pea epicotyls it could be demonstrated that this elicitor rapidly induced the biosynthesis of two enzymes of the phytoalexin biosynthetic pathway, phenylalanine ammonia-lyase (PAL) and chalcone synthase (Yamada *et al.*, 1989). This process was delayed in the simultaneous presence of elicitor and suppressor. Even after the induction of phytoalexin biosynthesis by the elicitor, addition of the suppressor caused a decrease in pisatin accumulation. Subsequent analysis of pisatin accumulation kinetics in the presence of different concentrations of elicitor and suppressor indicated that both compounds may not compete for the same target in the plant (Shiraishi *et al.*, 1991b).

A similar delay of pisatin accumulation as observed for the suppressor was also found in the presence of *ortho*-vanadate, an inhibitor of the plant plasma membrane $H^+$-ATPase (Yoshioka *et al.*, 1990). Therefore, in the search for the primary site of suppressor action, the influence of supprescins A and B and their peptide moieties on this membrane enzyme was analyzed (Kato *et al.*, 1993). *In vitro*, supprescin A had no effect on pea plasma membrane $H^+$-ATPase activity, whereas supprescin B was a noncompetitive inhibitor. In contrast, the peptide moieties of both supprescins inhibited $H^+$-ATPase activity in a competitive manner. In addition, a more detailed analysis of the inhibitory effect using different peptides suggested that the hexapeptide of supprescin B may include two active parts, one affecting the ATP-binding site, the other the phosphatase domain of the $H^+$-ATPase (Kato *et al.*, 1993). Since supprescin A reduced the inhibitory effect of supprescin B on enzyme activity, but both showed an additive effect on elicitor-induced pisatin accumulation (Shiraishi *et al.*, 1992), the presence of more than one site of suppressor action was suggested. Alternatively, the suppressor activity may affect fundamental functions in pea cells such as the polyphosphoinositide metabolism (Toyoda *et al.*, 1992, 1993).

Interestingly, inhibition of $H^+$-ATPase activity by the supprescins also occurred in isolated membranes from four nonhost species. However, after treating leaves with the suppressors cytochemical observations employing the lead precipitation technique in combination with electron microscopy revealed that the enzyme was only inhibited in the host plant, pea (Shiraishi *et al.*, 1991a). Taken together, these data indicate that the suppressors may not function simply by inhibiting elicitor binding to a receptor in pea membranes, but rather by affecting the signaling pathway that leads to the activation of the resistance response. This was further substantiated by the observation that the supprescins rapidly but transiently reverse the elicitor-induced transcription of the PAL gene resulting in a delayed defense response of the plant (Wada *et al.*, 1995).

Experimental evidence from the work on suppressors supports the hypothesis that these fungal factors may indeed be involved in mediating susceptibility of the host. The suppressors characterized so far appear to act at the species level, counteracting the activity of general elicitors. This would be in agreement with the model that basic susceptibility of a plant species is obtained through specific adaptation by a particular fungal species (Heath, 1981). Hence, suppressors are the products of fungal pathogenicity genes. Unfortunately, nothing is known about the genetic background of suppressor formation.

### 3.3. Detoxification of Antifungal Compounds

Not only the fungal pathogens, but also their hosts utilize toxins in their confrontation. To confine fungal development, many plants produce structurally widely diverse compounds with antifungal activity. If accumulated in unchallenged plants, they are termed phytoanticipins; if primarily synthesized upon infection, they are called phytoalexins (VanEtten *et al.*, 1994b). These compounds usually show a low level of specificity toward particular pathogens. To obstruct their impact, pathogens need to develop tolerance to these compounds or detoxification mechanisms.

The steroidlike saponins that occur in more than 100 plant families represent an example of phytoanticipins. The toxicity of these antibiotics is thought to originate from their ability to form complexes with membrane sterols, thereby causing a loss of membrane integrity (Bangham and Horne, 1962). However, several fungal species are capable of detoxifying saponins. This was recently shown for the root-infecting causal agent of the take-all disease of cereals, *Gaeumannomyces graminis* (Bowyer *et al.*, 1995). Most oat species accumulate the glycosylated triterpenoid saponin, avenacin A-1, in the epidermal cells of their roots, whereas one oat species, *Avena longiglumis*, as well as wheat are saponin-noncontaining plants. Wheat isolates of the fungus (*G. g.* var. *tritici*) are nonpathogenic on saponin-containing oats and sensitive to avenacin A-1. In contrast, oat isolates (*G. g.* var. *avenae*) are insensitive due to the production of an enzyme, avenacinase, that detoxifies the saponin by hydrolytic cleavage of terminal glucose residues. Disruption of the avenacinase gene in *G. g.* var. *avenae* yielded transformants that were no longer able to infect oats, but maintained their pathogenicity to wheat.

A homologous mechanism was found in the tomato pathogen *Septoria lycopersici*. This fungus produces an enzyme, $\beta_2$-tomatinase, which detoxifies the tomato steroidal alkaloid saponin, $\alpha$-tomatine, by removal of a single glucose residue to yield $\beta_2$-tomatine (Sandrock *et al.*, 1995; Osbourn *et al.*, 1995). However, transformation of the nonpathogenic fungus *Aspergillus nidulans* with the $\beta_2$-tomatinase gene did not yield transformants that were insensitive to $\alpha$- and $\beta_2$-tomatine. This indicates that $\beta_2$-tomatinase expression is not the only mechanism to cope with $\alpha$-tomatine and that *S. lycopersici*, but not *A. nidulans*, possesses additional nondegradative tolerance mechanisms (Sandrock *et al.*, 1995). In summary, the interactions of *G. graminis* and different cereals and of *S. lycopersici* and tomato provide examples of the role of fungal phytoanticipin inactivation. By gaining a new function, a detoxifying enzyme, the fungus obliterates an important plant defense strategy and opens a plant species for colonization.

A detoxification strategy can also be envisaged for the inactivation of phytoalexins which also have been associated with restricting the spectrum of fungal pathogens of a given plant species. For a long time it was believed that detoxification of the pea phytoalexin pisatin by pisatin demethylase (PDA) from *Nectria haematococca* constitutes the decisive step in fungal pathogenicity to this plant (VanEtten *et al.*, 1989). All naturally occurring isolates that lacked PDA and all Pda$^-$ progeny were essentially nonpathogenic on pea. However, Pda$^-$ mutants generated by disruption of the *PDA* gene remained virulent on pea. Instead, it turned out that the *PDA* gene is located on a small dispensable chromosome. Only if this chromosome is lost do the fungal isolates become nonpathogenic. This indicates that the chromosome carries, in addition to *PDA*, other, as-yet-unknown genes required for pathogenicity and that naturally occurring Pda$^-$ isolates lack the entire chromosome (VanEtten *et al.*, 1994a). Although the original hypothesis of a phytoalexin-detoxifying enzyme as determinant of compatibility no longer holds, these findings point to another possible speciation strategy of phytopathogenic fungi. Fungal dispensable chromosomes may be functionally analogous to bacterial plasmids which allow these microorganisms to occupy specific biological niches. The dispensable chromosome of *N. haematococca* contains genes that permit individual isolates of this broad-host-range pathogen to colonize different host plants (VanEtten *et al.*, 1994a).

## 4. GENETICALLY CONTROLLED PLANT RESISTANCE

This review has mainly dealt with characteristics of fungal pathogenicity on their host plants. Disease is the exception, however, rather than the rule and plants are resistant to the vast majority of potential pathogenic microorganisms. Some of the involved resistance mechanisms, such as the production of phytoanticipins or the activation of defense reactions in response to the interaction with general elicitors, are rather unspecific. There is, however, a highly specific resistance to particular races of pathogenic bacterial or fungal species. Data obtained with many such pathosystems have shown that these specific plant–pathogen interactions are under the direct genetic control of complementary gene pairs, avirulence genes in the pathogen and resistance genes in the host (gene-for-gene interactions) (Flor, 1955, 1971; Keen, 1990, 1992).

The elicitor–receptor model provides a biochemical interpretation of these genetic findings by presuming an avirulence gene to encode a race-specific elicitor of the plant defense response that interacts with a resistance gene-encoded receptor (Keen, 1982). The first avirulence gene was cloned more than 10 years ago from *Pseudomonas syringae* pv. *glycinea,* a bacterial pathogen of soybean, through genetic complementation of virulent strains (Staskawicz *et al.*, 1984). Since then, many such genes have been isolated from different bacteria in a similar way (for review see Dangl, 1994). However, none has been shown to encode an extracellular elicitor and their mode of action in triggering the plant defense remains unknown in almost all cases. In the light of recent results, the gene-for-gene phenomenon is now interpreted in more general terms; the products of pathogen avirulence genes interact directly or indirectly with host resistance genes, thus triggering the plant defense response. The avirulence gene *AvrD* from *Pseudomonas syringae* exemplifies that the gene product itself may not be the eliciting factor (Keen *et al.*, 1990). In addition, several lines of evidence point to the fact that the interaction of the gene products may occur at the plant plasma membrane, but also inside the plant cells (see below). The following section will focus on fungal avirulence genes cloned to date. However, the discussion on resistance genes will also include those involved in protection against bacteria.

### 4.1. Fungal Avirulence Genes

Genetic complementation using a shotgun approach as used for the isolation of bacterial avirulence genes is not feasible with fungi due to their bigger genome sizes. In addition, efficient transformation protocols are not available for many fungal plant pathogens, in particular for the biotrophic fungi. An alternative would be a positional cloning approach that was recently successfully used to isolate avirulence genes from *M. grisea* (Kang *et al.*, 1995; Sweigard *et al.*,

1995). However, many important phytopathogenic fungi belong to the Deuteromycetes (Imperfect Fungi). Their lack of a known sexual stage makes these fungi inaccessible by genetic analysis, thus preventing a map-based cloning strategy. Therefore, the first fungal avirulence genes were cloned from the tomato leaf mold pathogen, *Cladosporium fulvum* (Joosten *et al.*, 1994; Van Kan *et al.*, 1991), and the barley scald pathogen, *Rhynchosporium secalis* (Rohe *et al.*, 1995), only after identifying their products, race-specific elicitors, that are active exclusively in host cultivars of a specific resistance genotype.

### 4.1.1. Cultivar-Specific Avirulence

From *C. fulvum,* two avirulence genes, *Avr4* and *Avr9,* were cloned (Joosten *et al.*, 1994; Van Kan *et al.*, 1991; Van den Ackerveken *et al.*, 1992). Both encode elicitors of the hypersensitive response in tomato cultivars carrying the complementary resistance genes *Cf-4* and *Cf-9*. The gene products, cysteine-rich pre-pro-proteins, are processed by fungal and/or plant proteases to yield active proteins of 105 (AVR4) and 28 (AVR9) amino acids, respectively (Joosten *et al.*, 1994; Van den Ackerveken *et al.*, 1993). *Avr9* is a unique gene in fungal races that are avirulent on *Cf-9* tomato plants (Van Kan *et al.*, 1991). In contrast, all fungal races contain the *Avr4* gene. However, while this gene is identical in all races avirulent on *Cf-4* tomato plants, the virulent races carry alleles with single nucleotide alterations frequently affecting cysteine residues and, in one race, with a frameshift mutation leading to a truncated gene product. The *Avr4* gene is transcribed in all races, but the products of the mutated alleles were not detectable, indicating that these proteins are either unstable or not secreted (Joosten *et al.*, 1994; De Wit, 1995).

The *nip1* gene from *R. secalis* encodes an 82-amino acid pro-protein including a 22-amino acid secretory signal peptide. The mature protein is also cysteine-rich and elicits defense reactions in barley plants possessing the resistance gene *Rrs1* in a manner very similar to avirulent fungal races (Hahn *et al.*, 1993; Rohe *et al.*, 1995). A hypersensitive response is not included in the response of barley to *R. secalis* infection (Lehnackers and Knogge, 1990) or NIP1 treatment. All avirulent fungal races carry the *nip1* gene, whereas virulent races either lack the gene or contain an allele with single nucleotide exchanges resulting in elicitor inactivity of the gene product (Rohe *et al.*, 1995).

In the rice blast fungus, *M. grisea,* a gene family was identified that is involved in the gene-for-gene relationship with rice cultivars (Valent and Chumley, 1994). The *AVR2-YAMO* gene determining avirulence on cultivar Yashiro-mochi encodes a 223-amino acid protein that contains a sequence motif homologous to the active center of neutral $Zn^{2+}$-proteases (De Wit, 1995). Fungal virulence on cultivar Yashiro-mochi results from DNA insertion into

*AVR2-YAMO* or from deletions in the gene. In addition, point mutations in the putative protease active site lead to a loss of function, indicating that an enzymatically active gene product may be required for avirulence expression. Although direct evidence for protease activity of AVR2-YAMO is missing, this avirulence gene product may not be an elicitor itself, but may function by releasing an active elicitor from a plant or fungal precursor molecule.

#### 4.1.2. Species-Specific Avirulence

The gene-for-gene hypothesis was originally formulated for the subspecies interaction of flax and flax rust (Flor, 1995, 1971). However, there is no reason to assume that a similar type of interaction cannot be found at the species level (Heath, 1991). Consequently, single-gene-inherited avirulence was recently identified in the interaction of *M. grisea* and different grasses (Valent and Chumley, 1991, 1994). The *PWL* (pathogenicity to weeping lovegrass) genes from this fungus encode glycine-rich, hydrophilic proteins with putative secretory signal sequences that appear to trigger the defense response in different grass species.

*M. grisea* is pathogenic to more than 50 grass species. However, the fungus exists in genetically distinct, asexually reproducing populations that colonize different species. In a cross of two isolates, one from weeping lovegrass (*Eragrostis curvula*), the other from finger millet (*Eleusine coracana*), a single segregating gene, *PWL1*, was identified. Another gene, *PWL2*, was found in a cross between two laboratory parental strains that both infect rice, but only one infecting weeping lovegrass (Valent and Chumley, 1991, 1994). *PWL2* was isolated by positional cloning (Sweigard *et al.*, 1995) and shown to be a member of a gene family with homologs in many fungal strains isolated from various host species. Subsequently, three additional genes, *PWL1*, and *PWL3* from a finger millet pathogen and *PWL4* from a weeping lovegrass pathogen, were isolated by their homology to *PWL2* (Kang *et al.*, 1995).

*PWL3* and *PWL4* map to the same locus and appear to be allelic. In contrast, *PWL1* and *PWL2* are localized at different genomic positions. Transformation experiments revealed that *PWL1* and *PWL2*, but not *PWL3* and *PWL4*, abolish the ability of *M. grisea* to infect weeping lovegrass without losing pathogenicity to other hosts. The inactivity of *PWL4* appears to be caused by its improper expression, because it became functional when under the control of the *PWL1* or *PWL2* promoters, while *PWL3* remained nonfunctional. Furthermore, the pathogenic allele *pwl2-2* was found to code for a protein with a single amino acid exchange resulting in the loss of function (Sweigard *et al.*, 1995). Therefore, nontranscription or introduction of point mutations into the primary sequence appear to be ways to retain pathogenicity in *PWL*-expressing fungal strains.

### 4.1.3. Genuine Functions of Avirulence Genes

Restriction of the host range by the expression of avirulence genes is disadvantageous for fungal pathogens. Therefore, their role as avirulence determinants may be coincidental and rather a function of the plant defense-related recognition system. However, the genuine function of avirulence genes remains poorly understood and may not become evident under laboratory conditions.

In the case of the *PWL* genes, sequence comparison and mutation frequencies higher than in the rest of the *M. grisea* genome strongly indicate that this gene family is highly dynamic and rapidly evolving (Kang *et al.*, 1995). Although spontaneous *PWL2* deletions did not impair fungal growth and development under laboratory conditions, only one field isolate was detected that lacked *PWL2*-homologous DNA (Sweigard *et al.*, 1995). From these findings it was concluded that the *PWL* genes may provide some beneficial function to the fungus growing under natural conditions in the field.

The *Avr4* gene of *C. fulvum* is induced in plants and expressed during pathogenesis in compatible interactions. However, instability or nonsecretion of the mutated *Avr4* gene products had no visible effect on fungal development in susceptible tomato plants, indicating that the gene is dispensable (De Wit, 1995; Joosten *et al.*, 1994). The *Avr9* gene is activated after fungal hyphae have passed through the stomata and highly expressed in hyphae growing in the vicinity of vascular tissue. Several lines of evidence suggest a possible function of the gene product under conditions of nitrogen limitation. *In vitro*, *Avr9* is only expressed in the presence of low nitrogen concentrations (Van den Ackerveken *et al.*, 1994). Such conditions are possibly found in the apoplast of tomato leaves. The *Avr9* promoter contains several copies of a sequence motif that was identified as the recognition site of regulatory proteins in *Neurospora crassa* (Caddick, 1992) and *Aspergillus nidulans* (Fu and Marzluf, 1990; Marzluf *et al.*, 1992). Deletion of a number of these elements from the *Avr9* promoter abolishes the inducibility of the gene under low nitrogen conditions, suggesting that they are functional in the transcriptional regulation of the *C. fulvum* gene (De Wit, 1995). Nevertheless, under laboratory conditions the *Avr9* gene appears to be dispensable for fungal development in the plant, as indicated by the lack of the gene in races virulent on Cf-9 tomato and by disruption mutants. As with the *M. grisea PWL* genes, a possible role of the gene for fungal development under field conditions has been discussed (De Wit, 1995).

In contrast to these avirulence genes, the *nip1* gene from *R. secalis* appears to encode a factor that, in addition to its role as avirulence determinant, appears to be involved in the expression of virulence. Evidence comes from a fungal *nip1* disruption transformant that gained virulence on resistant *Rrs1* barley, but displays a reduced level of virulence on susceptible *rrs1* plants. This phenotype is similar, however, to that of wild-type $Nip1^-$ races, regardless of the presence or absence of the *Rrs1* gene in the host (W. Knogge, unpublished data). A contribution of NIP1 to fungal virulence is further substantiated by the finding that it induces disease symptoms (hence the name, necrosis-inducing protein, NIP) in all barley cultivars tested, although at higher concentrations than those required for elicitor activity. Moreover, it is toxic to other mono- and to several dicotyledonous plants (Wevelsiep *et al.*, 1991; W. Knogge, unpublished data). At least in part, the toxic activity appears to be based on an indirect stimulation of the plasma membrane $H^+$-ATPase (Wevelsiep *et al.*, 1993), reminiscent of the fusicoccin effect (see above). Therefore, the host plant, barley, apparently utilizes a fungal virulence factor in the recognition process resulting in plant resistance (Knogge, 1991, 1996).

## 4.2. Structure and Function of Plant Disease Resistance Genes

In recent years, several disease resistance genes conferring resistance to viral, bacterial, fungal and nematode pathogens have been isolated from a number of plants, using positional cloning or transposon tagging strategies (for reviews see Boyes *et al.*, 1996; Dangl, 1995; Staskawicz *et al.*, 1995; Baker *et al.*, 1997). The deduced amino acid sequences suggest that the products of these genes have common structural features and fall into different classes. Some may be surface receptors for signal molecules of pathogen origin, whereas others appear to be components of intracellular signaling pathways. There is no reason to assume that resistance genes acting against the various classes of pathogens are fundamentally different. Therefore, the following discussion will not be restricted to those genes controlling resistance to fungal pathogens.

The genes *Cf-2* and *Cf-9* confer resistance of tomato to races of *C. fulvum* expressing the avirulence genes *Avr2* and *Avr9*, respectively (Dixon *et al.*, 1996; Jones *et al.*, 1994). They encode structurally similar proteins with N-terminal putative signal peptides, 28 (*Cf-2*) and 38 (*Cf-9*) leucine-rich repeats (LRRs), and C-terminal transmembrane domains, indicating their membrane-anchored extracytoplasmic localization. A similar protein with imperfect LRRs is encoded by the $HS1^{pro\text{-}1}$ gene conferring resistance of sugar beet to the cyst nematode *Heterodera schachtii* (Cai *et al.*, 1997). LRRs are structural motifs that have been found in many proteins, including plant receptor-like protein kinases (Braun and Walker, 1996). The LRR domains of several receptors are involved in protein–protein interactions (Kobe and Deisenhofer, 1994, 1995; Jones and Jones, 1997) suggesting a function of the *Cf* gene products as receptors of protein ligands such as the fungal avirulence gene products.

Binding studies using the AVR9 elicitor peptide from *C. fulvum* as a ligand revealed, however, similar affinities to membranes from *Cf-9* and *Cf-0* tomato cultivars, although the latter lacks resistance genes and thus does not respond to elicitor treatment, as well as from other solanaceous species (Kooman-Gersmann *et al.*, 1996). It is therefore an open question whether the elicitor indeed interacts directly with the *Cf-9* gene product or whether additional molecules are involved. Alternatively, the presence of inactive *Cf-9* homologs in *Cf-0* plants may indicate that the receptor is nonfunctional in *Cf-0* tomato plants and other species, thus preventing AVR9-specific signal transduction (De Wit, 1995; Staskawicz *et al.*, 1995).

The predicted Cf-9 protein does not have any obvious cytoplasmic signaling capacity (Jones *et al.*, 1994). A cytoplasmic signal may therefore be generated by extracytoplasmic interaction of Cf-9 with a transmembrane protein that is attached to cytoplasmic signaling processes. Such a component, showing similarity with transmembrane receptor-like kinases from *Arabidopsis thaliana* such as RLK5 (Walker, 1993), TMK1 (Chang *et al.*, 1992), TMKL1 (Valon *et al.*, 1993), and PRK1 from *Petunia inflata* (Mu *et al.*, 1994), is encoded by the *Xa21* gene conferring resistance of rice to the bacterial pathogen *Xanthomonas oryzae* pv. *oryzae*, race 6. Following a putative signal sequence the protein has 23 LRRs, a transmembrane region, and a cytoplasmic serine/threonine protein kinase (STK) domain (Song *et al.*, 1995).

An STK domain with high similarity to Xa21 but no LRRs was also found in the predicted product of the *Pto* gene conferring resistance of tomato to *Pseudomonas syringae* pv. *tomato* (Martin *et al.*, 1993). The deduced structure of the Pto protein indicates its cytoplasmic localization with putative membrane attachment through N-terminal myristoylation. The Pto kinase (Loh and Martin, 1995) may therefore be a signal transducer interacting with the cytoplasmic domain of an as-yet-unknown transmembrane receptor whose extracellular domain binds the avrPto signal. The *Cf* gene products may also be able to interact with Pto-like proteins. Alternatively, the bacterial signal may directly interact intracellularly with the *Pto* gene product (see below).

All other resistance genes cloned to date are likely to be cytoplasmic. *RPS2* and *RPM1* confer resistance of *Arabidopsis thaliana* to *Pseudomonas syringae* isolates expressing the *avrRpt2* (Bent *et al.*, 1994; Mindrinos *et al.*, 1994) and the *avrRpml* or *avrB* genes (Bisgrove *et al.*, 1994; Grant *et al.*, 1995), respectively. The predicted products of these genes are highly similar, each possessing 14 LRRs, a putative N-terminal leucine zipper-type dimerization domain, and a nucleotide-binding site (Saraste *et al.*, 1990). The same number of LRRs and a nucleotide-binding site are also found in the product of the *N* gene conferring resistance of tobacco to tobacco mosaic virus (Dinesh-Kumar *et al.*, 1995; Whitham *et al.*, 1994).

Other proteins with similarity to *RPS2*, *RPM1*, and *N* are encoded by the *L6* and M genes conferring resistance of flax to the flax rust fungus, *Melampsora lini* (Lawrence *et al.*, 1995; Anderson *et al.*, 1997), and by the *RPP5* gene encoding resistance of *A. thaliana* to *Peronospora parasitica* (Parker *et al.*, 1997). The gene products contain hydrophobic N-termini with homology to the cytoplasmic domains of the interleukin-1 receptor and the *Drosophila* Toll protein. These domains are known to be involved in mediating downstream signaling in their respective systems (Dangl, 1995; Hultmark, 1994; Lamb, 1994). In contrast, no such domains, but 6–7 membrane spanning regions are found in the product of the barley *Mlo* gene (Büschges *et al.*, 1997). Recessive alleles of this gene mediate a broad, non-race-specific resistance reaction to powdery mildew (*Erysiphe graminis*).

The deduced amino acid sequences of the Cf-2, Cf-9, $HS1^{pro-1}$, and Xa21 proteins indicate their extracellular localization, which would be compatible with a possible role of resistance gene products in the perception of extracellular signals such as the AVR9 elicitor from *C. fulvum* or the NIP1 elicitor from *R. secalis*. The products of all other resistance genes appear to be intracellular proteins. This is not surprising in the case of the *N* gene product, because the intracellularly expressed viral replicase gene may be involved in the recognition process (Padgett and Beachy, 1993). It remains to be shown, however, how the cytoplasmic resistance gene products against fungi interact with their respective elicitors.

Until recently, very little information existed on the site and mode of action of bacterial avirulence gene products. Now data obtained with interactions involving the resistance genes *Pto*, *RPS2*, and *RPM1* provide evidence for intracellular recognition events, thus explaining the cytoplasmic localization of their products. The nonpathogenic bacteria *Pseudomonas fluorescens* and *E. coli* are capable of eliciting an HR that is specified by these resistance genes in *Arabidopsis* or tomato if they express the Hrp protein secretion system, the HrpZ harpin (He *et al.*, 1993), and the appropriate avirulence genes from *P. syringae* in the same cell. Furthermore, stable and transient expression of *avrB* in *Arabidopsis* resulted in *RPM1*-dependent necrosis, suggesting that the site of action of ArvB is indeed inside the plant cells (Gopalan *et al.*, 1996). Additional support that bacterial avirulence proteins are recognized inside host cells comes from the pepper-*Xanthomonas campestris* pv. *vesicatoria* pathosystem. Transient expression of the *avrBs3* gene in pepper leaves, using *Agrobacterium tumefaciens* for gene delivery, resulted in hypersensitive cell death, specifically on plants carrying the resistance gene *Bs3*. In addition, for its intracellular recognition, the AvrBs3 protein requires nuclear localization signals that are present in its C-terminal region. These results provide strong evidence that AvrBs3 is translocated into plant cells in a *Xanthomonas* Hrp type III secretion system and that nuclear factors are involved in AvrBs3 perception (Van den Ackerveken *et al.*, 1996).

### 4.3. Resistance-Related Signaling Pathways

Protein phosphorylation/dephosphorylation cascades have been shown to play a crucial role in regulating many cellular processes in such eukaryotes as animals and yeast (Hunter, 1995). Not surprisingly, changes in the level of protein phosphorylation are also included in the defense response of plants (Dietrich *et al.*, 1990; Felix *et al.*, 1991; Ranjeva and Boudet, 1987). For example, an STK was identified in tomato cells and inhibitor studies suggest this enzyme to be part of the signal transduction pathway that is activated by a fungal elicitor (Suzuki and Shinshi, 1995). In animals and yeast, a variety of substrates for signaling kinases is known including phosphatases, other kinases, cytoskeletal elements, and transcription factors. The recent finding that at least one of the components of a multimeric transcription factor that binds to the promoter of a defense-responsive gene, *PR-10a*, in potato appears to be positively regulated by phosphorylation upon treatment with the elicitor arachidonic acid or upon wounding (Després *et al.*, 1995).

The *avrPto*-specific defense response pathway currently provides the most advanced data on signaling pathways subsequent to resistance gene function and the role of protein phosphorylation. The resistance gene *Pto* is a member of a small gene family in tomato that also includes the *Fen* gene conferring sensitivity to the insecticide Fenthion (Martin *et al.*, 1994). Thus, the Pto and Fen proteins, which were demonstrated to be functional protein kinases with 80% identity (Loh and Martin, 1995), appear to interact with different ligands to initiate a signaling pathway resulting in both cases in cell death. Two additional signal transduction components, Prf (Salmeron *et al.*, 1994) and Pti (Zhou *et al.*, 1995), were identified. The *Prf* gene encodes a protein with a leucine zipper, a nucleotide-binding domain, and LRR motifs and is also within the *Pto* gene cluster, 24 kb from the *Pto* gene and 500 bp from the *Fen* gene (Salmeron *et al.*, 1996). Whereas *pto* mutants retained sensitivity to Fenthion, mutations in the *prf* locus resulted in both complete susceptibility to *P. s. tomato* and insensitivity to Fenthion, suggesting that *Prf* plays a role in signal transduction in response to both the pathogen elicitor and Fenthion.

Using the yeast two-hybrid system, another STK, Pti1 (Pto-interacting 1) was identified that physically interacts with Pto (Zhou *et al.*, 1995). In cross-phosphorylation assays Pto specifically phosphorylated Pti1, but Pti1 did not phosphorylate Pto, whereas Fen was unable to phosphorylate Pti1 and was not phosphorylated by Pti1. These findings indicate that Pti1 is part of the Pto-mediated signaling pathway, probably by acting as a component downstream of Pto in a phosphorylation cascade. It remains to be shown whether one of the *Pti* homologs in the tomato genome encodes an STK involved in the Fen pathway (Boyes *et al.*, 1996).

Evidence for recognition-specific signaling pathways has also been obtained in other plants. In mutation experiments additional genes were identified that have an impact on the function of particular resistance genes. In barley carrying the *Mla*$_{12}$ gene conferring resistance to powdery mildew (*Erysiphe graminis* f.sp. *hordei*), two unlinked genes, *Rar-1* and *Rar-2* (required for *Mla* -specified resistance; originally termed *Nar-1* and *Nar-2*) were detected (Freialdenhoven *et al.*, 1994). Both genes in addition to $Mla_{12}$ are required for triggering HR in attacked host cells as well as for high-level transcript accumulation of barley defense-related genes encoding chitinase, peroxidase, and pathogenesis-related protein 1.

Similarly, genes were detected that are required for the function of the barley *mlo* resistance gene. Mutation analysis identified two unlinked genes, *Ror-1* and *Ror-2* (required for *mlo*-specified resistance) that cause an intermediate phenotype compared to *mlo*-resistant and *Mlo*-susceptible barley upon inoculation with powdery mildew (Freialdenhoven *et al.*, 1996). However, the *mlo*-specified resistance is not affected by the *rar* mutations. Results from quantitative cytological analyses were interpreted such that the *Ror* genes may have a regulatory function in *mlo*-based resistance. A model was proposed in which the *Mlo* wild-type allele functions as a negative regulator and the *Ror* genes act as positive regulators of the non-race-specific resistance response.

The mutants in the barley *Rar* and *Ror* genes and possibly similar mutations detected in the *Rcr-1* and *Rcr-2* genes in Cf-9 tomato (required for full resistance to *C. fulvum* expressing the *Avr9* gene; Hammond-Kosack *et al.*, 1994) suggest a function of these genes in resistance gene-specific signaling pathways. Furthermore, different resistance genes appear to trigger different downstream responses, as was recently demonstrated using isogenic strains of *Pseudomonas syringae* pv. *maculicola* carrying either *avrRpt2* and/or *avrRpm1* or *avrB*, respectively. Inoculation of *Arabidopsis* ecotype Columbia (Col-0; genotype *RPS2/RPM1*) with bacteria carrying *avrRpt2* resulted in an early transcription of the genes *AIG1* and *AIG2* (*avrRpt2*-induced genes), but not of a different defense-related gene, *ELI3* (Reuber and Ausubel, 1996). Conversely, these *Arabidopsis* plants responded to inoculation with *avrRPM1*- or *avrB*-expressing bacteria with an early induction of *ELI3*, but not of *AIG1* and *AIG2*. Mixed-inoculum experiments using *avrRpt2* and *avrRpm1* bacteria revealed an interference between the *RPS2*- and *RPM1*-mediated resistance reactions, with the *avrRpm1* signal apparently blocking the *avrRpt2*-dependent induction of *AIG1*.

In another study using different bacterial strains harboring the same avirulence genes, however, the *RPS2*-specified reaction was found to be epistatic to that of *RPM1* (Ritter and Dangl, 1996). This was seen by analyzing HR timing (rapid induction mediated by *RPM1*, as opposed to slow induction by *RPS2*), bacterial growth in *planta*, and induction of *AIG1* and *ELI3*. This interference was demonstrated to occur outside the bacteria. An *Arabidopsis* Col-0-derived *rps2* mutant (encoding an *RPS2* gene product with a single amino acid exchange in the LRR region, but expressing a functional RPM1 gene) did not recognize bacteria expressing both avirulence genes. In addition, no HR was visible in mixed-inoculum experiments, suggesting that an *avrRpt2*-dependent signal represses the triggering of *RPM1*-specific resistance reactions even in the absence of a wild-type *RPS2* product. However, the interference could be overcome by a numerical excess of *avrRpm1*-expressing bacteria, indicating that the avirulence genes compete for some step in the signaling process.

Models for this interference and possible explanations for the discrepancy in the results from both groups have been discussed (Innes, 1996; Reuber and Ausubel, 1996; Ritter and Dangl, 1996). For the purpose of this chapter it suffices to note that these data strongly suggest the occurrence of qualitatively different resistance responses. These responses are mediated through distinct resistance gene-specific signaling pathways that may, however, connect at some point. Such cross-talk between signaling pathways originating from different receptors has already been demonstrated in animals and yeast. For the expression of plant resistance this may imply that adaptation to recognize the multitude of possible pathogen signal molecules may be based on diversification of the signaling pathways (Boyes *et al.*, 1996).

## 5. PERSPECTIVES

Recent years have seen tremendous progress in unraveling the molecular bases of plant–pathogen interactions. Not only the characterization of a number of plant resistance genes and of a multitude of plant defense reactions, but also the molecular analysis of fungal pathogenicity and virulence have substantially improved our understanding of the mechanisms that are responsible for plant disease or resistance. Many questions remain unanswered. However, fungal toxins undoubtedly have the potential to be crucial tools for future detailed analyses of fundamental cellular processes. Furthermore, the elucidation of the defense-related signal perception and signal transduction pathways of plants has come into reach through molecular genetic and biochemical analyses of resistance gene functions. Most excitingly, however, one can now address the questions of how new recognition specificities evolve in plants, whether there are indeed ancient relationships between the mechanisms of natural, or innate, immunity of mammals and insects and of polarity determination in the *Drosophila* embryo on the one hand and of plant resistance on the other hand (Briggs, 1995; Hultmark, 1994; Dangl, 1992; Dinesh-Kumar *et al.*, 1995), and how these findings may lead to the development of efficient strategies to improve disease resistance of crop plants.

ACKNOWLEDGMENTS. Drs. Jeff Dangl and Thorsten Nürnberger and my colleagues in the Department of Biochemistry at the Max-Planck-Institut für Züchtungsforschung are gratefully acknowledged for many extremely fruitful discussions on this topic over the years. The work in my laboratory was supported by grant No. 0136101 A from the Bundesministerium für Forschung und Technologie and by grant Kn225/3 from the Deutsche Forschungsgemeinschaft.

## 6. REFERENCES


Aducci, P., Marra, M., Fogliano, V., and Fullone, M. R., 1995, Fusicoccin receptors: Perception and transduction of the fusicoccin signal. *J. Exp. Bot.* **46:**1463–1478.

Ahn, J. H., and Walton, J. D., 1996, Chromosomal organization of *TOX2*, a complex locus controlling host-selective toxin biosynthesis in *Cochliobolus carbonum*, *Plant Cell* **8:**887–897.

Aitken, A., Collinge, D. B., Van Heusden, B. P. H., Isobe, T., Roseboom, P. H., Rosenfeld, G., and Soll, J., 1992, 14-3-3 proteins: A highly conserved, widespread family of eukaryotic proteins, *Trends Biochem. Sci.* **17:**498–501.

Alexander, D., Lawton, K., Uknes, S., Ward, E., and Ryals, J., 1994, Defense-related gene induction in plants, in *Genetic Engineering* (J. K. Setlow, ed.), pp. 195–212, Plenum Press, New York.

Allen, E. A., Hazen, B. E., Hoch, H. C., Kwon, Y., Leinhos, G. M. E., Staples, R. C., Stumpf, M. A., and Terhune, B. T., 1991a, Appressorium formation in response to topographical signals by 27 rust species, *Phytopathology* **81:**323–331.

Allen, E. A., Hoch, H. C., Stavely, J. R., and Steadman, J. R., 1991b, Uniformity among races of *Uromyces appendiculatus* in response to topographic signaling for appressorium formation, *Phytopathology* **81:**883–887.

Anderson, P. A., Lawrence, G. J., Morrish, B. C., Ayliffe, M. A., Finnegan, E. J., and Ellis, J. G., 1997, Inactivation of the flax rust resistance gene *M* associated with loss of a repeated unit within the leucine-rich repeat coding region, *Plant Cell* **9:**641–651.

Avni, A., Anderson, J. D., Holland, N., Rochaix, J. D., Gromet-Elhanan, Z., and Edelman, M., 1992, Tentoxin sensitivity of chloroplasts determined by codon 83 of β-subunit of proton-ATPase, *Science* **257:**1245–1247.

Baker, B., Zambryski, P., Staskawicz, B., and Dinesh-Kumar, S. P., 1997, Signaling in plant–microbe interactions, *Science* **276:**726–733.

Ballio, A., 1991, Non-host-selective fungal pathogens: Biochemical aspects and their mode of action, *Experientia* **47:**783–790.

Bangham, A. D., and Horne, R. W., 1962, Action of saponin on biological membranes, *Nature* **196:**952–953.

Basse, C. W., Bock, K., and Boller, T., 1992, Elicitors and suppressors of the defense response in tomato cells, *J. Biol. Chem.* **267:**10258–10265.

Beckerman, J. L., and Ebbole, D. J., 1996, *MPG1*, a gene encoding a fungal hydrophobin of *Magnaporthe grisea*, is involved in surface recognition, *Mol. Plant–Microbe Interact.* **9:**450–456.

Beissmann, B., and Kogel, K. H., 1992, Identification and characterization of suppressors, in *Modern Methods of Plant Analysis: Plant Toxin Analysis* (H. F. Linskens and J. F. Jackson, eds.), pp. 259–275, Springer-Verlag, Berlin.

Bent, A. F., Kunkel, B. N., Dahlbeck, D., Brown, K. L., Schmidt, R., Giraudet, J., Leung, J., and Staskawicz, B. J., 1994, *RPS2* of *Arabidopsis thaliana:* A leucine-rich repeat class of plant disease resistance genes, *Science* **265:**1856–1860.

Bisgrove, S. R., Simonich, M. T., Smith, N. M., Sattler, A., and Innes, R. W., 1994, A disease resistance gene in *Arabidopsis* with specificity for two different pathogen avirulence genes, *Plant Cell* **6:**927–933.

Bowden, C. G., Hintz, W. E., Jeng, R., Hubbes, M., and Horgen, P. A., 1994, Isolation and characterization of the cerato-ulmin toxin gene of the Dutch elm disease pathogen, *Ophiostoma ulmi*, *Curr. Genet.* **25:**323–329.

Bowden, C. G., Smalley, E., Guries, R. P., Hubbes, M., Temple, B., and Horgen, P. A., 1996, Lack of association between cerato-ulmin production and virulence in *Ophiostoma novo-ulmi*, *Mol. Plant–Microbe Interact.* **7:**556–564.

Bowles, D. J., 1990, Defense-related proteins in higher plants, *Annu. Rev. Biochem.* **59:**873–907.

Bowyer, P., Clarke, B. R., Lunness, P., Daniels, M. J., and Osbourn, A. E., 1995, Host range of a plant pathogenic fungus determined by a saponin detoxifying enzyme, *Science* **267:**371–374.

Boyes, D. C., McDowell, J. M., and Dangl, J. L., 1996, Plant pathology: Many roads lead to resistance, *Curr. Biol.* **6:**634–637.

Braun, C. J., Siedow, J. N., and Levings, C. S., III, 1990, Fungal toxins bind to the URF-13 protein in maize mitochondria and *Escherichia coli*, *Plant Cell* **2:**153–162.

Braun, D. M., and Walker, J. C., 1996, Plant transmembrane receptors: New pieces in the signaling puzzle, *Trends Biochem. Sci.* **21:**70–73.

Briggs, S. P., 1995, Grand unification theory in sight, *Curr. Biol.* **5:**128–131.

Briskin, D. P., and Hanson, J. B., 1992, How does the plant plasma membrane $H^+$-ATPase pump protons?, *J. Exp. Bot.* **43:**269–289.

Brosch, G., Ransom, R., Lechner, T., Walton, J. D., and Loidl, P., 1995, Inhibition of maize histone deacetylases by HC toxin, the host-selective toxin of *Cochliobolus carbonum*, *Plant Cell* **7:**1941–1950.

Burbelo, P. D., and Hall, A., 1995, Hot numbers in signal transduction, *Curr. Biol.* **5:**95–96.

Büschges, R., Hollricher, K., Panstruga, R., Simons, G., Wolter, M., Frijters, A., van Daelen, R., van der Lee, T., Diergarde, P., Groenendijk, J., Topsch, S., Vos, P., Salamini, F., and Schulze-Lefert, P., 1997, The barley *Mlo* gene: a novel control element of plant pathogen resistance, *Cell* **88:**695–705.

Bushnell, W. R., and Rowell, J. B., 1981, Suppressors of defense reactions: A model for roles in specificity, *Phytopathology* **71:**1012–1014.

Caddick, M. X., 1992, Characterization of a major *Aspergillus* regulatory gene, *areA*, in *Molecular Biology of Filamentous Fungi* (U. Stahl and P. Tudzinski, eds.), pp. 141–152, VCH, Weinheim, Germany.

Cai, D., Kleine, M., Kifle, S., Harloff, H. J., Sandal, N. N., Marcker, K. A., Klein-Lankhorst, R. M., Salentijn, E. M. J., Lange, W. Stiekema, W. J., Syss, U., Grundler, F. M. W., and Jung, C., 1997, Positional cloning of a gene for nematode resistance in sugar beet, *Science* **275:**832–834.

Carpenter, C. E., Mueller, R. J., Kazmierczak, P., Zhang, L., Villalon, D. K., and Van Alfen, N. K., 1992, Effect of a virus on accumulation of a tissue-specific cell-surface protein of the fungus *Cryphonectria parasitica*, *Mol. Plant–Microbe Interact.* **5:**55–61.

Chang, C., Schaller, G. E., Patterson, S. E., Kwok, S. F., Meyerowitz, E. M., and Bleecker, A. B., 1992, The *TMK1* gene from *Arabidopsis* codes for a protein with structural and biochemical characteristics of a receptor protein kinase, *Plant Cell* **4:**1263–1271.

Cheong, J. J., and Hahn, M. G., 1991, A specific, high affinity binding site for the hepta-β-glucoside elicitor exists in soybean membranes, *Plant Cell* **3:**137–147.

Corrêa, A., Jr., and Hoch, H. C., 1995, Identification of thigmoresponsive loci for cell differentiation in *Uromyces* germlings, *Protoplasma* **186:**34–40.

Cosio, E. G., Frey, T., and Ebel, J., 1992, Identification of a high-affinity binding protein for a hepta-β-glucoside phytoalexin elicitor in soybean, *Eur. J. Biochem.* **204:**1115–1123.

Côté, F., and Hahn, M. G., 1994, Oligosaccharins: Structures and signal transduction, *Plant Mol. Biol.* **26:**1379–1411.

Dangl, J. L., 1992, The major histocompatibility complex à la carte: Are there analogies to plant disease resistance genes on the menu? *Plant J.* **2:**3–11.

Dangl, J. L., 1994, The enigmatic avirulence genes of phytopathogenic bacteria, in *Bacterial Pathogenesis of Plants and Animals* (J. L. Dangl, ed.), pp. 99–118, Springer-Verlag, Berlin.

Dangl, J. L., 1995, Pièce de résistance: Novel classes of plant disease resistance genes, *Cell* **80:**363–366.

Danko, S. J., Kono, Y., Daly, J. M., Suzuki, Y., Takeuchi, S., and McCrery, D. A., 1984, Structure and biological activity of a host-specific toxin produced by the fungal corn pathogen *Phyllosticta maydis*, *Biochemistry* **23**:759–766.

Després, C., Subramaniam, R., Matton, D. P., and Brisson, N., 1995, The activation of the potato *PR-10a* gene requires the phosporylation of the nuclear factor PBF-1, *Plant Cell* **7**:589–598.

Dewey, R. E., Timothy, D. H., and Levings, C. S., III, 1987, A mitochondrial protein associated with cytoplasmic male sterility in the T cytoplasm of maize, *Proc. Natl. Acad. Sci. USA* **84**:5374–5378.

De Wit, P. J. G. M., 1995, Fungal avirulence genes and plant resistance genes: Unravelling the molecular basis of gene-for-gene interactions, in *Advances in Botanical Research*, Vol. 21 (J. H. Andrews and I. C. Tommerup, eds.), pp. 147–185, Academic Press, London.

Dickman, M. B., Podila, G. K., and Kolattukudy, P. E., 1989, Insertion of cutinase gene into a wound pathogen enables it to infect intact host, *Nature* **342**:446–448.

Dietrich, A., Mayer, J. E., and Hahlbrock, K., 1990, Fungal elicitors trigger rapid, transient and specific protein phosphorylation in parsley cell suspension cultures, *J. Biol. Chem.* **265**:6360–6368.

Dinesh-Kumar, S. P., Whitham, S., Choi, D., Hehl, R., Corr, C., and Baker, B., 1995, Transposon tagging of tobacco mosaic virus resistance gene *N:* Its possible role in the TMV-*N*-mediated signal transduction pathway, *Proc. Natl. Acad. Sci. USA* **92**:4175–4180.

Dixon, M. S., Jones, D. A., Keddie, J. S., Thomas, C. M., Harrison, K., and Jones, J. D. G., 1996, The tomato *Cf-2* disease resistance locus comprises two functional genes encoding leucine-rich repeat proteins, *Cell* **84**:451–459.

Dodd, J. L., and Hooker, A. L., 1990, Previously undescribed pathotype of *Bipolaris zeicola* on corn, *Plant Dis.* **74**:530.

Doke, N., and Tomiyama, K., 1980, Effect of hyphal wall components from *Phytophthora infestans*, *Physiol. Plant Pathol.* **16**:169–176.

Doke, N., Garas, N. A., and Kuc, J., 1979, Partial characterization and aspects of the mode of action of the hypersensitivity-inhibiting factor (HIF) isolated from *Phytophthora infestans*, *Physiol. Plant Pathol.* **15**:127–140.

Doke, N., Garas, N. A., and Kuc, J., 1980, Effect on host hypersensitivity of suppressors released during the germination of *Phytophthora infestans* cystospores, *Phytopathology* **70**:35–39.

Durbin, R. D., and Uchytil, T. F., 1977, A survey of plant insensitivity to tentoxin, *Phytopathology* **67**:602–603.

Ebel, J., and Cosio, E. G., 1994, Elicitors of plant defense responses, *Int. Rev. Cytol.* **148**:1–36.

Ebel, J., and Scheel, D., 1997, Signals in host–parasite interactions, in *Plant Relationships*, Vol. VI (G. C. Carroll and P. Tudzinsky, eds.), pp. 85–105, Springer-Verlag, Berlin.

Ebel, J., Bhagwat, A. A., Cosio, E. G., Feger, M., Kissel, U., Mithöfer, A., and Waldmüller, T., 1995, Elicitor-binding proteins and signal transduction in the activation of a phytoalexin defense response, *Can. J. Bot.* **73**:S506–S510.

Felix, G., Grosskopf, D. G., Regenass, M., and Boller, T., 1991, Rapid changes of protein phosphorylation are involved in transduction of the elicitor signal in plant cells, *Proc. Natl. Acad. Sci. USA* **88**:8831–8834.

Ferl, R. J., 1996, 14-3-3 proteins and signal transduction, *Annu. Rev. Plant Physiol. Plant Mol. Biol.* **47**:49–73.

Flor, H. H., 1955, Host–parasite interactions in flax rust—Its genetics and other implications, *Phytopathology* **45**:680–685.

Flor, H. H., 1971, Current status of the gene-for-gene concept, *Annu. Rev. Phytopathol.* **9**:275–296.

Freialdenhoven, A., Scherag, B., Hollricher, K., Collinge, D. B., Thordal-Christensen, H., and Schulze-Lefert, P.., 1994, *Nar-1* and *Nar-2*, two loci required for $Mla_{12}$-specified race-specific resistance to powdery mildew in barley, *Plant Cell* **6**:983–994.

Freialdenhoven, A., Peterhänsel, C., Kurth, J., Kreuzaler, F., and Schulze-Lefert, P., 1996, Identification of genes required to the function of non-race-specific *mlo* resistance to powdery mildew in barley, *Plant Cell* **8**:5–14.

Fu, Y.-H., and Marzluf, G. A., 1990, *Nit-2*, the major positive-acting nitrogen regulatory gene of *Neurospora crassa*, encodes a sequence-specific DNA-binding protein, *Proc. Natl. Acad. Sci. USA* **87**:5331–5335.

Garas, N. A., Doke, N., and Kuc, J., 1979, Suppression of the hypersensitive reaction in potato tubers by mycelial components from *Phytophthora infestans*, *Physiol. Plant Pathol.* **15**:117–126.

Gopalan, S., Bauer, D. W., Alfano, J. R., Loniello, A. O., He, S. Y., and Collmer, A., 1996, Expression of the *Pseudomonas syringae* avirulence protein AvrB in plant cells alleviates its dependence on the hypersensitive response and pathogenicity (Hrp) secretion system in eliciting genotype-specific hypersensitive cell death, *Plant Cell* **8**:1095–1105.

Grant, M. R., Godiard, L., Straube, E., Ashfield, T., Lewald, J., Sattler, A., Innes, R. W., and Dangl, J. L., 1995, Structure of the *Arabidopsis RPM1* gene enabling dual specificity disease resistance, *Science* **269**:843–846.

Hahn, M., Jüngling, S., and Knogge, W., 1993, Cultivar-specific elicitation of barley defense reactions by the phytotoxic peptide NIP1 from *Rhynchosporium secalis*, *Mol. Plant–Microbe Interact.* **6**:745–754.

Hammond-Kosack, K. E., Jones, D. A., and Jones, J. D. G., 1994, Identification of two genes required in tomato for full *Cf*-9-dependent resistance to *Cladosporium fulvum*, *Plant Cell* **6**:361–374.

He, S. Y., Huang, H. C., and Collmer, A., 1993, *Pseudomonas syringae* pv. *syringae* harpin$_{Pss}$: A protein that is secreted via the Hrp pathway and elicits the hypersensitive response in plants, *Cell* **73**:1255–1266.

Heath, M. C., 1981, A generalized concept of host–parasite specificity, *Phytopathology* **71**:1121–1123.

Heath, M. C., 1982, The absence of active defense mechanisms in compatible host–pathogen interactions, in *Active Defense Mechanisms in Plants* (R. K. S. Wood, ed.), pp. 143–156, Plenum Press, New York.

Heath, M. C., 1991, The role of gene-for-gene interactions in the determination of host species specificity, *Phytopathology* **81**:127–130.

Hoch, H. C., Staples, R. C., Whitehead, B., Comeau, J., and Wolf, E. D., 1987, Signaling for growth orientation and cell differentiation by surface topography in *Uromyces*, *Science* **235**:1659–1662.

Howard, R. J., Ferrari, M. A., Roach, D. H., and Money, D. H., 1991, Penetration of hard substances by a fungus employing enormous turgor pressures, *Proc. Natl. Acad. Sci. USA* **88**:11281–11284.

Hu, N., Mills, D. A., Huchzermeyer, B., and Richter, M. L., 1993, Inhibition by tentoxin of cooperativity among nucleotide binding sites on chloroplast coupling factor 1, *J. Biol. Chem.* **268**:8536–8540.

Hultmark, D., 1994, Ancient relationships, *Nature* **367**:116–117.

Hunter, T., 1995, Protein kinases and phosphatases: the Yin and Yang of protein phosphorylation and signaling, *Cell* **80**:225–236.

Innes, R. W., 1996, Plant–pathogen interactions: Unexpected findings on signal input and output, *Plant Cell* **8**:133–136.

Johal, G. S., and Briggs, S. P., 1992, Reductase activity encoded by the *HM1* disease resistance gene in maize, *Science* **258**:985–987.

Jones, D. A., and Jones, J. D. G., 1997, The role of leucine-rich repeat proteins in plant defences, *Adv. Bot. Res.* **24**:89–167.

Jones, D. A., Thomas, C. M., Hammond-Kosack, K. E., Balint-Kurti, P. J., and Jones, J. D. G.,

1994, Isolation of the tomato *Cf-9* gene for resistance to *Cladosporium fulvum* by transposon tagging, *Science* **266:**789–793.

Joosten, M. H. A. J., Cozijnsen, A. J., and De Wit, P. J. G. M., 1994, Host resistance to a fungal tomato pathogen lost by a single base-pair change in an avirulence gene, *Nature* **367:**384–387.

Kang, S., Sweigard, J. A., and Valent, B., 1995, The *PWL* host specificity gene family in the blast fungus *Magnaporthe grisea, Mol. Plant–Microbe Interact.* **8:**939–948.

Kato, T., Shiraishi, T., Toyoda, K., Saitoh, K., Satoh, Y., Tahara, M., Yamada, T., and Oku, H., 1993, Inhibition of ATPase activity in pea plasma membranes by fungal suppressors from *Mycosphaerella pinodes* and their peptide moieties, *Plant Cell Physiol.* **34:**439–445.

Keen, N. T., 1982, Specific recognition in gene-for-gene host-parasite systems, *Adv. Plant Pathol.* **1:**35–82.

Keen, N. T., 1986, Pathogenic strategies of fungi, in *Recognition in Microbe–Plant Symbiotic and Pathogenic Interactions,* Vol. H 4 (B. Lugtenberg, ed.), pp. 171–188, Springer-Verlag, Berlin.

Keen, N. T., 1990, Gene-for-gene complementarity in plant–pathogen interactions, *Annu. Rev. Genet.* **24:**447–463.

Keen, N. T., 1992, The molecular biology of disease resistance, *Plant Mol. Biol.* **19:**109–122.

Keen, N. T., Yoshikawa, M., and Wang, M. C., 1983, Phytoalexin elicitor activity of carbohydrates from *Phytophthora megasperma* f.sp. *glycinea* and other sources, *Plant Physiol.* **71:**466–471.

Keen, N. T., Tamaki, S., Kobayashi, D., Gerhold, D., Stayton, M., Shen, H., Gold, S., Lorang, J., Thordal-Christensen, H., Dahlbeck, D., and Staskawicz, B., 1990, Bacteria expressing avirulence gene *D* produce a specific elicitor of the soybean hypersensitive reaction, *Mol. Plant–Microbe Interact.* **3:**112–121.

Kim, S.-D., Knoche, H. W., and Dunkle, L. D., 1987, Essentiality of the ketone function for toxicity of the host-selective toxin produced by *Helminthosporium carbonum, Physiol. Mol. Plant Pathol.* **30:**433–440.

Klotz, M. G., 1988, The action of tentoxin on membrane processes in plants, *Physiol. Plant.* **74:**575–582.

Knogge, W., 1991, Plant resistance genes for fungal pathogens—Physiological models and identification in cereal crops, *Z. Naturforsch.* **46c:**969–981.

Knogge, W., 1996, Fungal infection of plants, *Plant Cell* **8:**1711–1722.

Knogge, W., 1997, Elicitors and suppressors of the resistance response, in *Resistance of Crop Plants Against Fungi* (R. Heitefuss, H. H. Hoppe and H. Hartleb, eds.), pp. 159–182, Fischer Verlag, Jena, Germany.

Kobe, B., and Deisenhofer, J., 1994, The leucine-rich repeat: A versatile binding motif, *Trends Biochem. Sci.* **19:**415–421.

Kobe, B., and Deisenhofer, J., 1995, Proteins with leucine-rich repeats, *Curr. Opin. Struct. Biol.* **5:**409–416.

Kogel, G., Beissmann, B., Reisener, H. J., and Kogel, K. H., 1988, A single glycoprotein from *Puccinia graminis* f.sp. *tritici* cell walls elicits the hypersensitive lignification response in wheat, *Physiol. Mol. Plant Pathol.* **33:**173–185.

Kolattukudy, P. E., 1985, Enzymatic penetration of the plant cuticle by fungal pathogens, *Annu. Rev. Phytopathol.* **23:**223–250.

Kolattukudy, P. E., Rogers, L. M., Li, D., Hwang, C. S., and Flaishman, M. A., 1995, Surface signaling in pathogenesis, *Proc. Natl. Acad. Sci. USA* **92:**4080–4087.

Kombrink, E., and Somssich, I. E., 1995, Defense responses of plants to pathogens, in *Advances in Botanical Research,* Vol. 21 (J. H. Andrews and I. C. Tommerup, eds.), pp. 1–34, Academic Press, London.

Kono, Y., and Daly, J. M., 1979, Characterization of the host-specific pathotoxin produced by

*Helminthosporium maydis*, race T, affecting corn with Texas male sterile cytoplasm, *Bioorg. Chem.* **8**:391–397.

Kono, Y., Knoche, H. W., and Daly, J. M., 1981, Structure: Fungal host-specific, in *Toxins in Plant Disease* (E. D. Durbin, ed.), pp. 221–257, Academic Press, New York.

Kooman-Gersmann, M., Honée, G., Bonnema, G., and De Wit, P. J. G. M., 1996, A high affinity binding site for the AVR9 peptide elicitor of *Cladosporium fulvum* is present on plasma membranes of tomato and other solanaceous plants, *Plant Cell* **8**:929–939.

Korthout, H. A. A. J., and De Boer, A. H., 1994, A fusicoccin binding protein belongs to the family of 14-3-3 brain protein homologs, *Plant Cell* **6**:1681–1692.

Kubo, Y., and Furusawa, I., 1991, Melanin biosynthesis: Prerequisite for successful invasion of the plant host by appressoria of *Colletotrichum* and *Pyricularia*, in *The Fungal Spore and Disease Initiation in Plants and Animals* (G. T. Cole and H. C. Hoch, eds.), pp. 205–218, Plenum Press, New York.

Kubo, Y., Nakamura, H., Kobayashi, K., Okuno, T., and Furusawa, I., 1991, Cloning of a melanin biosynthetic gene essential for appressorial penetration of *Colletotrichum lagenarium*, *Mol. Plant–Microbe Interact.* **4**:440–445.

Lamb, C. J., 1994, Plant disease resistance genes in signal perception and transduction, *Cell* **76**:419–422.

Lamb, C. J., Lawton, M. A., Dron, M., and Dixon, R. A., 1989, Signals and transduction mechanisms for activation of plant defenses against microbial attack, *Cell* **56**:215–224.

Lawrence, G. J., Finnegan, E. J., Ayliffe, M. A., and Ellis, J. G., 1995, The *L6* gene for flax rust resistance is related to the *Arabidopsis* bacterial resistance gene *RPS2* and the tobacco viral resistance gene *N*, *Plant Cell* **7**:1195–1206.

Lee, Y. H., and Dean, R. A., 1993, cAMP regulates infection structure formation in the plant pathogenic fungus *Magnaporthe grisea*, *Plant Cell* **5**:693–700.

Lehnackers, H., and Knogge, W., 1990, Cytological studies on the infection of barley cultivars with known resistance genotypes by *Rhynchosporium secalis*, *Can. J. Bot.* **68**:1953–1961.

Levings, C. S., III, and Siedow, J. N., 1992, Molecular basis of disease susceptibility in the Texas cytoplasm of maize, *Plant Mol. Biol.* **19**:135–147.

Levings, C. S., III, Rhoads, D. M., and Siedow, J. N., 1995, Molecular interactions of *Bipolaris maydis* T-toxin and maize, *Can. J. Bot.* **73**:S483–S489.

Loh, Y.-T., and Martin, G. B., 1995, The *Pto* bacterial resistance gene and the *Fen* insecticide sensitivity gene encode functional protein kinases with serine/threonine specificity, *Plant Physiol.* **108**:1735–1739.

Marrè, E., 1979, Fusicoccin: A tool in plant physiology, *Annu. Rev. Plant Physiol.* **30**:273–288.

Marrè, E., and Ballarin-Denti, A., 1985, The proton pump of the plasmalemma and the tonoplast of higher plants, *J. Bioenerg. Biomembranes* **17**:1–21.

Martin, G. B., Brommonschenkel, S. H., Chunwongse, J., Frary, A., Ganal, M. W., Spivey, R., Wu, T., Earle, E. D., and Tanksley, S. D., 1993, Map-based cloning of a protein kinase gene conferring resistance in tomato, *Science* **262**:1432–1436.

Martin, G. B., Frary, A., Wu, T., Brommonschenkel, S., Chunwongse, J., Earle, E. D., and Tanksley, S. D., 1994, A member of the tomato *Pto* gene family confers sensitivity to fenthion resulting in rapid cell death, *Plant Cell* **6**:1543–1552.

Marzluf, G. A., Kanaan, M., and Fu, Y.-H., 1992, Molecular analysis of DNA-binding, *trans*-acting regulatory proteins of *Neurospora crassa*, in *Molecular Biology of Filamentous Fungi* (U. Stahl and P. Tudzynski, eds.), pp. 153–166, VCH, Weinheim, Germany.

Matthews, D. E., Gregory, P., and Gracen, V. E., 1979, *Helminthosporium maydis* race T toxin induces leakage of $NAD^+$ from T cytoplasm corn mitochondria, *Plant Physiol.* **63**:1149–1153.

Meeley, R. G., and Walton, J. D., 1991, Enzymatic detoxification of HC-toxin, the host-selective cyclic peptide from *Cochliobolus carbonum*, *Plant Physiol.* **97:**1080–1086.

Meeley, R. B., and Walton, J. D., 1993, Molecular biology and biochemistry of *Hm1*, a maize gene for fungal resistance, in *Advances in Molecular Genetics of Plant–Microbe Interactions*, Vol. 2 (E. W. Nester and D. P. S. Verma, eds.), pp. 463–467, Kluwer, Dordrecht, The Netherlands.

Meeley, R. B., Johal, G. S., Briggs, S. P., and Walton, J. D., 1992, A biochemical phenotype for a disease resistance gene of maize, *Plant Cell* **4:**71–77.

Mendgen, K., and Deising, H., 1993, Infection structures of fungal plant pathogens—A cytological and physiological evaluation, *New Phytol.* **124:**192–213.

Meyer, C., Waldkotter, K., Sprenger, A., Schlosser, U. G., Luther, M., and Weiler, E. W., 1993, Survey of the taxonomic and tissue distribution of microsomal binding sites for the non-host selective fungal phytotoxin, fusicoccin, *Z. Naturforsch.* **48c:** 595–602.

Meyer, W. L., Templeton, G. E., Grable, C. I., Jones, R., Kuiper, L. F., Lewis, R. B., Sigel, C. W., and Woodhead, S. M., 1975, Use of $^1$H nuclear magnetic resonance spectroscopy for sequence and configuration analysis of cyclic tetrapeptides. The structure of tentoxin, *J. Am. Chem. Soc.* **97:**3802–3809.

Michelet, B., and Boutry, M., 1995, The plasma membrane $H^+$-ATPase, *Plant Physiol.* **108:**1–6.

Mindrinos, M., Katagiri, F., Yu, G.-L., and Ausubel, F. M., 1994, The *A. thaliana* disease resistance gene *RPS2* encodes a protein containing a nucleotide-binding site and leucine-rich repeats, *Cell* **78:**1089–1099.

Mitchell, T. K., and Dean, R. A., 1995, The cAMP-dependent protein kinase catalytic subunit is required for appressorium formation and pathogenesis by the rice blast pathogen *Magnaporthe grisea*, *Plant Cell* **7:**1869–1878.

Mu, J. H., Lee, H. S., and Kao, T. H., 1994, Characterization of a pollen-expressed receptor-like kinase gene of *Petunia inflata* and the activity of its encoded kinase, *Plant Cell* **6:**709–721.

Nelson, O. E., and Ullstrup, A. J., 1964, Resistance to leaf spot in maize, *J. Hered.* **55:**195–199.

Nicholson, R. L., and Epstein, L., 1991, Adhesion of fungi to the plant surface, in *The Fungal Spore and Disease Initiation in Plants and Animals* (G. T. Cole and H. C. Hoch, eds.), pp. 3–23, Plenum Press, New York.

Nikolskaya, A. N., Panaccione, D. G., and Walton, J. D., 1995, Identification of peptide synthetase-encoding genes from filamentous fungi producing host-selective phytotoxins or analogs, *Gene* **165:**207–211.

Nürnberger, T., Nennstiel, D., Hahlbrock, K., and Scheel, D., 1995, Covalent cross-linking of the *Phytophthora megasperma* oligopeptide elicitor to its receptor in parsley membranes, *Proc. Natl. Acad. Sci. USA* **92:**2338–2342.

Oecking, C., Eckerskorn, C., and Weiler, E. W., 1994, The fusicoccin receptor of plants is a member of the 14-3-3 superfamily of eukaryotic regulatory proteins, *FEBS Lett.* **352:**163–166.

Oku, H., Shiraishi, T., Ouchi, S., and Ishiura, M., 1980, A new determinant of pathogenicity in plant disease, *Naturwissenschaften* **67:**310.

Osbourn, A., Bowyer, P., Lunness, P., Clarke, B., and Daniels, M., 1995, Fungal pathogens of oat roots and tomato leaves employ closely related enzymes to detoxify different host plant saponins, *Mol. Plant–Microbe Interact.* **8:**971–978.

Ouchi, S., Hibino, C., Oku, H., Fujiwara, M., and Nakabayashi, H., 1979, The induction of resistance or susceptibility, in *Recognition and Specificity in Plant Host–Parasite Interactions* (J. M. Daly and I. Uritani, eds.), pp. 49–65, University Park Press, Baltimore, Maryland.

Padgett, H. S., and Beachy, R. N., 1993, Analysis of a tobacco mosaic virus strain capable of overcoming *N* gene-mediated resistance, *Plant Cell* **5:**577–586.

Panaccione, D. G., Scott-Craig, J. S., Pocard, J.-A., and Walton, J. D., A cyclic peptide synthetase gene required for pathogenicity of the fungus *Cochliobolus carbonum*, *Proc. Natl. Acad. Sci. USA* **89**:6590–6594.

Parker, J. E., Coleman, M. J., Szabo, V., Frost, L. N., Schmidt, R., Van der Biezen, E. A., Moores, T., Dean, C., Daniels, M. J., and Jones, J. D. G., 1997, The *Arabidopsis* downy mildew resistance gene *RPP5* shares similarity to the Toll and interleukin-1 receptors with *N* and *L6*, *Plant Cell* **9**:879–894.

Podila, G. K., Rogers, L. M., and Kolattukudy, P. E., 1993, Chemical signals from avocado surface wax trigger germination and appressorium formation in *Colletotrichum gloeosporioides*, *Plant Physiol.* **103**:267–272.

Ranjeva, R., and Beudet, A. M., 1987, Phosphorylation of proteins in plants: Regulatory effects and potential involvement in stimulus/response coupling, *Annu. Rev. Plant Physiol.* **38**:73–93.

Reuber, T. L., and Ausubel, F. M., 1996, Isolation of *Arabidopsis* genes that differentiate between resistance responses mediated by the *RPS2* and *RPM1* disease resistance genes, *Plant Cell* **8**:241–249.

Rhoads, D. M., Levings, C. S., III, and Siedow, J. N., 1995, URF13, a ligand-gated, pore-forming receptor for T-toxin in the inner membrane of cms-T mitochondria, *J. Bioenerg. Biomembranes* **27**:437–445.

Ritter, C., and Dangl, J. L., 1996, Interference between two specific pathogen recognition events mediated by distinct plant disease resistance genes, *Plant Cell* **8**:251–257.

Rogers, L. M., Flaishman, M. A., and Kolattukudy, P. E., 1994, Cutinase gene disruption in *Fusarium solani* f.sp. *pisi* decreases its virulence on pea, *Plant Cell* **6**:935–945.

Rohe, M., Gierlich, A., Hermann, H., Hahn, M., Schmidt, B., Rosahl, S., and Knogge, W., 1995, The race-specific elicitor, NIP1, from the barley pathogen, *Rhynchosporium secalis*, determines avirulence on host plants of the *Rrs1* resistance genotype, *EMBO J.* **14**:4168–4177.

Ryals, J., Uknes, S., and Ward, E., 1994, Systemic acquired resistance, *Plant Physiol.* **104**:1109–1112.

Salmeron, J. M., Barker, S. J., Carland, F. M., Mehta, A. Y., and Staskawicz, B. J., 1994, Tomato mutants altered in bacterial disease resistance provide evidence for a new locus controlling pathogen recognition, *Plant Cell* **6**:511–520.

Salmeron, J. M., Oldroyd, G. E. D., Rommens, C. M. T., Scofield, S. R., Kim, H. S., Lavelle, D. T., Dahlbeck, D., and Staskawicz, B. J., 1996, Tomato *prf* is a member of the leucine-rich repeat class of plant disease resistance genes and lies embedded within the *Pto* kinase gene cluster, *Cell* **86**:123–133.

Sandrock, R. W., DellaPenna, D., and VanEtten, H., 1995, Purification and characterization of $\beta_2$-tomatinase, an enzyme involved in the degradation of $\alpha$-tomatine and isolation of the gene encoding $\beta_2$-tomatinase from *Septoria lycopersici*, *Mol. Plant–Microbe Interact.* **8**:960–970.

Saraste, M., Sibbald, P. R., and Wittinghofer, A., 1990, The P-loop—A common motif in ATP- and GTP-binding proteins, *Trends Biochem. Sci.* **15**:430–434.

Scheffer, R. P., and Livingston, R. S., 1984, Host-selective toxins and their role in plant diseases, *Science* **223**:17–21.

Scheffer, R. P., Nelson, R. R., and Ullstrup, A. J., 1967, Inheritance of toxin production and pathogenicity in *Cochliobolus carbonum* and *Cochliobolus victoriae*, *Phytopathology* **57**:1288–1291.

Scott-Craig, J. S., Panaccione, D. G., Pocard, J. A., and Walton, J. D., 1992, The cyclic peptide synthetase catalyzing HC-toxin production in the filamentous fungus *Cochliobolus carbonum* is encoded by a 15.7-kilobase open reading frame, *J. Biol. Chem.* **267**:26044–26049.

Serrano, R., 1989, Structure and function of plasma membrane ATPase, *Annu. Rev. Plant Physiol. Plant Mol. Biol.* **40**:61–94.

Shaw, D. S., 1991, Genetics, *Adv. Plant Pathol.* **7**:131–170.

Shiraishi, T., Oku, H., Yamashita, M., and Ouchi, S., 1978, Elicitor and suppressor of pisatin induction in spore germination fluid of pea pathogen, *Mycosphaerella pinodes*, *Ann. Phytopathol. Soc. Japan* **44**:659–665.

Shiraishi, T., Araki, M., Yoshioka, H., Kobayashi, I., Yamada, T., Ichinose, Y., Kunoh, H., and Oku, H., 1991a, Inhibition of ATPase activity in pea plasma membranes *in situ* by a suppressor from a pea pathogen, *Mycosphaerella pinodes*, *Plant Cell Physiol.* **32**:1067–1075.

Shiraishi, T., Yamada, T., Oku, H., and Yoshioka, H., 1991b, Suppressor production as a key factor for fungal pathogenesis, in *Molecular Strategies of Pathogens and Host Plants* (S. S. Patil, S. Ouchi, D. Mills, and C. Vance, eds.), pp. 151–162, Springer-Verlag, New York.

Shiraishi, T., Saitoh, K., Kim, H. M., Kato, T., Tahara, M., Oku, H., Yamada, T., and Ichinose, Y., 1992, Two suppressors, suppresscins A and B, secreted by a pea pathogen, *Mycosphaerella pinodes*, *Plant Cell Physiol.* **33**:663–667.

Siedow, J. N., Rhoads, D. M., Ward, G. C., and Levings, C. S., III, 1995, The relationship between the mitochondrial gene T-*urf13* and fungal pathotoxin sensitivity in maize, *Biochim. Biophys. Acta* **1271**:235–240.

Song, W.-Y., Wang, G.-L., Chen, L.-L., Kim, H.-S., Pi, L.-Y., Hosten, T., Gardner, J., Wang, B., Zhai, W.-X., Zhu, L.-H., Fauquet, C., and Ronald, P., 1995, A receptor kinase-like protein encoded by the rice disease resistance gene, *Xa21*, *Science* **270**:1804–1806.

Stahl, D. J., and Schäfer, W., 1992, Cutinase is not required for fungal pathogenicity on pea, *Plant Cell* **4**:621–629.

Staskawicz, B. J., Dahlbeck, D., and Keen, N. T., 1984, Cloned avirulence gene of *Pseudomonas syringae* pv. *glycinea* determines race-specific incompatibility on *Glycine max* (L.) Merr., *Proc. Natl. Acad. Sci. USA* **81**:6024–6028.

Staskawicz, B. J., Ausubel, F. M., Baker, B. J., Ellis, J. G., and Jones, J. D. G., 1995, Molecular genetics of plant disease resistance, *Science* **268**:661–667.

Suzuki, K., and Shinshi, H., 1995, Transient activation and tyrosine phosphorylation of a protein kinase in tobacco cells treated with a fungal elicitor, *Plant Cell* **7**:639–647.

Sweigard, J. A., Chumley, F. G., and Valent, B., 1992, Disruption of a *Magnaporthe grisea* cutinase gene, *Mol. Gen. Genet.* **232**:183–190.

Sweigard, J. A., Carroll, A. M., Kang, S., Farrall, L., Chumley, F. G., and Valent, B., 1995, Identification, cloning and characterization of *PWL2*, a gene for host-species specificity in the rice blast fungus, *Plant Cell* **7**:1221–1233.

Talbot, N. J., Ebbole, D. J., and Hamer, J. E., 1993, Identification and characterization of *MPG1*, a gene involved in pathogenicity from the rice blast fungus *Magnaporthe grisea*, *Plant Cell* **5**:1575–1590.

Talbot, N. J., Kershaw, M. J., Wakley, G. E., De Vries, O. M. H., Wessels, J. G. H., and Hamer, J. E., 1996, *MPG1* encodes a fungal hydrophobin involved in surface interactions during infection-related development of *Magnaporthe grisea*, *Plant Cell* **8**:985–999.

Toyoda, K., Shiraishi, T., Yoshioka, H., Yamada, T., Ichinose, Y., and Oku, H., 1992, Regulation of polyphosphoinositide metabolism in pea plasma membranes by elicitor and suppressor from a pea pathogen, *Mycosphaerella pinodes*, *Plant Cell Physiol.* **33**:445–452.

Toyoda, K., Shiraishi, T., Yamada, T., Ichinose, Y., and Oku, H., 1993, Rapid changes in polyphosphoinositide metabolism in pea in response to fungal signals, *Plant Cell Physiol.* **34**:729–735.

Tucker, B. E., Hoch, H. C., and Staples, R. C., 1986, The involvement of F actin in *Uromyces* cell differentiation: The effects of cytochalasin E and phalloidin, *Protoplasma* **135**:88–101.

Turgeon, B. G., Kodama, M., Yang, G., Rose, M. S., Lu, S. W., and Yoder, O. C., 1995, Function and chromosomal location of the *Cochliobolus heterostrophus Tox1* locus, *Can. J. Bot.* **73**:S1071–S1076.

Tzeng, D. D., and DeVray, J. E., 1993, Role of oxygen radicals in plant disease development, *Adv. Plant Pathol.* **10:**1–34.

Tzeng, T. H., Lyngholm, L. K., Ford, C. F., and Bronson, C. R., 1992, A restriction fragment length polymorphism map and electrophoretic daryotype of the fungal maize pathogen *Cochliobolus heterostrophus*, *Genetics* **130:**81–96.

Valent, B., and Chumley, F. G., 1991, Molecular genetic analysis of the rice blast fungus, *Magnaporthe grisea*, *Annu. Rev. Phytopathol.* **29:**443–467.

Valent, B., and Chumley, F. G., 1994, Avirulence genes and mechanisms of genetic instability in the rice blast fungus, in *Rice Blast Diseases* (R. Zeigler, P. S. Teng, and S. Leong, eds.), pp. 111–134, CAB International, Wallingford, Oxon, U.K.

Valon, C., Smalle, J., Goodman, H. M., and Giraudat, J., 1993, Characterization of an *Arabidopsis thaliana* gene (*TMKL1*) encoding a putative transmembrane protein with an unusual kinase-like domain, *Plant Mol. Biol.* **23:**415–421.

Van den Ackerveken, G. F. J. M., Van Kan, J. A. L., and De Wit, P. J. G. M., 1992, Molecular analysis of the avirulence gene *avr9* of the fungal tomato pathogen *Cladosporium fulvum* fully supports the gene-for-gene hypothesis, *Plant J.* **2:**359–366.

Van den Ackerveken, G. F., Vossen, P., and De Wit, P. J. G. M., 1993, The AVR9 race-specific elicitor of *Cladosporium fulvum* is processed by endogenous and plant proteases, *Plant Physiol.* **103:**91–96.

Van den Ackerveken, G. F., Dunn, R. M., Cozijnsen, A. J., Vossen, J. P., Van den Broek, H. W., and De Wit, P. J. G. M., 1994, Nitrogen limitation induces expression of the avirulence gene *avr9* in the tomato pathogen *Cladosporium fulvum*, *Mol. Gen. Genet.* **243:**277–85.

Van den Ackerveken, G., Marois, E., and Bonas, U., 1996, Recognition of the bacterial avirulence protein AvrBs3 occurs inside the host plant cell, *Cell* **87:**1307–1316.

Van der Hoeven, P. C. J., Siderius, M., Korthout, H. A. A. J., Drabkin, A. V., and De Boer, A. H., 1996, A calcium and free fatty acid-modulated protein kinase as putative effector of the fusicoccin 14-3-3 receptor, *Plant Physiol.* **111:**857–865.

VanEtten, H. D., Matthews, D. E., and Matthews, P. S., 1989, Phytoalexin detoxification: Importance for pathogenicity and practical implications, *Annu. Rev. Phytopathol.* **27:**143–164.

VanEtten, H., Funnell-Baerg, D., Wasmann, C., and McCluskey, K., 1994a, Location of pathogenicity genes on dispensable chromosomes in *Nectria haematococca* MPVI, *Ant. Leeuwenhoek* **65:**263–267.

VanEtten, H. D., Mansfield, J. W., Bailey, J. A., and Farmer, E. E., 1994b, Two classes of plant antibiotics: Phytoalexins versus "phytoanticipins," *Plant Cell* **6:**1191–1192.

Van Kan, J. A. L., Van den Ackerveken, G. F. J. M., and De Wit, P. J. G. M., 1991, Cloning and characterization of cDNA of avirulence gene *avr9* of the fungal pathogen *Cladosporium fulvum*, causal agent of tomato leaf mold, *Mol. Plant–Microbe Interact.* **4:**52–59.

Vera-Estrella, R., Barkla, B. J., Higgins, V. J., and Blumwald, E., 1994, Plant defense response to fungal pathogens. Activation of host plasma membrane $H^+$-ATPase by elicitor induced enzyme dephosphorylation, *Plant Physiol.* **104:**209–215.

Wada, M., Kato, H., Malik, K., Sriprasertsak, P., Ichinose, Y., Shiraishi, T., and Yamada, T., 1995, A supprescin from a phytopathogenic fungus deactivates transcription of a plant defense gene encoding phenylalanine ammonia-lyase, *J. Mol. Biol.* **249:**513–519.

Walker, J. C., 1993, Receptor-like protein kinase genes of *Arabidopsis thaliana*, *Plant J.* **3:**451–456.

Walton, J. D., 1993, Molecular basis of specificity in maize leaf spot disease, in *Advances in Molecular Genetics of Plant–Microbe Interactions*, Vol. 2 (E. W. Nester and D. P. S. Verma, eds.), pp. 313–323, Kluwer, Dordrecht, The Netherlands.

Walton, J. D., and Panaccione, D. G., 1993, Host-selective toxins and disease specificity: Perspectives and progress, *Annu. Rev. Phytopathol.* **31:**275–303.

Walton, J. D., Ahn, J.-H., Akimitsu, K., Pitkin, J. W., and Ransom, R., 1994a, Leaf-spot disease of maize: Chemistry, biochemistry, and molecular biology of a host-selective cyclic peptide, in *Advances in Molecular Genetics of Plant–Microbe Interactions*, Vol. 3 (M. J. Daniels, J. A. Downie, and A. E. Osbourn, eds.), pp. 231–237, Kluwer, Dordrecht, The Netherlands.

Walton, J. D., Akimitsu, K., Ahn, J. H., and Pitkin, J. W., 1994b, Towards an understanding of the *TOX2* gene of *Cochliobolus carbonum*, in *Host-Specific Toxin: Biosynthesis, Receptor and Molecular Biology* (K. Kohmoto and O. C. Yoder, eds.), pp. 227–237, Tottori University, Tottori, Japan.

Welz, H. G., and Leonard, K. J., 1988, Genetic variation in field populations of race 0, 2, and 3 of *Bipolaris zeicola* in 1988, *Phytopathology* **78:**1574.

Welz, H. G., and Leonard, K. J., 1993, Phenotypic variation and parasitic fitness of races of *Cochliobolus carbonum* on corn in North Carolina, *Phytopathology* **83:**593–601.

Wessels, J. G. H., 1994, Developmental regulation of fungal wall formation, *Annu. Rev. Phytopathol.* **32:**413–437.

Wevelsiep, L., Kogel, K. H., and Knogge, W., 1991, Purification and characterization of peptides from *Rhynchosporium secalis* inducing necrosis in barley, *Physiol. Mol. Plant Pathol.* **39:**471–482.

Wevelsiep, L., Rüpping, E., and Knogge, W., 1993, Stimulation of barley plasmalemma $H^+$-ATPase by phytotoxic peptides from the fungal pathogen *Rhynchosporium secalis*, *Plant Physiol.* **101:**297–301.

Whitham, S., Dinesh-Kumar, S. P., Choi, D., Hehl, R., Corr, C., and Baker, B., 1994, The product of the tobacco mosaic virus resistance gene *N:* Similarity to Toll and the interleukin-1 receptor, *Cell* **78:**1101–1115.

Woloshuk, C. P., and Kolattukudy, P. E., 1986, Mechanism by which contact with plant cuticle triggers cutinase gene expression in the spores of *Fusarium solani* f.sp. *pisi*, *Proc. Natl. Acad. Sci. USA* **83:**1704–1708.

Yamada, T., Hashimoto, H., Shiraishi, T., and Oku, H., 1989, Suppression of pisatin, phenylalanine ammonia-lyase mRNA, and chalcone synthase mRNA accumulation by a putative pathogenicity factor from the fungus *Mycosphaerella pinodes*, *Mol. Plant–Microbe Interact.* **2:**256–261.

Yoshikawa, M., and Sugimoto, K., 1993, A specific binding site in soybean membranes for a phytoalexin elicitor released from fungal cells walls by β-1,3-endoglucanase, *Plant Cell. Physiol.* **34:**1229–1237.

Yoshioka, H., Shiraishi, T., Yamada, T., Ichinose, Y., and Oku, H., 1990, Suppression of pisatin production and ATPase activity in pea plasma membranes by orthovanadate, Verapamil and a suppressor from *Mycosphaerella pinodes*, *Plant Cell Physiol.* **31:**1139–1146.

Zhang, L., Villalon, D., Sun, Y., Kazmierczak, P., and Van Alfen, N. K., 1994, Virus-associated down-regulation of the gene encoding cryparin, an abundant cell-surface protein from the chestnut blight fungus, *Cryphonectria parasitica*, *Gene* **139:**59–64.

Zhou, X. L., Stumpf, M. A., Hoch, H. C., and Kung, C., 1991, A mechanosensitive channel in whole cells and in membrane patches of the fungus *Uromyces*, *Science* **253:**1415–1417.

Zhou, J., Loh, Y. T., Bressan, R. A., and Martin, G. B., 1995, The tomato gene *Pti1* encodes a serine/threonine kinase that is phosphorylated by *Pto* and is involved in the hypersensitive response, *Cell* **83:**925–935.

Ziegler, E., and Albersheim, P., 1977, Host–pathogen interactions. XIII. Extracellular invertases secreted by three races of a plant pathogen are glycoproteins which possess different carbohydrate structures, *Plant Physiol.* **59:**1104–1110.

Ziegler, E., and Pontzen, R., 1982, Specific inhibition of glucan-elicited glyceollin accumulation on soybeans by an extracellular mannan-glycoprotein of *Phytophthora megasperma* f.sp. *glycinea*, *Physiol. Plant Pathol.* **20:**321–331.

*Chapter 8*

# *Arabidopsis thaliana*

# A Robust Model System for Studying Plant–Pathogen Interactions

Keith R. Davis

## 1. *ARABIDOPSIS* AS A MODEL SYSTEM

*Arabidopsis thaliana* is a rather unimposing crucifer that has rapidly developed into a major model system in experimental plant biology over the past 15 years. *A. thaliana* was first used experimentally almost 100 years ago, although it was not until the 1940s that its development into a bona fide model system began (Rédei, 1992). Numerous attributes contribute to the utility of *A. thaliana* as an experimental system. These include its rapid life cycle (4–6 weeks) and small size, which allow large numbers of plants to be expeditiously grown in soil or defined media, a small genome of ~1 Mbp with little repetitive DNA distributed among five chromosomes, the ability to generate large numbers of seeds by either self-fertilization or outcrossing, the relative ease of generating large mutant populations using a variety of mutagens, and facile methods of generating transgenic *A. thaliana* plants. These attributes make *A. thaliana* a powerful system for classical and molecular genetic studies of plant growth and development as well as plant responses to the environment (Rédei, 1975; Meyerowitz, 1987; Meyerowitz and Somerville, 1994).

---

**Keith R. Davis** Plant Biotechnology Center and Department of Plant Biology, The Ohio State University, Columbus, Ohio 43210-1002.

*Subcellular Biochemistry, Volume 29: Plant–Microbe Interactions*, edited by Biswas and Das. Plenum Press, New York, 1998.

In addition to these long-recognized experimental advantages, the research efforts of numerous laboratories have resulted in the availability of significant new resources that greatly facilitate *A. thaliana* research and provide the basis of novel strategies that can be applied to diverse experimental questions. These resources include extensive genetic and physical maps of the *A. thaliana* genome; recombinant inbred lines that can be used to rapidly map newly cloned genes; mutant populations generated by insertional mutagenesis with either T-DNA or transposons, including enhancer and gene trap lines; numerous cDNA and genomic libraries, including large insert BAC, YAC, and P1 phage libraries; a large collection of cDNAs representing expressed sequence tags (ESTs); and a newly initiated multinational genome sequencing project. An additional resource that is of particular relevance to studies of *A. thaliana*–pathogen interactions is the availability of several hundred *A. thaliana* accessions collected from around the world that can be screened for differences in response to specific pathogens. Many of these resources are readily available to researchers through the Arabidopsis Biological Resource Center at Ohio State University and the Nottingham Arabidopsis Stock Centre at Nottingham University. In addition, increasing amounts of information concerning *A. thaliana* resources and research are available through internet access (http://aims.cps.msu.edu/aims/, http://genome-www.stanford.edu/Arabidopsis/, http://lenti.med.umn.edu/arabidopsis/arab—top—page.html, and associated links), thus facilitating the rapid dissemination of significant new developments. Taken together, the experimental advantages of *A. thaliana* in conjunction with these resources make this small weed an extremely robust system to address fundamental questions in plant biology.

## 2. *ARABIDOPSIS* PATHOSYSTEMS

Given its inherent experimental advantages, it is not surprising that attempts have been made to use *A. thaliana* to investigate plant–pathogen interactions (Davis, 1992). To a large extent, these studies have been successful and a number of pathogens that can infect *A. thaliana* have been identified. These include numerous bacterial, fungal, and viral pathogens as well as nematodes and mycoplasmas. The following sections sumarize the salient features of these pathosystems.

### 2.1. Bacterial Pathogens of *A. thaliana*

#### 2.1.1. *Xanthomonas campestris*

*Xanthomonas campestris* pv. *campestris*, a well-characterized pathogen of *Brassica* spp., was one of the first bacterial pathogens tested for virulence on *A.*

*thaliana*. Several groups found that the majority of *X. c.* pv. *campestris* isolates caused the development of typical black rot symptoms and multiplied several logs *in planta* in inoculated leaves (Tsuji and Somerville, 1988; Simpson and Johnson, 1990; Daniels *et al.*, 1991; Davis *et al.*, 1991). Surveys of the susceptibility of *A. thaliana* accessions to specific *X. c.* pv. *campestris* strains revealed that some accessions are resistant to *X. c.* pv. *campestris* infection. Initial studies by Simpson and Johnson (1990) showed differential responses of three *A. thaliana* accessions to two different *X. c.* pv. *campestris* strains and suggested that in one case resistance may be associated with a hypersensitive response (HR; see Section 3). Genetic analyses of these differential interactions were not presented. Daniels and co-workers (Daniels *et al.*, 1991) also reported differential interactions between *A. thaliana* accessions and provided some evidence that resistance was mediated by "gene-for-gene" interactions typical of plant–pathogen resistance responses (Keen, 1990; Staskawicz *et al.*, 1995). Further evidence that gene-for-gene interactions occur between *A. thaliana* and *X. campestris* strains includes the demonstration that an avirulence gene isolated from *X. c.* pv. *raphani* is recognized by most, but not all *A. thaliana* accessions (Parker *et al.*, 1993). More recent studies confirm that some *A. thaliana* accessions exhibit differential resistance phenotypes to specific *X. c.* pv. *campestris* strains and that resistance is associated with the activation of a putative defense gene (Aufsatz and Grimm, 1994). Although these results suggest the possibility that *A. thaliana* may have resistance genes specific for *X. campestris* strains, none of these genes has been clearly identified.

A distinct resistance response to *X. c.* pv. *campestris* was described by Tsuji *et al.* (1991). They found that accession Col-0 inoculated with *X. c.* pv. *campestris* strain 2D520 remained asymptomatic, whereas accession Pr-0 developed spreading chlorotic lesions typical of *X. c.* pv. *campestris* infection. Interestingly, the multiplication of this strain in both Col-0 and Pr-0 was identical, indicating that the phenotypic differences between these accessions is due to the tolerance of Col-0 to the presence of the bacteria. Genetic studies of this interaction indicate that this tolerance is due to a single, dominant nuclear gene, designated *RXC1*. Recent studies indicate that the tolerance to Col-0 to infection is not due to the activation of a typical HR and the associated expression of defense genes (Buell and Somerville, 1995).

A number of different *X. campestris* pathovars, including pathovars *aberrans, amoraciae, graminis, malvacearum, phaseoli, raphani, translucens, vasculorum, vesicatoria,* and *vitians,* have been tested on *A. thaliana*. Several virulent strains of *armoraciae, aberrans,* and *raphani* have been identified; however, symptom severity with most of these strains is reduced compared to the more virulent *X. c.* pv. *campestris* (Davis *et al.*, 1991; Parker *et al.*, 1993). The other *X. campestris* pathovars were found to be avirulent and either induced no obvious symptoms or a weak HR (Daniels *et al.*, 1991).

### 2.1.2. *Pseudomonas*

*Pseudomonas* species include a large number of strains that are pathogens of numerous plant species. Pseudomonads that were initially assessed for virulence on *A. thaliana* include *P. aeruginosa, P. cichorii,* and *P. syringae* pvs. *atrofaciens, delphinii, maculicola, morsprunorum, pisi, savastanoi, syringae,* and *tomato* (Davis *et al.*, 1991; Debener *et al.*, 1991; Dong *et al.*, 1991; Whalen *et al.*, 1991). These studies demonstrated that several strains of *P. s.* pvs. *maculicola* and *tomato* were virulent on some *A. thaliana* accessions and have become the focus of the majority of work currently being done on *Pseudomonas–A. thaliana* interactions. The *P. s.* pvs. *maculicola* and *tomato* are closely related pathovars that are pathogenic on various *Brassica* spp. and tomato, respectively. Virulent strains cause typical leaf speck disease on *A. thaliana*, chlorotic lesions that often develop necrotic centers. Previous studies of these two pathovars indicated that the production of the phytotoxin coronatine is a major virulence determinant in most hosts (Bender *et al.*, 1991; Durbin, 1991; Gross, 1991; Cuppels and Ainsworth, 1995) and recent results indicate that this is also the case with *A. thaliana*. It has been shown that coronatine production is required for *P. s.* pv. *tomato* to infect susceptible *A. thaliana* accessions and that the toxin may function to suppress the activation of defense gene expression (Mittal and Davis, 1995). Interestingly, an *A. thaliana* mutant resistant to coronatine is resistant to a *P. s.* pv. *atropurpurea* strain that also produces coronatine as a virulence factor (Feys *et al.*, 1994).

Detailed studies of the interaction of *A. thaliana* accessions with different *P. s.* pvs. *maculicola* and *tomato* strains in conjunction with cloned avirulence (*avr*) genes from other *P. syringae* pathovars have demonstrated that clear gene-for-gene relationships apply in these interactions and a number of resistance (*R*) gene specificities have been identified. These specificities include recognition of *avrRpt2* from *P. s.* pv. *tomato* (Dong *et al.*, 1991; Whalen *et al.*, 1991), *avrRmp1* from *P. s.* pv. *maculicola* (Debener *et al.*, 1991), and the closely related *avrPpiA1* gene from *P. s.* pv. *pisi* (Dangl *et al.*, 1992), *avrB* from *P. s.* pv. *glycinea* (Innes *et al.*, 1993; Wanner *et al.*, 1993), *avrPph3* from *P. s.* pv. *phaseolicola* (Simonich and Innes, 1995), and *avrRps4* from *P. s.* pv. *pisi* (Hinsch and Staskawicz, 1996). The recognition specificity toward *avrB* and *avrRpm1* is particularly interesting since it appears that a single *R* gene, *RPS3* (*RPM1*), confers recognition of both of these distinct avirulence genes in *A. thaliana* (Bisgrove *et al.*, 1994; Grant *et al.*, 1995), whereas in soybean, specific alleles or closely linked genes exist at the *RPG1* locus that can discriminate between these two *avr* genes (Ashfield *et al.*, 1995). Mutant screens based on the identification of plants that have lost resistance and/or do not express an HR in response to a strain containing a specific *avr* gene have been successful (Kunkel *et al.*, 1993; Yu *et al.*, 1993; Bisgrove *et al.*, 1994), and in several cases the *R* genes corresponding to specific *avr* genes have been cloned (see Section 3.1).

The interaction of *A. thaliana* with other *Pseudomonas* strains has also been investigated in some detail. A number of *P. cichorii* strains have been tested on a number of *A. thaliana* accessions (Dangl *et al.*, 1991; Davis *et al.*, 1991) and in all cases these strains induced HR. Since no virulent strains were identified and attempts to identify *avr* genes were unsuccessful (E. Schott, K. R. Davis, and F. M. Ausubel, unpublished results), *P. cichorii* has not proved useful for investigating questions concerning specificity. However, it does serve as a model for investigating nonhost resistance. *A. thaliana* leaves inoculated with *P. cichorii* develop a rapid HR that is associated with the induction of defense-related genes (Davis *et al.*, 1991). Exposure of cultured *Arabidopsis* cells with *P. cichorii* causes a rapid and dramatic alkalization of the medium that is characteristic of the 'XR' response (Atkinson and Baker, 1989; Baker *et al.*, 1991). Another example of a likely nonhost resistance response is the interaction of *A. thaliana* with certain *P. s.* pv. *syringae* strains. Studies of this interaction led to the identification of an *A. thaliana* phytoalexin, camalexin (Tsuji *et al.*, 1992).

A recently described *A. thaliana–Pseudomonas* system that is particularly interesting involves *P. aeruginosa* (Rahme *et al.*, 1995). Specific *P. aeruginosa* strains can infect both *A. thaliana* and mouse using a full-thickness skin burn model. Analysis of specific *P. aeruginosa* mutants in both host systems revealed that several pathogenicity genes, such as *toxA*, *plcS*, and *gacA*, encode virulence factors that are required for infection of both the plant and animal hosts. These results are of major significance since they clearly demonstrate that some aspects of pathogenicity are highly conserved in animal and plant pathogens and that studies of plant disease resistance may have real significance for understanding some human diseases.

## 2.2. Fungal Pathogens of *A. thaliana*

A variety of fungal pathogens that can infect *A. thaliana* have been identified and include obligate biotrophs that exhibit very specific interactions with different *A. thaliana* accessions as well as less specialized necrotrophs that can infect a number of different plant species. Fungi that have been shown to be pathogens of *A. thaliana* in laboratory and/or natural settings include the oomycetes *Albugo candida*, *Peronospora parasitica*, and *Pythium* species; the ascomycetes *Alternaria brassicae*, *Botrytis cinerea*, *Cladosporium* species, *Erysiphe* species, *Fusarium oxysporum*, *Leptosphaeria maculans*, and *Sclerotinia sclerotiorum;* the basidiomycetes *Rhizoctonia solani* and *Puccinia thlaspeos;* the plasmodiophoromycete *Plasmodiophora brassicae;* the deuteromycete *Chromelosporium ollare;* and the hyphomycete *Geniculosporium ser-*

*pens* (reviewed in Davis, 1992; Dangl, 1993; Mauch-Mani and Slusarenko, 1993; Crute *et al.*, 1994). As was observed for some bacterial pathogens, specific interactions between fungal isolates and *A. thaliana* accessions consistent with gene-for-gene recognition have been documented for the fungal pathogens *Albugo candida, Erysiphe cichoracearum,* and *Peronospora parasitica* (Koch and Slusarenko, 1990a; Holub *et al.*, 1994, 1995; Adam and Somerville, 1996). The identification of specific interactions between *A. thaliana* and these fungi is of some agronomic importance since it is likely that resistance mechanisms, e.g., specific *R* genes, identified in *A. thaliana* can be applied to developing resistance in crop plants that are also susceptible to these or closely related fungal pathogens. Given this possibility, several groups are beginning to elucidate the genetic and molecular basis of resistance to these pathogens and nicely illustrate the advantages of using *A. thaliana* as a model host system.

### 2.2.1. *Peronospora parasitica*

*P. parasitica* infection of *A. thaliana* represents the best-characterized interaction between *A. thaliana* and a fungal pathogen. As pointed out by Slusarenko and co-workers, reports of infection of *A. thaliana* by *P. parasitica* date back to the early 1900s (Joos *et al.*, 1996), although it was not until 1990 that the first publication appeared that suggested the possibility of gene-for-gene interactions in this plant–pathogen combination (Koch and Slusarenko, 1990a). *P. parasitica* is an obligate biotroph and is known to infect a variety of cruciferous plant species. Symptoms observed in *A. thaliana* plants infected with *P. parasitica* vary from the development of a profuse mat of conidiophores on the surface of infected tissues that gives the plant the appearance of being covered in down (hence the common name for this disease is downy mildew) in highly susceptible accessions, to a rapid HR in resistant accessions. In an elegant series of experiments, Crute and co-workers investigated the interaction phenotypes of numerous *A. thaliana* accessions with *P. parasitica* isolates (Holub *et al.*, 1994). By classifying aspects of both fungal development and host response in different *A. thaliana* accessions and in progeny derived from crosses of accessions with distinct interaction phenotypes, specific loci were identified that correlated with specific types of interactions. This system offers great potential for identifying specific genes that are important in determining resistance phenotypes and providing new insight into the molecular basis for the observed differences in symptom development.

The genetic analysis of *R* genes that confer resistance to *P. parasitica* (*RPP* genes) has advanced rapidly over the past several years. Over 20 *RPP* loci have been identified using a relatively small number of *P. parasitica* isolates and *A. thaliana* accessions (Holub *et al.*, 1994; Joos *et al.*, 1996; Reignault *et al.*, 1996). This suggests that *A. thaliana* has a significant capability to recognize this pathogen and that additional *RPP* loci remain to be identified. *RPP*-mediated resistance has been found to be a semidominant trait in many cases, suggesting that gene dosage may be very important for the expression of resistance. In addition, *RPP* loci are found in clusters throughout the genome, and in some cases are associated with *R* genes that confer resistance to other pathogens. Several groups are currently attempting to clone *RPP* genes, primarily by positional cloning methods. The availability of specific *RPP* genes will allow detailed studies of the mechanisms whereby *RPP* gene products trigger a resistance response, as well as interactions between different *RPP* genes that presumably result in the various interaction phenotypes that have been observed.

### 2.2.2. *Erysiphe* Species

*Erysiphe* species, the causal agent of powdery mildew, are important pathogens of several crop species. An initial study demonstrated that an isolate of *E. cruciferarum* originally obtained from infected oilseed rape could infect *A. thaliana* accession Wei-0 (Koch and Slusarenko, 1990b). The symptoms observed in infected *A. thaliana* plants were typical of powdery mildew in other plant hosts, namely, the development of white colonies on leaf surfaces that eventually coalesce and cover large regions of the leaf surface. More recent studies using 11 different *E. cruciferarum* isolates and 50 *A. thaliana* accessions have demonstrated that three accessions exhibit resistance to powdery mildew infection (Shunyan, Ellwood, Oliver, and Turner, unpublished data, cited in Adam and Somerville, 1996). The genetic basis for this resistance in these examples remains to be published.

Since resistance to powdery mildew is conferred by single *R* genes in other plant hosts, it was expected that this will also be the case in *A. thaliana*. This expectation has been confirmed by recent studies testing different accessions with an *E. cichoracearum* field isolate that was originally obtained from an infected *A. thaliana* plant (Adam and Somerville, 1996). Of the 59 accessions tested, 11 showed some resistance to infection. Resistance phenotypes varied from little or no fungal growth to delayed or reduced fungal growth. This is similar to the various levels of resistance previously described for the interaction of *P. parasitica* with *A. thaliana* accessions. More detailed studies of six of the resistant accessions revealed that single, nonallelic semidominant loci conferred resistance in four accessions (Kas-1, Stw-0, Su-0, and Wa-1). Resistance in accessions Te-0 and Sl-0 was recessive and was conferred by a single locus in Te-0 and by two unlinked loci in Sl-0. Five of these resistance loci, designated *RPW1–5*, were mapped; one locus each was found on chromosomes II, IV, and V, while two loci mapped to chromosome III. The eventual cloning of the *RPW* genes and the elucidation of their function will no doubt provide important insights into the interaction of powdery mildew with plant hosts.

### 2.2.3. *Albugo candida*

*A. candida* is a fungal pathogen that is closely related to *P. parasitica* and is the causal agent of white blister. Wild populations of *A. thaliana* infected with *A. candida* have been noted throughout Europe and this is often closely associated with *P. parasitica* infections (Holub *et al.*, 1995, and references therein). Detailed analyses of the interaction of an extensive panel of *A. thaliana* accessions with two *A. candida* isolates revealed a range of interaction phenotypes reminiscent of that observed in studies using *P. parasitica* (Holub *et al.*, 1995). Accessions were identified that exhibited resistance to specific *A. candida* isolates; however, the frequency at which resistant accessions were identified was significantly less than that observed using *P. parasitica*. Genetic studies of three accessions exhibiting three distinct resistance phenotypes resulted in the identification of three loci responsible for recognition of *A. candida*. Two of these loci behaved as partially dominant, or, under some conditions, recessive, loci, while the third was completely dominant. This latter dominant locus, designated *RAC1*, has been mapped to the top arm of chromosome I.

As is the case with *P. parasitica* and *Erysiphe* species, the establishment of defined interactions of *A. thaliana* with *A. candida* offers great potential for elucidating the underlying basis for the more subtle variations in resistance that are often observed in crop plants infected with these fungal pathogens. It is likely that gene dosage effects, as well as the interaction of *R* genes with other unlinked 'modifier' genes, can be important in determining what type of resistance phenotype will be expressed. A clear understanding of these interactions will be greatly facilitated by the isolation of these various *R* genes and the characterization of how their gene products interact with pathogen components and/or other host factors involved in the development of a resistance response. Given the experimental advantages of *A. thaliana* in conjunction with the array of genetic and molecular tools available, it is likely that progress will be more rapid in this model host compared to similar studies using crop species.

However, all of the fungal pathogen systems described have a major deficiency, namely, the lack of genetics for these pathogens. A formal demonstration that an interaction represents a true gene-for-gene interaction requires that specific *avr* genes are identified in fungal isolates that correspond to specific *R* genes. The availability of cloned *avr* genes has proven extremely useful in studies of resistance to bacterial pathogens since it allows the production of nearly isogenic strains that differ only by a specific *avr* gene. Thus, any differences in a susceptible and resistant response in a specific host genotype can be attributed to recognition of a specific *avr* gene. The development of genetics in these fungal pathogens will significantly improve these already valuable plant–pathogen systems.

## 2.3. Viral Pathogens of *A. thaliana*

Viruses represent a unique class of pathogenic agents because of the extremely intimate associations formed between the virus and its host. In plant systems, it seems clear that host factors play critical roles throughout the entire virus life cycle, including the expression of viral genes, viral genome replication, and the movement of virus throughout the plant. However, essentially nothing is known about the specific host factors that are involved in supporting virus infection. To address this deficiency, several groups have identified specific *A. thaliana*–virus interactions that should prove useful in the identification of plant host factors important for viral pathogenesis. The early development of *A. thaliana*–virus interactions has been previously reviewed (Davis, 1992; Dangl, 1993; Simon, 1994). The following subsections describe recent results concerning several of the better developed *A. thaliana*–virus interaction systems. The results obtained using these systems clearly indicate that *A. thaliana* is an excellent host plant for studying plant–virus interactions.

### 2.3.1. Cauliflower Mosaic Virus (CaMV)

CaMV, a double-stranded DNA virus in the caulimovirus group of plant viruses, was one of the first clearly recognized viral pathogens of *A. thaliana* (Susnova and Polak, 1975; Balàzs and Lebeurier, 1981). This interaction is proving useful for studies of both host resistance and symptom development as well as analysis of viral functions (Zijlstra and Hohn, 1992; Leisner *et al.*, 1993). *A. thaliana* inoculated with virulent CaMV isolates are stunted and develop chlorotic patches and vein clearing that causes the plants to have a mosaic appearance typical of CaMV infection of other plant hosts. Variations in the timing and severity of symptom development have been observed after inoculating different *A. thaliana* accessions with various CaMV isolates (Melcher, 1989; Leisner and Howell, 1992). The observed differences in symptom development could be correlated with differences in the accumulation of virus particles as well as differences in the timing of floral induction. Early-flowering accessions were generally found to develop less severe symptoms (Leisner and Howell, 1992). Studies in which a 'skeleton hybridization' method (Melcher *et al.*, 1981) was used to monitor CaMV movement in whole plants (Leisner *et al.*, 1993) demonstrated that virus movement was directly correlated with the growth rate of the plants. For example, *A. thaliana* plants grown under short-day conditions (which slows growth rates and delays the initiation of flowering) exhibit more widespread movement of the virus compared to plants with faster growth rates and earlier flowering times. This 'developmental resistance' is most likely due to the fact that rapid-flowering plants mature before the virus has time to spread; it is known that CaMV typically can only move systemically and cause symptoms in developing tissues, presumably because of the changes in source–sink relationships between tissues that occur during maturation.

Resistance of *A. thaliana* to CaMV infection that apparently does not relate to the developmental resistance described above has been described (Leisner and Howell, 1992; Leisner *et al.*, 1993). Accession En-2 did not exhibit any symptom development after inoculation with three different CaMV isolates and skeleton hybridization studies demonstrated that CaMV does not move systemically in these plants. Genetic analyses indicate that resistance in En-2 is due to a single, semidominant gene located on chromosome 1. Analysis of CaMV-infected En-2 plants demonstrated that some pathogenesis-related (PR) proteins are induced, suggesting that resistance could be due to the activation of an active defense response (Callaway *et al.*, 1996).

### 2.3.2. Beet Curly Top Virus (BCTV)

BCTV is a second DNA virus that is proving useful for basic studies of the interaction of *A. thaliana* with a viral host and provides a system for studying both resistance and the molecular basis of symptom development. BCTV is a single-stranded DNA virus in the geminivirus subgroup II and exhibits a broad host range (Stanley, 1991; Lazarowitz, 1992). Over the past several years we have characterized the interaction of different *A. thaliana* accessions with two cloned isolates of BCTV, CFH, and Logan. These two isolates have seven common open-reading frames (ORFs), three rightward virion sense-strand ORFs designated L1–L4. BCTV–Logan and BCTV–CFH show greater than 95% amino acid similarity in the three rightward ORFs; however, the four leftward ORFs differ significantly, with amino acid similarities of 58–87% (Stenger, 1994).

Our initial studies demonstrated that BCTV–Logan and BCTV–CFH differed significantly in their ability to infect and cause symptoms in different *A. thaliana* accessions (Davis, 1992; S. Lee *et al.*, 1994). Symptoms appeared on susceptible plants 2–3 weeks after inoculation with BCTV–Logan and after 10–15 days with BCTV–CFH. Symptoms were more severe in BCTV–CFH-infected plants and included leaf curling, the formation of stunted, deformed inflorescence structures, and the accumulation of anthocyanin pigments in symptomatic tissues. Analysis of viral DNA accumulation indicated that symptom development and severity were correlated with the amount of viral DNA present in the plants. Two accessions, Ms-0 and Pr-0, were found to be resistant to BCTV–Logan in that they did not develop symptoms nor accumulate detectable levels of viral DNA. Studies of viral DNA replication in excised inflorescence pieces demonstrated that BCTV–Logan could replicate in tissues from these resistant accessions, suggesting that resistance was due to a block in viral movement. Genetic studies of these two accessions indicate that resistance is due to two distinct single, recessive loci (S. Lee *et al.*, 1994). These resistance loci represent the first examples of single loci conferring resistance to any geminivirus. We have recently implemented positional cloning strategies to isolate these resistance loci.

In addition to identifying accessions that were resistant to infection by BCTV–Logan, several accessions were found to develop more severe systems compared to many of the other accessions tested. One accession in particular (Sei-0) was found to be 'hypersusceptible" to the Logan strain, in that it developed very severe symptoms, including severely deformed inflorescences with callus-like structures, and accumulated high levels of viral DNA. Further studies demonstrated that Logan mutants that are deficient in the production of proteins required for viral DNA replication and symptom development (ORFs R2 and R3) could initiate symptomatic infections of Sei-0, whereas these mutants did not cause symptoms in all other accessions tested. These results indicate that Sei-0 is in some way more compatible with the Logan strain. Genetic analyses of $F_2$ progeny derived from a cross of Sei-0 and Col-0 indicate that the hypersusceptible phenotype may be due to a single, dominant locus (S. Lee, J. Park, and K. R. Davis, unpublished observations).

Microscopic studies of the BCTV-induced cell division that was dramatically expressed in Sei-0 demonstrated that the activation of cell division is preceded by the disruption of the phloem. Phloem is a major conducting tissue; thus, disruption of phloem tissue by BCTV infection is likely to cause changes in the balance of some metabolites and hormones. There was a strong correlation between the amount of phloem disruption and the severity of symptoms observed at the whole-plant level. BCTV infection of Sei-0 resulted in the complete disruption of phloem bundles, whereas in Col-0, phloem cells necrosis was much more limited. The mechanism of phloem disruption is not clear, but it is likely that virus multiplication and accumulation in nuclei may disrupt cellular functions, resulting in cell death. It is also possible that the virus directly triggers programmed cell death in infected cells. We are currently investigating this very intriguing possibility.

An important aspect of phloem disruption observed in BCTV-infected *A. thaliana* was the induction of cell division within the phloem, and in the case of accession Sei-0, the induction of cell division in the phloem and surrounding cortex cells. Analysis of the expression of GUS reporter gene activity in transgenic plants containing constructs with promoters of the cell-cycle genes *cdc2* and *cyc1* and the auxin-induced *saur* promoter showed that *saur* promoter activity was induced concomitantly with cell-cycle gene promoter activities during BCTV infection (S. Lee, J. Park, J. Hur, and K. R. Davis, unpublished). Histochemical staining for GUS activity showed that cells in symptomatic tissues at the inflorescence shoot tip of the three transgenic lines expressed high levels of GUS activity. This strongly suggests that changes in auxin concentration are involved in the induction of cell division in BCTV-infected plants. The kinetics of induction of *saur, cdc2,* and *cyc1* promoter activities after virus inoculation did not show any clear differences. Thus, the activation of these promoter activities is tightly linked in symptomatic tissues. Auxin is believed to be produced in shoot tips and leaf margins and then translocated through problem vessels or by basipetal cell-to-cell movement. In virus-infected plants, auxin may not move from source tissues, because of phloem disruption. Therefore, high concentrations of auxin may accumulate in the shoot tips and leaves. This suggestion is consistent with the high GUS activities found in inflorescence shoot tips and leaves of the BCTV-infected SAUR:GUS plants. RNA blot analyses of *cdc2* and *saur* transcript accumulation were consistent with the expression patterns observed in transgenic plants expressing the *cdc2* and *saur* reporter genes. Accumulation of both transcripts induced by BCTV infection were similar with respect to both the timing and magnitude of induction. These studies taken together clearly demonstrated a strong correlation between auxin-induced gene expression and the activation of cell cycle genes. This suggests that the activation of cell division is caused by an increase in the local auxin concentration rather than a direct activation of the cell cycle by BCTV.

We have further defined the factors important for symptom development caused by BCTV using a molecular genetic approach based on expressing BCTV-encoded proteins in transgenic plants. Results of these studies indicate that the BCTV ORF L4 is a primary symptom determinant. Expression of L4 from either BCTV–CFH or BCTV–Logan in transgenic plants results in a range of phenotypes that include stunting, leaf curling and deformation, abnormal development of inflorescence structures, and the inhibition of flowering. More recent studies aimed at isolating the plant host factors that interact with L4 have utilized the yeast two-hybrid system for investigating protein–protein interactions (Chien *et al.*, 1991). Several *A. thaliana* cDNAs that encode proteins that specifically interact with L4 have been identified and are currently being characterized (K. Buckley, D. Ware, and K. R. Davis, unpublished). These studies will no doubt lead to a more detailed explanation of the molecular basis of symptom expression in this plant–virus interaction.

### 2.3.3. Turnip Crinkle Virus (TCV)

TCV is a single-stranded RNA virus in the carmovirus group that also exhibits a rather broad host range and represents one of the simplest RNA viruses known. It is also interesting from the point of view that it is often associated with subviral RNAs that can have profound effects on symptom development in susceptible hosts (Simon, 1994). Infection of *A. thaliana* accessions with TCV in the presence or absence of virulent subviral RNAs indicated that most accessions were susceptible to TCV and that the presence of a virulent subviral RNA was required for maximum symptom development (Simon *et al.*, 1992; Dempsey *et al.*, 1993). *A. thaliana* plants inoculated with a TCV genomic RNA alone typically develop moderate symptoms that include mild chlorosis, stunting, and inflorescence curling. In contrast, plants inoculated with TCV and a virulent subviral RNA developed spreading chlorotic lesions within 1–2 weeks that subsequently become necrotic, usually leading to the death of the plants by 3–4 weeks after inoculation. However, one accession, Dijon (Di-0), appeared to be resistant to TCV infection.

Dijon plants were normally asymptomatic (a few plants developed mild symptoms several weeks after infection) or in some cases developed localized necrotic lesions reminiscent of HR (Simon *et al.*, 1992; Dempsey *et al.*, 1993; Uknes *et al.*, 1993). The demonstration that TCV infection of Dijon caused the accumulation of salicylic acid, increased accumulation of several pathogenesis-related (PR) protein transcripts, and the expression of systemic acquired resistance suggests that this may indeed represent a case of HR-mediated resistance (Dempsey *et al.*, 1993; Uknes *et al.*, 1993). Protoplast infection and whole-plant hybridization experiments revealed that the resistance exhibited by Dijon plants was due to an inhibition of systemic (long-distance) movement rather than to a block in viral replication or cell-to-cell movement (Simon *et al.*, 1992; Simon, 1994). More recent studies in which recombinant viruses were made between TCV and an isolate of cardamine chlorotic fleck virus (CCFV) that can infect *A. thaliana* (Skotnicki *et al.*, 1993) suggest that the coat protein (or its corresponding ORF) is a primary determinant for resistance in Dijon and may function by affecting virus accumulation (Kong *et al.*, 1995; Oh *et al.*, 1995).

No clear genetic basis for the resistance of Dijon to TCV has been published. This may be due to the fact that expression of resistance can be significantly affected by the growth conditions used to rear plants and that even under ideal conditions a small percentage of the plants exhibit some symptoms (Simon *et al.*, 1992; Dempsey *et al.*, 1993). It has been suggested that the original Di-0 accession represents a heterogeneous population containing both susceptible and resistant individuals. Two lines have been generated from the original Di-0 population that are more uniformly susceptible or resistant. In this case, approximately 70% of the TCV-inoculated plants of the resistant line exhibit complete resistance (Dempsey *et al.*, 1993). Although the environmental effects and variability in disease phenotypes certainly complicate the genetic analysis of this resistance locus, it is likely that strategies will be developed to overcome these difficulties.

### 2.3.4. Tobacco Mosaic Virus (TMV)

TMV, the type member of the tobamovirus group, is one of the best known plant RNA viruses and its interaction with solanaceous hosts has been well characterized (Dawson, 1992). *A. thaliana* was first noted as a host for TMV by Sasnova and Polak (1975). Later studies confirmed this observation by demonstrating that a crucifer-infecting strain of TMV, TMV-Cg, could infect *A. thaliana* (Ishikawa *et al.*, 1991). Symptoms observed in *A. thaliana* plants inoculated with TMV-Cg exhibited varying degrees of stunting, dark-greening of rosette leaves, and abnormal inflorescence development that was associated with reduced fertility. It was noted that younger plants (15–20 days after sowing) developed symptoms with a few days after inoculation, whereas older plants (>3 weeks after sowing) generally remained asymptomatic.

Most of the characterization of TMV–*A. thaliana* interactions has focused on accumulation of the coat protein (CP) in infected plants. Ishikawa *et al.* (1991) used enzyme-linked immunosorbent assay (ELISA)-based detection of CP as the basis for a screen to identify mutant *A. thaliana* plants exhibiting alterations in TMV multiplication. Two independent mutants were identified that exhibited reduced CP accumulation. Genetic analyses of these *TOM* (tobamovirus multiplication) mutants revealed that they both contained a single recessive locus conferring the reduced CP accumulation phenotype and that the mutations were allelic. This *tom-1* mutation did not have any effect on the replication of the unrelated RNA viruses, TCV, or turnip yellow mosaic virus (TYMV). Subsequent studies demonstrated that the *tom-1* mutation caused a concomitant reduction of TMV genomic RNA and CP in a protoplast assay. In addition, it was observed that coinoculation of TMV and cucumber mosaic virus (CMV) in *tom1* protoplasts greatly reduced the accumulation of TMV RNA, whereas coinoculation of TMV with CMV had no effect in protoplasts derived from wild-type plants (Ishikawa *et al.*, 1993). Coinoculation of TMV and TCV had no effect on TMV accumulation in either wild-type or *tom-1* protoplasts. These results suggest that the *TOM1* gene encodes a protein required for the efficient replication of TMV RNA and that at least one other gene exists that is required for both TMV and CMV multiplication.

A recent study has been published concerning the expression of the TMV 30-kD movement protein (P-30) in *A. thaliana* (Hughes *et al.*, 1995). It was found that in TMV-infected plants or transgenic plants expressing the movement protein a truncated form of P-30 accumulated, suggesting that P-30 is proteolytically processed in this host. The inability of transgenic *A. thaliana* expressing P-30 to complement a TMV P-30 mutant indicates that the processed form of P-30 is not functional. The ability of TMV to infect *A. thaliana* despite this processing was attributed to the fact that some unprocessed P-30 is present during early times after inoculation. These results provide additional support for previous suggestions that the N-terminal portion of P-30 is required for its function. Additional experiments demonstrated that P-30 is not phosphorylated in *A. thaliana*. This result is of particular interest since previous studies have suggested that phosphorylation of P-30 may be important for inactivating P-30 in tobacco (Citovsky *et al.*, 1993). Thus, *A. thaliana* appears to have an alternative pathway for the inactivation of P-30. In addition, these studies point out a potential difficulty in using TMV–*A. thaliana* as a model for studying plant–virus interactions.

### 2.3.5. Other Viral Pathogens of *A. thaliana*

Given the fact that the majority of known plant viruses of agronomic importance are single-stranded RNA viruses, it is not surprising that most studies of *A. thaliana*–virus interactions have focused on this broad class of plant pathogens. Several other RNA viruses have been shown to infect *A. thaliana;* however, these interactions are currently in the initial stages of development. As mentioned previously, TYMV (Ramachandra *et al.*, 1988; Ishikawa *et al.*, 1991; Bozarth *et al.*, 1992; Dangl, 1993), CMV (Ishikawa *et al.*, 1993), and CCFV (Skotnicki *et al.*, 1993) have been shown to infect *A. thaliana*. Recent reports have documented that tomato spotted wilt virus (German *et al.*, 1995), Chinese (oilseed) rape mosaic virus (Aguilar *et al.*, 1996), and tobacco ringspot virus (Lee *et al.*, 1996) are also capable of causing systemic infections in *A. thaliana*. In the case of tobacco ringspot virus, tolerance to virus infection observed in several ecotypes was found to be due to a single incompletely dominant locus, designated *TTR1*. In addition, several studies have demonstrated that *A. thaliana* tissues or cells can support some aspects of virus replication or assembly using viruses that do not cause systemic infections (Stenger *et al.*, 1992; Wellink *et al.*, 1993). It is possible that studies of these nonhost interactions using *A. thaliana* mutants with specific alterations in various cellular processes will provide useful information concerning host proteins required for virus multiplication.

## 2.4. Nematodes as Pathogens of *A. thaliana*

Nematodes represent an interesting class of eukaryotic parasites that cause significant crop losses throughout the world. However, in spite of their economic importance, little is known about the interaction of plant pathogenic nematodes with their hosts, particularly at the molecular level (Sijmons *et al.*, 1994). This is primarily due to the fact that all known root-parasitic nematodes are obligate biotrophs for which axenic culture conditions have not been identified. In addition, nematodes have a long and complex life cycle compared to viral and microbial plant pathogens; thus, many investigators have shied away from using these organisms as 'model' plant pathogens. Nonetheless, several groups have attempted to take the advantages that *A. thaliana* offers as a model host plant to increase our understanding of nematode–plant interactions. Progress in characterizing the interaction of *A. thaliana* has been recently reviewed (Sijmons *et al.*, 1994), so only a few highlights will be discussed here.

The ability to examine *A. thaliana*–nematode interactions has been greatly facilitated by the development of culture conditions using a defined agar medium that allow high infection rates of several nematode species (Sijmons *et al.*, 1991). This system allows quick access to the root systems and provides for facile visualization of *in vivo* nematode behavior and plant responses. Nematode species that are known to parasitize *A. thaliana* include the sedentary root knot nematode *Meloidogyne incognita*, the sedentary cyst-forming nematodes *Heterodera schachtii, H. trifolii,* and *H. cajani,* and the migratory nematode species *Pratylenchus penetrans* (Sijmons *et al.*, 1991; Wyss *et al.*, 1992). Both *H. goettingiana* and *Globoder rostochiensis* initiate infections of *A. thaliana roots;* however, they do not complete their life cycles (Sijmons *et al.*, 1994).

Detailed observations of the infection of *A. thaliana* by several of these nematode species have been published (Sijmons *et al.*, 1991, 1994; Wyss *et al.*, 1992; Böckenhoff and Grundler, 1994). These studies demonstrate that the root knot nematode *Meloidogyne incognita* and the sedentary cyst-forming *Heterodera* species form feeding structures typical of those observed in other plant hosts. Additional studies of *Heterodera schachtii* infection of *A. thaliana* demonstrated that nematode infection induced phloem unloading at infection sites (Böckenhoff *et al.*, 1996). Effects of nematode infection on patterns of root gene expression have been documented using transgenic *A. thaliana* plants expressing defined promoter-GUS reporter genes of random 'promoter-trap' lines (Goddijn *et al.*, 1993). These studies demonstrated that the activities of several promoters that are typically high in uninfected roots, such as the *PAL1* and CaMV 35S promoters, are downregulated in nematode feeding structures. Both root knot and cyst nematodes were found to induce expression of *cdc2a*-GUS and *cyc1*-GUS reporter genes (Niebel *et al.*, 1996). Analysis of a panel of promoter-trap lines having strong expression in uninfected roots demonstrated that the downregulation of genes normally expressed in roots is a general phenomenon. Comparison of the promoter sequences of these downregulated genes suggests that the *as-1* element may be involved; however, functional studies are required to clearly determine if this is indeed the case. Screens of promoter trap lines for plants that exhibited GUS activity in feeding structures resulted in the isolation of several lines that exhibited GUS expression in feeding structures, but not in adjacent root cells. These lines also exhibited some GUS activity in the vascular cylinder of uninfected roots; thus expression was not totally specific to feeding structures. Further analysis of these promoter trap lines will provide important clues as to possible mechanisms responsible for the dramatic effects of nematode infection of root gene expression and the role of these changes in the development of feeding structures.

Genetic approaches have also been applied to characterizing several of these defined nematode interactions. These studies involved evaluating various *A. thaliana* accessions for differences in response to nematode infection; screening mutagenized *A. thaliana* populations for plants that are resistant to infection, or that exhibit altered symptom development at specific infection stages; and examining any changes in the infection process in previously characterized *A. thaliana* mutants that would be expected to potentially affect the nematode infection, e.g., hormone or root development mutants (Sijmons *et al.*, 1991, 1994). Initial surveys of different accessions using several different *H. schachtii* populations failed to find any resistant lines, although some differences in female populations in roots were noted. Screens of ethyl methanesulfonate (EMS)-mutagenized *A. thaliana* with either *H. schachtii* or *M. incognita* resulted in the isolation of one mutant that was affected in root hair penetration by *M. incognita*. Preliminary data indicated that this mutation is inherited as a single locus, but no further information is available about this potentially interesting mutation. Initial studies of nematode infection of 14 different *A. thaliana* hormone mutants yielded the surprising result that none of these mutations have major effects on nematode infection or development. Some minor effects were noted, including relatively early invasion in the ethylene mutant *etr1*, increased nematode-induced lateral root formation in a gibberellin mutant (*ga-5*), and decreased nematode-induced lateral root formation in one auxin-resistant mutant (*axr-1*). More detailed analyses of these more subtle phenotypes and additional studies of other classes of mutants that affect root development may provide important clues to the mechanisms involved in nematode infection and the development of differentiated feeding structures.

## 2.5. Other Pathogens of *A. thaliana*

Additional pathogens of *A. thaliana* include the mollicutes, an interesting group of cell-wall-less prokaryotes that include spiroplasmas and mycoplasmalike organisms (MLOs). *Spiroplasma citir*, the causal agent of horseradish brittle root, and beet leafhopper-transmitted virescence agent (BLTVA) have both been shown to infect *A. thaliana* using beet leafhoppers as an insect vector (Golino *et al.*, 1988; Fletcher and Eastman, 1992; Fletcher *et al.*, 1993). Screens for *S. citri*- and *BLTVA-resistant A. thaliana* accessions have been negative (Fletcher *et al.*, 1993). One potential problem with the utility of this pathosystem for genetic studies is that leafhopper feeding is quite damaging and infection rates are 80–90%. However, since many of the overall effects of mollicute infection of *A. thaliana* are similar to those observed in other plant hosts, it is likely that further studies of this interaction will provide new clues concerning the basis of symptom development. For example, BLVTA causes premature induction of flowering in several hosts, including *A. thaliana*. Studies utilizing BLVTA infection of the available floral induction and development mutants may prove useful in characterizing this phenomenon.

A major area in which there is currently little information is the suitability of

*A. thaliana* for studies of plant–insect interactions. Anyone who has grown *A. thaliana* for any length of time is aware that a variety of insect pests, including aphids, thrips, fungus gnat larvae, and spider mites, consider it a suitable host. As mentioned above, some species of leafhoppers will also feed on *A. thaliana* and it is likely that other insects that feed on cruciferous plants will also utilize *A. thaliana* as a host. Studies of these interactions may provide useful clues concerning the roles of specific plant compounds as insect attractants or deterrents and as a model to test whether overexpression of insecticidal proteins or secondary metabolites provides protection against insect attack.

## 3. *ARABIDOPSIS* DEFENSE RESPONSES

A major question concerning plant disease resistance that continues to plague investigators is what the molecular and biochemical basis of resistance really is and precisely how the resistance response is triggered. A voluminous body of literature has been published that describes a number of diverse responses that are associated with a resistance response and potential signals that may activate these responses (Bell, 1981; Collinge and Slusarenko, 1987; Matern and Kneusel, 1988; Hahlbrock and Scheel, 1989; Dixon and Harrison, 1990; Côté and Hahn, 1994; Ryan, 1994; Dixon and Paiva, 1995; Kuc, 1995; Low and Merida, 1996; Schaller and Ryan, 1996). Most of these responses are thought to be involved in producing toxic metabolites or proteins that directly inhibit pathogen growth or in generating physical barriers that prevent pathogens from easily moving from the infection site. Unfortunately, because of the overall complexity of resistance responses, it has been difficult to demonstrate clearly by correlative approaches that a particular response is required for resistance. Thus, a major impetus for the development of *A. thaliana* as a model system for plant–pathogen studies was the hope that the combination of biochemical and molecular studies with genetic analyses of mutant plants with specific alterations in disease resistance would provide some clear insights into the basis of plant disease resistance (Davis, 1992).

A major defense reaction that is commonly associated with disease resistance is HR (Keen, 1990). HR is distinguished by rapid, localized necrosis at the site of infection and the activation of a large number of defense-related genes. The induction of HR is thought to be triggered by the recognition of molecular signals produced by the pathogen by receptors present in the responding plant cells. In 'gene-for-gene' systems, a highly evolved recognition system has evolved that is highlighted by highly specific interactions between pathogen *avr* genes and plant *R* genes (Keen, 1990, 1992; Heath, 1991; Crute, 1994; Staskawicz *et al.*, 1995). Another resistance response that has gained much attention which is often associated with the induction of HR is systemic acquired resistance (SAR) (Madamanchi and Kuc, 1991). SAR is characterized by the induction of resistance throughout the plant and is correlated with the expression of a number of PR proteins (Ryals *et al.*, 1995; Bol *et al.*, 1996). SAR-mediated resistance is rather broad spectrum and often prevents infection by viral, bacterial, and fungal pathogens. The following subsections will briefly summarize some of the details of HR and SAR in *A. thaliana*, with an emphasis on how genetic and molecular approaches applied in this model system have provided new insights into the regulation of these defense responses.

### 3.1. Identification of Resistance (*R*) Genes in *A. thaliana*

A major contribution that studies of *A. thaliana*–pathogen interactions have made is in the isolation and characterization of *R* genes that confer recognition to microbial pathogens. As described in previous sections, a number of *R* gene specificities have been defined for several *A. thaliana* pathosystems. Although gene-for-gene interactions between plants and their pathogens and the genetic identification of *R* genes have been known since the pioneering work of Flor during the 1940s (Flor, 1971), it was not until 1992 that the first *R* gene was cloned (Johal and Briggs, 1992). Shortly thereafter, several other *R* genes were isolated that conferred specificity to bacterial, fungal, and viral pathogens (reviewed in Staskawicz *et al.*, 1995; J. Jones, 1996), including two *A. thaliana R* genes, *RPS2* and *RPS3* (*RPM1*) (Bent *et al.*, 1994; Mindrinos *et al.*, 1994; Grant *et al.*, 1995). Both of these *A. thaliana R* genes were cloned using positional cloning methods.

The *A. thaliana R* genes *RPS2* and *RPS3* (*RPM1*) are similar in structure, both having a N-terminal leucine zipper domain, an internal nucleotide-binding site (NBS), and a C-terminal leucine-rich repeat (LRR) domain. Similar NBS and LRR domains are also found in the tobacco *N* gene and the flax *L6* gene which confer resistance to TMV and flax rust, respectively (Whitham *et al.*, 1994; Lawrence *et al.*, 1995). Ectopic expression of bacterial avirulence genes *avrB* and *avrRpt2* in plant tissues expressing the corresponding resistance genes resulted in the induction of a resistance response (Leister *et al.*, 1996; Gopalan *et al.*, 1996). These observations suggest that recognition requires only the presence of the avirulence gene product and that the recognition event may occur in the cytoplasm. Current models predict that the LRR domain is responsible for recognizing race-specific pathogen elicitors and that this interaction may trigger a protein kinase cascade and/or the activation of a NADPH oxidase complex that initiates an oxidative burst (Staskawicz *et al.*, 1995; J. Jones, 1996). Further functional studies of these *A. thaliana R* genes will provide important tests of these models of *R* gene function.

### 3.2. Defense-Related Gene Expression in *A. thaliana*

As described in the previous sections, *A. thaliana* exhibits HR in response to infection with specific bacterial, fungal, and viral pathogens. The first studies of *A. thaliana* defense responses utilized cultured cells treated with a bacterial pectinase elicitor that was known to induce defense reactions in parsley cell cultures and soybean cotyledons (Davis and Ausubel, 1989). These studies demonstrated that several enzyme activities and/or mRNAs of the defense-related genes phenylalanine ammonia-lyase (PAL), 4-coumarate:CoA ligase (4CL), caffeic acid *O*-methyl transferase (CMT), and β-1,3-glucanase were induced by elicitor treatment. These studies gave the first indication that the defense responses of *A. thaliana* were similar to those reported for other plants.

Since this initial study, detailed studies of defense gene expression during the interaction of *A. thaliana* with a variety of virulent and avirulent pathogens have been completed (reviewed in Crute *et al.*, 1994). These studies have demonstrated that transcripts for several genes, including *PAL1, PAL2, 4CL,* those for glutathione-*S*-transferase (*GST1*) and lipoxygenase (*LOX1*), the elicitor-induced gene *ELI3,* and that for 3-deoxy-D-arabinoheptulosonate 7-phosphate synthase (*DHS1*), are induced more rapidly and/or to higher levels in interactions with avirulent pathogens that result in HR compared to compatible interactions with virulent pathogens (Davis *et al.*, 1991; Dong *et al.*, 1991; Keith *et al.*, 1991; Kiedrowski *et al.*, 1992; Lummerzheim *et al.*, 1993; Melan *et al.*, 1993; Wanner *et al.*, 1993; Greenberg *et al.*, 1994; D. Lee *et al.*, 1995; Mittal and Davis, 1995). Other studies have identified additional genes of unknown function that are also differentially induced by avirulent pathogens (Aufsatz and Grimm, 1994; Reuber and Ausubel, 1996). The induction of genes involved in phenylpropanoid biosynthesis, such as *PAL* and *4CL,* is a common feature of resistance responses in many plants. In *A. thaliana,* the induction of this pathway is likely to be involved in the production of lignin and other phenolic compounds, such as salicylic acid (Mauch-Mani and Slusarenko, 1996; S. Lee and K. R. Davis, unpublished observations). Given recent studies that suggest that the induction of salicylic acid accumulation is required for the establishment of HR (Delaney *et al.*, 1994), the induction of the phenylpropanoid pathway may be a critical component of the resistance response. Besides genes that are preferentially expressed by avirulent bacteria, several genes have been identified that are induced to similar extents by both virulent and avirulent pathogen strains or genes that are preferentially induced by virulent pathogens (Dong *et al.*, 1991; Crute *et al.*, 1994; Sharma and Davis, 1995). The fact that a number of distinct gene expression patterns are induced by pathogen infection suggests that multiple signaling pathways are involved in the response of *A. thaliana* to microbial attack. This is supported by experiments in which plants deficient in the production of salicylic acid or that have a mutation affecting disease resistance are altered in some aspects of pathogen-induced gene expression, but not others (Greenberg *et al.*, 1994; Sharma *et al.*, 1996).

The analysis of defense gene expression during *A. thaliana*–pathogen interactions has served several important purposes. First, the results have clearly demonstrated that the overall response of *A. thaliana* is very similar to that of other plants. Thus, new information gained using this model system will likely be directly applicable to other plant–pathogen systems. Second, characterized defense genes can serve as defined probes that can be used to access the effects of specific mutations on disease resistance. A third use for analyses of defense gene expression is to compare the overall pattern of gene expression in plants infected with different pathogens or nearly isogenic pathogens that contain different *avr* genes. This type of analysis provides some indication as to potential overlap between resistance responses induced by different pathogens and has been used to demonstrate that there can be interference between *avr* genes in inducing resistance responses (Reuber and Ausubel, 1996). Finally, the isolation of novel genes specifically induced during pathogen infection may provide additional clues about specific signaling steps or biochemical responses that are important for the expression of resistance.

### 3.3. Activation of Programmed Cell Death and the Expression of HR and SAR

A hallmark of HR is the rapid death of plant cells at the infection site. A number of studies have suggested that in some cases this cell death is actually an example of programmed cell death (reviewed in Johal *et al.*, 1995; A. M. Jones and Dangl, 1996). One major piece of evidence that supports this proposition is that it has been possible to identify cell death mutants in several plant species, including *A. thaliana*. Some of these "accelerated cell death" (*acd*) or "lesion-simulating disease resistance" (*lsd*) mutants develop necrotic lesions in the absence of pathogen infection concomitant with the induction of defense-related gene expression and the accumulation of antimicrobial phytoalexins (Greenberg and Ausubel, 1993; Dietrich *et al.*, 1994; Greenberg *et al.*, 1994; Weymann *et al.*, 1995). These mutations are likely to be in genes encoding factors important for controlling the expression of HR.

The activation of HR and the often accompanying cell death is thought to be dependent on the production of active oxygen species (AOS) that serve as signals that activate downstream events required for the establishment of resistance (Lamb, 1994; Mehdy, 1994; Low and Merida, 1996). In addition to AOS, salicylic acid may play a signaling role in HR as well, since it has been shown to induce defense gene expression and potentiate responses to AOS (Kauss and

Jeblick, 1995, 1996; Mur *et al.*, 1996). The observations that several cell death mutants accumulate high levels of salicylic acid (Greenberg *et al.*, 1994; Weymann *et al.*, 1995) and that the *lsd1* mutant accumulates abnormally high levels of superoxide (Jabs *et al.*, 1996) provide evidence for the role of these compounds in controlling some aspects of HR.

The induction of SAR is also often associated with localized cell death in one part of the plant and the accumulation of salicylic acid (Klessig and Malamy, 1994). *A. thaliana* has been shown to exhibit SAR in response to microbial infection (Uknes *et al.*, 1993; Cameron *et al.*, 1994; Lawton *et al.*, 1995), by treatment with salicylic acid or other chemicals (Delaney *et al.*, 1994; Vernooij *et al.*, 1995; Lawton *et al.*, 1996), and by ozone exposure (Sharma *et al.*, 1996). It is noteworthy that cell death mutants that accumulate salicylic acid also are highly resistant to pathogen infection (Dietrich *et al.*, 1994; Greenberg *et al.*, 1994; Weymann *et al.*, 1995). Mutant screens based on the failure of *A. thaliana* plants to exhibit SAR or SAR-related gene expression have resulted in the isolation of interesting mutants, *npr1* (nonexpresser of PR genes; Cao *et al.*, 1994) and *nim1* (noninducible immunity; Delaney *et al.*, 1995). The *npr1* and *nim1* phenotypes are due to single recessive nuclear mutations that abolish SAR by affecting a step downstream of salicylic acid. These mutations also have some effects on local resistance, adding support to the conjecture that some aspects of HR and SAR rely on common signaling pathways. The *NPR1* gene has recently been cloned and shown to encode a novel protein containing ankyrin repeats (Cao *et al.*, 1997). Based on the structure of NPR1, it has been suggested that it may function in a manner similar to that of other stress-related transcriptional regulators, such as NF-κB and I-κB, which also contain ankyrin repeats. Another class of SAR mutants was identified by screening mutagenized *A. thaliana* for the constitutive expression of PR gene expression (Bowling *et al.*, 1994). This *cpr1* mutation caused the constitutive expression of SAR and PR gene expression and was correlated with the presence of elevated salicylic acid levels. Genetic studies demonstrated that *cpr1* phenotype is due to a single, recessive nuclear mutation. The effects of the *cpr1* mutation were ameliorated by introduction of a bacterial gene encoding salicylate hydroxylase, which converts salicylic acid to catechol. This suggests that the *CPR1* encodes a product that is a negative regulator of SAR at a step upstream of salicylic acid production. The identification of HR and SAR mutants clearly demonstrates the utility of applying *A. thaliana* genetics to studies of resistance responses.

### 3.4. Additional *A. thaliana* Disease Resistance Mutants

A number of groups have exploited mutational analysis of *A. thaliana* in an attempt to identify key steps in the induction and expression of disease resistance. Initial studies tended to focus on identifying mutant plants that had lost resistance to a particular avirulent pathogen or did not exhibit an observable HR. These studies were greatly facilitated by the availability of nearly isogenic bacterial strains that differed only by the presence of a single *avr* gene that was recognized by the *A. thaliana* accession used for mutagenesis. The majority of mutants that were identified using this strategy had mutations in *R* genes corresponding to the *avr* gene present in the strain used for screening (see Section 3.1). However, this strategy has yielded one disease resistance mutation that does not correspond to an alteration of an *R* gene. This mutation, designated *ndr1* (non-race-specific disease resistance) renders this mutant susceptible to infection by normally avirulent *P. s.* pv. *tomato* containing any of four *avr* genes as well as several avirulent isolates of *P. parasitica* (Century *et al.*, 1995). Interestingly, *ndr1* plants expressed HR in response to several of the bacterial strains, even though the plants were clearly susceptible based on increased bacterial multiplication. In addition, resistance to phytopathogens for which *A. thaliana* is not a host (the bacterial pathogen *P. s.* pv. *glycinia* and the fungal pathogen *P. parasitica* race P-006) was not affected by this mutation. The fact that *ndr1* plants are altered in resistance to two distinct pathogens suggests that *NDR1* may represent a step in signal transduction pathway that is generally activated in response to pathogen recognition. Further studies of defense gene activation and other defense reactions in the *ndr1* mutant will no doubt provide important clues as to key components of disease resistance.

Mutant screens based on the lack of expression of specific defense responses have also been successful. *A. thaliana* has been shown to accumulate the phytoalexin camalexin in response to pathogen infection (Tsuji *et al.*, 1992). Glazebrook and Ausubel (1994) conducted a brute force biochemical screen and identified camalexin-deficient *A. thaliana* mutants. These phytoalexin-deficient (*pad*) mutants fell into three distinct complementation groups and may represent mutations in specific steps in camalexin biosynthesis. Resistance of the *pad* mutants to infection by avirulent *P. syringae* strains was not affected, demonstrating that camalexin accumulation is not required for this resistance response. However, *pad1* and *pad2* were found to be more susceptible to infection with virulent *P. syringae* strains, suggesting that camalexin accumulation may play a role in limiting the growth of virulent bacteria.

The observation that some *pad* defense response mutants exhibited increased susceptibility to infection by virulent bacteria prompted an additional screen based on the identification of mutant *A. thaliana* plants that display increased symptom development after inoculation with low doses of the virulent *P. syringae* pv. *maculicola* strain ES4326 (Glazebrook *et al.*, 1996). Twelve enhanced disease susceptibility (*eds*) mutants were identified. Two of these mutants represented camalexin-deficient mutants, an additional *pad2* mutant and a new mutant class designated as *pad4*. Two mutants represented additional alleles of the previously described *npr1* gene, while the remaining mutants defined seven novel genes of unknown function.

Detailed studies of the expression of the defense-related genes *PR1*, *PR5*, and *BGL2* in the newly defined *npr1* mutants demonstrated that *PR1* mRNA accumulation was greatly reduced in leaves inoculated with the virulent *P. s.* pv. *maculicola* ES4326 strain, whereas expression of *PR5* and *BGL2* was largely unaffected. This result suggests that *PR1* may play a role in limiting the growth of virulent bacterial pathogens and is consistent with the observation that *PR1* expression is required for SAR and ozone-induced disease resistance (Cao *et al.*, 1994; Delaney *et al.*, 1994; Sharma *et al.*, 1996). The fact that the pathogen induction of *PR1* mRNA accumulation in the *npr1* mutants was reduced, whereas accumulation of *PR5* and *BGL2* transcripts was normal, also suggests that there are at least two distinct signaling pathways involved in defense gene activation by virulent pathogens. Similar conclusions were drawn from studies where it was shown that ozone-induced resistance to infection with *P. s.* pv. *maculicola* 4326 in an *npr1* mutant was correlated with greatly reduced expression of *PR1*, but not in *PAL* or *GST1* (Sharma *et al.*, 1996).

The fact that several *eds* mutants were identified that corresponded to disease resistance mutants identified by screening for loss of resistance suggests that the expression of resistance to avirulent pathogens and the limitation of virulent pathogens require common resistance mechanisms. The success of the screen for *eds* mutants demonstrates that mutant hunts based on identifying plants that are either altered in resistance to avirulent pathogens or have increased sensitivity to virulent pathogens can both be fruitful. Given that mutant screens based on identifying mutants that have lost resistance to pathogens containing specific avirulence determinants tend to identify mutations in *R* genes, it is possible the screens for *eds*-like mutants may prove more useful for identifying mutations in genes involved in downstream signaling steps or the elaboration of specific biochemical responses important for conferring resistance.

## 4. CONCLUSIONS

The development of *A. thaliana* as a model host for studying plant–pathogen interactions has progressed rapidly. Initial predictions that the application of genetic and molecular genetic approaches that are feasible with this model plant to studies of plant–pathogen interactions would be successful have been borne out and the data leave little doubt that studies using *A. thaliana* as a host plant are productive and relevant. The establishment of numerous diverse *A. thaliana* pathosystems and the exponentially increasing amount of information becoming available about the basic biology of *A. thaliana* will ensure that this small weed will continue to play a major role in deciphering the complex processes that underlie disease resistance.

ACKNOWLEDGMENTS. I would like to take this opportunity to thank the numerous colleagues who have contributed to the development of the study of *Arabidopsis*–pathogen interactions and made this the stimulating and exciting area it is today. I gratefully acknowledge my former and current students and postdocs for their efforts and contributions to my research. Special thanks go to Fred Ausubel, who afforded me the opportunity to get started in this field and continues to provide helpful advice. Research concerning *A. thaliana*–pathogen interactions in my laboratory is funded by the National Institute of General Medical Sciences, Grant No. GM45570.

## 5. REFERENCES

Adam, L., and Somerville, S. C., 1996, Genetic characterization of five powdery mildew disease resistance loci in *Arabidopsis thaliana*, *Plant J.* **9**:341–356.

Aguilar, I., Sanchez, A., Martin, A. M., Martinez-Herrera, D., and Ponz, F., 1996, Nucleotide sequence of Chinese rape mosaic virus (oilseed rape mosaic virus), a crucifer tobamovirus infectious on *Arabidopsis thaliana*, *Plant Mol. Biol.* **30**:191–197.

Ashfield, T., Keen, N. T., Buzzell, R. I., and Innes, R. W., 1995, Soybean resistance genes specific for different *Pseudomonas syringae* avirulence genes are allelic, or closely linked, at the RPG1 locus, *Genetics* **141**:1597–1604.

Atkinson, M. M., and Baker, C. J., 1989, Role of the plasmalemma H+-ATPase in *Pseudomonas syringae*-induced K+/H+ exchange in suspension-cultured tobacco cells, *Plant Physiol.* **91**:298–303.

Aufsatz, W., and Grimm, C., 1994, A new, pathogen-inducible gene of *Arabidopsis* is expressed in an ecotype-specific manner, *Plant Mol. Biol.* **25**:229–239.

Baker, C. J., O'Neill, N. R., Keppler, L. D., and Orlandi, E. W., 1991, Early responses during plant bacteria interactions in tobacco cell suspensions, *Phytopathology* **81**:1504–1507.

Balàzs, E., and Lebeurier, G., 1981, *Arabidopsis* is a host of cauliflower mosaic virus, *Arabidopsis Inf. Serv.* **18**:130–134.

Bell, A. A., 1981, Biochemical mechanisms of disease resistance, *Annu. Rev. Plant Physiol.* **32**:21–81.

Bender, C. L., Young, S. A., and Mitchell, R. E., 1991, Conservation of plasmid DNA sequences in coronatine-producing pathovars of *Pseudomonas syringae*, *Appl. Environ. Microbiol.* **57**:993–999.

Bent, A. F., Kunkel, B. N., Dahlbeck, D., Brown, K. L., Schmidt, R., Giraudat, J., Leung, J., and Staskawicz, B., 1994, *RPS2* of *Arabidopsis thaliana:* A leucine-rich repeat class of plant disease resistance genes, *Science* **265**:1856–1860.

Bisgrove, S. R., Simonich, M. T., Smith, N. M., Sattler, A., and Innes, R. W., 1994, A disease resistance gene in *Arabidopsis* with specificity for two different avirulence genes, *Plant Cell* **6**:927–933.

Böckenhoff, A., and Grundler, F. M. W., 1994, Studies on the nutrient uptake by the beet cyst nematode *Heterodera schachtii* by *in situ* microinjection of fluorescent probes into the feeding structures in *Arabidopsis thaliana*, *Parasitology* **109**:249–254.

Böckenhoff, A., Prior, D. A. M., Grundler, F. M. W., and Oparka, K. J., 1996, Induction of phloem unloading in *Arabidopsis thaliana* roots by the parasitic nematode *Heterodera schachtii*, *Plant Physiol.* **112:**1421–1427.

Bol, J. F., Buchel, A. S., Knoester, M., Baladin, T., Van Loon, L. C., and Linthorst, H. J. M., 1996, Regulation of the expression of plant defence genes, *Plant Growth Regul.* **18:**87–91.

Bowling, S. A., Guo, A., Cao, H., Gordon, A. S., Klessig, D. F., and Dong, X., 1994, A mutation in *Arabidopsis* that leads to constitutive expression of systemic acquired resistance, *Plant Cell* **6:**1845–1857.

Bozarth, C., Weiland, J., and Dreher, T., 1992, Expression of ORF-69 of turnip yellow mosaic virus is necessary for viral spread in plants, *Virology* **187:**124–130.

Buell, C. R., and Somerville, S. C., 1995, Expression of defense-related and putative signaling genes during tolerant and susceptible interactions of *Arabidopsis* with *Xanthomonas campestris* pv. *campestris*, *Mol. Plant–Microbe Interact.* **8:**435–443.

Callaway, A., Liu, W. N., Andrianov, V., Stenzler, L., Zhao, J. M., Wettlaufer, S., Jayakumar, P., and Howell, S. H., 1996, Characterization of cauliflower mosaic virus (CaMV) resistance in virus-resistant ecotypes of *Arabidopsis*, *Mol. Plant Microbe Interact.* **9:**810–818.

Cameron, R. K., Dixon, R. A., and Lamb, C. J., 1994, Biologically induced systemic acquired resistance in *Arabidopsis thaliana*, *Plant J.* **5:**715–725.

Cao, H., Bowling, S. A., Gordon, A. S., and Dong, X., 1994, Characterization of an *Arabidopsis* mutant that is nonresponsive to inducers of systemic acquired resistance, *Plant Cell* **6:**1583–1592.

Cao, H., Glazebrook, J., Clarke, J. D., Volko, S., and Dong, X. N., 1997, The *Arabidopsis NPR1* gene that controls systemic acquired resistance encodes a novel protein containing ankyrin repeats, *Cell* **88:**57–63.

Century, K. S., Holub, E. B., and Staskawicz, B. J., 1995, *NDR1*, a locus of *Arabidopsis thaliana* that is required for disease resistance to both a bacterial and a fungal pathogen, *Proc. Natl. Acad. Sci. USA* **92:**6597–6601.

Chien, C.-T., Bartel, P. L., Sternglanz, R., and Fields, S., 1991, The two-hybrid system: A method to identify and clone genes for proteins that interact with a protein of interest, *Proc. Natl. Acad. Sci. USA* **88:**9578–9582.

Citovsky, V., Mclean, B. G., Zupan, J. R., and Zambryski, P., 1993, Phosphorylation of tobacco mosaic virus cell-to-cell movement protein by a developmentally regulated plant cell wall-associated protein kinase, *Gene Dev.* **7:**904–910.

Collinge, D. B., and Slusarenko, A. J., 1987, Plant gene expression in response to pathogens, *Plant Mol. Biol.* **9:**389–410.

Côté, F., and Hahn, M. G., 1994, Oligosaccharins: Structures and signal transduction, *Plant Mol. Biol.* **26:**1379–1411.

Crute, I. R., 1994, Gene-for-gene recognition in plant–pathogen interactions, *Philos. Trans. R. Soc. Lond. Biol.* **346:**345 –349.

Crute, I., Beynon, J., Dangl, J., Holub, E., Mauch-Mani, B., Slusarenko, A., Staskawicz, B., and Ausubel, F., 1994, Microbial pathogenesis of *Arabidopsis*, in *Arabidopsis* (E. M. Meyerowitz and C. R. Somerville, eds.), pp. 705–747, CSHL Press, New York.

Cuppels, D. A., and Ainsworth, T., 1995, Molecular and physiological characterization of *Pseudomonas syringae* pv. *tomato* and *Pseudomonas syringae* pv. *maculicola* strains that produce the phytotoxin coronatine, *Appl. Environ. Microbiol.* **61:**3530–3536.

Dangl, J. L., 1993, Applications of *Arabidopsis thaliana* to outstanding issues in plant–pathogen interactions, *Int. Rev. Cytol.* **144:**53–83.

Dangl, J., Lehnackers, H., Kiedrowski, S., Debener, T., Rupprecht, C., Arnold, M., and Somssich, I., 1991, Interactions between *Arabidopsis thaliana* and phytopathogenic *Pseudomonas* pathovars: A model for the genetics of disease resistance, in *Advances in Molecular Genetics of*

*Plant–Microbe Interactions*, Vol. 1 (H. Hennecke and D. P. S. Verma, eds.), pp. 78–83, Kluwer, Dordrecht, The Netherlands.

Dangl, J. L., Ritter, C., Gibbon, M. J., Mur, L. A. J., Wood, J. R., Goss, S., Mansfield, J., Taylor, J. D., and Vivian, A., 1992, Functional homologs of the *Arabidopsis RPM1* disease resistance gene in bean and pea, *Plant Cell* **4:**1359–1369.

Daniels, M. J., Fan, M.-J., Barber, C. E., Clarke, B. R., and Parker, J. E., 1991, Interaction between *Arabidopsis thaliana* and *Xanthomonas campestris*, in *Advances in Molecular Genetics of Plant–Microbe Interactions*, Vol. 1 (H. Hennecke and D. P. S. Verma, eds.), pp. 84–89, Kluwer, Dordrecht, The Netherlands.

Davis, K. R., 1992, *Arabidopsis thaliana* as a model host for studying plant–pathogen interactions, in *Molecular Signals in Plant–Microbe Communication* (D. P. S. Verma, ed.), pp. 393–406, CRC Press, Boca Raton, Florida.

Davis, K. R., and Ausubel, F. M., 1989, Characterization of elicitor-induced defense responses in suspension-cultured cells of *Arabidopsis*, *Mol. Plant–Microbe Interact.* **2:**363–368.

Davis, K. R., Schott, E., and Ausubel, F. M., 1991, Virulence of selected phytopathogenic pseudomonads in *Arabidopsis thaliana*, *Mol. Plant–Microbe Interact.* **4:**477–488.

Dawson, W. O., 1992, Tobamovirus–plant interactions, *Virology* **186:**359–367.

Debener, T., Lehnackers, H., Arnold, M., and Dangl, J. L., 1991, Identification and molecular mapping of a single *Arabidopsis thaliana* locus determining resistance to a phytopathogenic *Pseudomonas syringae* isolate, *Plant J.* **1:**289–302.

Delaney, T. P., Uknes, S., Vernooij, B., Friedrich, L., Weymann, K., Negrotto, D., Gaffney, T., Gut, R. M., Kessmann, H., Ward, E., and Ryals, J., 1994, A central role of salicylic acid in plant disease resistance, *Science* **266:**1247–1250.

Delaney, T. P., Friedrich, L., and Ryals, J. A., 1995, *Arabidopsis* signal transduction mutant defective in chemically and biologically induced disease resistance, *Proc. Natl. Acad. Sci. USA* **92:**6602–6606.

Dempsey, D. A., Wobbe, K. K., and Klessig, D. F., 1993, Resistance and susceptible responses of *Arabidopsis thaliana* to turnip crinkle virus, *Phythpathology* **83:**1021–1029.

Dietrich, R. A., Delaney, T. P., Uknes, S. J., Ward, E. R., Ryals, J. A., and Dangl, J. L., 1994, *Arabidopsis* mutants simulating disease resistance response, *Cell* **77:**565–577.

Dixon, R. A., and Harrison, M. J., 1990, Activation, structure and organization of genes involved in microbial defense in plants, *Adv. Genet.* **28:**165–234.

Dixon, R. A., and Paiva, N. L., 1995, Stress-induced phenylpropanoid metabolism, *Plant Cell* **7:**1085–1097.

Dong, X., Mindrinos, M., Davis, K. R., and Ausubel, F. M., 1991, Induction of *Arabidopsis* defense genes by virulent and avirulent *Pseudomonas syringae* strains and by a cloned avirulence gene, *Plant Cell* **3:**61–72.

Durbin, R. D., 1991, Bacterial phytotoxins—Mechanisms of action, *Experientia* **47:**776–783.

Feys, B. J. F., Benedetti, C. E., Penfold, C. N., and Turner, J. G., 1994, *Arabidopsis* mutants selected for resistance to the phytotoxin coronatine are male sterile, insensitive to methyl jasmonate, and resistant to a bacterial pathogen, *Plant Cell* **6:**751–759.

Fletcher, J., and Eastman, C. E., 1992, *Arabidopsis thaliana* as an experimental host of the mollicute *Spiroplasma citri*, *Plant Dis.* **76:**862–864.

Fletcher, J., Golino, D. A., and Eastman, C. E., 1993, *Arabidopsis* as an experimental host plant of the phytopathogenic mollicutes, in *Arabidopsis thaliana as a Model for Plant–Pathogen Interactions* (K. R. Davis and R. Hammerschmidt, eds.), pp. 99–108, APS Press, St. Paul.

Flor, H., 1971, Current status of the gene-for-gene concept, *Annu. Rev. Phytopathol.* **9:**275–296.

German, T. L., Adkins, S., Witherell, A., Richmond, K. E., Knaack, W. R., and Willis, D. K., 1995, Infection of *Arabidopsis thaliana* ecotype Columbia by tomato spotted wilt virus, *Plant Mol. Biol. Rep.* **13:**110–117.

Glazebrook, J., and Ausubel, F. M., 1994, Isolation of phytoalexin-deficient mutants of *Arabidopsis thaliana* and characterization of their interactions with bacterial pathogens, *Proc. Natl. Acad. Sci. USA* **91**:8955–8959.

Glazebrook, J., Rogers, E. E., and Ausubel, F. M., 1996, Isolation of *Arabidopsis* mutants with enhanced disease susceptibility by direct screening, *Genetics* **143**:973–982.

Goddijn, O. J. M., Lindsey, K., van der Lee, F. M., Klap, J. C., and Sijmons, P. C., 1993, Differential gene expression in nematode-induced feeding structures of transgenic plants harbouring promoter-*gusA* fusion constructs, *Plant J.* **4**:863–873.

Golino, D. A., Shaw, M., and Rappaport, L., 1988, Infectin of *Arabidopsis thaliana* (L.) Heyhn. with a mycoplasma-like organism, the beet leafhopper transmitted virescence agent, *Arabidopsis Inf. Serv.* **26**:9–14.

Gopalan, S., Bauer, D. W., Alfano, J. R., Loniello, A. O., He, S. Y., and Collmer, A., 1996, Expression of the *Pseudomonas syringae* avirulence protein AvrB in plant cells alleviates its dependence on the hypersensitive response and pathogenicity (Hrp) secretion system in eliciting genotype-specific hypersensitive cell death, *Plant Cell* **8**:1095–1105.

Grant, M. R., Godiard, L., Straube, E., Ashfield, T., Lewald, J., Sattler, A., Innes, R. W., and Dangl, J. L., 1995, Structure of the *Arabidopsis RPM1* gene enabling dual specificity disease resistance, *Science* **269**:843–846.

Greenberg, J. T., and Ausubel, F. M., 1993, *Arabidopsis* mutants compromised for the control of cellular damage during pathogenesis and aging, *Plant J.* **4**:327–342.

Greenberg, J. T., Guo, A., Klessig, D. F., and Ausubel, F. M., 1994, Programmed cell death in plants: A pathogen-triggered response activated coordinately with multiple defense functions, *Cell* **77**:551–563.

Gross, D. C., 1991, Molecular and genetic analysis of toxin production by pathovars of *Pseudomonas syringae*, *Annu. Rev. Phytopathol.* **29**:247–278.

Hahlbrock, K., and Scheel, D., 1989, Physiology and molecular biology of phenylpropanoid metabolism, *Annu. Rev. Plant Physiol. Plant Mol. Biol.* **40**:347–369.

Heath, M. C., 1991, The role of gene-for-gene interactions in the determination of host species specificity, *Phytopathology* **81**:127–130.

Hinsch, M., and Staskawicz, B., 1996, Identification of a new *Arabidopsis* disease resistance locus, *RPS4*, and cloning of the corresponding avirulence gene, *avrRps4*, from *Pseudomonas syringae* pv. *pisi*, *Mol. Plant–Microbe Interact.* **9**:55–61.

Holub, E. B., Beynon, J. L., and Crute, I. R., 1994, Phenotypic and genotypic characterization of interactions between isolates of *Peronospora parasitica* and accessions of *Arabidopsis thaliana*, *Mol. Plant–Microbe Interact.* **7**:223–239.

Holub, E. B., Brose, E., Tör, M., Clay, C., Crute, I. R., and Beynon, J. L., 1995, Phenotypic and genotypic variation in the interaction between *Arabidopsis thaliana* and *Albugo candida*, *Mol. Plant–Microbe Interact.* **8**:916–928.

Hughes, R. K., Perbal, M. C., Maule, A. J., and Hull, R., 1995, Evidence for proteolytic processing of tobacco mosaic virus movement protein in *Arabidopsis thaliana*, *Mol. Plant–Microbe Interact.* **8**:658–665.

Innes, R., Bisgrove, S., Smith, N., Bent, A., Staskawicz, B., and Liu, Y., 1993, Identification of a disease resistance locus in *Arabidopsis* that is functionally homologous to the *RPG1* locus of soybean, *Plant J.* **4**:813–820.

Ishikawa, M., Obata, F., Kumagai, T., and Ohno, T., 1991, Isolation of mutants of *Arabidopsis thaliana* in which accumulation of tobacco mosaic virus coat protein is reduced to low levels, *Mol. Gen. Genet.* **230**:33–38.

Ishikawa, M., Naito, S., and Ohno, T., 1993, Effects of the *tom1* mutation of *Arabidopsis thaliana* on the multiplication of tobacco mosaic virus RNA in protoplasts, *J. Virol.* **67**:5328–5338.

Jabs, T., Dietrich, R. A., and Dangl, J. L., 1996, Initiation of runaway cell death in an *Arabidopsis* mutant by extracellular superoxide, *Science* **273:**1853–1856.

Johal, G. S., and Briggs, S. P., 1992, Reductase activity encoded by the *HM1* disease resistance gene in maize, *Science* **258:**985–987.

Johal, G. S., Hulbert, S. H., and Briggs, S. P., 1995, Disease lesion mimics of maize: A model for cell death in plants, *BioEssays* **17:**685–692.

Jones, A. M., and Dangl, J. L., 1996, Logjam at the Styx: Programmed cell death in plants, *Trends Plant Sci.* **1:**114–119.

Jones, J., 1996, Plant disease resistance genes: Structure, function and evolution, *Curr. Opin. Biotechnol.* **7:**155–160.

Joos, H. J., Mauch-Mani, B., and Slusarenko, A. J., 1996, Molecular mapping of the *Arabidopsis* locus *RPP11* which conditions isolate-specific hypersensitive resistance against downy mildew in ecotype RLD, *Theor. Appl. Genet.* **92:**281–284.

Kauss, H., and Jeblick, W., 1995, Pretreatment of parsley suspension cultures with salicylic acid enhances spontaneous and elicited production of $H_2O_2$, *Plant Physiol.* **108:**1171–1178.

Kauss, H., and Jeblick, W., 1996, Influence of salicylic acid on the induction of competence for $H_2O_2$ elicitation—Comparison of ergosterol with other elicitors, *Plant Physiol.* **111:**755–763.

Keen, N. T., 1990, Gene-for-gene complementarity in plant–pathogen interactions, *Annu. Rev. Genet.* **24:**447–463.

Keen, N. T., 1992, The molecular biology of disease resistance, *Plant Mol. Biol.* **19:**109–122.

Keith, B., Dong, X. N., Ausubel, F. M., and Fink, G. R., 1991, Differential induction of 3-deoxy-D-ararbino-heptulosonate 7-phosphate synthase genes in *Arabidopsis thaliana* by wounding and pathogenic attack, *Proc. Natl. Acad. Sci. USA* **88:**8821–8825.

Kiedrowski, S., Kawalleck, P., Hahlbrock, K., Somssich, I. E., and Dangl, J. L., 1992, Rapid activation of a novel plant defense gene is strictly dependent on the *Arabidopsis-RPM1* disease resistance locus, *EMBO J.* **11:**4677–4684.

Klessig, D. F., and Malamy, J., 1994, The salicylic acid signal in plants, *Plant Mol. Biol.* **26:**1439–1458.

Koch, E., and Slusarenko, A., 1990a, *Arabidopsis* is susceptible to infection by a downy mildew fungus, *Plant Cell* **2:**437–445.

Koch, E., and Slusarenko, A. J., 1990b, Fungal pathogens of *Arabidopsis thaliana* (L.) Heyn., *Bot. Helv.* **100:**257–268.

Kong, Q., Oh, J.-W., and Simon, A. E., 1995, Symptom attenuation by a normally virulent satellite RNA of turnip crinkle virus is associated with the coat protein open reading frame, *Plant Cell* **7:**1625–1634.

Kuc, J., 1995, Phytoalexins, stress metabolism, and disease resistance in plants, *Annu. Rev. Phytopathol.* **33:**275–297.

Kunkel, B. N., Bent, A. F., Dahlbeck, D., Innes, R. W., and Staskawicz, B. J., 1993, *RPS2*, an *Arabidopsis* disease resistance locus specifying recognition of *Pseudomonas syringae* strains expressing the avirulence gene *avrRpt2*, *Plant Cell* **5:**865–875.

Lamb, C. J., 1994, Plant disease resistance genes in signal perception and transduction, *Cell* **76:**419–422.

Lawrence, G. J., Finnegan, E. J., Ayliffe, M. A., and Ellis, J. G., 1995, The *L6* gene for flax rust resistance is related to the *Arabidopsis* bacterial resistance gene *RPS2* and the tobacco viral resistance gene *N*, *Plant Cell* **7:**1195–1206.

Lawton, K., Weymann, K., Friedrich, L., Vernooij, B., Uknes, S., and Ryals, J., 1995, Systemic acquired resistance in *Arabidopsis* requires salicylic acid but not ethylene, *Mol. Plant–Microbe Interact.* **8:**863–870.

Lawton, K. A., Friedrich, L., Hunt, M., Weymann, K., Delaney, T., Kessmann, H., Staub, T., and

Ryals, J., 1996, Benzothiadiazole induces disease resistance in *Arabidopsis* by activation of the systemic acquired resistance signal transduction pathway, *Plant J.* **10**:71–82.

Lazarowitz, S. G., 1992, Geminiviruses: Genome structure and function, *Crit. Rev. Plant Sci.* **11**:327–349.

Lee, D., Ellard, M., Wanner, L. A., Davis, K. R., and Douglas, C. J., 1995, The *Arabidopsis thaliana* 4-coumarate:CoA ligase (*4CL*) gene: Stress and developmentally regulated expression and nucleotide sequence of its cDNA, *Plant Mol. Biol.* **28**:871–884.

Lee, J. M., Hartman, G. L., Domier, L. L., and Bent, A. F., 1996, Identification and map location of TTR1, a single locus in *Arabidopsis thaliana* that confers tolerance to tobacco ringspot nepovirus, *Mol. Plant Microbe Interact.* **9**:729–735.

Lee, S., Stenger, D. C., Bisaro, D. M., and Davis, K. R., 1994, Identification of loci in *Arabidopsis* that confer resistance to geminivirus infection, *Plant J.* **6**:525–535.

Leisner, S. M., and Howell, S. H., 1992, Symptom variation in different *Arabidopsis thaliana* ecotypes produced by cauliflower mosaic virus, *Phytopathology* **82**:1042–1046.

Leisner, S. M., Turgeon, R., and Howell, S. H., 1993, Effects of host plant development and genetic determinants on the long-distance movement of cauliflower mosaic virus in *Arabidopsis, Plant Cell* **5**:191–202.

Leister, R. T., Ausubel, F. M., and Katagiri, F., 1996, Molecular recognition of pathogen attach occurs inside of plant cells in plant disease resistance specified by the *Arabidopsis* genes RPS2 and RPM1, *Proc. Natl. Acad. Sci. USA* **93**:15497–15502.

Low, P. S., and Merida, J. R., 1996, The oxidative burst in plant defense: Function and signal transduction, *Physiol. Plant.* **96**:533–542.

Lummerzheim, M., de Oliveira, D., Castresana, C., Miguens, F. C., Louzada, E., Roby, D., Van Montagu, M., and Timmerman, B., 1993, Identification of compatible and incompatible interactions between *Arabidopsis thaliana* and *Xanthomonas campestris* pv. *campestris* and characterization of the hypersensitive response, *Mol. Plant–Microbe Interact.* **6**:532–544.

Madamanchi, N. R., and Kuc, J., 1991, Induced systemic resistance in plants, in *The Fungal Spore and Disease Initiation in Plants and Animals* (G. T. Cole and H. C. Hoch, eds.), pp. 347–362, Plenum Press, New York.

Matern, U., and Kneusel, R. E., 1988, Phenolic compounds in plant disease resistance, *Phytoparasitica* **16**:153–170.

Mauch-Mani, B., and Slusarenko, A. J., 1993, *Arabidopsis* as a model host for studying plant–pathogen interactions, *Trends Microbiol.* **1**:265–270.

Mauch-Mani, B., and Slusarenko, A. J., 1996, Production of salicylic acid precursors is a major function of phenylalanine ammonia-lyase in the resistance of *Arabidopsis* to *Peronospora parasitica, Plant Cell* **8**:203–212.

Mehdy, M. C., 1994, Active oxygen species in plant defense against pathogens, *Plant Physiol.* **105**:467–472.

Melan, M. A., Dong, X., Endara, M. E., Davis, K. R., Ausubel, F. M., and Peterman, T. K., 1993, An *Arabidopsis thaliana* lipoxygenase gene can be induced by pathogens, abscisic acid, and methyl jasmonate, *Plant Physiol.* **101**:441–450.

Melcher, U., 1989, Symptoms of cauliflower mosiac virus infection in *Arabidopsis thaliana* and turnip, *Bot. Gaz.* **150**:139–147.

Melcher, U., Gardner, C. O., and Essenberg, R. C., 1981, Clones of cauliflower mosaic virus identified by molecular hybridization in turnip leaves, *Plant Mol. Biol.* **1**:63–73.

Meyerowitz, E. M., 1987, *Arabidopsis thaliana, Annu. Rev. Genet.* **21**:93–111.

Meyerowitz, E. M., and Somerville, C. R., eds., 1994, *Arabidopsis*, Cold Spring Harbor Laboratory Press, Cold Spring Harbor, New York.

Mindrinos, M., Katagiri, F., Yu, G. L., and Ausubel, F. M., 1994, The *A. thaliana* disease resistance gene *RPS2* encodes a protein containing a nucleotide-binding site and leucine-rich repeats, *Cell* **78:**1089–1099.

Mittal, S., and Davis, K. R., 1995, Role of the phytotoxin coronatine in the infection of *Arabidopsis thaliana* by *Pseudomonas syringae* pv. *tomato*, *Mol. Plant–Microbe Interact.* **8:**165–171.

Mur, L., Naylor, G., Warner, S., Sugars, J. M., White, R. F., and Draper, J., 1996, Salicylic acid potentiates defence gene expression in tissue exhibiting acquired resistance to pathogen attack, *Plant J.* **9:**559–571.

Niebel, A., Engler, J. D., Hemerly, A., Ferreira, P., Inzé, D., Van, M. M., and Gheysen, G., 1996, Induction of cdc2a and cyc1At expression in *Arabidopsis thaliana* during early phases of nematode-induced feeding cell formation, *Plant J.* **10:**1037–1043.

Oh, J. W., Kong, Q. Z., Song, C. Z., Carpenter, C. D., and Simon, A. E., 1995, Open reading frames of turnip crinkle virus involved in satellite symptom expression and incompatibility with *Arabidopsis thaliana* ecotype Dijon, *Mol. Plant–Microbe Interact.* **8:**979–987.

Parker, J. E., Barber, C. E., Mi-jiao, F., and Daniels, M. J., 1993, Interaction of *Xanthomonas campestris* with *Arabidopsis thaliana:* Characterization of a gene from *X. c.* pv. *raphani* that confers avirulence to most *A. thaliana* accessions, *Mol. Plant–Microbe Interact.* **6:**216–224.

Rahme, L. G., Stevens, E. J., Wolfort, S. F., Shao, J., Tompkins, R. G., and Ausubel, F. M., 1995, Common virulence factors for bacterial pathogenicity in plants and animals, *Science* **268:**1899–1902.

Ramachandra, S., Lommel, S., and White, F., 1988, Interaction of turnip yellow mosaic virus (TYMV) with *Arabidopsis thaliana*, *Phytopathology* **78:**1559.

Rédei, G. P., 1975, *Arabidopsis* as a genetic tool, *Annu. Rev. Genet.* **9:**111–127.

Rédei, G., 1992, A heuristic glance at the past of *Arabidopsis* genetics, in *Methods in Arabidopsis Research* (C. Koncz, N.-H. Chua, and J. Schell, eds.), pp. 1–15, World Scientific Singapore.

Reignault, P., Frost, L. N., Richardson, H., Daniels, M. J., Jones, J., and Parker, J. E., 1996, Four *Arabidopsis RPP* loci controlling resistance to the Noco2 isolate of *Peronospora parasitica* map to regions known to contain other *RPP* recognition specificities, *Mol. Plant–Microbe Interact.* **9:**464–473.

Reuber, T. L., and Ausubel, F. M., 1996, Isolation of *Arabidopsis* genes that differentiate between resistance responses mediated by the *RPS2* and *RPM1* disease resistance genes, *Plant Cell* **8:**241–249.

Ryals, J., Lawton, K. A., Delaney, T. P., Friedrich, L., Kessmann, H., Neuenschwander, U., Uknes, S., Vernooij, B., and Weymann, K., 1995, Signal transduction in systemic acquired resistance, *Proc. Natl. Acad. Sci. USA* **92:**4202–4205.

Ryan, C. A., 1994, Oligosaccharide signals: From plant defense to parasite offense, *Proc. Natl. Acad. Sci. USA* **91:**1–2.

Schaller, A., and Ryan, C. A., 1996, Systemin—A polypeptide defense signal in plants, *BioEssays* **18:**27–33.

Sharma, Y. K., and Davis, K. R., 1995, Isolation of a novel *Arabidopsis* ozone-induced cDNA by differential display, *Plant Mol. Biol.* **29:**91–98.

Sharma, Y. K., León, J., Raskin, I., and Davis, K. R., 1996, Ozone-induced responses in *Arabidopsis thaliana:* The role of salicylic acid in the accumulation of defense-related transcripts and induced resistance, *Proc. Natl. Acad. Sci. USA* **93:**5099–5104.

Sijmons, P. C., Atkinson, H. J., and Wyss, U., 1994, Host cell responses to plant-parasitic nematodes, *Annu. Rev. Phytopathol.* **32:**235–259.

Sijmons, P. C., Grundler, F. M. W., Vonmende, N., Burrows, P. R., and Wyss, U., 1991, *Arabidopsis thaliana* as a new model host for plant-parasitic nematodes, *Plant J.* **1:**245–254.

Sijmons, P. C., von Mende, N., and Grundler, F. M. W., 1994, Plant-parasitic nematodes, in *Arabidopsis* (E. M. Meyerowitz and C. R. Somerville, eds.), pp. 749–767, CSHL Press, New York.

Simon, A. E., 1994, Interactions between *Arabidopsis thaliana* and viruses, in *Arabidopsis* (E. M. Meyerowitz and C. R. Somerville, eds.), pp. 685–704, CSHL Press, New York.

Simon, A. E., Li, X. H., Lew, J. E., Stange, R., Zhang, C., Polacco, M., and Carpenter, C. D., 1992, Susceptibility and resistance of *Arabidoposis thaliana* to turnip crinkle virus, *Mol. Plant–Microbe Interact.* **5:**496–503.

Simonich, M. T., and Innes, R. W., 1995, A disease resistance gene in *Arabidopsis* with specificity for the *avrPph3* gene of *Pseudomonas syringae* pv. *phaseolicola, Mol. Plant–Microbe Interact.* **8:**637–640.

Simpson, R. B., and Johnson, L. J., 1990, *Arabidopsis thaliana* as a host for *Xanthomonas campestris* pv. *campestris, Mol. Plant–Microbe Interact.* **3:**233–237.

Skotnicki, M. L., Mackenzie, A. M., Torronen, M., and Gibbs, A. J., 1993, The genomic sequence of cardamine chlorotic fleck carmovirus, *J. Gen. Virol.* **74:**1933–1937.

Stanley, J., 1991, The molecular determinants of geminivirus pathogenesis, *Semin. Virol.* **2:**139–149.

Staskawicz, B. J., Ausubel, F. M., Baker, B. J., Ellis, J. G., and Jones, J., 1995, Molecular genetics of plant disease resistance, *Science* **268:**661–667.

Stenger, D. C., 1994, Complete nucleotide sequence of the hypervirulent CFH strain of beet curly top virus, *Mol. Plant–Microbe Interact.* **7:**154–157.

Stenger, D. C., Davis, K. R., and Bisaro, D. M., 1992, Limited replication of tomato golden mosaic virus DNA in explants of nonhost species, *Mol. Plant–Microbe Interact.* **5:**525–527.

Susnova, V., and Polak, Z., 1975, Susceptibility of *Arabidopsis thaliana* to infection with some plant viruses, *Biol. Planta* **17:**156–158.

Tsuji, J., and Somerville, S. C., 1988, *Xanthomonas campestris* pv. *campestris*-induced chlorosis in *Arabidopsis thaliana, Arabidopsis Inf. Serv.* **26:**1–8.

Tsuji, J., Somerville, S. C., and Hammerschmidt, R., 1991, Identification of a gene in Arabidopsis thaliana that controls resistance to *Xanthomonas campestris* pv. *campestris, Physiol. and Mol. Plant Pathol.* **38:**57–65.

Tsuji, J., Jackson, E. P., Gage, D. A., Hammerschmidt, R., and Somerville, S. C., 1992, Phytoalexin accumulation in *Arabidopsis thaliana* during the hypersensitive reaction to *Pseudomonas syringae* pv. *syringae, Plant Physiol.* **98:**1304–1309.

Uknes, S., Winter, A. M., Delaney, T., Vernooij, B., Morse, A., Friedrich, L., Nye, G., Potter, S., Ward, E., and Ryals, J., 1993, Biological induction of systemic acquired resistance in *Arabidopsis, Mol. Plant–Microbe Interact.* **6:**692–698.

Vernooij, B., Friedrich, L., Goy, P. A., Staub, T., Kessmann, H., and Ryals, J., 1995, 2,6-Dichloroisonicotinic acid-induced resistance to pathogens without the accumulation of salicylic acid, *Mol. Plant–Microbe Interact.* **8:**228–234.

Wanner, L. A., Mittal, S., and Davis, K. R., 1993, Recognition of the avirulence gene *avrB* from *Pseudomonas syringae* pv. *glycinea* by *Arabidopsis thaliana, Mol. Plant–Microbe Interact.* **6:**582–591.

Wellink, J., van Lent, J. W., Verver, J., Sijen, T., Goldbach, R. W., and van Kammen, A., 1993, The cowpea mosaic virus M RNA-encoded 48-kilodalton protein is responsible for induction of tubular structures in protoplasts, *J. Virol.* **67:**3660–3664.

Weymann, K., Hunt, M., Uknes, S., Neuenschwander, U., Lawton, K., Steiner, H. Y., and Ryals, J., 1995, Suppression and restoration of lesion formation in *Arabidopsis* lsd mutants, *Plant Cell* **7:**2013–2022.

Whalen, M. C., Innes, R. W., Bent, A. F., and Staskawicz, B. J., 1991, Identification of *Pseudomonas syringae* pathogens of *Arabidopsis* and a bacterial locus determining avirulence on both *Arabidopsis* and soybean, *Plant Cell* **3:**49–59.

Whitham, S., Dinesh, K. S., Choi, D., Hehl, R., Corr, C., and Baker, B., 1994, The product of the tobacco mosaic virus resistance gene *N:* Similarity to Toll and the interleukin-1 receptor, *Cell* **78:**1101–1115.

Wyss, U., Grundler, F. M. W., and Munch, A., 1992, The parasitic behaviour of 2nd-stage juveniles of *Meloidogyne incognita* in roots of *Arabidopsis thaliana*, *Nematologica* **38:**98–111.

Yu, G. L., Katagiri, F., and Ausubel, F. M., 1993, *Arabidopsis* mutations at the RPS2 locus result in loss of resistance to *Pseudomonas syringae* strains expressing the avirulence gene *avrRpt2*, *Mol. Plant–Microbe Interact.* **6:**434–443.

Zijlstra, C., and Hohn, T., 1992, Cauliflower mosaic virus gene VI controls translation from dicistronic expression units in transgenic *Arabidopsis* plants, *Plant Cell* **4:**1471–1484.

*Chapter 9*

# Engineering Resistance against Viral Diseases in Plants

Carlos A. Malpica, Maria Teresa Cervera, Chris Simoens, and Marc Van Montagu

## 1. INTRODUCTION

Viral diseases can result in severe losses to many agriculturally important crops. Genetic engineering proposes an interesting alternative to more traditional techniques intended to preserve plants from becoming infected. This review addresses proven genetic engineering strategies devised to control virus infection and also points to other schemes which are currently being tested. Although it has no ambition of covering the entire field, an effort has been made to point out as many different elements as possible to understand the molecular mechanisms on which the strategies are based, their degree of effectiveness, and their applicability.

Plant viruses consist of a nucleic acid genome (RNA or DNA) wrapped in a coat made of protein. By contrast, viroids consist of only a small RNA molecule.

---

**Carlos A. Malpica, Maria Teresa Cervera, Chris Simoens, and Marc Van Montagu** Laboratorium voor Genetica, Department of Genetics, Flanders Interuniversity Institute for Biotechnology (VIB), Universiteit Gent, B-9000 Gent, Belgium. *Present address for C.A.M.:* Instituto de Medio Ambiente (IMA), Centro de Investigaciones Energéticas Medioambientales y Tecnológicas (CIEMAT), E-28040 Madrid, Spain. *Present address for M.T.C.:* Instituto Nacional de Investigación y Tecnología Agraria y Alimentaria (CIT-INIA), E-28040 Madrid, Spain.

*Subcellular Biochemistry, Volume 29: Plant–Microbe Interactions,* edited by Biswas and Das. Plenum Press, New York, 1998.

Both viruses and viroids need the host cellular machinery to replicate. They are generally dependent on other living organisms, such as fungi, nematodes, and diverse insects (e.g., aphids and hoppers) to enter into host plants. As they make use of host cellular functions for expressing their own messenger RNA and trigger self-protein synthesis, viral pathogens weaken the plants and reduce crop yields. Visible symptoms include yellowing, mottling, and mosaic on leaves, distortion and necrosis of leaves, ringspots and line patterns on leaves, color breaking of flowers, general stunting, and loss of vigor. Fruits may also be mottled, misshapen, or necrotic and may drop prematurely. Distortions and discoloration can make a product unmarketable.

The genetic pool of natural plant populations may contain resistance genes to possible new pathogens. Plant viruses exist in many strains and evolve rapidly. It may take only a few nucleotide/amino acid substitutions to change an avirulent virus into a virulent form which overcomes natural resistance. When a new virulent strain comes in contact with a susceptible plant population, the result can be catastrophic. Intensive agricultural systems and the trend toward monocultures of genetically uniform varieties have made modern agriculture extremely vulnerable to viral diseases.

Another cause for enhanced impact of viral diseases on agricultural and horticultural crops is the increased worldwide movement of germplasm and food stocks. Banana bunchy top virus (in banana), papaya ringspot virus (in papaya), and tomato spotted wilt virus (in many crops) could all extend dramatically their infection area. Special regulations and technical guidelines have been worked out by national and international authorities (such as the Food and Agricultural Organization of the United Nations and the International Board for Plant Genetic Resources) to guarantee phytosanitary safety of germplasm transfer. Even when the imported propagative material is disease-free, it may encounter new, locally existing viruses and strains in its area, leading to an unexpected epidemic and the failure of the new variety.

The exact economic losses caused by plant viral diseases are very difficult to estimate, especially in the endemic cases. According to Henry *et al.* (1995), the annual loss due to potato viruses is £75 million in the United Kingdom, while the crop value was £431 million in 1992. In general, losses are more important in tropical and subtropical regions, perhaps due to the great diversity of potentially virulent pathogens at these latitudes. Other examples illustrate the magnitude of the problem when diseases occur in less-developed areas and where certification and control measures are not available or cannot be implemented. This is the case of rice tungro disease in East Asia (Herdt, 1991) or African cassava mosaic virus (ACMV) in Africa (Fauquet and Fargette, 1990). A worldwide example is that of citrus tristeza virus (Hull, 1984).

Among traditional strategies available for crop protection against viruses, breeding virus resistance is the most straightforward approach. However, it has been very difficult to achieve. Of the rare sources of natural resistance for viruses of major crops, most are confined to single gene loci, which are easily overcome by mutant virus isolates (for reviews see Fraser, 1990, 1992). The virus resistance present in wild species is often cumbersome to introduce into agriculturally important cultivars. Other sources of resistance can be directed to the virus vectors. For instance, the resistance bred into rice against rise tungro disease has been shown to be a resistance against the leafhopper vector of that disease (Dahal *et al.*, 1990). In this case, too, resistance can be temporary, due to genetic variation in vector populations. Indirect approaches, such as plant virus-avoiding strategies, have been the most successful. One of these approaches consists in using only certified virus-free planting material, obtained from micropropagation via meristem culture, for example. This material may be applied together with the eradication of diseased plants. In many temperate climate countries, for instance, special schemes exist in which virus-free potatoes are produced in regions where vectors are rare, to be cultivated in the more productive areas. Similar schemes are followed for many other crops, such as carnation, strawberry, and citrus.

Genetic engineering strategies to protect crops against viral diseases derive from a better knowledge of the molecular mechanisms involved in important steps of the life cycle of the viruses. Understanding the role of structural and functional viral proteins during plant–virus interactions has helped to refine a number of strategies such as so-called coat protein-mediated resistance, movement protein-mediated resistance, and replicase-mediated resistance. Together, they belong to the broad class of pathogen-derived strategies. On the other hand, plants have a general stress defense system and plant pathogens induce specific gene expression from which plant gene-derived strategies have been developed. Other genes have been imported from other organisms into transgenic plants to achieve virus control.

## 2. PATHOGEN-DERIVED STRATEGIES

Plant viruses enter the host cell carried by a vector or by mechanical inoculation, thereby starting the virus life cycle. Upon virion disassembly, the viral genome is uncovered and replication is initiated. There are many different ways by which a plant virus may replicate, depending on the nature of nucleic acid content, RNA or DNA, and specific features of the genome, which could, for example, be multipartite or code for a polyprotein (for review see Matthews, 1991). Viral structural and functional proteins are synthesized which make possible virus gene expression, virus genome replication, virus cell-to-cell movement, and virus encapsidation. These represent many steps at which pathogen-derived strategies have been achieved by expressing viral wild-type or modified proteins into transgenic plants.

## 2.1. Structural Proteins: Coat Proteins

The use of the viral coat protein (CP) was the first example of pathogen-derived resistance. In 1986, resistance to tobacco mosaic virus (TMV) infection was reported in transgenic tobacco plants expressing the viral CP (Powell Abel *et al.*, 1986). Since this first report, resistance to other families of virus has been reported and so-called coat protein-mediated resistance (CPMR) has proven to be a widespread and efficient tool against several tobamoviruses (Nejidat and Beachy, 1990; R. S. Nelson *et al.*, 1987, 1988; Powell Abel *et al.*, 1986), alfalfa mosaic virus (AlMV) (Taschner *et al.*, 1994; Loesch-Fries *et al.*, 1987), beet necrotic yellow vein virus (Kallerhoff *et al.*, 1990), cucumber mosaic virus (CMV) (Gielen *et al.*, 1996; Gonsalves *et al.*, 1992, 1994; Okuno *et al.*, 1993; Yoshioka *et al.*, 1993; Yie *et al.*, 1992; Namba *et al.*, 1991; Quemada *et al.*, 1991; Cuozzo *et al.*, 1988), potato virus X (PVX) (Fehér *et al.*, 1992; Lawson *et al.*, 1990; Hoekema *et al.*, 1989; Hemenway *et al.*, 1988), potato aucuba mosaic virus (Leclerc and AbouHaidar, 1995), tobacco rattle virus (Ploeg *et al.*, 1993; Angenent *et al.*, 1990; van Dun and Bol, 1988; van Dun *et al.*, 1987), tomato spotted wilt virus (TSWV) (Pang *et al.*, 1994; MacKenzie and Ellis, 1992), grapevine chrome mosaic (Brault *et al.*, 1993), and the related grapevine fanleaf virus (Bardonnet *et al.*, 1994), as well as a number of potyviruses (Cassidy and Nelson, 1995; Fuchs and Gonsalves, 1995; Scorza *et al.*, 1995; Malnoë *et al.*, 1994; Silva-Rosales *et al.*, 1994; Dinant *et al.*, 1993; Fang and Grumet, 1993; Farinelli and Malnoë, 1993; Laimer da Câmara Machado *et al.*, 1993; Lindbo and Dougherty, 1992a; Namba *et al.*, 1992; Regner *et al.*, 1992). The only DNA virus known to be protected by CPMR is the geminivirus tomato yellow leaf curl virus (Kunik *et al.*, 1994).

Many mechanisms have been proposed to be involved in CPMR. A first possibility is that the protecting event takes place prior to virus uncoating. In many cases, the resistance is attenuated or can be overcome by higher virus titer inoculum (R. S. Nelson *et al.*, 1987, 1988; Powell Abel *et al.*, 1986) or inoculation with viral RNA (R. S. Nelson *et al.*, 1988; Register and Beachy, 1988; Loesch-Fries *et al.*, 1987). Virion disassembly could perturb the decapsidation/reencapsidation equilibrium. Once RNA transcription has started, the transgenic CP cannot encapsidate the intruding RNA. This mechanism may explain why viral RNA overcomes the resistance. In the case of TMV, ribosomes bind to the exposed 5′ end and translation proceeds along the exposed RNA while displacing the CP in a so-called striptosome system (Shaw *et al.*, 1986; Wilson, 1984). A chimeric construct which carried the reporter gene for β-glucuronidase (GUS) and the TMV origin of assembly (Osbourn *et al.*, 1989a,b) was made into a pseudovirion by coating with TMV CP and used for transfection of transgenic protoplasts. GUS activity in protoplasts which had been transformed with TMV CP was only about 1% of that observed in nontransformed protoplasts. Transfecting the protoplasts with naked, uncoated constructs resulted in a 30- to 50-fold increase in GUS activity in the transgenic protoplasts compared with the coated pseudovirions, indicating a possible inhibition of their decapsidation. The resistance-conferring stage occurred earlier than the cotranslational stripping, because a partially stripped TMV also overcomes CPMR (Register and Beachy, 1988). Transgenic TMV CP plants were crossed with plants expressing RNA transcripts carrying the chloramphenicol acetyltransferase (CAT) reporter gene and the TMV origin of assembly. CAT transcript coating by endogenous CP was very low, although CPMR was detected (Osbourne *et al.*, 1989b). This experiment suggests that prevention of TMV decapsidation and not reencapsidation is the mechanism interfering with the infection cycle. In another experiment, protoplasts transformed with TMV U1 CP were resistant to TMV U1 infection; they were, however, susceptible to infection by the cowpea strain of TMV or its naked RNA (Osbourn *et al.*, 1989a). When infected with heterovirions made of TMV U1 CP and cowpea strain RNA, the protoplasts remained unprotected. These results indicate the encapsidation/decapsidation equilibrium is not the only mechanism involved in CPMR. The possible mechanisms of TMV CPMR are discussed in detail by Reimann-Philipp and Beachy (1993).

However, viral CPs may have functions other than purely structural ones that may account for CPMR. For example, inoculation of transgenic plants expressing potato virus S (MacKenzie *et al.*, 1991), arabis mosaic virus (Bardonnet *et al.*, 1994; Bertioli *et al.*, 1992), and AlMV (Turner *et al.*, 1987) CP genes with viral RNA did not always overcome the resistance. A mixture of AlMV RNA1 and AlMV RNA2 is infectious only when supplemented with some molecules of the viral CP or with AlMV RNA4, which encodes the viral CP (Bol *et al.*, 1971). Clearly, AlMV CP is essential for its replication and could interfere against virus infection when expressed in transgenic plants. A single amino acid change in AlMV CP was shown to abolish its capacity to confer CPMR (Turner *et al.*, 1991). Other experiments demonstrate that AlMV CP is actively involved in a later stage of viral replication (Bol *et al.*, 1993).

TMV CP may also have an effect on the local and long-distance movement of the virus within the plant. A systemically spreading virus with a mutated CP conferred a hypersensitive reaction in *Nicotiana sylvestris* (Knorr and Dawson, 1988). A gene-for-gene interaction was involved between the viral CP gene and a host gene. In addition, TMV CP mutations did not overcome the localizing effect conferred by the tobacco N gene (a mutation at the 126-kDa TMV gene has been implicated in N gene-mediated resistance; Padgett and Beachy, 1993). In any particular genetic environment, the CP may influence viral movement. Such an effect has indeed been observed when CP transgenic plants were grafted onto unmodified plants and the root stocks were inoculated with TMV: the movement of the virus into the graft was reduced as compared with the control (Wisniewski *et al.*, 1990). Furthermore, when a section of a CP-bearing transgenic plant was grafted between two sections of wild-type tobacco, viral spread from the lower part to the apical part was also reduced, provided the intermediate transgenic graft carried a transgenic leaf. A similar observation has also been reported for potato virus S (MacKenzie *et al.*, 1991). The CP of tobacco etch virus (TEV) has also been shown to be involved in cell-to-cell and long-distance movement (Dolja *et al.*, 1995).

The range of CPMR has been extensively studied. The proposed mechanisms of CPMR are based on interactions between transgenic viral CP and a specific target, viral RNA or viral host protein. CPMR is generally of a relatively narrow range. In some cases, it protects against its own virus and closely related strains, but not against distantly related strains (Turner *et al.*, 1991; Nejidat and Beachy, 1990; Osbourn *et al.*, 1989a,b). Potyvirus CPMR is an exception in that it generally exhibits a wider range of protection from homologous resistance (to its own virus) to heterologous resistance (to other viruses from within the family) (Dinant *et al.*, 1993; Ling *et al.*, 1991; Stark and Beachy, 1989).

A positive correlation between the level of expression of the CP and the degree of CPMR in transgenic plants has been reported in many cases (Fitchen and Beachy, 1993), indicating that the protein itself must be present for resistance. Untranslatable constructs of TMV CP and AlMV CP genes did not confer resistance in transgenic plants (Powell *et al.*, 1989; van Dun *et al.*, 1987). In cases where a correlation between CP accumulation and resistance was not observed, e.g., CMV (Namba *et al.*, 1991), potyviruses (Lawson *et al.*, 1990; Stark and Beachy, 1989), and TSWV (MacKenzie and Ellis, 1992; Gielen *et al.*, 1991), a mechanism based solely on the CP may be coinciding with another mechanism, such as RNA-mediated protection (see Section 2.3.1).

## 2.2. Functional Proteins

### 2.2.1. Replicase-Associated Genes

Viral replication is a very crucial step in the development of viral infection. A slight modification in the rate of virus replication may strongly affect the success of virus colonization of the host plant. In recent years, many reports have described that replication-association genes in transgenic plants have conferred resistance to virus infection (for review see Baulcombe, 1994; Carr and Zaitlin, 1993). The number of virus-encoded genes required to ensure viral replication may vary from one family to another. Plant viral gene products are considered to be part of a larger replicating complex, which also includes host proteins, as, for example, described for AlMV (Quadt *et al.*, 1991). The term replicase comprises all the components needed for virus replication. The so-called replicase-mediated resistance therefore involves expression of virus-encoded proteins that actively participate in replication, whether they have a regulatory or a targeting role. It has not been demonstrated that gene constructs used in replicase-mediated resistance produce functional proteins involved in replication.

Replicase-mediated resistance was first reported in 1990 (Golemboski *et al.*, 1990) and has been reported for TMV (Donson *et al.*, 1993; Golemboski *et al.*, 1990), CMV (Carr *et al.*, 1994; Zaitlin *et al.*, 1994; Anderson *et al.*, 1992), PVX (Longstaff *et al.*, 1993; Braun and Hemenway, 1992), potato virus Y (PVY) (Audy *et al.*, 1994), cowpea mosaic virus (CoMV) (Sijen *et al.*, 1995), pea early browning virus (PEBV) (MacFarlane and Davis, 1992), pepper mild mottle virus (Tenllado *et al.*, 1995), cymbidium ringspot virus (CymbRSV) (Rubino and Russo, 1995; Lupo *et al.*, 1994), and ACMV (Hong and Stanley, 1996). The resistance observed is in many cases stronger than CPMR, as it is maintained even when higher titers of viral inoculum are applied. The resistance is also quite specific to the original virus and does not extend to other viruses of the same group, even closely related ones. Only one case of broad-spectrum resistance to tobamoviruses has been described (Donson *et al.*, 1993). Transgenic expression of a full-length replication-associated gene did not always yield a resistant phenotype. Full-length replicase genes were used with success in the case of PVX, CoMV, and CymbRSV (Rubino and Russo, 1995; Sijen *et al.*, 1995; Braun and Hemenway, 1992), whereas AlMV and brome mosaic virus (BMV) (van Dun *et al.*, 1988; Mori *et al.*, 1992) did not confer resistance. Deliberately defective constructs have been made. A considerable gene insertion was placed in a TMV replication construct (Donson *et al.*, 1993), a deletion was made that allowed expression of 75% of the CMV gene product (Anderson *et al.*, 1992), and a dysfunctional mutant of PVX replicase gene was used for transformation (Longstaff *et al.*, 1993; Braun and Hemenway, 1992). Resistance conferred by a construct which is unable to produce a functional protein might be explained by RNA-mediated protection (see Section 2.3.1), at least until further data are obtained. The case of readthrough proteins such as TMV or PEBV is, however, more difficult to explain. The TMV replicase sequence, encompassing the open-reading frame (ORF) for a putative 54-kDa protein that was never found in infected or engineered plants, could nonetheless confer resistance in transgenic tobacco (Golemboski *et al.*, 1990). Known TMV-encoded proteins of the replication system of this virus are the 126-kDa and 183-kDa proteins, the larger resulting from a readthrough of the smaller. None of these proteins confers replicase-mediated resistance in transgenic plants (Donson *et al.*, 1993; Golemboski *et al.*, 1990). The 54-kDa protein can be considered a truncated 183-kDa protein, as it results from the readthrough part of this protein. This protein alone could be thought to be a dysfunctional protein and, as such, might lead to replicase-mediated resistance. Another explanation is that the protein may be produced at levels which make it undetectable with available methods. The only indirect indication that the protein is synthesized *in vivo* is the presence of its mRNA and double-stranded RNA (dsRNA) in host plants (Sulzinski *et al.*, 1985). Therefore, it can be postulated that it plays a regulatory role in wild-type conditions and perturbs normal virus replication in transgenic plants, leading to the resistance phenotype. Other experiments have been made to assess whether the protein had to be synthesized. A slightly modified gene was capable of inducing resistance, whereas a frameshift mutation, directing synthesis of only parts of the protein, did not confer resistance in protoplasts (Carr *et al.*, 1992). These results suggest an involvement of the protein and not the RNA in the resistance phenomenon.

### 2.2.2. Movement Proteins

Plant viruses encode a protein that facilitates its movement from cell to cell. TMV (Deom *et al.*, 1990), CoMV (Wellink and van Kammen, 1989), AlMV (Schoumacher *et al.*, 1992), tobacco rattle virus (Ziegler-Graff *et al.*, 1991), and cauliflower mosaic virus (CaMV) (Citovsky *et al.*, 1991) possess such a so-called movement protein (MP). The various aspects of viral MPs and mode of action have been described (for a review see Wolf and Lucas, 1994).

The 30-kDa gene product of TMV was the first to be directly identified as involved in movement. Several TMV temperature-sensitive mutants were found to be unable to spread at nonpermissive temperatures and coinfection with a wild-type virus enabled movement of the mutant, showing complementation of the movement function (Taliansky *et al.*, 1982). All transport-defective TMV strains were mutated at the 30-kDa protein gene and, reciprocally, a wild-type TMV was converted into a transport-defective one by *in vitro* mutagenesis of the 30-kDa gene. Movement of a transport-defective mutant was complemented and restored in transgenic plants expressing the nonmutated 30-kDa MP (Holt and Beachy, 1991). The amount of viral MP in transgenic plants influences several aspects of viral movement (Arce-Johnson *et al.*, 1995).

Plasmodesmata, the transmembrane structures that prolong the endoplasmic reticulum between adjacent plant cells, have been shown to be involved in viral transport. The MP of TMV accumulates near the plasmodesmata both in normally infected and in MP transgenic plants (Moore *et al.*, 1992; Atkins *et al.*, 1991; Deom *et al.*, 1990; Wolf *et al.*, 1989) and directly or indirectly changes the exclusion properties of the plasmodesmata. Under normal physiological conditions, plasmodesmata allow the transport of low-molecular-weight compounds (less than 1 kDa and about 2.0 nm in diameter). However, the size of TMV particles ($18 \times 300$ nm) or free-folded TMV RNA [10 nm in diameter according to Citovsky and Zambryski (1993)] indicates that the plasmodesmata gate size must increase to allow passage of these forms. Additionally, plasmodesmata in transgenic leaves were found to be permeable to indicator dyes 5–6 nm in diameter. Microinjection of TMV MP into cells increased permeability up to 9 nm (Citovsky *et al.*, 1990). Even at its highest permeability, the MP must change the viral transport function in another way to enable either virions or viral RNA to migrate to the adjacent cell. Several MPs have been characterized as single-stranded nucleic acid-binding proteins. Citovsky *et al.* (1992) have proposed a model in which TMV MP interacts with TMV RNA to make a 2-nm-thin transport complex.

TMV MP is proposed to have two distinct domains, one that binds to plasmodesmata and allows its size exclusion limit to increase, and another that binds unspecifically to single-strand nucleic acids. If this were a general movement strategy for RNA viruses, it could explain why a certain MP may complement a defective one, even of a quite distant virus. Since the MP interacts both with plasmodesmata and viral RNA, a modified MP would probably still be capable of interfering with transport of a wild-type virus and bring about resistance, as is the case with 30-kDa temperature-sensitive MP of TMV mutant Ni2519 (Malyshenko *et al.*, 1993). In the same experiment, wild-type BMV MP induces resistance to TMV presumably by trapping TMV viral RNA. Resistance using a dysfunctional TMV MP is found against distant viruses of the tobamovirus group (Lapidot *et al.*, 1993). The work of Cooper *et al.* (1995) and Beck *et al.* (1994) indicates that multivirus-resistant transgenic plants might be obtained that express a dysfunctional MP. Recent data by Rohfritsch *et al.* (1996) indicate that a modified MP may also alter plant metabolism.

### 2.2.3. Other Viral Proteins

Some viruses require processing of a large polyprotein into mature proteins. In that case, another important step of the virus life cycle involves control of virus gene expression by protease processing of the genome. The potyvirus genome encodes a single large polyprotein synthesized from an RNA in sense orientation. The primary polyprotein is self-processed to give seven mature viral proteins (Robaglia *et al.*, 1989; Hellmann *et al.*, 1986). The order (5′ to 3′) of cistrons is: P1, helper component (HC), P3, cylindrical inclusion (CI), NIa, NIb, and CP. NIa is the major potyvirus protease, cleaving the primary polyprotein into several mature proteins (Hellmann *et al.*, 1988). Proteins directed by the first two cistrons, P1 and HC, have also been shown to be proteases, participating in self-processing of the potyviral polyprotein (Carrington *et al.*, 1989). The P1 gene product cleaves between the first and second proteins. The HC protein, which owes its name to the fact that it is required for insect transmission of potyviruses, is also a protease cleaving at its own C terminus (Verchot *et al.*, 1991). It is also involved in long-distance movement (Cronin *et al.*, 1995) and genome amplification (Kasschau and Carrington, 1995). The processing of the proteins derived from the 5′ end is independent of NIa and does not require the rest of the potyviral genome.

Two reports of protease-mediated resistance, both against potyviruses, have been published. The resistance is comparatively stronger than that provided by CPMR. A small fraction of NIa-transformed plants (3 lines out of 50) were found to be fully protected from PVY for as much as 60 days after inoculation (Vardi *et al.*, 1993). NIa and CI of tobacco vein mottle virus (TVMV) were assayed in transgenic plants. NIa generated resistance, whereas CI did not (Maiti *et al.*, 1993). The NIa-induced resistance was strong, as it stood against inoculums of 100 $\mu g\ ml^{-1}$, but was specific only to the homologous virus.

It can be expected that transgenic plants carrying one or two of the genes coding for the proteases P1 and HC produce resistance lines in a similar way. HC, as insect transmissibility factor and protease, is of special interest, as it is also known to interact with the viral CP (Lecoq *et al.*, 1993) and is involved in long-distance movement (Cronin *et al.*, 1995).

In the reported cases of protease-mediated potyvirus resistance, the expressed proteases were proteolytically active. A mechanism by which the plant becomes protected due to cotranslational, immature processing of the primary protein of the invading virus by the endogenous protease might explain the observed resistance. It remains that only 5–10% of transgenic plants were shown to be resistant. The role of the N-terminal portion of the NIa protein that is a genome-linked protein called VPg and that is attached to the 5′ end of the viral RNA in protection was also evaluated. In this respect, in the case of TVMV, the engineered construct carried VPg sequences, and the possibility that the presence of VPg outside of its usual context interferes with postulated VPg functions (initiation of RNA replication, packaging) was discussed by Maiti *et al.* (1993). In the case of PVY, most of the VPg sequence was deleted, and resistance due to the lack of VPg was also discussed (Vardi *et al.*, 1993). In view of these results, one may conclude that VPg is apparently not involved in protection.

## 2.3. Nucleotide Sequence Expression

### 2.3.1. Viral Subgenomic Sequences: Sense and Antisense RNA

Genetically transformed plants expressing an untranslatable CP gene sequence of potyviruses (Cassidy and Nelson, 1995; Dougherty *et al.*, 1994; Silva-Rosales *et al.*, 1994; Smith *et al.*, 1994; Farinelli and Malnoë, 1993; Lindbo *et al.*, 1993b; Lindbo and Dougherty, 1992b) and TSWV (Pang *et al.*, 1993, 1994; de Haan *et al.*, 1992) exhibited resistance even in the absence of the protein itself. Of all possible mechanisms that may explain these observations, RNA-mediated silencing is the preferred hypothesis. Most of the relevant data come from studies with TEV (Lindbo *et al.*, 1993a). TEV CP-transformed plants may not initially become resistant. Resistance, however, was induced by TEV infection and developed gradually. When the transformed plant reached an antiviral state, the steady-state level of endogenous mRNA, as well as that of the viral RNA, was reduced to 5–10% of that of control plants. Transcription was not affected at all, and the steady-state level of the transgene primary transcripts remained unaltered (Lindbo et al., 1993b; Lindbo and Dougherty, 1992a). A proposed mechanism for RNA-mediated resistance is the formation of a dsRNA in the cytoplasm by hybridization of the transgene transcript to the virus strand of opposite polarity. In this model, the RNA duplex would presumably block viral or host-associated functions such as replication and translation. This mechanism involves the production of antisense RNA in the cytoplasm by an RNA-dependent RNA polymerase (RdRp). Short dsRNA segments become a target for RNA degradation. The observation that transgenic tobacco transcribing short antisense sequences to the 5′ end of TMV RNA (A. Nelson *et al.*, 1993) or the 5′ sense sequences of PVY RNA (Dougherty *et al.*, 1994) are resistant may also be attributed to a similar mechanism. The observations by Hammond and Kamo (1995) indicate, however, that viral dsRNA formed in the cytoplasm can be so stable that it can be purified from air-dried leaves. Also, it remains unclear why viral helicase activity should not simply separate the inhibitor from target strands.

Experimental evidence indicates that RNA-mediated resistance based on potyviral CP gene sequences is a very specific type of resistance and the broad resistance provided by CPMR is not observed. The plants become protected only from cognate viruses. This has led to the hypothesis that dsRNAs act as elicitors of specific host defense (Goodwin *et al.*, 1996). Huntley and Hall (1993b) suggested that some, but not all, viral RNA sequences could serve as templates for an endogenous host-encoded RdRp analogous to those described in animal cells. It has been suggested that a certain threshold level of a certain RNA molecule must be exceeded (by a summation of transgene transcript and viral RNA) to activate the cytoplasmic posttranscriptional RNA degradation process and elicit virus resistance in a homology-dependent resistance type (Goodwin *et al.*, 1996; Smith *et al.*, 1994; Lindbo *et al.*, 1993b). Some "unknown genomic feature" located at or adjacent to the transgene can influence cytoplasmic RNA turnover, as it has been observed that not all transgenes have "silencing" capacity and that this feature can be inherited (Mueller *et al.*, 1995). More recently, English *et al.* (1996) proposed that qualitative, in addition to the quantitative, aberrations (when exceeding a threshold level) may be involved in triggering the silencing mechanism. This might consist in the formation of aberrant transgene RNAs. The so-called asRNA would then serve as a template for a plant-encoded RdRp that would produce an antisense RNA. Subsequently, this asRNA would target transgene transcripts as well as viral RNA for destruction, by hybridizing with the homologous regions.

Subgenomic RNA regulatory sequences may be used to obtain resistant plants. Partial resistance against TYMV was obtained in transgenic rapeseed that express genes including the 3′-untranslated region of the TYMV genome (Zaccomer *et al.*, 1993). This 3′ region contained the tRNA-like structure acting as initiation site. Transgenic expression of antisense RNA, including the viral 3′-untranslated regulatory regions, in transgenic *N. benthamiana* led to very high levels of resistance against bean yellow mosaic potyvirus (Hammond and Kamo, 1995). Similarly, Morch *et al.* (1987) have shown that in *in vitro* experiments with turnip yellow mosaic virus (TYMV) sense fragments compete for replicase with genomic sense RNAs. The mechanism of interference with the viral life cycle is unclear. A sense tRNA-like structure, essential for initiation of RNA replication, probably acts as a decoy molecule competing both for host factors that interact with the virus genomic tRNA-like structure and, especially, for the limited pool of viral replicase.

Antisense RNA-mediated strategies have also been applied with success. In many cases, viral genomic sequences have been introduced into transgenic plants in such a way that the negative viral strand (antisense) is transcribed. The aim of this strategy has been to downregulate gene expression by antisense RNA in plants and thereby to obtain virus-resistant plants. Antisense regulation of gene expression is a natural mechanism in prokaryotes which is also employed in eukaryotes (Hélène and Toulmé, 1990). In the nucleus, antisense RNA may bind to the loop initiated by RNA polymerase and interfere with transcription initiation. It may also bind to the new transcript and prevent its elongation and correct termination. Most plant viruses replicate in the cytoplasm and do not go through any nuclear phase. Therefore, antisense viral sequences most probably exert resistance only through cytoplasmic events. In the cytoplasm, hybridization of antisense RNA to mRNA may lead to translation arrest and prevent ribosomes from binding to RNA or sliding along the transcript. Formation of dsRNA may also be a signal for degradation as described before, and expression may decline due to a lower steady-state level of the respective mRNA. Competition for viral replicase components may also occur.

Antisense-mediated resistance has been observed against CMV (Cuozzo *et al.*, 1988), PVX (Hemenway *et al.*, 1988), TMV (Powell *et al.*, 1989), and potyviruses (Fang and Grumet, 1993; Lindbo *et al.*, 1993b). In all these cases, antisense-mediated resistance was less frequent and milder than that mediated by CP. Strong antisense-mediated resistance was reported against potato leaf roll virus (Kawchuk *et al.*, 1991). Another example relates to DNA viruses: expression of the *AL1* gene, an essential gene for viral DNA replication of the geminivirus tomato golden mosaic virus (TGMV), in an antisense orientation induced resistance in transgenic plants (Day *et al.*, 1991). In transgenic plants with antisense RNA-induced resistance against TMV, the transgene complements the last 117 bases of TMV RNA (Powell *et al.*, 1989). Blocking of the replicase-binding site (at the 3′-untranslated region) was suggested as a possible mechanism. DNA geminiviruses replicate in the nucleus. Antisense inhibition of

TGMV was expected to take place in the nucleus and to confer resistance to a wider range of geminiviruses because the *AL1* sequence is conserved among this viral group. Indeed, the antisense expression of TGMV *AL1* sequence conferred resistance to beet curly top virus, which is 64% similar to TGMV (it contains a stretch of 280 bases which is over 80% similar to that of TGMV) (Day *et al.*, 1991). The *AL1* of ACMV is also 63% similar to that of TGMV, but the similarity is more dispersed and no protection by antisense TGMV *AL1* was observed (Bejarano and Lichtenstein, 1994). A further study of viral genomic sites which are more highly affected by antisense-mediated resistance will probably lead to more efficiently protective antisense-based systems.

### 2.3.2. Satellite RNAs

Satellite RNAs (satRNAs) are small RNA molecules, ranging approximately from 200 to 1500 nucleotides, that depend on a host virus for replication and encapsidation and do not share any significant sequence homology with their helper viruses. Several types can be distinguished depending on size and structure (Fritsch and Mayo, 1989; Roossinck *et al.*, 1992). Only the larger satRNAs (900–1500 nucleotides) code for proteins. Some of the smaller satRNAs have a circular, highly base-paired viroid-like structure and are therefore sometimes called virusoids. They typically have ribozyme activity (see Section 2.3.4). When a satRNA encodes its own capsid protein and encapsidates its own RNA, it is called a satellite virus. The ability of some satRNAs to attenuate the symptoms of their helper virus led to their early and widespread use in spray inoculations of greenhouse and field crops. In contrast, other satellites enhance symptom expression, and others have no effect.

A certain level of resistance of the transgenic plant toward the corresponding virus can be expected when copies of symptom-reducing satRNAs are introduced into the plant genome. This has indeed been shown for CMV on tobacco (Baulcombe *et al.*, 1986), tomato (McGarvey *et al.*, 1994; Saito *et al.*, 1992), and petunia (Kim *et al.*, 1995). A few reports have shown that the CMV satRNA could also give partial protection to other related viruses, namely tomato aspermy virus (TAV) (Harrison *et al.*, 1987) and TMV (Kim *et al.*, 1995). The satellite approach has also been successful for tobacco ringspot virus (TRSV) (Gerlach *et al.*, 1987). In general, disease symptoms do develop on the inoculated transgenic plants, but the plants recover completely. Symptoms never appeared after challenging perfect CMV satRNA-monomer-expressing tobacco lines (Peña *et al.*, 1994). Transgenic tobacco plants expressing both CMV CP and satRNA genes were shown to have a two-fold higher CMV resistance compared to plants with CMV CP or satRNA alone (Yie *et al.*, 1992), indicating possible additive effects between satRNA-mediated resistance and CPMR.

Competition with the viral genomic RNA for a limiting amount of replicating enzyme has been proposed as a mechanism for satRNA-mediated disease attenuation. This mechanism would lead to inhibition of virus replication. However, in the case of CMV satRNA resulting in tolerance to TAV infection (Harrison *et al.*, 1987), no inhibition of TAV RNA replication or production of intact virus particles was observed. This phenomenon is not clearly understood. It might be that the satRNA slightly delays virus replication, interacts directly with the viral RNA, or encodes and produces small proteins that interfere with symptom induction. After transforming *Nicotiana benthamiana* with CymbRSV satRNA, no disease tolerance could be noticed (Rubino and Russo, 1995). Rather, replication of satRNA was found to suppress accumulation of defective interfering (DI) RNA (see Section 2.3.3) which acts as an attenuator of virus replication and disease in the absence of satRNA. In this case, satRNA may outcompete DI RNA for replication and/or packaging virions. Because both DI RNA and satRNA require the helper virus for replication, it is reasonable that both molecules utilize at least some virus-encoded factors in common.

The satellite strategy is limited because not many viruses have satellites, and also raises some concerns, such as the relative ease by which satRNAs can mutate to more severe forms (a single nucleotide change suffices) and the fact that new satellite–virus combinations may result in more severe symptoms (Sleat and Palukaitis, 1992; Devic *et al.*, 1990; Jaegle *et al.*, 1990).

### 2.3.3. Defective Interfering Sequences

Defective interfering (DI) sequences are deletion mutants of viral genomes that have generally lost all essential viral genes needed for movement, replication, and encapsidation. DI RNAs or DNAs require the presence of a helper virus for *trans*-acting factors necessary for replication and often multiply and accumulate at the expense of the helper virus from which they are derived. Interference with the helper genome may result in significant symptom attenuation. Their function in the field may be to modulate virus levels in natural infections and so prevent the development of severe symptoms that might otherwise have an adverse effect on the virus population. As for satRNAs, DI sequences can be introduced into plants and thus confer virus resistance.

While common in animal viruses, DI sequences have a limited occurrence in plant viruses, mainly the genus tombusvirus and carmovirus, and some geminiviruses (Burgyan *et al.*, 1991; Li *et al.*, 1989; Hillman *et al.*, 1987). DNA copies of DI RNA of CymbRSV were introduced into *Nicotiana benthamiana,* resulting in CymbRSV disease protection (Kollàr *et al.*, 1993). Another successful example was the introduction of subgenomic DNA B, derived from the native Kenyan isolate ACMV in *N. benthamiana,* which behaves like a DI sequence, leading to partial resistance toward ACMV (Stanley *et al.*, 1990). Similar positive results were obtained with monopartite beet curly top geminivirus (Frischmuth and Stanley, 1994).

Although the approach offers a valuable alternative to CPMR, the limited availability of naturally occurring DI sequences is a serious drawback. This may, however, be overcome by the use of artificial constructions of DI RNAs as in the case of TMV (Raffo and Dawson, 1991) and BMV (Huntley and Hall, 1993a,b; Marsh *et al.*, 1991a,b). Coinoculation experiments have demonstrated that several different derivatives of BMV RNAs function as DI RNAs with viral RNA and that all probably contain *cis*-acting elements involved in BMV RNA replication. Some of these *cis*-acting elements—an artificial DI RNA derived from RNA2, a sense tRNA-like structure corresponding to the 3′ end of RNA2, and an antisense sequence corresponding to the intercistronic region of BMV RNA3—have been introduced into rice, where they indeed conferred high levels of resistance against BMV. A reduction up to 95% in accumulation of progeny viral RNAs was observed when plants were inoculated with virion RNA (Huntley and Hall, 1996). Artificial DI sequences have a great potential as an alternative tool for obtaining plant virus resistance, but the requirement for specific sequences, hybridization or host enzyme interaction, makes it a very specific type of resistance.

### 2.3.4. Ribozymes

Ribozymes are small RNA molecules, derived from certain satRNAs and viroids, which are capable of highly specific, intramolecular catalytic cleavage of RNA (Symons, 1989). Haseloff and Gerlach (1988) reported simple rules for the design of synthetic hammerhead ribozymes that could specifically bind and cleave other RNA molecules in a bimolecular reaction. These ribozymes typically consist of a 22-nucleotide catalytic domain flanked by sequences that are complementary to the target RNA. The flanking sequences, or "hybridizing arms," align the catalytic domain to juxtapose with the target RNA and provide specificity in the interaction with the target. The length of the "arms" can vary from six nucleotides to much longer antisense sequences. In most naturally occurring hammerhead ribozymes, cleavage occurs 3′ of a GUC trinucleotide. Synthetic hammerhead ribozymes can also catalyze efficiently *in vitro* cleavage adjacent to GUA, GUU, CUC, and UUC trinucleotides (Shimayama *et al.*, 1995; Perriman *et al.*, 1992).

The use of ribozymes has been much more common in animal systems. In *Drosophila*, ribozymes linked to a region-specific or inducible promoter have been used as a tool for mutagenesis, to follow gene function in a specific region or at a particular developmental stage (Zhao and Pick, 1993). Ribozymes are also a new pharmaceutical class of reagents that offer potential in treating a number of different medical disorders, including infectious diseases such as viral hepatitis, and cancer (Lieber and Kay, 1996). However, there are only a few reports of ribozyme activity in plant cells. Ribozyme-mediated gene inactivation strategies have been successfully employed against the bacterial neomycin phosphotransferase II (NPTII) gene in protoplasts of *Nicotiana tabacum* (Steinecke *et al.*, 1992) and TMV replication in protoplasts (Edington and Nelson, 1992). In the transient expression assay of Steinecke *et al.* (1992) a 10:1 ratio of ribozyme to *nptII*-expressing vector was needed to achieve a reduction of active NPT enzyme of 58% and a 100:1 ratio was required to eliminate the active NPT enzyme completely. Surprisingly, when the same chimeric ribozyme gene was stably expressed in transgenic tobacco, a 2:1 ratio was sufficient to reduce the amount of NPT enzyme by about 80% (Wegener *et al.*, 1994). Exact compartmentalization of both ribozyme and target might have helped to reduce the need for high levels of ribozyme. Alternatively, host factors not expressed in protoplasts might interfere. Atkins *et al.* (1995) introduced ribozyme genes to target the negative strand of citrus exocortis viroid (CEVd) in tomato and obtained a moderate reduction in the accumulation of CEVd RNA. Unexpectedly, when targeting the positive strand of CEVd, an increase in the rate of CEVd RNA accumulation was observed. As viroids are very difficult to tackle by gene engineering strategies, ribozymes might be an interesting tool, but more has to be known about their mechanism. For instance, the inclusion of extra hammerhead motifs failed to have an apparent additive effect on the antisense RNA (Atkins *et al.*, 1995). Also, de Feyter *et al.* (1996) noticed that the ribozyme targeted against TMV conferred resistance in tobacco primarily by an antisense mechanism, although involvement of gene silencing cannot be excluded. Otherwise, antisense and ribozyme RNA may act by binding directly to the positive-sense TMV RNA to inhibit replication and translation, or indirectly by competing for viral or host factors needed for replication or spread.

In any case, higher ribozyme levels might improve virus or viroid resistance. High levels of ribozyme RNAs in plant cells were obtained by embedding hammerhead ribozyme motifs in an actively transcribed tRNA sequence combined with the use of an autonomously ACMV-derived replicating vector (Perriman *et al.*, 1995). Obviously, much more has to be known about the mechanism of ribozyme activity in plants to make it an efficient alternative for engineered virus and viroid resistance.

## 3. HOST-GENE-DERIVED STRATEGIES

In conventional breeding, native genes which confer some protection (tolerance or resistance) and are derived from a resistant cultivar, a related species, or a sexually compatible wild relative are introduced into commercial plants of interest. In most phytopathological cases, the introduced resistance is broken within several years due to the appearance and natural selection of resistance-breaking pathogens. Thanks to the development of gene tagging technology by restriction fragment length polymorphism (RFLP), random amplified polymorphic DNA (RAPD), amplified fragment length polymorphism (AFLP), transposons, microsatellites, and other molecular marker technology it is possible to mark a position on the chromosome within or near the gene of interest. From that point on, methods such as chromosome walking enable the identification and the characterization of the gene itself (Michelmore, 1995). This technique has been successful for the *N* gene. Indeed, introduction of the *N* gene into a sensitive tobacco cultivar resulted in a resistant transgenic plant (Whitham *et al.*, 1994). Other stable resistance genes which may be exploited, once isolated and characterized, are the *Tm-2a* gene in tomato and *Rx* gene in potato. Chromosome walking from RFLP and RAPD markers is already under way for *Tm-2a* (Ohmori *et al.*, 1995; Dax *et al.*, 1994). Additional information on the *Rx* gene (Köhm *et al.*, 1993) and its mode of action may lead to deliberate manipulation of the gene or of the *Rx*-derived metabolic pathways for the construction of virus-resistant plants.

Expression of a whole set of genes is stimulated in plants after pathogen infections and other stresses, such as air pollution, osmotic stress, or heat shock. The products of these genes have been termed "pathogenesis related" (PR) proteins. PR proteins are associated with general defense mechanisms and are expressed during lignin deposition or plant cell death. Several of them have been found to be hydrolytic enzymes, such as chitinases and glucanases, and two were recognized by human interferon antibodies (see Section 4.2). The expression of many of these genes is stimulated in tobacco by TMV infection particularly in *N* gene-carrying genotypes. It was hoped that some of these genes could be used to generate virus-resistant transgenic plants, but this has not been the case (Cutt *et al.*, 1989; Linthorst *et al.*, 1989). Interestingly, however, Beffa *et al.* (1996) observed decreased symptom severity in transgenic tobacco plants carrying an antisense β-glucanase construct (Beffa *et al.*, 1996). Recently, a model has been put forward suggesting that a low steady-state level of defense mechanism proteins in normal, noninfected plants is due to ubiquitin-mediated protein degradation rather than to low expression. Transgenic plants expressing a ubiquitin mutant that interferes with ubiquitin-dependent protein degradation exhibited resistance to pathogen infection, including TMV. These transgenics were also phenotypically similar to resistant plants in their development of necrotic lesions (Becker *et al.*, 1993). Although the basic mechanism of the observed phenomenon is not clearly understood, a new type of protection from a host gene could be developed from this basis.

Plant extracts of pokeweed (*Phytolacca* spp.) has been shown to protect against viral infection, initiating a search for a purified plant virus inhibitor. The pokeweed antiviral protein (PAP) consists of three forms that are active against a range of viruses (Barbieri *et al.*, 1982). PAP was shown to be a ribosomal-binding protein, deglycosylating a specific base on eukaryotic 28S rRNA, thereby preventing binding of elongation factor 2 and arresting protein synthesis. It is easily extractable and is believed to enter the cytoplasm whenever the cell wall is damaged (Ready *et al.*, 1986). PAP has recently been cloned in plants (Lodge *et al.*, 1993). Transgenic plants (tobacco, *N. benthamiana*, and potato) expressing less than 5 ng PAP per mg appeared normal. These plants were protected from PVX, CMV, and PVY (as well as aphid-inoculated PVY). This approach offers the possibility of developing broad-spectrum resistance by cloning a single gene. Work is underway to knock down PAP phytotoxicity and transform transgenic plants with a modified gene.

## 4. IMPORTED STRATEGIES

### 4.1. Prokaryotic Genes

Very little has been reported regarding the use of prokaryotic genes to achieve virus resistance in plants. Wilson (1993) cites the use of PVX subgenomic RNA promoter and diphtheria toxin mRNA to obtain transgenic tobacco plants protected against systemic spread of PVX. The use of bacterial toxins to induce virus resistance may be explained by the triggering of a general defense mechanism in plants. Watanabe *et al.* (1995) succeeded in expressing fusion yeast dsRNAse which, as proposed, can account for a broad type of resistance against a range of nonrelated plant RNA viruses based on degradation of dsRNA intermediates during virus replication.

### 4.2. Eukaryotic Genes

Expression of diverse forms of recombinant antibodies in plants, single-chain fragment and full-length antibodies, has become possible due to the development of hybridoma-derived monoclonal antibodies and gene cloning techniques (Plückthun, 1991). Expression and assembly of functional antibodies in transgenic tobacco were first demonstrated for a murine antibody (Hiatt *et al.*, 1989). Since then other reports have followed (De Neve *et al.*, 1993; Owen *et al.*, 1992; Benvenuto *et al.*, 1991). However, the plant-produced antibodies, also called plantibodies, are fairly new and antibody-mediated protection from only two plant viruses, artichoke mottle crinkle virus (Tavladoraki *et al.*, 1993) and TMV (Voss *et al.*, 1995), has been reported. Antiviral antibodies appear to act via the neutralization of viral surface proteins in a way that intervenes with the initial establishment of infection. Many possible improvements of this type of antibody-mediated protection are yet to be worked out. It seems that the level of protein expression may help as well as proper targeting. Single-chain fragment antibody technology has increased the possibilities. The choice of viral epitopes is very important. A conserved CP motif may lead to a broader type of resistance to many strains of the same virus family. Anti-CP antibodies may also interfere with insect transmission. Functional viral proteins accumulate in infected cells to lower levels than structural proteins. Antibody to a functional viral gene product might be more efficient than an antibody against a structural protein antibody. In addition, a broader spectrum of resistance may potentially be achieved when the antibody is directed to a conserved region of a functional protein that is shared by viruses of many groups. Replication-associated genes are a good example because several motifs are rather well conserved.

Among plant mechanisms of natural defense against viral infection, the synthesis of an antiviral substance in virus-infected plants has been suggested (Sela and Applebaum, 1962). The antivirally active fraction was designated "antiviral factor" (AVF). Technical advances, such as high-performance liquid chromatography for separation of proteins, have allowed AVF to be resolved into several different proteins (Sela, 1986). A major AVF protein was purified to homogeneity and characterized as a 22-kDa glycoprotein. AVF activity was particularly stimulated in host plants that confined viral infection to local lesions, such as tobacco varieties carrying the *N* gene. In this respect, they resemble a number of PR proteins. Some biological characteristics of AVF were reminiscent of animal interferon. Human interferon is transiently antivirally active in plants, where it was found to protect against TMV infection (Orchansky *et al.*, 1982). The protective activity of interferon in plants has been controversial for many years and several reports have indicated an absence of protection (Huisman *et al.*, 1985; Loesch-Fries *et al.*, 1985). These inconsistencies seem to be due to the applied interferon concentration, of which only low levels are active in plants (Kaplan *et al.*, 1988). In addition, two plant glycoproteins with antiviral activity, gp22 and gp35, were isolated from TMV-infected *N* gene tobacco by immunoaffinity chromatography using antibodies raised to human β-interferon (Edelbaum *et al.*, 1990). Partial amino acid sequencing identified gp35 as a 1,3-glucanase PR protein. gp22 seems to belong to another family of PR proteins typified by osmotin (which belongs to the PR5 group) (Edelbaum *et al.*, 1991).

Experiments with interferon indicate that it stimulates several cellular pathways, including that of the dsRNA-dependent 2′–5′ oligoadenylate synthase in animal cells. This enzyme polymerases ATP to a series of oligoadenylates with an unusual phosphodiester link (2′–5′ rather than 3′–5′), cumulatively designated 2′–5′ A (Samuel, 1991). 2′–5′ A activates a ribonuclease which plays a role in activating an antiviral state. 2′–5′ A and related derivatives have been found to be antivirally active in plants (Truve *et al.*, 1993, 1994; Devash *et al.*, 1982, 1985). Viral infection in plants elicited an enzyme analogous to the 2′–5′ A enzyme which polymerizes ATP to a short oligonucleotide, similar to 2′–5′ A

(Sher *et al.*, 1990). Interferon and 2′–5′ A were reported to have other physiological activities in plants (Kulaeva *et al.*, 1992). Transgenic plants expressing mammalian interferon have been obtained (Nakamura *et al.*, 1994; Truve *et al.*, 1993, 1994; Edelbaum *et al.*, 1992). Transgenic potato plants expressing rat oligoadenylate synthase were shown to become PVX-resistant (Truve *et al.*, 1993). The occurrence of an interferon-analogous plant system has therefore been suggested. This may open the way to manipulation of an interferon-like system which may prove to be effective for broad-spectrum plant virus protection.

## 5. CLOSING REMARKS

The identification and better understanding of different important steps in the life cycle of viruses has enabled the development of transgenic virus control strategies. The source of protecting genes can be as diverse as of viral, plant host, or nonhost origin. This observation stresses the importance of environmental studies from the applied side, in the form of field tests, for the fundamental knowledge required to better understand the risks which may be involved in the release of genetically engineered organisms.

Several field tests have mainly been conducted with CP-expressing transgenic plants (Gielen *et al.*, 1996; Clough and Hamm, 1995; Conner *et al.*, 1994; Malnoë *et al.*, 1994; Vallejo *et al.*, 1994; Gonsalves *et al.*, 1992; Jongedijk *et al.*, 1992; Kaniewski *et al.*, 1990; Lawson *et al.*, 1990; R. S. Nelson *et al.*, 1988). The overall conclusion is quite encouraging, because most of the plants in the transgenic population maintained at least a tolerance level of protection under field conditions (Kaniewski and Thomas, 1993). Crop yields from inoculated transgenic fields were comparable to those of their respective healthy control fields, indicating the feasibility of using transgenic resistant plants in agriculture. In most cases, the transgenic plants were tested for resistance by mechanical inoculation with purified virus or infected crude sap. Vector-mediated inoculation has been tested in only a few cases. For example, CMV-resistant transgenic plants were found to be protected from aphid-transmitted CMV (Gonsalves *et al.*, 1992). Similarly, transgenic tobacco carrying the CP gene of PVY were also protected from aphid-borne PVY (van der Vlugt and Goldbach, 1993). Tobacco plants transformed by tobacco rattle virus CP were protected from the aphid-transmitted virus, but were susceptible to the nematode-transmitted virus (Ploeg *et al.*, 1993).

If the release of resistant transgenic plants in the environment is being envisaged, heterologous encapsidation and viral recombination are to be considered when evaluating possible risks (de Zoeten, 1991). Heterologous encapsidation occurs in nature as exemplified by acquired insect transmissibility. The effect of CP in insect transmissibility has been documented in the case of potyviruses (Gal-On *et al.*, 1992; Atreya *et al.*, 1990; Rochow, 1970). Infection of a transgenic plant by a slightly related virus is quite possible and the virus may be transencapsidated by the endogenous transgenic CP and its insect transmissibility characteristics altered (Lecoq *et al.*, 1993; Farinelli *et al.*, 1992; Bourdin and Lecoq, 1991). Heteroencapsidation has also been demonstrated between the unrelated CMV and AlMV viruses (Candelier-Harvey and Hull, 1993). Greenhouse and field experiments are quite widespread and undesirable traits have not been reported. It can be said that since the newly created phenotypes only take one generation, as the next generation reverts to the original viral phenotype, transencapsidation does not seem to pose a really lasting problem. Viral RNA recombination is also quite common and has been implicated in virus evolution. RNA recombination has also been reported for plant viruses (Greene and Allison, 1994). The new phenotypes may become stable in this case. Falk and Bruening (1994) have found little foreseeable risk to agriculture from RNA recombination, since it is an infrequent event and occurs far less frequently than normal mutation events. The recombinant would only survive under considerable selective pressure and then only in the unlikely event that it is more viable than the competing natural viruses. It is believed that naturally mixed infections readily occur in nature and provide better opportunities for RNA recombination than transgenic gene–virus genome recombination.

ACKNOWLEDGMENTS. The authors wish to thank Laurent Farinelli for a critical reading of the manuscript and Martine De Cock for help in preparing it. Part of this review was supported by grants from the Belgian Programme on Interuniversity Poles of Attraction (Prime Minister's Office, Science Policy Programming, #38) and the Vlaams Actieprogramma Biotechnologie (ETC 002). M.T.C. is indebted to the European Union for an individual fellowship from the Human Capital Mobility Program (41AS8694). C.S. was a Postdoctoral Fellow of the Fund for Scientific Research (Flanders).

## 6. REFERENCES

Anderson, J. M., Palukaitis, P., and Zaitlin, M., 1992, A defective replicase gene induces resistance to cucumber mosaic virus in transgenic tobacco plants, *Proc. Natl. Acad. Sci. USA* **89:**8759–8763.

Angenent, G. C., van den Ouweland, J. M. W., and Bol, J. F., 1990, Susceptibility to virus infection of transgenic tobacco plants expressing structural and nonstructural genes of tobacco rattle virus, *Virology* **175:**191–198.

Arce-Johnson, P., Kahn, T. W., Reimann-Philipp, U., Rivera-Bustamante, R., and Beachy, R. N., 1995, The amount of movement protein produced in transgenic plants influences the establishment, local movement, and systemic spread of infection by movement protein-deficient tobacco mosaic virus, *Mol. Plant–Microbe Interact.* **8:**415–423.

Atkins, D., Hull, R., Wells, B., Roberts, K., Moore, P., and Beachy, R. N., 1991, The tobacco mosaic virus 30K movement protein in transgenic tobacco plants is localized to plasmodesmata, *J. Gen. Virol.* **72**:209–211.

Atkins, D., Young, M., Uzzell, S., Kelly, L., Fillatti, J., and Gerlach, W. L., 1995, The expression of the antisense and ribozyme genes targeting citrus exocortis viroid in transgenic plants, *J. Gen. Virol.* **76**:1781–1790.

Atreya, C. D., Raccah, B., and Pirone, T. P., 1990, A point mutation in the coat protein abolishes aphid transmissibility of a potyvirus, *Virology* **178**:161–165.

Audy, P., Palukaitis, P., Slack, S. A., and Zaitlin, M., 1994, Replicase-mediated resistance to potato virus Y in transgenic tobacco plants, *Mol. Plant–Microbe Interact.* **7**:15–22.

Barbieri, L., Aron, G. M., Irvin, J. D., and Stirpe, F., 1982, Purification and partial characterization of another form of the antiviral protein from the seeds of *Phytolacca americana* L. (pokeweed), *Biochem. J.* **203**:55–59.

Bardonnet, N., Hans, F., Serghini, M. A., and Pinck, L., 1994, Protection against virus infection in tobacco plants expressing the coat protein of grapevine fanleaf nepovirus, *Plant Cell Rep.* **13**:357–360.

Baulcombe, D., 1994, Replicase-mediated resistance: A novel type of virus resistance in transgenic plants? *Trends Microbiol.* **2**:60–63.

Baulcombe, D. C., Saunders, G. R., Bevan, M. W., Mayo, M. A., and Harrison, B. D., 1986, Expression of biologically active viral satellite RNA from the nuclear genome of transformed plants, *Nature* **321**:446–449.

Beck, D. L., Van Dolleweerd, C. J., Lough, T. J., Balmori, E., Voot, D. M., Andersen, M. T., O'Brien, I. E. W., and Forster, R. L. S., 1994, Disruption of virus movement confers broad-spectrum resistance against systemic infection by plant viruses with a triple gene block, *Proc. Natl. Acad. Sci. USA* **91**:10310–10314.

Becker, F., Buschfeld, E., Schell, J., and Bachmair, A., 1993, Altered response to viral infection by tobacco plants perturbed in ubiquitin system, *Plant J.* **3**:875–881.

Beffa, R. S., Hofer, R.-M., Thomas, M., and Meins, F., Jr, 1996, Decreased susceptibility to viral disease of β-1,3-glucanase-deficient plants generated by antisense transformation, *Plant Cell* **8**:1001–1011.

Bejarano, E. R., and Lichtenstein, C. P., 1994, Expression of TGMV antisense RNA in transgenic tobacco inhibits replication of BCTV but not ACMV geminiviruses, *Plant Mol. Biol.* **24**:241–248.

Benvenuto, E., Ordàs, R. J., Tavazza, R., Ancora, G., Biocca, S., Cattaneo, A., and Galeffi, P., 1991, 'Phytoantibodies': A general vector for the expression of immunoglobulin domains in transgenic plants, *Plant Mol. Biol.* **17**:865–874.

Bertioli, D. J., Cooper, J. I., Edwards, M. L., and Hawes, W. S., 1992, Arabis mosaic nepovirus coat protein in transgenic tobacco lessens disease severity and virus replication, *Ann. Appl. Biol.* **120**:47–54.

Bol, J. F., Van Vloten-Doting, L., and Jaspars, E. M. J., 1971, A functional equivalence of top component *a* RNA and coat protein in the initiation of infection by alfalfa mosaic virus, *Virology* **46**:73–85.

Bol, J. F., Brederode, F. T., Neeleman, L., Taschner, P. E. M., and Tumer, N. E., 1993, Complementation and disruption of viral processes in transgenic plants, *Phil. Trans. R. Soc. Lond. B* **342**:259–263.

Bourdin, D., and Lecoq, H., 1991, Evidence that heteroencapsidation between two potyviruses is involved in aphid transmission of a non-aphid-transmissible isolate from mixed infections, *Phytopathology* **81**:1459–1464.

Brault, V., Candresse, T., le Gall, O., Delbos, R. P., Lanneau, M., and Dunez, J., 1993, Genetically engineered resistance against grapevine chrome mosaic nepovirus, *Plant Mol. Biol.* **21**:89–97.

Braun, C. J., and Hemenway, C. L., 1992, Expression of amino-terminal portions or full-length viral replicase genes in transgenic plants confers resistance to potato virus X infection, *Plant Cell* **4**:735–744.

Burgyan, J., Rubino, L., and Russo, M., 1991, *De novo* generation of cymbidium ringspot virus defective interfering RNA, *J. Gen. Virol.* **72**:505–509.

Candelier-Harvey, P., and Hull, R., 1993, Cucumber mosaic virus genome is encapsidated in alfalfa mosaic virus coat protein expressed in transgenic tobacco plants, *Transgenic Res.* **2**:277–285.

Carr, J. P., and Zaitlin, M., 1993, Replicase mediated resistance, *Semin. Virol.* **4**:339–347.

Carr, J. P., Marsh, L. E., Lomonossoff, G. P., Sekiya, M. E., and Zaitlin, M., 1992, Resistance to tobacco mosaic virus induced by the 54-kDa gene sequence requires expression of the 54-kDa protein, *Mol. Plant–Microbe Interact.* **5**:397–404.

Carr, J. P., Gal-On, A., Palukaitis, P., and Zaitlin, M., 1994, Replicase-mediated resistance to cucumber mosaic virus in transgenic plants involves suppression of both virus replication in the inoculated leaves and long-distance movement, *Virology* **199**:439–447.

Carrington, J. C., Cary, S. M., Parks, T. D., and Dougherty, W. G., 1989, A second proteinase encoded by a plant potyvirus genome, *EMBO J.* **8**:365–370.

Cassidy, B. G., and Nelson, R. S., 1995, Differences in protection phenotypes in tobacco plants expressing coat protein genes from peanut stripe potyvirus with or without an engineered ATG, *Mol. Plant–Microbe Interact.* **8**:357–365.

Citovsky, V., and Zambryski, P., 1993, Transport of nucleic acids through membrane channels: Snaking through small holes, *Annu. Rev. Microbiol* **47**:167–197.

Citovsky, V., Knorr, D., Schuster, G., and Zambryski, P., 1990, The P30 movement protein of tobacco mosaic virus is a single-strand nucleic acid binding protein, *Cell* **60**:637–647.

Citovsky, V., Knorr, D., and Zambryski, P., 1991, Gene I, a potential cell-to-cell movement locus of cauliflower mosaic virus, encodes an RNA binding protein, *Proc. Natl. Acad. Sci. USA* **88**:2476–2480.

Citovsky, V., Wong, M. L., Shaw, A. L., Prasad, B. V. V., and Zambryski, P., 1992, Visualization and characterization of tobacco mosaic virus movement protein binding to single-stranded nucleic acids, *Plant Cell* **4**:397–411.

Clough, G. H., and Hamm, P. B., 1995, Coat protein transgenic resistance to watermelon mosaic and zucchini yellows mosaic virus in squash and cantaloupe, *Plant Dis.* **79**:1107–1109.

Conner, A. J., Williams, M. K., Abernethy, D. J., Fletcher, P. J., and Genet, R. A., 1994, Field performance of transgenic potatoes. *N. Z. J. Crop Hortic. Sci.* **22**:361–371.

Cooper, B., Lapidot, M., Heick, J. A., Dodds, J. A., and Beachy, R. N., 1995, A defective movement protein of TMV in transgenic plants confers resistance to multiple viruses whereas the functional analog increases susceptibility, *Virology* **206**:307–313.

Cronin, S., Verchot, J., Haldeman-Cahill, R., Schaad, M. C., and Carrington, J. C., 1995, Long-distance movement factor: A transport function of the potyvirus helper component proteinase, *Plant Cell* **7**:549–559.

Cuozzo, M., O'Connell, K. M., Kaniewski, W., Fang, R.-X., Chua, N.-H., and Tumer, N. E., 1988, Viral protection in transgenic tobacco plants expressing the cucumber mosaic virus coat protein or its antisense RNA, *Bio/technology* **6**:549–557.

Cutt, J. R., Harpster, M. H., Dixon, D. C., Carr, J. P., Dunsmuir, P., and Klessig, D. F., 1989, Disease response to tobacco mosaic virus in transgenic tobacco plants that constitutively express the pathogenesis-related PR1b gene, *Virology* **173**:89–97.

Dahal, G., Hibino, H., and Saxena, R. C., 1990, Association of leafhopper feeding behavior with transmission of rice tungro to susceptible and resistant rice cultivars, *Phytopathology* **80**:371–377.

Dax, E., Livneh, O., Edelbaum, O., Kedar, N., Gavish, N., Karchi, H., Milo, J., Sela, I., and Rabinowitch, H., 1994, A random amplified polymorphic DNA (RAPD) molecular marker for the *Tm-2-a* gene in tomato, *Euphytica* **74**:159–163.

Day, A. G., Bejarano, E. R., Buck, K. W., Burrell, M., and Lichtenstein, C. P., 1991, Expression of an antisense viral gene in transgenic tobacco confers resistance to the DNA virus tomato golden mosaic virus, *Proc. Natl. Acad. Sci. USA* **88:**6721–6725.

de Feyter, R., Young, M., Schroeder, K., Dennis, E. S., and Gerlach, W., 1996, A ribozyme gene and an antisense gene are equally effective in conferring resistance to tobacco mosaic virus on transgenic tobacco, *Mol. Gen. Genet.* **250:**329–338.

de Haan, P., Gielen, J. J. L., Prins, M., Wijkamp, I. G., van Schepen, A., Peters, D., van Grinsven, M. Q. J. M., and Goldbach, R., 1992, Characterization of RNA-mediated resistance to tomato spotted wilt virus in transgenic tobacco plants, *Bio/technology* **10:**1133–1137.

De Neve, M., De Loose, M., Jacobs, A., Van Houdt, H., Kaluza, B., Weidle, U., Van Montagu, M., and Depicker, A., 1993, Assembly of an antibody and its derived antibody fragment in *Nicotiana* and *Arabidopsis, Transgenic Res.* **2:**227–237.

Deom, C. M., Schubert, K. R., Wolf, S., Holt, C. A., Lucas, W. J., and Beachy, R. N., 1990, Molecular characterization and biological function of the movement protein of tobacco mosaic virus in transgenic plants, *Proc. Natl. Acad. Sci. USA* **87:**3284–3288.

Devash, Y., Biggs, S., and Sela, I., 1982, Multiplication of tobacco mosaic virus in tobacco leaf disks is inhibited by (2′–5′) oligoadenylate, *Science* **216:**1415–1416.

Devash, Y., Reichman, M., Sela, I., Reichenbach, N. L., and Suhadolnik, R. J., 1985, Plant oligoadenylates: Enzymatic synthesis, isolation and biological activities, *Biochemistry* **24:**593–599.

Devic, M., Jaegle, M., and Baulcombe, D., 1990, Cucumber mosaic virus satellite RNA (strain Y): Analysis of sequences which affect systemic necrosis on tomato, *J. Gen. Virol.* **71:**1443–1449.

de Zoeten, G. A., 1991, Risk assessment: Do we let history repeat itself? *Phytopathology* **81:**585–586.

Dinant, S., Blaise, F., Kusiak, C., Astier-Manifacier, S., and Albouy, J., 1993, Heterologous resistance to potato virus Y in transgenic tobacco plants expressing the coat protein gene of lettuce mosaic potyvirus, *Phytopathology* **83:**818–824.

Dolja, V. V., Haldeman-Cahill, R., Montgomery, A. E., Vandenbosch, K. A., and Carrington, J. C., 1995, Capsid protein determinants involved in cell-to-cell and long distance movement of tobacco etch potyvirus, *Virology* **206:**1007–1016.

Donson, J., Kearney, C. M., Turpen, T. H., Khan, I. A., Kurath, G., Turpen, A. M., Jones, G. E., Dawson, W. O., and Lewandowski, D. J., 1993, Broad resistance to tobamoviruses is mediated by a modified tobacco mosaic virus replicase transgene, *Mol. Plant–Microbe Interact.* **6:**635–642.

Dougherty, W. G., Lindbo, J. A., Smith, H. A., Parks, T. D., Swaney, S., and Proebsting, W. M., 1994, RNA-mediated virus resistance in transgenic plants: Exploitation of a cellular pathway possibly involved in RNA degradation, *Mol. Plant–Microbe Interact.* **7:**544–552.

Edelbaum, O., Ilan, N., Grafi, G., Sher, N., Stram, Y., Novick, D., Tal, N., Sela, I., and Rubinstein, M., 1990, Two antiviral proteins from tobacco: Purification and characterization by monoclonal antibodies to human β-interferon, *Proc. Natl. Acad. Sci. USA* **87:**588–592.

Edelbaum, O., Sher, N., Rubinstein, M., Novick, D., Tal, N., Moyer, M., Ward, E., Ryals, J., and Sela, I., 1991, Two antiviral proteins, gp35 and gp22, correspond to β-1,3-glucanase and an isoform of PR-5, *Plant Mol. Biol.* **17:**171–173.

Edelbaum, O., Stein, D., Holland, N., Gafni, Y., Livneh, O., Novick, D., Rubinstein, M., and Sela, I., 1992, Expression of active human interferon β in transgenic plants, *J. Interferon Res.* **12:**449–453.

Edington, B. V., and Nelson, R. S., 1992, Utilisation of ribozymes in plants: Plant viral resistance, in *Gene Regulation: Biology of Antisense RNA and DNA* (R. P. Erickson and J. G. Izant, eds.), pp. 209–222, Raven Press, New York, NY.

English, J. J., Mueller, E., and Baulcombe, D. C., 1996, Suppression of virus accumulation in transgenic plants exhibiting silencing of nuclear genes, *Plant Cell* **8:**179–188.

Falk, B. W., and Bruening, G., 1994, Will transgenic crops generate new viruses and new diseases, *Science* **263:**1395–1396.

Fang, G., and Grumet, R., 1993, Genetic engineering of potyvirus resistance using constructs derived from the zucchini yellow mosaic virus coat protein gene, *Mol. Plant–Microbe Interact.* **6:**358–367.

Farinelli, L., and Malnoë, P., 1993, Coat protein gene-mediated resistance to potato virus Y in tobacco: Examination of the resistance mechanisms—is the transgenic coat protein required for protection? *Mol. Plant–Microbe Interact.* **6:**284–292.

Farinelli, L., Malnoë, P., and Collet, G. F., 1992, Heterologous encapsidation of potato virus Y strain O ($PVY^o$) with the transgenic coat protein of PVY strain N ($PVY^n$) in *Solanum tuberosum* cv. Bintje, *Bio/technology* **10:**1020–1025.

Fauquet, C., and Fargette, D., 1990, African cassava mosaic virus: Etiology, epidemiology, and control, *Plant Dis.* **74:**404–411.

Fehér, A., Skryabin, K. G., Balázs, E., Preiszner, J., Shulga, O. A., Zakharyev, V. M., and Dudits, D., 1992, Expression of PVX coat protein gene under the control of extensin-gene promoter confers virus resistance on transgenic potato plants, *Plant Cell Rep.* **11:**48–52.

Fitchen, J. H., and Beachy, R. N., 1993, Genetically engineered protection against viruses in transgenic plants, *Annu. Rev. Microbiol.* **47:**739–763.

Fraser, R. S. S., 1990, The genetics of resistance to plant viruses, *Annu. Rev. Phytopathol.* **28:**179–200.

Fraser, R. S. S., 1992, The genetics of plant–virus interactions: Implications for plant breeding, *Euphytica* **63:**175–185.

Frischmuth, T., and Stanley, J., 1994, Beet curly top virus symptom amelioration in *Nicotiana benthamiana* transformed with a naturally occurring viral subgenomic DNA, *Virology* **200:**826–830.

Fritsch, C., and Mayo, M. A., 1989, Satellites of plant viruses, in *Plant Viruses, Vol. I: Structure and Replication* (C. L. Mandahar, ed.), pp. 289–321, CRC Press, Boca Raton, Florida.

Fuchs, M., and Gonsalves, D., 1995, Resistance of transgenic hybrid squash ZW-20 expressing the coat protein genes of zucchini yellow mosaic virus and watermelon mosaic virus 2 to mixed infections by both potyviruses, *Bio/technology* **13:**1466–1473.

Gal-On, A., Antignus, Y., Rosner, A., and Raccah, B., 1992, A zucchini yellow mosaic virus coat protein gene mutation restores aphid transmissibility but has no effect on multiplication, *J. Gen. Virol.* **73:**2183–2187.

Gerlach, W. L., Llewellyn, D., and Haseloff, J., 1987, Construction of a plant disease resistance gene from the satellite RNA of tobacco ringspot virus, *Nature* **328:**802–805.

Gielen, J. J. L., de Haan, P., Kool, A. J., Peters, D., van Grinsven, M. Q. J. M., and Goldbach, R. W., 1991, Engineered resistance to tomato spotted wilt virus, a negative-strand RNA virus, *Bio/technology* **9:**1363–1367.

Gielen, J., Ultzen, T., Bontems, S., Loots, W., Van Schepen, A., Westerbroek, A., de Haan, P., and Van Grinsven, M., 1996, Coat protein-mediated protection to cucumber mosaic virus infections in cultivated tomato, *Euphytica* **88:**139–149.

Golemboski, D. B., Lomonossoff, G. P., and Zaitlin, M., 1990, Plants transformed with a tobacco mosaic virus nonstructural gene sequence are resistant to the virus, *Proc. Natl. Acad. Sci. USA* **87:**6311–6315.

Gonsalves, D., Chee, P., Provvidenti, R., Seem, R., and Slightom, J. L., 1992, Comparison of coat protein-mediated and genetically-derived resistance in cucumbers to infection by cucumber mosaic virus under field conditions with natural challenge inoculations by vectors, *Bio/technology* **10:**1562–1570.

Gonsalves, C., Xue, B., Yepes, M., Fuchs, M., Ling, K., Namba, S., Chee, P., Slightom, J. L., and Gonsalves, D., 1994, Transferring cucumber mosaic virus-white leaf strain coat protein gene into *Cucumis melo* L., and evaluating transgenic plants for protection against infections, *J. Am. Soc. Hortic. Sci.* **119:**345–355.

Goodwin, J., Chapman, K., Swaney, S., Parks, T. D., Wernsman, E. A., and Dougherty, W. G., 1996, Genetic and biochemical dissection of transgenic RNA-mediated virus resistance, *Plant Cell* **8:**95–105.

Greene, A. E., and Allison, R. F., 1994, Recombination between viral RNA and transgenic plant transcripts, *Science* **263:**1423–1425.

Hammond, J., and Kamo, K. K., 1995, Effective resistance to potyvirus infection conferred by expression of antisense RNA in transgenic plants, *Mol. Plant–Microbe Interact.* **8:**674–682.

Harrison, B. D., Mayo, M. A., and Baulcombe, D. C., 1987, Virus resistance in transgenic plants that express cucumber mosaic virus satellite RNA, *Nature* **328:**799–802.

Haseloff, J., and Gerlach, W. L., 1988, Simple RNA enzymes with new and highly specific endoribonuclease activities, *Nature* **334:**585–591.

Hélène, C., and Toulmé, J.-J., 1990, Specific regulation of gene expression by antisense, sense and antigene nucleic acids, *Biochim. Biophys. Acta* **1049:**99–125.

Hellmann, G. M., Hiremath, S. T., Shaw, J. G., and Rhoads, R. E., 1986, Cistron mapping of tobacco vein mottling virus, *Virology* **151:**159–171.

Hellmann, G. M., Shaw, J. G., and Rhoads, R. E., 1988, *In vitro* analysis of tobacco vein mottling virus $NI_a$ cistron: Evidence for a virus-encoded protease, *Virology* **163:**554–562.

Hemenway, C., Fang, R.-X., Kaniewski, W. K., Chua, N.-H., and Tumer, N. E., 1988, Analysis of the mechanisms of protection in transgenic plants expressing the potato virus X coat protein or its antisense RNA, *EMBO J.* **7:**1273–1280.

Henry, C. M., Barker, I., Pratt, M., Pemberton, A. W., Farmer, M. J., Cotten, J., Ebbels, D., Coates, D., and Stratford, R., 1995, *Risks Associated with the Use of Genetically Modified Virus Tolerant Plants*, Ministry of Agriculture, Fisheries and Food, London, UK.

Herdt, R. W., 1991, Research priorities for rice biotechnology, in *Rice Biotechnology* (G. S. Khush and G. H. Toenniessen, eds.), pp. 19–54, CAB International, Wallingford, Oxon, U.K.

Hiatt, A., Cafferkey, R., and Bowdish, K., 1989, Production of antibodies in transgenic plants, *Nature* **342:**76–78.

Hillman, B. I., Carrington, J. C., and Morris, T. J., 1987, A defective interfering RNA that contains a mosaic of a plant virus genome, *Cell* **51:**427–433.

Hoekema, A., Huisman, M. J., Molendijk, L., van den Elzen, P. J. M., and Cornelissen, B. J. C., 1989, The genetic engineering of two commercial potato cultivars for resistance to potato virus X, *Bio/technology* **7:**273–278.

Holt, C. A., and Beachy, R. N., 1991, *In vivo* complementation of infectious transcripts from mutant tobacco mosaic virus cDNAs in transgenic plants, *Virology* **181:**109–117.

Hong, Y., and Stanley, J., 1996, Virus resistance in *Nicotiana benthamiana* conferred by African cassava mosaic virus replication-associated protein (AC1) transgene, *Mol. Plant–Microbe Interact.* **9:**219–225.

Huisman, M. J., Broxterman, H. J. G., Schellekens, H., and van Vloten-Doting, L., 1985, Human interferon does not protect cowpea plant cell protoplasts against infection with alfalfa mosaic virus, *Virology* **143:**622–625.

Hull, R., 1984, Rapid diagnosis of plant virus infections by spot hybridization, *Trends Biotechnol.* **2:**88–91.

Huntley, C. C., and Hall, T. C., 1993a, Interference with brome mosaic virus replication by targeting the minus strand promoter, *J. Gen. Virol.* **74:**2445–2452.

Huntley, C. C., and Hall, T. C., 1993b, Minus sense transcripts of brome mosaic virus RNA-3 intercistronic region interfere with viral replication, *Virology* **192:**290–297.

Huntley, C. C., and Hall, T. C., 1996, Interference with brome mosaic virus replication in transgenic rice, *Mol. Plant–Microbe Interact.* **9:**164–170.

Jaegle, M., Devic, M., Longstaff, M., and Baulcombe, D., 1990, Cucumber mosaic virus satellite RNA (Y strain): Analysis of sequences which affect yellow mosaic symptoms on tobacco, *J. Gen. Virol.* **71:**1905–1912.

Jongedijk, E., de Schutter, A. A. J. M., Stolte, T., van den Elzen, P. J. M., and Cornelissen, B. J. C., 1992, Increased resistance to potato virus X and preservation of cultivar properties in transgenic potato under field conditions, *Bio/technology* **10:**422–429.

Kallerhoff, J., Perez, P., Bouzoubaa, S., Ben Tahar, S., and Perret, J., 1990, Beet necrotic yellow vein virus coat protein-mediated protection in sugarbeet (*Beta vulgaris* L.) protoplasts, *Plant Cell Rep.* **9:**224–228.

Kaniewski, W. K., and Thomas, P. E., 1993, Field testing of virus resistant transgenic plants, *Semin. Virol.* **4:**389–396.

Kaniewski, W., Lawson, C., Sammons, B., Haley, L., Hart, J., Delannay, X., and Tumer, N. E., 1990, Field resistance of transgenic Russet Burbank potato to effects of infection by potato virus X and potato virus Y, *Bio/technology* **8:**750–754.

Kaplan, I. B., Taliansky, M. E., Malyshenko, S. I., Ogarkov, V. I., and Atabekov, J. G., 1988, Effect of human interferon on reproduction of plant and myxoviruses, *Arch. Phytopath. Pflanzenschutz* **24:**3–8.

Kasschau, K. D., and Carrington, J. C., 1995, Requirement for HC-Pro processing during genome amplification of tobacco etch potyvirus, *Virology* **209:**268–273.

Kawchuk, L. M., Martin, R. R., and McPherson, J., 1991, Sense and antisense RNA-mediated resistance to potato leafroll virus in Russet Burbank potato plants, *Mol. Plant–Microbe Interact.* **4:**247–253.

Kim, S. J., Paek, K. H., and Kim, B. D., 1995, Delay of disease development in transgenic petunia plants expressing cucumber mosaic virus I-17N-satellite RNA, *J. Am. Soc. Hortic. Sci.* **120:**353–359.

Knorr, D. A., and Dawson, W. O., 1988, A point mutation in the tobacco mosaic virus capsid protein gene induces hypersensitivity in *Nicotiana sylvestris*, *Proc. Natl. Acad. Sci. USA* **85:**170–174.

Köhm, B. A., Goulden, M. G., Gilbert, J. E., Kavanagh, T. A., and Baulcombe, D. C., 1993, A potato virus X resistance gene mediates an induced, nonspecific resistance in protoplasts, *Plant Cell* **5:**913–920.

Kollàr, À., Dalmay, T., and Burgyàn, J., 1993, Defective interfering RNA-mediated resistance against cymbidium ringspot tombusvirus in transgenic plants, *Virology* **193:**313–318.

Kulaeva, O. N., Fedina, A. B., Burkhanova, E. A., Karavaiko, N. N., Karpeisky, M. Y., Kaplan, I. B., Taliansky, M. E., and Atabekov, J. G., 1992, Biological activities of human interferon and 2′–5′ oligoadenylates in plants, *Plant Mol. Biol.* **20:**383–393.

Kunik, T., Salomon, R., Zamir, D., Navor, N., Zeidan, M., Michelson, I., Gafni, Y., and Czosnek, H., 1994, Transgenic tomato plants expressing the tomato yellow leaf curl virus capsid protein are resistant to the virus, *Bio/technology* **12:**500–504.

Laimer da Câmara Machado, M., da Câmara Machado, A., Hanzer, V., Weiss, H., Regner, F., Steinkellner, H., Plail, R., Knapp, E., and Katinger, H., 1993, Coat protein-mediated protection against plum pox virus, *Acta Hortic.* **336:**85–92.

Lapidot, M., Gafny, R., Ding, B., Wolf, S., Lucas, W. J., and Beachy, R. N., 1993, A dysfunctional movement protein of tobacco mosaic virus that partially modifies the plasmodesmata and limits virus spread in transgenic plants, *Plant J.* **4:**959–970.

Lawson, C., Kaniewski, W., Haley, L., Rozman, R., Newell, C., Sanders, P., and Tumer, N. E., 1990, Engineering resistance to mixed virus infection in a commercial potato cultivar: Resistance to potato virus X and potato virus Y transgenic Russet Burbank, *Bio/technology* **8:**127–134.

Leclerc, D., and AbouHaidar, M. G., 1995, Transgenic tobacco plants expressing a truncated form of the PAMV capsid protein (CP) gene show CP-mediated resistance to potato aucuba mosaic virus, *Mol. Plant–Microbe Interact.* **8:**58–65.

Lecoq, H., Ravelonandro, M., Wipf-Scheibel, C., Monsion, M., Raccah, B., and Dunez, J., 1993, Aphid transmission of a non-aphid-transmissible strain of zucchini yellow mosaic potyvirus from transgenic plants expressing the capsid protein of plum pox potyvirus, *Mol. Plant–Microbe Interact.* **6:**403–406.

Li, X. H., Heaton, L. A., Morris, T. J., and Simon, A. E., 1989, Turnip crinkle virus defective interfering RNAs intensify viral symptoms and are generated *de novo*, *Proc. Natl. Acad. Sci. USA* **86:**9173–9177.

Lieber, A., and Kay, M. A., 1996, Adenovirus-mediated expression of ribozymes in mice, *J. Virol.* **70:**3153–3158.

Lindbo, J. A., and Dougherty, W. G., 1992a, Pathogen-derived resistance to a potyvirus: Immune and resistant phenotypes in transgenic tobacco expressing altered forms of a potyvirus coat protein nucleotide sequence, *Mol. Plant–Microbe Interact.* **5:**144–153.

Lindbo, J. A., and Dougherty, W. G., 1992b, Untranslatable transcripts of the tobacco etch virus coat protein gene sequence can interfere with tobacco etch virus replication in transgenic plants and protoplasts, *Virology* **189:**725–733.

Lindbo, J. A., Silva-Rosales, L., Proebsting, W. M., and Dougherty, W. G., 1993a, Induction of a highly specific antiviral state in transgenic plants: Implications for regulation of gene expression and virus resistance, *Plant Cell* **5:**1749–1759.

Lindbo, J. A., Silva-Rosales, L., and Dougherty, W. G., 1993b, Pathogen derived resistance to potyviruses: Working, but why? *Semin. Virol.* **4:**369–379.

Ling, K., Namba, S., Gonsalves, C., Slightom, J. L., and Gonsalves, D., 1991, Protection against detrimental effects of potyvirus infection in transgenic tobacco plants expressing the papaya ringspot virus coat protein gene, *Bio/technology* **9:**752–758.

Linthorst, H. J. M., Meuwissen, R. L. J., Kauffmann, S., and Bol, J. F., 1989, Constitutive expression of pathogenesis-related proteins PR-1, GRP, and PR-S in tobacco has no effect on virus infection, *Plant Cell* **1:**285–291.

Lodge, J. K., Kaniewski, W. K., and Tumer, N. E., 1993, Broad-spectrum virus resistance in transgenic plants expressing pokeweed antiviral protein, *Proc. Natl. Acad. Sci. USA* **90:**7089–7093.

Loesch-Fries, L. S., Halk, E. L., Nelson, S. E., and Krahn, K. J., 1985, Human leukocyte interferon does not inhibit alfalfa mosaic virus in protoplasts or tobacco tissue, *Virology* **143:**626–629.

Loesch-Fries, L. S., Merlo, D., Zinnen, T., Burhop, L., Hill, K., Krahn, K., Jarvis, N., Nelson, S., and Halk, E., 1987, Expression of alfalfa mosaic virus RNA 4 in transgenic plants confers virus resistance, *EMBO J.* **6:**1845–1851.

Longstaff, M., Brigneti, G., Boccard, F., Chapman, S., and Baulcombe, D., 1993, Extreme resistance to potato virus X infection in plants expressing a modified component of the putative viral replicase, *EMBO J.* **12:**379–386.

Lupo, R., Rubino, L., and Russo, M., 1994, Immunodetection of the 33 K/92 K polymerase proteins in cymbidium ringspot virus-infected and in transgenic plant tissue extracts, *Arch. Virol.* **138:**135–142.

MacFarlane, S. A., and Davis, J. W., 1992, Plants transformed with a region of the 201-kilodalton replicase gene from pea early browning virus RNA1 are resistant to virus infection, *Proc. Natl. Acad. Sci. USA* **89:**5829–5833.

MacKenzie, D. J., and Ellis, P. J., 1992, Resistance to tomato spotted wilt virus infection in transgenic tobacco expressing the viral nucleocapsid gene, *Mol. Plant–Microbe Interact.* **5:**34–40.

MacKenzie, D. J., Tremaine, J. H., and McPherson, J., 1991, Genetically engineered resistance to potato virus S in potato cultivar Russet Burbank, *Mol. Plant–Microbe Interact.* **4:**95–102.

Maiti, I. B., Murphy, J. F., Shaw, J. G., and Hunt, A. G., 1993, Plants that express a potyvirus proteinase gene are resistant to virus infection, *Proc. Natl. Acad. Sci. USA* **90**:6110–6114.

Malnoë, P., Farinelli, L., Collet, G. F., and Reust, W., 1994, Small-scale field tests with transgenic potato, cv Bintje, to test resistance to primary and secondary infections with potato virus y, *Plant Mol. Biol* **25**:963–975.

Malyshenko, S. I., Kondakova, O. A., Nazarova, J. V., Kaplan, I. B., Taliansky, M. E., and Atabekov, J. G., 1993, Reduction of tobacco mosaic virus accumulation in transgenic plants producing non-functional viral transport proteins, *J. Gen. Virol.* **74**:1149–1156.

Marsh, L. E., Pogue, G. P., Szybiak, U., Connell, J. P., and Hall, T. C., 1991a, Non-replicating deletion mutants of brome mosaic virus RNA-2 interfere with viral replication, *J. Gen. Virol* **72**:2367–2374.

Marsh, L. E., Pogue, G. P., Connell, J. P., and Hall, T. C., 1991b, Artificial defective interfering RNAs derived from brome mosaic virus, *J. Gen. Virol.* **72**:1787–1792.

Matthews, R. E. F., 1991, *Plant Virology*, 3rd ed., Academic Press, San Diego, California.

McGarvey, P. B., Montasser, M. S., and Kaper, J. M., 1994, Transgenic tomato plants expressing satellite RNA are tolerant to some strains of cucumber mosaic virus, *J. Am. Soc. Hortic. Sci.* **119**:642–647.

Michelmore, R., 1995, Molecular approaches to manipulation of disease resistance genes, *Annu. Rev. Phytopathol.* **33**:393–427.

Moore, P. J., Fenczik, C. A., Deom, C. M., and Beachy, R. N., 1992, Developmental changes in plasmodesmata in transgenic tobacco expressing the movement protein of tobacco mosaic virus, *Protoplasma* **170**:115–127.

Morch, M. D., Joshi, R. L., Denial, T. M., and Haenni, A. L., 1987, A new 'sense' RNA approach to block viral RNA replication *in vitro*, *Nucleic Acids Res.* **15**:4123–4130.

Mori, M., Mise, K., Okuno, T., and Furusawa, I., 1992, Expression of brome mosaic virus-encoded replicase genes in transgenic tobacco plants, *J. Gen. Virol* **73**:169–172.

Mueller, E., Gilbert, J., Davenport, G., Brigneti, G., and Baulcombe, D. C., 1995, Homology-dependent resistance: Transgenic virus resistance in plants related to homology-dependent gene silencing, *Plant J.* **7**:1001–1013.

Nakamura, S., Yoshikawa, M., Taira, H., and Ehara, Y., 1994, Plants transformed with mammalian 2′–5′ oligoadenilate synthase gene show resistance to virus infections, *Ann. Phytopathol. Soc. Japan* **60**:691–693.

Namba, S., Ling, K., Gonsalves, C., Gonsalves, D., and Slightom, J. L., 1991, Expression of the gene encoding the coat protein of cucumber mosaic virus (CMV) strain WL appears to provide protection to tobacco plants against infection by several different CMV strains, *Gene* **107**:181–188.

Namba, S., Ling, K., Gonsalves, C., Slightom, J. L., and Gonsalves, D., 1992, Protection of transgenic plants expressing the coat protein gene of watermelon mosaic virus II or zucchini yellow mosaic virus against six potyviruses, *Phytopathology* **82**:940–946.

Nejidat, A., and Beachy, R. N., 1990, Transgenic tobacco plants expressing a coat protein gene of tobacco mosaic virus are resistant to some other tobamoviruses, *Mol. Plant–Microbe Interact.* **3**:247–251.

Nelson, A., Roth, D. A., and Johnson, J. D., 1993, Tobacco mosaic virus infection of transgenic *Nicotiana tabacum* plants is inhibited by antisense constructs directed at the 5′ region of viral RNA, *Gene* **127**:227–232.

Nelson, R. S., Powell Abel, P., and Beachy, R. N., 1987, Lesions and virus accumulation in inoculated transgenic tobacco plants expressing the coat protein gene of tobacco mosaic virus, *Virology* **158**:126–132.

Nelson, R. S., McCormick, S. M., Delannay, X., Dubé, P., Layton, J., Anderson, E. J., Kaniewska, M., Proksch, R. K., Horsch, R. B., Rogers, S. G., Fraley, R. T., and Beachy, R. N., 1988, Virus tolerance, plant growth, and field performance of transgenic tomato plants expressing coat protein from tobacco mosaic virus, *Bio/technology* **6**:403–409.

Ohmori, T., Murata, M., and Motoyoshi, F., 1995, Identification of RAPD markers linked to the *Tm-2* locus in tomato, *Theor. Appl. Genet.* **90**:307–311.

Okuno, T., Nakayama, M., Yoshida, S., Furusawa, I., and Komiya, T., 1993, Comparative susceptibility of transgenic tobacco plants and protoplasts expressing the coat protein gene of cucumber mosaic virus to infection with virions and RNA, *Phytopathology* **83**:542–547.

Orchansky, P., Rubinstein, M., and Sela, I., 1982, Human interferons protect plants from virus infection, *Proc. Natl. Acad. Sci. USA* **79**:2278–2280.

Osbourn, J. K., Watts, J. W., Beachy, R. N., and Wilson, T. M., 1989a, Evidence that nucleocapsid disassembly and a later step in virus replication are inhibited in transgenic tobacco protoplasts expressing TMV coat protein, *Virology* **172**:370–373.

Osbourn, J. K., Plaskitt, K. A., Watts, J. W., and Wilson, T. M. A., 1989b, Tobacco mosaic virus coat protein and reporter gene transcripts containing the TMV origin-of-assembly sequence do not interact in double-transgenic tobacco plants: Implications for coat protein-mediated protection, *Mol. Plant–Microbe Interact.* **2**:340–345.

Owen, M., Gandecha, A., Cockburn, B., and Whitelam, G., 1992, Synthesis of a functional antiphytochrome single-chain $F_v$ protein in transgenic tobacco, *Bio/technology* **10**:790–794.

Padgett, H. S., and Beachy, R. N., 1993, Analysis of a tobacco mosaic virus strain capable of overcoming *N* gene-mediated resistance, *Plant Cell* **5**:577–586.

Pang, S.-Z., Slightom, J. L., and Gonsalves, D., 1993, Different mechanisms protect transgenic tobacco against tomato spotted wilt and impatiens necrotic spot *Tospoviruses*, *Bio/technology* **11**:819–824.

Pang, S.-Z., Bock, J. H., Gonsalves, C., Slightom, J. L., and Gonsalves, D., 1994, Resistance of transgenic *Nicotiana benthamiana* plants to tomato spotted wilt and impatiens necrotic spot tospoviruses: Evidence of involvement of the N protein and N gene RNA in resistance, *Phytopathology* **84**:243–249.

Peña, L., Trad, J., Díaz-Ruiz, J. R., McGarvey, P. B., and Kaper, J. M., 1994, Cucumber mosaic virus protection in transgenic tobacco plants expressing monomeric, dimeric or partial sequences of a benign satellite RNA, *Plant Sci.* **100**:71–81.

Perriman, R., Delves, A., and Gerlach, W. L., 1992, Extended target-site specificity for a hammerhead ribozyme, *Gene* **113**:157–163.

Perriman, R., Bruening, G., Dennis, E. S., and Peacock, W. J., 1995, Effective ribozyme delivery in plant cells, *Proc. Natl. Acad. Sci. USA* **92**:6175–6179.

Ploeg, A. T., Mathis, A., Bol, J. F., Brown, D. J. F., and Robinson, D. J., 1993, Susceptibility of transgenic tobacco plants expressing tobacco rattle virus coat protein to nematode-transmitted and mechanically inoculated tobacco rattle virus, *J. Gen. Virol* **74**:2709–2715.

Plückthun, A., 1991, Antibody engineering, *Curr. Opin. Biotechnol.* **2**:238–246.

Powell Abel, P., Nelson, R. S., De, B., Hoffmann, N., Rogers, S. G., Fraley, R. T., and Beachy, R. N., 1986, Delay of disease development in transgenic plants that express the tobacco mosaic virus coat protein gene, *Science* **232**:738–743.

Powell, P. A., Stark, D. M., Sanders, P. R., and Beachy, R. N., 1989, Protection against tobacco mosaic virus in transgenic plants that express tobacco mosaic virus antisense RNA, *Proc. Natl. Acad. Sci. USA* **86**:6949–6952.

Quadt, R., Rosdorff, H. J. M., Hunt, T. W., and Jaspars, E. M. J., 1991, Analysis of the protein composition of alfalfa mosaic virus RNA-dependent RNA polymerase, *Virology* **182**:309–315.

Quemada, H. D., Gonsalves, D., and Slightom, J. L., 1991, Expression of coat protein gene from cucumber mosaic virus strain C in tobacco: Protection against infections by CMV strains transmitted mechanically or by aphids, *Phytopathology* **81**:794–802.

Raffo, A. J., and Dawson, W. O., 1991, Construction of tobacco mosaic virus subgenomic replicons that are replicated and spread systemically in tobacco plants, *Virology* **184**:277–289.
Ready, M. P., Brown, D. T., and Robertus, J. D., 1986, Extracellular localization of pokeweed antiviral protein, *Proc. Natl. Acad. Sci. USA* **83**:5053–5056.
Register III, J. C., and Beachy, R. N., 1988, Resistance to TMV in transgenic plants results from interference with an early event in infection, *Virology* **166**:524–532.
Regner, F., da Câmara Machado, A., Laimer da Câmara Machado, M., Steinkellner, H., Mattanovich, D., Hanzer, V., Weiss, H., and Katinger, H., 1992, Coat protein mediated resistance to Plum Pox Virus in *Nicotiana clevelandii* and *N. benthamiana*, *Plant Cell Rep.* **11**:30–33.
Reimann-Philipp, U., and Beachy, R. N., 1993, Plant resistance to virus diseases through genetic engineering: Can a similar approach control plant-parasitic nematodes? *J. Nematol.* **25**:541–547.
Robaglia, C., Durand-Tardif, M., Tronchet, M., Boudazin, G., Astier-Manifacier, S., and Casse-Delbart, F., 1989, Nucleotide sequence of potato virus Y (N strain) genomic RNA, *J. Gen. Virol.* **70**:935–947.
Rochow, W. F., 1970, Barley yellow dwarf virus: Phenotypic mixing and vector specificity, *Science* **167**:875–878.
Rohfritsch, O., Poirson, A., Turner, A., Gagey, M.-J., Roberts, K., Stussi-Garaud, C., and Godefroy-Colburn, T., 1996, A modified form of the alfalfa mosaic virus movement protein induces stressed phenotypes in transgenic tobacco, *Can. J. Bot.* **74**:939–951.
Roossinck, M. J., Sleat, D., and Palukaitis, P., 1992, Satellite RNAs of plant viruses: Structure and biological effects, *Microbiol. Rev.* **56**:265–279.
Rubino, L., and Russo, M., 1995, Characterization of resistance to cymbidium ringspot virus in transgenic plants expressing a full-length viral replicase gene, *Virology* **212**:240–243.
Saito, Y., Komari, T., Masuta, C., Hayashi, Y., Kumashiro, T., and Takanami, Y., 1992, Cucumber mosaic virus-tolerant transgenic tomato plants expressing a satellite RNA, *Theor. Appl. Genet.* **83**:679–683.
Samuel, C. E., 1991, Antiviral actions of interferon. Interferon regulated cellular proteins and their surprisingly selective antiviral activities, *Virology* **183**:1–11.
Schoumacher, F., Erny, C., Berna, A., Godefroy-Colburn, T., and Stussi-Garaud, C., 1992, Nucleic acid-binding properties of alfalfa mosaic virus movement protein produced in yeast, *Virology* **188**:896–899.
Scorza, R., Levy, L., Damsteegt, V., Yepes, L. M., Cordts, J., Hadidi, A., Slightom, J., and Gonsalves, D., 1995, Transformation of plum with the papaya ringspot virus coat protein gene and reaction of transgenic plants to plum pox virus, *J. Am. Soc. Hortic. Sci* **120**:943–952.
Sela, I., 1986, Preparation and measurement of an antiviral protein found in tobacco cells after infection with tobacco mosaic virus, in *Interferons, Part C* (S. Pestka, ed.), pp. 734–744, Academic Press, San Diego, California.
Sela, I., and Applebaum, S. W., 1962, Occurrence of antiviral factor in virus-infected plants, *Virology* **17**:543–548.
Shaw, J. G., Plaskitt, K. A., and Wilson, T. M. A., 1986, Evidence that tobacco mosaic virus particles disassemble cotranslationally *in vivo*, *Virology* **148**:326–336.
Sher, N., Edelbaum, O., Barak, Z., Grafi, G., Stram, Y., and Sela, I., 1990, Induction of an ATP-polymerizing enzyme in TMV-infected tobacco and its homology to human 2′–5′ A synthase, *Virus Genes* **4**:27–39.
Shimayama, T., Nishikawa, S., and Taira, K., 1995, Generality of the NUX rule: Kinetic analysis of the results of systematic mutations in the trinucleotide at the cleavage site of hammerhead ribozymes, *Biochemistry* **34**:3649–3654.
Sijen, T., Wellink, J., Hendriks, J., Verver, J., and van Kammen, A., 1995, Replication of cowpea mosaic virus RNA1 or RNA2 is specifically blocked in transgenic *Nicotiana benthamiana* plants

expressing the full-length replicase or movement protein genes, *Mol. Plant–Microbe Interact.* **8**:340–347.

Silva-Rosales, L., Lindbo, J. A., and Dougherty, W. G., 1994, Analysis of transgenic tobacco plants expressing a truncated form of a potyvirus coat protein nucleotide sequence, *Plant Mol. Biol.* **24**:929–939.

Sleat, D. E., and Palukaitis, P., 1992, A single nucleotide change within a plant virus satellite RNA alters the host specificity of disease induction, *Plant J.* **2**:43–49.

Smith, H. A., Swaney, S. L., Parks, T. D., Wernsman, E. A., and Dougherty, W. G., 1994, Transgenic plant virus resistance mediated by untranslatable sense RNAs: Expression, regulation, and fate of nonessential RNAs, *Plant Cell* **6**:1441–1453.

Stanley, J., Frischmuth, T., and Ellwood, S., 1990, Defective viral DNA ameliorates symptoms of geminivirus infection in transgenic plants, *Proc. Natl. Acad. Sci. USA* **87**:6291–6295.

Stark, D. M., and Beachy, R. N., 1989, Protection against potyvirus infection in transgenic plants: Evidence for broad spectrum resistance, *Bio/technology* **7**:1257–1262.

Steinecke, P., Herget, T., and Schreier, P. H., 1992, Expression of a chimeric ribozyme gene results in endonucleolytic cleavage of target mRNA and a concomitant reduction of gene expression *in vivo, EMBO J.* **11**:1525–1530.

Sulzinski, M. A., Gabard, K. A., Palukaitis, P., and Zaitlin, M., 1985, Replication of tobacco mosaic virus. VIII. Characterization of a third subgenomic TMV RNA, *Virology* **145**:132–140.

Symons, R. H., 1989, Self-cleavage of RNA in the replication of small pathogens of plants and animals, *Trends Biochem. Sci.* **14**:445–450.

Taliansky, M. E., Malyshenko, S. I., Pshennikova, E. S., Kaplan, I. B., Ulanova, E. F., and Atabekov, J. G., 1982, Plant virus-specific transport function. I. Virus genetic control required for systemic spread, *Virology* **122**:318–326.

Taschner, P. E. M., van Marle, G., Brederode, F. T., Tumer, N. E., and Bol, J. F., 1994, Plants transformed with a mutant alfalfa mosaic virus coat protein gene are resistant to the mutant but not to wild-type virus, *Virology* **203**:269–276.

Tavladoraki, P., Benvenuto, E., Trinca, S., De Martinis, D., Cattaneo, A., and Galeffi, P., 1993, Transgenic plants expressing a functional single-chain Fv antibody are specifically protected from virus attack, *Nature* **366**:469–472.

Tenllado, F., García-Luque, I., Serra, M. T., and Díaz-Ruíz, J. R., 1995, *Nicotiana benthamiana* plants transformed with the 54-kDa region of the pepper mild mottle tobamovirus replicase gene exhibit two types of resistance responses against viral infection, *Virology* **211**:170–183.

Truve, E., Aaspôllu, A., Honkanen, J., Puska, R., Mehto, M., Hassi, A., Teeri, T. H., Kelve, M., Seppänen, P., and Saarma, M., 1993, Transgenic potato plants expressing mammalian 2′–5′ oligoadenylate synthetase are protected from potato virus X infection under field conditions, *Bio/technology* **11**:1048–1052.

Truve, E., Kelve, M., Aaspôllu, A., Kuuskalu, A., Seppänen, P., and Saarma, M., 1994, Principles and background for the construction of transgenic plants displaying multiple virus resistance, *Arch. Virol. Suppl.* **9**:41–50.

Tumer, N. E., O'Connell, K. M., Nelson, R. S., Sanders, P. R., Beachy, R. N., Fraley, R. T., and Shah, D. M., 1987, Expression of alfalfa mosaic virus coat protein gene confers cross-protection in transgenic tobacco and tomato plants, *EMBO J.* **6**:1181–1188.

Tumer, N. E., Kaniewski, W., Haley, L., Gehrke, L., Lodge, J. K., and Sanders, P., 1991, The second amino acid of alfalfa mosaic virus coat protein is critical for coat protein-mediated protection, *Proc. Natl. Acad. Sci. USA* **88**:2331–2335.

Vallejo, R. L., Collins, W. W., Schiavone, R. D., Lommel, S. A., and Young, J. B., 1994, Extreme resistance to infection by potato virus Y and potato virus X in an advanced hybrid *Solanum phureja–S. stenotomum* diploid potato population, *Am. Potato J.* **71**:617–628.

van der Vlugt, R. A. A., and Goldbach, R. W., 1993, Tobacco plants transformed with the potato virus $Y^N$ coat protein gene are protected against different PVY isolates and against aphid-mediated infection, *Transgenic Res.* **2:**109–114.

van Dun, C. M. P., and Bol, J. F., 1988, Transgenic tobacco plants accumulating tobacco rattle virus coat protein resist infection with tobacco rattle virus and pea early browning virus, *Virology* **167:**649–652.

van Dun, C. M. P., Bol, J. F., and van Vloten-Doting, L., 1987, Expression of alfalfa mosaic virus and tobacco rattle virus coat protein genes in transgenic tobacco plants, *Virology* **159:**299–305.

van Dun, C. M. P., van Vloten-Doting, L., and Bol, J. F., 1988, Expression of alfalfa mosaic virus cDNA1 and 2 into transgenic tobacco plants, *Virology* **163:**572–578.

Vardi, E., Sela, I., Edelbaum, O., Livneh, O., Kuznetsova, L., and Stram, Y., 1993, Plants transformed with a cistron of a potato virus Y protease (NIa) are resistant to virus infection, *Proc Natl. Acad. Sci. USA* **90:**7513–7517.

Verchot, J., Koonin, E. V., and Carrington, J. C., 1991, The 35-kDa protein from the N-terminus of the potyviral polyprotein functions as a third virus-encoded proteinase, *Virology* **185:**527–535.

Voss, A., Niersbach, M., Hain, R., Hirsch, H. J., Liao, Y. C., Kreuzaler, F., and Fischer, R., 1995, Reduced virus infectivity in *N. tabacum* secreting a TMV-specific full-size antibody, *Mol. Breeding* **1:**39–50.

Watanabe, Y., Ogawa, T., Takahashi, H., Ishida, I., Takeuchi, Y., Yamamoto, M., and Okada, Y., 1995, Resistance against multiple plant viruses in plants mediated by double stranded-RNA specific ribonuclease, *FEBS Lett.* **372:**165–168.

Wegener, D., Steinecke, P., Herget, T., Petereit, I., Philip, C., and Schreier, P. H., 1994, Expression of a reporter gene is reduced by a ribozyme in transgenic plants, *Mol. Gen. Genet.* **245:**465–470.

Wellink, J., and van Kammen, A., 1989, Cell-to-cell transport of cowpea mosaic virus requires both the 58K/48K proteins and the capsid proteins, *J. Gen. Virol.* **70:**2279–2286.

Whitham, S., Dinesh-Kumar, S. P., Choi, D., Hehl, R., Corr, C., and Baker, B., 1994, The product of the tobacco mosaic virus resistance gene *N:* Similarity to Toll and the interleukin-1 receptor, *Cell* **78:**1101–1115.

Wilson, T. M. A., 1984, Cotranslational disassembly of tobacco mosaic virus *in vitro*, *Virology* **137:**255–265.

Wilson, T. M. A., 1993, Strategies to protect crop plants against viruses: Pathogen-derived resistance blossoms, *Proc. Natl. Acad. Sci. USA* **90:**3134–3141.

Wisniewski, L. A., Powell, P. A., Nelson, R. S., and Beachy, R. N., 1990, Local and systemic spread of tobacco mosaic virus in transgenic tobacco, *Plant Cell* **2:**559–567.

Wolf, S., and Lucas, W. J., 1994, Virus movement proteins and other molecular probes of plasmodesmal function, *Plant Cell Environ.* **17:**573–585.

Wolf, S., Deom, C. M., Beachy, R. N., and Lucas, W. J., 1989, Movement protein of tobacco mosaic virus modifies plasmodesmatal size exclusion limit, *Science* **246:**377–379.

Yie, Y., Zhao, F., Zhao, S. Z., Liu, Y. Z., Liu, Y. L., and Tien, P., 1992, High resistance to cucumber mosaic virus conferred by satellite RNA and coat protein in transgenic commercial tobacco cultivar G-140, *Mol. Plant–Microbe Interact.* **5:**460–465.

Yoshioka, K., Hanada, K., Harada, T., Minobe, Y., and Oosawa, K., 1993, Virus resistance in transgenic melon plants that express the cucumber mosaic virus coat protein gene and in their progeny, *Jpn. J. Breeding* **43:**629–634.

Zaccomer, B., Cellier, F., Boyer, J.-C., Haenni, A.-L., and Tepfer, M., 1993, Transgenic plants that express genes including the 3′ untranslated region of the turnip yellow mosaic virus (TYMV) genome are partially protected against TYMV infection, *Gene* **136:**87–94.

Zaitlin, M., Anderson, J. M., Perry, K. L., Zhang, P., and Palukaitis, P., 1994, Specificity of replicase-mediated resistance to cucumber mosaic virus, *Virology* **201**:200–205.

Zhao, J. J., and Pick, L., 1993, Generating loss-of-function phenotypes of the *fushi tarazu* gene with a targeted ribozyme in *Drosophila*, *Nature* (London) **365**:448–451.

Ziegler-Graff, V., Guilford, P. J., and Baulcombe, D. C., 1991, Tobacco rattle virus RNA-1 29K gene product potentiates viral movement and also affect symptoms induction in tobacco, *Virology* **182**:145–155.

*Chapter 10*

# Biosynthesis and Regulation of Coronatine, a Non-Host-Specific Phytotoxin Produced by *Pseudomonas syringae*

Carol L. Bender, David A. Palmer, Alejandro Peñaloza-Vázquez, Vidhya Rangaswamy, and Matthias Ullrich

## 1. INTRODUCTION

### 1.1. *Pseudomonas syringae:* Biology and Pathogenicity

*Pseudomonas syringae* induces a wide variety of symptoms on plants, including blights (rapid death of tissue), leaf spots, and galls. The species *P. syringae* is subdivided into pathogenic variants (pathovars; pvs.) which vary in host range. Two distinct reactions are possible when *P. syringae* cells are infiltrated into plant tissue. One potential outcome is a compatible, susceptible interaction

**Carol L. Bender, David A. Palmer, Alejandro Peñaloza-Vázquez, Vidhya Rangaswamy, and Matthias Ullrich** Department of Plant Pathology, 110 Noble Research Center, Oklahoma State University, Stillwater, Oklahoma 74078-3032. *Present address for D.A.P.:* Zeneca Ag Products, Eastern Regional Technical Center, Whitakers, North Carolina 27891. *Present address for M.U.:* Max-Planck-Insitut für Terrestrische Mikrobiologie, D-35043 Marburg, Germany.

*Subcellular Biochemistry, Volume 29: Plant–Microbe Interactions*, edited by Biswas and Das. Plenum Press, New York, 1998.

which is characterized by a symptom called "water-soaking," a reaction which is followed by pathogen proliferation and advanced symptom development. In contrast, resistant host cells undergo a reaction known as the hypersensitive response (HR) and become necrotic 12–24 hr after inoculation; this severely restricts multiplication of the pathogen and results in host resistance. A cluster of genes termed the *hrp* region (for *h*ypersensitive *r*esponse and *p*athogenicity) is conserved in phytopathogenic prokaryotes and controls the ability of a bacterium to induce an HR in a resistant host and pathogenicity in a susceptible host (Willis *et al.*, 1991). The *hrp* genes are now known to encode genes for the regulation and biosynthesis of a type III secretion pathway that is used by both plant and animal pathogens to secrete virulence proteins (Salmond, 1994).

## 1.2. Phytotoxin Production in *P. syringae*

It is important to note that the *hrp* gene cluster is required for pathogenicity of *P. syringae* on plant hosts. However, in addition to the *hrp* genes, phytopathogenic pseudomonads encode gene products that significantly enhance pathogen virulence, including extracellular polysaccharides, phytotoxins, and cell wall-degrading enzymes (Denny, 1995; Boucher *et al.*, 1992, Gross, 1991). Phytotoxins are important virulence factors in many diseases incited by *P. syringae*. In a general sense, a phytotoxin is a product of a pathogen or host–pathogen interaction that directly injures living host protoplasts and influences the course of disease development or symptomology. Both fungal and bacterial pathogens produce a number of secondary metabolites which are toxic to plant cells. However, this does not necessarily mean that they are important in plant disease. Consequently, phytopathologists have developed criteria for assessing the involvement of toxins in plant disease. Factors which suggest that a toxic metabolite has a role in disease include the following: (1) reproduction of disease symptoms with the purified toxin; (2) a correlation between yield of toxin and pathogenicity; (3) production of the toxin during active growth of the pathogen *in planta;* and (4) reduced virulence or lack of virulence in nontoxigenic strains. Phytotoxins may be host-specific and exhibit the same specificity as the producing pathogen, or non-host-specific with a wider host range of activity than the producing pathogen. Most toxins produced by *P. syringae* are non-host-specific and cause symptoms on many plants which cannot be infected by the toxin-producing pathogen.

Although phytotoxins are not required for pathogenicity in *P. syringae*, they generally function as virulence factors in this pathogen, and their production results in increased disease severity. Several reports have indicated that *P. syringae* phytotoxins contribute to pathogen virulence. For example, phaseolotoxin-deficient mutants of *P. syringae* pv. *phaseolicola* were unable to move systemically in bean (Patil *et al.*, 1974). Also, lesions produced by a syringomycin-deficient mutant of *P. syringae* pv. *syringae* were significantly smaller than those produced by the wild-type toxigenic strain (Xu and Gross, 1988a). Gnanamanickman *et al.* (1982) found that nontoxigenic strains of *P. syringae* pv. *glycinea* did not induce stunting, systemic chlorosis, and leaf elongation, symptoms which are associated with production of the phytotoxin coronatine by this bacterium. Furthermore, coronatine-defective mutants of *P. syringae* pv. *tomato* produced lesions significantly smaller than the coronatine-producing strain and were unable to achieve the population levels attained by the toxigenic wild type (Bender *et al.*, 1987). Therefore, we concluded that coronatine synthesis is important in the virulence of *P. syringae* pv. *tomato* and contributes significantly to both lesion expansion and multiplication of the bacterium in tomato leaves (Bender *et al.*, 1987).

### 1.2.1. Mode of Action

Although several *P. syringae* phytotoxins cause similar symptoms (necrosis or chlorosis), they have completely different modes of action. For example, the *P. syringae* phytotoxins phaseolotoxin, tagetitoxin, and tabtoxin inhibit ornithine carbamoyl transferase, RNA polymerase, and glutamine synthetase, respectively (Thomas *et al.*, 1983; Matthews and Durbin, 1990; Patil *et al.*, 1970). The phytotoxin syringomycin is known to form transmembrane pores which alter ion flux in plasma membranes of both plants and fungi (Hutchison *et al.*, 1995). Because of their antimicrobial activity, phaseolotoxin, tabtoxin, and syringomycin can be detected in assays utilizing toxin-sensitive bacteria or fungi (Gasson, 1980; Staskawicz and Panopoulos, 1979; Gross and DeVay, 1977). In contrast, the phytotoxin coronatine is not antimicrobial and instead possesses biological activity directed primarily at plant cells.

### 1.2.2. Biosynthetic Origins

The toxins produced by *P. syringae* are varied in origin and include monocyclic beta-lactam (tabtoxin), sulfodiaminophosphinyl peptide (phaseolotoxin), peptidolactone (syringomycins), hemithioketal (tagetitoxin), and polyketide (coronatine) structures (Gross, 1991). It has been difficult to obtain information on biosynthetic intermediates in some of these pathways. One major reason for the lack of characterized intermediates in these diverse pathways is that synthesis may be catalyzed by multifunctional proteins or polypeptide complexes where intermediates are transferred between enzymatic domains and not released into the cytoplasm. Alternatively, conversion of intermediates to the final product may occur very rapidly and impede qualitative analyses. Recent analysis of genes required for syringomycin production indicate that biosynthesis proceeds via a nonribosomal, thiotemplate mechanism which is commonly utilized for peptide antibodies (J.-H. Zhang *et al.*, 1995). A similar mechanism is likely to be required for biosynthesis of phaseolotoxin, since this molecule contains a tripeptide moiety consisting of ornithine, alanine, and homoarginine.

### 1.3. Genetics of Phytotoxin Production

Progress has been made on the cloning of genes required for the synthesis of *P. syringae* phytotoxins. Mutants defective in phaseolotoxin synthesis have been generated, and genes required for production of this toxin have been isolated (Kamdar *et al.*, 1991; Y. Zhang *et al.*, 1993; Peet *et al.*, 1986), but specific steps in the phaseolotoxin biosynthetic pathway have not been identified or characterized. The cloning of the complete biosynthetic cluster for tabtoxin synthesis was recently reported by Kinscherf *et al.* (1991). A subsequent study showed that *tblA*, which mapped within the tabtoxin biosynthetic cluster, was required for tabtoxin biosynthesis (Barta *et al.*, 1992). The nucleotide sequence of *tblA* did not reveal its function; however, similarity between *tblA* and *tabA* was noted (Barta *et al.*, 1993). The *tabA* gene was sequenced in another tabtoxin-producing strain of *P. s. tabaci* and was shown to be required for tabtoxin production, but a function for *tabA* was not demonstrated (Engst and Shaw, 1992).

Biosynthesis of the syringomycins is thought to occur via a multienzyme system (Gross, 1991) and several large proteins (100–500 kDa) have been correlated with the production of this class of phytotoxin (Mo and Gross, 1991a; Morgan and Chatterjee, 1988; Xu and Gross, 1988b). Several genes which are essential for syringomycin biosynthesis have been identified, and nucleotide sequencing has provided some hints regarding their possible function. When translated, the *syrB* gene shows relatedness to adenylate-forming enzymes involved in peptide antibiotic and siderophore synthesis (J.-H. Zhang *et al.*, 1995). *syrC*, a gene adjacent to *syrB*, shows similarity to thioesterases, which further substantiates the role of a thiotemplate mechanism in syringomycin biosynthesis. *syrD* showed relatedness to ATP-binding secretion proteins, and the implication is that SyrD functions as an ATP-driven efflux pump for the secretion of syringomycin (Quigley *et al.*, 1993).

### 1.4. Regulation of Phytotoxin Production

Some progress has been made on elucidating factors which regulate phytotoxin biosynthesis in *P. syringae*. Barta *et al.* (1992) showed that *tblA*, a gene required for tabtoxin synthesis, is regulated by the *lemA* (*le*sion *m*anifestation) gene in *P. syringae* pv. *coronafaciens*, which indicates the importance of *lemA* in modulating tabtoxin biosynthesis in this pathovar. The *lemA* locus was shown to be conserved among *P. syringae* pathovars and contained conserved domains found in two-component regulatory proteins (Hrabak and Willis, 1992; Rich *et al.*, 1992). Interestingly, *lemA* was also shown to be required for syringomycin production in *P. syringae* pv. *syringae*, but was not required for lesion formation or phaseolotoxin biosynthesis in *P. syringae* pv. *phaseolicola* (Hrabak and Willis, 1993; Rich *et al.*, 1992). Apparently, *lemA* has differential effects in *P.*

*syringae* pathovars which may have developed during divergence and adaptation to specific hosts (Rich *et al.*, 1992).

Phaseolotoxin production in *P. syringae* pv. *phaseolicola* is regulated by temperature. Goss (1940) showed that the chlorosis associated with phaseolotoxin production occurred at cooler temperatures (18–20°C), but was absent at warmer temperatures (28–32°C). Rowley *et al.* (1993) recently described the production of a "repressor-like protein" when *P. syringae* pv. *phaseolicola* was grown at 28°C, a temperature unfavorable for phaseolotoxin synthesis. Production of this putative repressor at the nonpermissive temperature may explain thermoregulation of phaseolotoxin biosynthesis.

The effect of different nutritional parameters on syringomycin production was previously investigated; syringomycin synthesis was repressed at phosphate concentrations higher than 1 mM and positively regulated by the addition of iron *in vitro* (Gross, 1985). A *lacZ* fusion was used to show that the *syrB* locus is strongly induced by the plant phenolics arbutin and salicin (Mo and Gross, 1991b). These results may have a broader significance, since other investigators have shown that genes required for pathogenicity (*hrp* loci) and host recognition (*avr* genes) in *P. syringae* are also modulated by environmental and host factors (Rahme *et al.*, 1992; Xiao *et al.*, 1992; Huynh *et al.*, 1989).

## 1.5. Transgenic Plants with Resistance to Phytotoxins

Phaseolotoxin has been shown to competitively inhibit ornithine carbamoyl transferase (OCTase), a critical enzyme in the urea cycle. Inhibition of OCTase causes an accumulation of ornithine and a deficiency in intracellular pools of arginine, leading to chlorosis. Since the urea cycle is important in both prokaryotes and eukaryotes, one wonders how the producing organism, *P. syringae* pv. *phaseolicola*, deals with phaseolotoxin. *P. syringae* pv. *phaseolicola* actually synthesizes two forms of OCTase; one form is resistant to phaseolotoxin (ROCTase) and the second form is sensitive (SOCTase). During conditions favorable to toxin production, *P. syringae* pv. *phaseolicola* synthesizes the ROCTase isomer. The gene encoding the ROCTase (*argK*) has been cloned and sequenced (Hatziloukas and Panopoulos, 1992; Mosqueda *et al.*, 1990). Two laboratories (de la Fuente-Martinez *et al.*, 1992; Hatziloukas and Panopoulos, 1992) have utilized the ROCTase gene as a source of resistance to engineer phaseolotoxin-resistant tobacco plants. In plant cells, OCTase is produced in the chloroplast. This was done by fusing the *argK* coding region to the transit peptide of the small subunit of ribulose bisphosphate carboxylase. This construct was cloned into a plant transformation vector, introduced into *Agrobacterium tumefaciens*, and then introduced into tobacco using the leaf disk transformation method. Rooted plants were then transplanted to soil and screened for OCTase activity and phaseolotoxin sensitivity. Control tobacco plants which contained SOCTase showed chlorosis in response to phaseolotoxin treatment and had a reduced chlorophyll content. Transgenic plants with the ROCTase (*argK*) did not turn chlorotic when treated with phaseolotoxin and had no change in chlorophyll content. Control tobacco plants were systemically infected when inoculated with *P. syringae* pv. *phaseolicola,* but the transgenics expressing ROCTase showed a localized HR in response to infection. Therefore, a level of host resistance was expressed in the transgenics to both the pathogen and toxin. A current priority is to produce transgenic bean plants with resistance to phaseolotoxin, since bean is the natural host of *P. syringae* pv. *phaseolicola*. This will present a greater challenge since bean is more difficult to transform than tobacco.

## 2. THE PHYTOTOXIN CORONATINE

### 2.1. Mode of Action for Coronatine

The structure of coronatine (Fig. 1A) is unusual and consists of two distinct parts: coronafacic acid (CFA; Fig. 1C) is of polyketide origin, whereas coronamic acid (CMA: Fig. 1B) is an ethylcyclopropyl amino acid derived from isoleucine (Parry *et al.*, 1994; Mitchell, 1985; Ichihara *et al.*, 1977). In addition to eliciting chlorosis, COR is known to induce hypertrophy, inhibit root elongation, and stimulate ethylene biosynthesis (Feys *et al.*, 1994; Kenyon and Turner, 1992; Bender *et al.*, 1987; Sakai *et al.*, 1979). Recently, several reports have noted the striking structural and functional homologies between COR and jas-

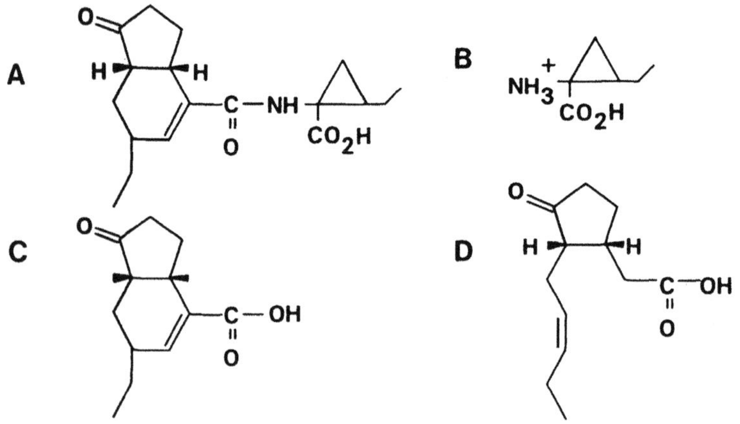


**FIGURE 1.** Structure of coronatine (A), coronamic acid (B), coronafacic acid (C), and jasmonate (D).

monic acid (Fig. 1D), a compound which is produced in a variety of plants in response to biological stress. The relatedness of these two molecules had led several investigators to suggest that COR may function as a molecular mimic of the octadecanoid signaling molecules of higher plants (Koda *et al.*, 1996; Feys *et al.*, 1994; Weiler *et al.*, 1994). Recently, Feys *et al.* (1994) generated a coronatine-insensitive (*coi1*) mutant of *Arabidopsis thaliana* which was insensitive to the effects of both COR and methyl jasmonate, suggesting a similar mode of action.

We became interested in comparing the effects of COR, CFA, and methyl jasmonate in tomato tissue (Palmer and Bender, 1995). When light microscopy was used to compare healthy and coronatine-treated tomato tissue, several changes became evident in tissue exposed to coronatine. The epidermal wall was significantly thicker in coronatine-treated tissue and there were noticeable differences in the size and shape of chloroplasts. In coronatine-treated tissue, chloroplasts were significantly smaller and stained more intensely (Palmer and Bender, 1995). One of the most pronounced differences was the appearance of spherical and cubical proteinaceous structures in the vacuole of tomato tissue treated with coronatine. These structures were markedly similar to the proteinase inhibitors which had been previously reported in plant tissues exposed to various biological stresses (Akers and Hoff, 1980; Shumway *et al.*, 1970, 1976). Furthermore, the presence of proteinase inhibitors in the coronatine-treated tissue was confirmed by finding that this tissue significantly inhibited the activity of both chymopapain and chymotrypsin (Palmer and Bender, 1995). More recently, we obtained antisera to both chymopapain inhibitor (Bolter, 1993) and chymotrypsin inhibitor (Ryan, 1967) and used these antibodies to positively confirm the identity of the proteinaceous structures in coronatine-treated tomato tissue. When coronatine-treated tissue was incubated with antisera to chymopapain inhibitor and then with a secondary antibody conjugated to gold, we saw very high specific labeling of the cubical crystals with gold particles, indicating that these structures were chymopapain inhibitor (D. A. Palmer and C. L. Bender, unpublished data). A similar experiment indicated that the spherical crystals were chymotrypsin inhibitor; thus we concluded that both chymopapain and chymotrypsin inhibitor are specifically induced in response to coronatine in tomato tissue.

We found that COR, CFA, and methyl jasmonate all induced proteinase inhibitors in tomato tissue. However, only coronatine caused cell wall thickening, changes in chloroplast structure, and chlorosis. CFA and methyl jasmonate did not induce these changes in tomato tissue (Palmer and Bender, 1995). Consequently, we suspect that the CMA moiety, or perhaps the amide linkage between CFA and CMA, may impart additional biological activities to COR in tomato. Therefore we have concluded that COR does not function solely as a molecular mimic of methyl jasmonate in tomato tissue (Palmer and Bender, 1995). Consequently, the precise mode of action for coronatine may remain unclear until a receptor for the toxin is located in the plant.

## 2.2. Biosynthetic Pathway to Coronatine

Parry *et al.* (1994) initially reported that the CFA portion of coronatine is a polyketide derived from three acetate units, one butyrate unit, and one unit of pyruvate. Polyketide synthesis is biochemically very similar to fatty acid biosynthesis; the latter begins with the condensation of acetyl-CoA (starter unit) and malonyl-CoA (chain extender unit) and then proceeds with a cycle of modifications on the carbonyl group of malonyl-CoA (i.e., reduction to a hydroxyl, dehydration to produce a double bond, and further reduction to form a saturated fatty acid). Unlike fatty acid synthesis, polyketide synthetases (PKSs) can make additional choices in the starter and extender groups, and the products vary in the extent of reduction. Recent studies have suggested that the pyruvate used for CFA biosynthesis is actually converted into α-ketoglutarate before incorporation to CFA, and that α-ketoglutarate may serve as the starter unit for CFA assembly (Parry *et al.*, 1996). It is important to mention that PKSs are generally classified as type I or II systems and consist of protein complexes that act on covalently bound substrates which are attached as thioesters to an acyl carrier protein (ACP) (Katz and Donadio, 1993). Type I polyketides usually have a fairly reduced structure and are synthesized by large multifunctional proteins which consist of individual domains which catalyze specific and discrete reactions in a nonreiterative fashion (Hutchinson and Fujii, 1995). The functional activities catalyzed by the domains within the type I PKS are often apparent in the structure of the growing polyketide chain. Conversely, type II PKSs are most often associated with synthesis of aromatic polyketides, and biosynthesis occurs on monofunctional proteins which associate in a complex. Unlike the type I system, the type II PKS may utilize one or more enzymes in a reiterative fashion.

The biosynthetic origins of COR are truly "mixed" since only the CFA portion is derived via polyketide synthesis. The coronamic acid portion of COR originates from isoleucine via alloisoleucine and is cyclized by an unknown mechanism (Mitchell, 1985; Parry *et al.*, 1991). We have shown that both CFA and CMA function as distinct intermediates and are secreted by COR-producing strains at low levels (Mitchell *et al.*, 1994; Young *et al.*, 1992). The final step in the pathway to COR is presumed to be the coupling (CPL) or ligation of CFA and CMA via amide bond formation (Fig. 2).

## 2.3. Coronatine Biosynthesis Is Generally Plasmid Encoded

Coronatine production has been demonstrated in five pathovars of *P. syringae*, including *atropurpurea*, *glycinea*, *maculicola*, *morsprunorum*, and *tomato*, which infect ryegrass, soybean, crucifers, *Prunus* spp., and tomato, respectively. In several host–pathogen interactions, COR was shown to have a distinct role in virulence (Mittal and Davis, 1995; Bender *et al.*, 1987; Sato *et*

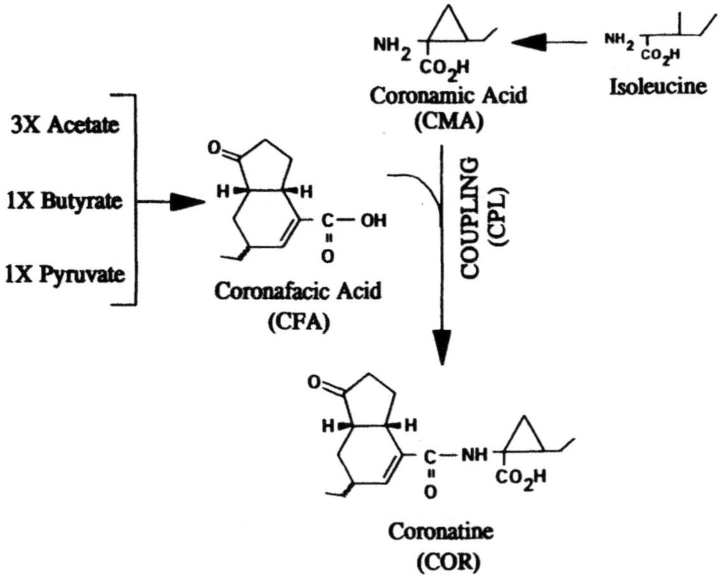


**FIGURE 2.** Biochemical pathways involved in the synthesis of coronatine (COR). COR consists of a polyketide component (CFA) which is coupled (CPL) via amide bond formation to an ethylcyclopropyl amino acid (CMA).

*al.*, 1983). However, it is interesting to note that production of COR is dispensable in many pathovars, since pathogenic strains of pvs. *glycinea, maculicola, morsprunorum,* and *tomato* have been isolated which no longer produce the toxin (Ullrich *et al.*, 1993; Mitchell *et al.*, 1983; Mitchell, 1982). Part of this variability in COR production might be caused by the occurrence of the COR biosynthetic cluster on large transmissible plasmids (Cuppels and Ainsworth, 1995; Bender *et al.*, 1991; Sato, 1988). Several reports have shown that the COR biosynthetic cluster occurs on indigenous plasmids in pvs. *atropurpurea, glycinea, morsprunorum,* and *tomato* (Liang *et al.*, 1994; Ullrich *et al.*, 1993; Bender *et al.*, 1991). The tendency of the COR biosynthetic cluster to occur as a plasmid-encoded trait leads to the speculation that these genes have been horizontally transferred by plasmids in *P. syringae* (Bender *et al.*, 1991).

## 2.4. Detection of Coronatine

Unlike phaseolotoxin, tabtoxin, and syringomycin, coronatine is not antimicrobial and cannot be detected in bioassays which utilize other bacteria or fungi. Although the toxin can be readily detected by its ability to induce chlorosis in a variety of plants, this assay is qualitative rather than quantitative. Völksch *et al.* (1989) described a semiquantitative assay in which a hypertrophic reaction on potato tissue is utilized to detect coronatine. Although this assay is sensitive, some variability can result depending on the cultivar utilized and the age of the potato tissue.

When derivatized to a methyl ester, coronatine can also be detected by gas chromatography (GC) (Bender *et al.*, 1989); this method is very sensitive and quantitative, but labor-intensive and slow when compared to biological assays. To facilitate quantitative detection of coronatine, we developed an improved extraction and fractionation method which utilizes high-performance (HPLC) liquid chromatography rather than GC. This scaled-down method involves the direct extraction of organic acids from 0.5 ml of culture supernatant, a 9-min fractionation on a reverse-phase $C_8$ column in a gradient of acetonitrile and water, and quantitative detection of coronafacoyl compounds at 208 nm (Palmer and Bender, 1993). The extraction can be performed in microfuge tubes, takes approximately 3 min to complete, and accurately separates and quantifies coronafacic acid, coronatine, coronafacoylvaline, and coronafacoylisoleucine (the latter two compounds contain CFA linked to valine and isoleucine, respectively).

## 2.5. Conservation of Coronatine Synthesis Genes in *P. syringae*

Our original goal was to characterize the COR biosynthetic cluster in *P. syringae* pv. *tomato* PT23.2 since $COR^-$ mutants were available, and the genes encoding coronatine were shown to be encoded by a 101-kb plasmid designated pPT23A (Bender *et al.*, 1987, 1989). However, several problems hindered the genetic analysis of COR production in *P. syringae* pv. *tomato* PT23.2: (1) this strain was recalcitrant to genetic manipulation, and neither conjugation nor electroporation was consistently reliable for the introduction of cloned DNA; (2) pPT23A was not stably maintained in PT23.2, which made it difficult to obtain mutants by homologous recombination; and (3) the low yields of coronatine produced by PT23.2 *in vitro* (0.3 mg/L) made it necessary to scale-up culture volumes for extraction and analysis by HPLC, thus increasing the time and expense involved in each analysis.

Because of these difficulties, we investigated whether other coronatine-producing ($COR^+$) strains of *P. syringae* would be more amenable to genetic studies. Four different $COR^+$ strains of *P. syringae* were evaluated: pv. *glycinea* PG4180, pv. *atropurpurea* PA1304, pv. *morsprunorum* PM3714, and pv. *morsprunorum* PM567 (Bender *et al.*, 1991). The results of this study indicated that the genes controlling COR biosynthesis in these strains were also encoded by large (90–105 kb) indigenous plasmids (Bender *et al.*, 1991). We resumed our genetic analysis of COR production in *P. syringae* pv. *glycinea* PG4180 because this strain is easy to manipulate genetically, consistently synthesizes large amounts of coronatine *in vitro* (20–40 mg/L), and infects soybean, a host which is easy to cultivate.

## 2.6. Coronatine Synthesis in *P. syringae* pv. *glycinea* PG4180

Tn*5* mutagenesis suggested that the COR biosynthesis genes in *P. syringae* pv. *glycinea* PG4180 reside on a 90-kb plasmid designated p4180A (Bender *et al.*, 1991). The involvement of p4180A in COR production was demonstrated by transforming this plasmid into two nonproducers of coronatine, *P. syringae* pv. *syringae* PS51 and PS61. Organic acids were then extracted from PS51 and PS61 transformants containing p4180A and analyzed by HPLC and combined gas chromatography/mass spectrometry. PS51 and PS61 transformants containing p4180A produced coronafacic acid and coronatine, providing unequivocal evidence for the role of p4180A in the synthesis of coronatine (Bender *et al.*, 1993).

We have used various approaches to characterize the COR biosynthetic cluster encoded by p4180A, including (1) saturation Tn*5* mutagenesis, (2) feeding studies using exogenously supplied CFA and CMA, (3) complementation of selected mutants with cloned DNA, (4) expression of selected regions of the COR gene cluster in COR-nonproducers, and (5) nucleotide sequence analysis (Liyanage *et al.*, 1995b; Ullrich and Bender, 1994; Ullrich *et al.*, 1994; Bender *et al.*, 1993; Young *et al.*, 1992). These approaches facilitated the construction of a functional map of the COR biosynthetic region (Fig. 3). Saturation Tn*5* mutagenesis indicated that a 32-kb contiguous region was absolutely required for COR biosynthesis (Ullrich *et al.*, 1994; Bender *et al.*, 1993; Young *et al.*, 1992). Tn*5* mutagenesis was especially helpful in defining the borders of the gene cluster; the left and right borders were defined by the mutations designated N53 and G4, respectively. The insertions at these positions and to the left and right, respectively, had no effect on COR production and resulted in the wild-type phenotype (Bender *et al.*, 1993; Ullrich *et al.*, 1994). Two regions in the COR biosynthetic cluster contained structural genes for CMA and CFA biosynthesis; these were separated by a 3.4-kb regulatory region (REG, Fig. 3). The functional area designated CPL was required for the coupling of CFA and CMA via amide bond formation (Bender *et al.*, 1993).

Ullrich and Bender (1994) showed that a 6.9-kb region contained the CMA biosynthetic gene cluster; the nucleotide sequence of this DNA revealed the presence of three genes designated *cmaA* (2.7 kb), *cmaT* (1.2 kb), and *cmaU* (0.9 kb) (Fig. 3). The deduced amino acid sequence of *cmaA* indicates that the enzyme contains an amino acid-activating domain and a putative iron-binding region; the latter is significant because of its conservation in the active site of certain enzymes which catalyze oxidative cyclizations. Furthermore, the deduced amino acid sequence of *cmaT* suggests that it functions as a thioesterase,

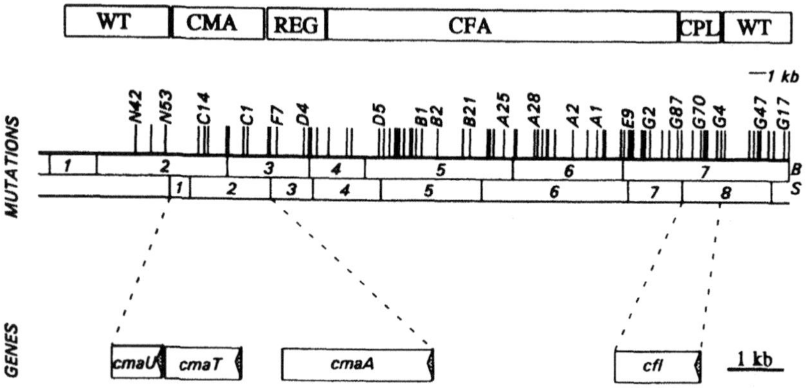


**FIGURE 3.** Functional and physical map of the COR biosynthetic gene cluster. Vertical lines with letter/number designations represent Tn*5* insertions; mutations in the regions designated wild type (WT) had no effect on COR production. Enzymes used for restriction mapping were *Bam*HI and *Sst* I. Functional regions of the COR biosynthetic cluster are shown in boxed areas; CMA, coronamic acid biosynthetic gene cluster; REG, regulatory region; CFA, coronafacic acid biosynthetic gene cluster; CPL, region required for coupling of CFA and CMA. The location of genes with a role in CMA biosynthesis (*cmaA*, *cmaT*, and *cmaU*) and the ligation of CFA and CMA (*cfl*) are indicated.

providing further support to the role of a thiotemplate mechanism for CMA biosynthesis (Ullrich and Bender, 1994).

The region designated CPL was shown to be required for the coupling of CFA and CMA via amide bond formation (Bender *et al.*, 1993). Liyanage *et al.* (1995b) sequenced this region and identified a 1.4-kb gene designated *cfl* (*co*ronafacate *l*igase; Fig. 3). Coronafacate ligase is most closely related to enzymes which activate carboxylic acids by adenylation, and we suspect that this enzyme adenylates CFA and the CFA-adenylate is then ligated to CMA. Coronafacate ligase has now been overproduced in *Escherichia coli* in a soluble form, and assays are currently underway to demonstrate its activity *in vitro*.

Complementation analyses, nucleotide sequencing, and exogenous feeding studies using blocked mutants (Bender *et al.*, 1993; Penfold *et al.*, 1996) have been used to define the location of the CFA biosynthetic gene cluster (Fig. 3). In *Sst*I fragment #7, Penfold *et al.* (1996) identified three open-reading frames (ORFs) which showed relatedness to genes involved in the synthesis of polyketide compounds. ORFs were identified with relatedness to acyl carrier proteins (ACP), acetyl dehydratases (AD), and β-keto synthases (KS). ACP and KS have structural roles in the assembly of precursors into the nascent polyketide, whereas AD is involved in the modification (dehydration) of keto groups. These homologies are consistent with a role for this region in the synthesis of CFA, which is a product of the polyketide pathway.

## 2.7. Regulation of Coronatine Biosynthesis

Since coronatine is derived from two distinct pathways (isoleucine and polyketide), it is likely that regulatory controls exist for these two pathway components (CFA and CMA). Environmental and nutritional factors were previously examined for their effects on coronatine production in *P. syringae* pv. *glycinea* PG4180 by varying the components of a defined minimal medium (Palmer and Bender, 1993). In this study, bacterial growth was recorded as dry weight, and coronatine production was measured by HPLC. In all experiments, coronatine production was reported in mg/L and reported as a function of dry weight to adjust for differences in bacterial growth.

Temperature had profound effects on coronatine biosynthesis with maximal production at 18°C and negligible yields at 30°C (Palmer and Bender, 1993). Bacterial growth was relatively unaffected over the range of temperatures tested (14–30°C). This response to temperature is consistent with symptom development in the field, since *P. syringae* pv. *glycinea* is predominantly a cool-weather pathogen. CFA and CMA were also subject to the same pattern of temperature control, with optimal production at 18°C (Palmer, 1995; Ullrich and Bender, 1994).

Recent data indicate that the production of both CFA and CMA is regulated at the transcriptional level by temperature. Transcriptional fusions of the CFA and CMA promoter regions to a promoterless glucuronidase gene indicated that transcriptional activity in both biosynthetic gene clusters was maximal at 18°C and fourfold less at 28°C (Liyanage *et al.*, 1995a; Ullrich and Bender, 1994). The higher level of transcriptional activity for the CMA and CFA biosynthetic promoters at 18°C helps explain why COR production is optimal at this temperature.

Several mutants (F7, P1, and D4 in Fig. 3) were isolated which did not produce CFA or CMA. Furthermore, all three mutants were defective in transcriptional activation of the CMA and CFA biosynthetic promoters, suggesting that these mutants were defective in regulatory genes which control both CMA and CFA biosynthesis. COR production was restored to F7, P1, and D4 by pMUH34, a clone containing a 3.4-kb *Hin*dIII/*Eco*RI fragment (Ullrich *et al.*, 1995). Furthermore, pMUH34 restored transcriptional activation and thermoregulation of the CMA and CFA promoters to F7, P1, and D4. The nucleotide sequence of pMUH34 revealed the presence of three genes, *corP*, *corS*, and *corR* (Ullrich *et al.*, 1995). The deduced amino acid sequence of *corP* and *corR* indicated relatedness to response regulators which function as members of two-component regulatory systems, whereas *corS* shows similarity to histidine protein kinases, which function as environmental sensors (Parkinson and Kofoid, 1992). The COR regulatory system is modified from the two-component paradigm since it contains two response regulator proteins together with a single sensor protein (Ullrich *et al.*, 1995).

### 2.8. Coronatine: Conclusions and Perspectives

We have shown that the production of coronatine requires two distinct biosynthetic gene clusters. Precursor feeding experiments and nucleotide sequence analyses indicate that the biosynthetic route to CFA requires the polyketide pathway. Schemes for the assembly of CFA from the precursor units remain speculative, but nucleotide sequencing (Penfold *et al.*, 1996) and the identification of a potential intermediate to CFA (Mitchell *et al.*, 1995) have provided new clues. Nucleotide sequence analysis of the CMA biosynthetic gene cluster has indicated that CMA assembly involves the thiotemplate mechanism utilized by nonribosomal peptide synthetases. Furthermore, the pathway to CMA involves an oxidative cyclization that represents a new route for the biosynthesis of cyclopropanes.

Perhaps one of the most novel evolutionary events in COR biosynthesis is the joining of the CFA and CMA pathways by a centrally located "modified" two-component regulatory region which is temperature-responsive (Ullrich *et al.*, 1995). The widespread documentation of two-components systems and the variety of traits regulated by these proteins in bacteria is astonishing. Despite the explosion of information on two-component systems in the past decade, many issues remain unresolved, such as the nature of the factor "sensed" (if any), how two-component systems have evolved, and mechanistic aspects of the phosphotransfer and "phosphatase" reactions (Magasanik, 1995; Parkinson, 1995). Although numerous two-component systems have been described in phytopathogenic bacteria (Grimm *et al.*, 1995; H.-C. Huang *et al.*, 1995; Willis *et al.*, 1994; Mills *et al.*, 1993; Charles *et al.*, 1992; Leigh and Coplin, 1992), only the VirA/VirG system of *Agrobacterium tumefaciens* has been investigated in depth.

Although we have demonstrated that the modified two-component system in the COR gene cluster responds to temperature, it is not yet clear whether temperature directly controls the functional activities of the regulatory proteins or if the temperature signal is transduced by a higher tier of regulatory proteins. The role of temperature in the expression of virulence genes in human and animal pathogens is well established (Mekalanos, 1992). Transcription of these genes is generally activated when microbes enter their warm-blooded hosts, and some of the mechanisms used to sense increased temperature have been documented (Dorman, 1995; Tobe *et al.*, 1995; Uhl and Miller, 1995; Hoe and Goguen, 1993). Although thermoregulated expression of virulence genes in plant-associated bacteria has been reported, the mechanisms responsible for the temperature response remain obscure (Hugovieux-Cotte-Pattat *et al.*, 1992; Lanham *et al.*, 1991). Pathogens like *Pseudomonas syringae* are subject to extreme temperature fluctuations in nature, and some temperatures are more conducive to disease development than others. Consequently, the thermoregulation of virulence genes in *P. syringae* could have broad significance.

## 3. SUMMARY

Many *P. syringae* pathovars are known to produce low-molecular-weight, diffusible toxins in infected host plants. These phytotoxins reproduce some of the symptoms of the relevant bacterial disease and are effective at very low concentrations. Phytotoxins generally enhance the virulence of the *P. syringae* pathovar which produces them, but are not required for pathogenesis.

Genes encoding phytotoxin production have been identified and cloned from several *P. syringae* pathovars. With the exception of coronatine, toxin biosynthetic gene clusters are generally chromosomally encoded. In several pathovars, the toxin biosynthetic gene cluster also contains a resistance gene which functions to protect the producing strain from the biocidal effects of the toxin. In the case of phaseolotoxin, a resistance gene (*argK*) has been utilized to engineer phaseolotoxin-resistant tobacco plants.

Although *P. syringae* phytotoxins can induce very similar effects in plants (chlorosis and necrosis), their biosynthesis and mode of action can be quite different. Knowledge of the biosynthetic pathways to these toxins and the cloning of the structural genes for their biosynthesis has relevance to the development of new bioactive compounds with altered specificity. For example, polyketides constitute a huge family of structurally diverse natural products including antibiotics, chemotherapeutic compounds, and antiparasitics. Most of the research on polyketide synthesis in bacteria has focused on compounds synthesized by *Streptomyces* or other actinomycetes. It is also important to note that it is now possible to utilize a genetic rather than synthetic approach to biosynthesize novel polyketides with altered biological properties (Hutchinson and Fujii, 1995; Kao *et al.*, 1994; Donadio *et al.*, 1993; Katz and Donadio, 1993). Most of the reprogramming or engineering of novel polyketides has been done using actinomycete PKSs, but much of this technology could also be applied to polyketides synthesized by *Pseudomonas* when sufficient sequence information is available. It is important to note that *Pseudomonas* produces a variety of antimicrobial compounds from the polyketide pathway, including mupirocin (pseudomonic acid) (Feline *et al.*, 1977), pyoluteorin (Cuppels *et al.*, 1986), and 2,4 diacetylphloroglucinol (Phl) (Bangera and Thomashow, 1996). Pseudomonic acid is valued for its pharmaceutical properties as an antibiotic (Aldridge, 1992), whereas pyoluteorin and Phl have antifungal properties (Howell and Stipanovic, 1980; Keel *et al.*, 1992). A thorough understanding of the biosynthetic pathway to polyketide phytotoxins such as coronatine may ultimately lead to the development of novel compounds with altered biological properties. Thus, specific genes in the biosynthetic pathways of *P. syringae* phytotoxins could be deployed in other systems to develop new compounds with a wide range of activities.

ACKNOWLEDGMENTS. C.B. acknowledges financial support for this work from the Oklahoma Agricultural Experiment Station and National Science Foundation grants DMB-8902561, EHR-9108771, INT-9220628, MCB-9603618, and MCB-9316488.

## 4. REFERENCES

Akers, C. P., and Hoff, J. E., 1980, Simultaneous formation of chymopapain inhibitor activity and cubical crystals in tomato leaves, *Can. J. Bot.* **58:**1000–1003.

Aldridge, K. E., 1992, *In vitro* antistaphylococcal activities of two investigative fluoroquinolones, CI-960 and WIN 57273, compared with those of ciprofloxacin, mupirocin (pseudomonic acid), and peptide-class antimicrobial agents, *Antimicrob. Agents Chemother.* **36:**851.

Bangera, M. G., and Thomashow, L. S., 1996, Characterization of a genomic locus required for synthesis of the antibiotic 2,4-diacetylphloroglucinol by the biological control agent *Pseudomonas fluorescens* Q2-87, *Mol. Plant–Microbe Interact.* **9:**83–90.

Barta, T. M., Kinscherf, T. G., and Willis, D. K., 1992, Regulation of tabtoxin production by the *lemA* gene in *Pseudomonas syringae, J. Bacteriol.* **174:**3021–3029.

Barta, T. M., Kinscherf, T. G., Uchytil, T. F., and Willis, D. K., 1993, DNA sequence and transcriptional analysis of the *tblA* gene required for tabtoxin biosynthesis by *Pseudomonas syringae, Appl. Environ. Microbiol.* **59:**458–466.

Bender, C. L., Stone, H. E., Sims, J. J., and Cooksey, D. A., 1987, Reduced pathogen fitness of *Pseudomonas syringae* pv. *tomato* Tn*5* mutants defective in coronatine production, *Physiol. Mol. Plant Pathol.* **30:**272–283.

Bender, C. L., Malvick, D. K., and Mitchell, R. E., 1989, Plasmid-mediated production of the phytotoxin coronatine in *Pseudomonas syringae* pv. *tomato, J. Bacteriol.* **171:**807–812.

Bender, C. L., Young, S. A., and Mitchell, R. E., 1991, Conservation of plasmid DNA sequences in coronatine-producing pathovars of *Pseudomonas syringae, Appl. Environ. Microbiol.* **57:**993–999.

Bender, C. L., Liyanage, H., Palmer, P., Ullrich, M., Young, S., and Mitchell, R., 1993, Characterization of the genes controlling biosynthesis of the polyketide phytotoxin coronatine including conjugation between coronafacic and coronamic acid, *Gene* **133:**31–38.

Bolter, C. J., 1993, Methyl jasmonate induces papain inhibitor(s) in tomato leaves, *Plant Physiol.* **103:**1347–1353.

Boucher, C. A., Gough, C. L., and Arlat, M., 1992, Molecular genetics of pathogenicity determinants of *Pseudomonas solanacearum* with special emphasis on *hrp* genes, *Annu. Rev. Phytopathol.* **30:**443–461.

Charles, T. C., Jin, S., and Nester, E. W., 1992, Two-component sensory transduction systems in phytobacteria, *Annu. Rev. Phytopathol.* **30:**463–484.

Cuppels, D. C., and Ainsworth, T., 1995, Molecular and physiological characterization of *Pseudomonas syringae* pv. *tomato* and *Pseudomonas syringae* pv. *maculicola* strains that produce the phytotoxin coronatine, *Appl. Environ. Microbiol.* **61:**3530–3536.

Cuppels, D. A., Howell, C. R., Stipanovic, R. D., Stoessl, A., and Stothers, J. B., 1986, Biosynthesis of pyoluteorin: A mixed polyketide-tricarboxylic acid cycle origin demonstrated by [1,2-$^{13}C_2$]acetate incorporation, *Z. Naturforsch.* **41:**532–536.

De la Fuente-Martinez, J., Mosqueda-Cano, G., Alvarez-Morales, A., and Herrara-Estrella, L., 1992, Expression of a bacterial phaseolotoxin-resistant ornithyl transcarbamylase in transgenic tobacco confers resistance to *Pseudomonas syringae* pv. *phaseolicola, Bio/technology* **10:**905–909.

Denny, T. P., 1995, Involvement of bacterial polysaccharides in plant pathogenesis, *Ann. Rev. Phytopathol.* **33:**173–197.

Donadio, S., McAlpine, J. B., Sheldon, P. J., Jackson, M., and Katz, L., 1993, An erythromycin analog produced by reprogramming of polyketide synthesis, *Proc. Natl. Acad. Sci. USA* **90:**7119–7123.

Dorman, C. J., 1995, DNA topology and the global control of bacterial gene expression: Implications for the regulation of virulence gene expression, *Microbiology* **141:**1271–1280.

Engst, K., and Shaw, P. D., 1992, Identification of a *lysA*-like gene required for tabtoxin biosynthesis and pathogenicity in *Pseudomonas syringae* pv. *tabaci* strain PTBR2.024, *Mol. Plant–Microbe Interact.* **5:**322–329.

Feline, T. C., Jones, R. B., Mellows, G., and Phillips, L., 1977, Pseudomonic acid. Part 2. Biosynthesis of pseudomonic acid A, *J. Chem. Soc. Trans. I* **1977:**309–318.

Feys, B. J. F., Benedetti, C. E., Penfold, C. N., and Turner, J. G., 1994, *Arabidopsis* mutants selected for resistance to the phytotoxin coronatine are male sterile, insensitive to methyl jasmonate, and resistant to a bacterial pathogen, *Plant Cell* **6:**751–759.

Gasson, M. J., 1980, Indicator technique for antimetabolic toxin production by phytopathogenic species of *Pseudomonas*, *Appl. Environ. Microbiol* **39:**25–29.

Gnanamanickman, S. S., Starratt, A. N., and Ward, E. W. B., 1982, Coronatine production *in vitro* and *in vivo* and its relation to symptom development in bacterial blight of soybean, *Can. J. Bot.* **60:**645–650.

Goss, R. W., 1940, The relation of temperature to common and halo blight of beans, *Phytopathology* **30:**258–264.

Grimm, C., Aufsatz, W., and Panopoulos, N. J., 1995, The *hrpRS* locus of *Pseudomonas syringae* pv. *phaseolicola* constitutes a complex regulatory unit, *Mol. Microbiol.* **15:**155–165.

Gross, D. C., 1985, Regulation of syringomycin synthesis in *Pseudomonas syringae* pv. syringae and defined conditions for its production, *J. Appl. Bacteriol.* **58:**167–174.

Gross, D. C., 1991, Molecular and genetic analysis of toxin production by pathovars of *Pseudomonas syringae*, *Annu. Rev. Phytopathol.* **29:**247–278.

Gross, D. C., and DeVay, J. E., 1977, Population dynamics and pathogenesis of *Pseudomonas syringae* in maize and cowpea in relation to the *in vitro* production of syringomycin, *Phytopathology* **67:**475–483.

Hatziloukas, E., and Panopoulos, N. J., 1992, Origin, structure, and regulation of *argK*, encoding the phaseolotoxin-resistant ornithine carbamoyltransferase in *Pseudomonas syringae* pv. *phaseolicola*, and functional expression of *argK* in transgenic tobacco, *J. Bacteriol.* **174:**5895–5909.

Hoe, N. P., and Goguen, J. D., 1993, Temperature sensing in *Yersinia pestis:* Translation of the LcrF activator protein is thermally regulated, *J. Bacteriol.* **175:**7901–7909.

Howell, C. R., and Stipanovic, R. D., 1980, Suppression of *Pythium ultimum*-induced damping-off of cotton seedlings by *Pseudomonas fluorescens* and its antibiotic, pyoluteorin, *Phytopathology* **70:**712–715.

Hrabak, E. M., and Willis, D. K., 1992, The *lemA* gene required for pathogenicity of *Pseudomonas syringae* pv. *syringae* on bean is a member of a family of two-component regulators, *J. Bacteriol.* **174:**3011–3020.

Hrabak, E. M., and Willis, D. K., 1993, Involvement of the *lemA* gene in production of syringomycin and protease by *Pseudomonas syringae* pv. *syringae*, *Mol. Plant–Microbe Interact.* **6:**368–375.

Huang, H.-C., He, S. Y., Bauer, D. W., and Collmer, A., 1992, The *Pseudomonas syringae* pv. *syringae* 61 *hrpH* product, an envelope protein required for elicitation of the hypersensitive response in plants, *J. Bacteriol.* **174:**6878–6885.

Huang, J., Carney, B. F., Denny, T. P., Weissinger, A. K., and Schell, M. A., 1995, A complex network regulates expression of *eps* and other virulence genes of *Pseudomonas solanacearum*, *J. Bacteriol.* **177:**1259–1267.

Hugovieux-Cotte-Pattat, N., Dominguez, H., and Robert-Baudouy, J., 1992, Environmental conditions affect transcription of the pectinase genes of *Erwinia chrysanthemi* 3937, *J. Bacteriol.* **174:**7807–7818.

Hutchinson, C. R., and Fujii, I., 1995, Polyketide synthase gene manipulation: A structure–function approach in engineering novel antibiotics, *Annu. Rev. Microbiol.* **49:**201–38.

Hutchison, M. L., Tester, M. A., and Gross, D. C., 1995, Role of biosurfactant and ion channel-forming activities of syringomycin in transmembrane ion flux: A model for the mechanism of action in the plant–pathogen interaction, *Mol. Plant–Microbe Interact.* **8:**610–620.

Huynh, T. V., Dahlbeck, D., and Staskawicz, B. J., 1989, Bacterial blight of soybean: Regulation of a pathogen gene determining host cultivar specificity, *Science* **245:**1374–1377.

Ichihara, A., Shiraishi, K., Sato, H., Sakamura, S., Nishiyama, K., Sakai, R., Furusaki, A., and Matsumoto, T., 1977, The structure of coronatine, *J. Am. Chem. Soc.* **99:**636–637.

Kamdar, H. V., Rowley, K. B., Clements, D., and Patil, S. S., 1991, *Pseudomonas syringae* pv. *phaseolicola* genomic clones harboring heterologous DNA sequences suppress the same phaseolotoxin-deficient mutants, *J. Bacteriol.* **173:**1073–1079.

Kao, C. M., Katz, L., and Khosla, C., 1994, Engineered biosynthesis of a complete macrolactone in a heterologous host, *Science* **265:**509–512.

Katz, L., and Donadio, S., 1993, Polyketide synthesis: Prospects for hybrid antibiotics, *Annu. Rev. Microbiol.* **47:**875–912.

Keel, C., Schnider, U., Maurhofer, M., Voisard, C., Laville, J., Burger, U., Wirthner, P., Haas, D., and Defago, D., 1992, Suppression of root diseases by *Pseudomonas fluorescens* CHAO: Importance of the bacterial secondary metabolite 2,4-diacetylphloroglucinol, *Mol. Plant–Microbe Interact.* **5:**4–13.

Kenyon, J. S., and Turner, J. G., 1992, The stimulation of ethylene synthesis in *Nicotiana tabacum* leaves by the phytotoxin coronatine, *Plant Physiol.* **100:**219–224.

Kinscherf, T. G., Coleman, R. H., Barta, T. M., and Willis, D. K., 1991, Cloning and expression of the tabtoxin biosynthetic region from *Pseudomonas syringae*, *J. Bacteriol.* **173:**4124–4132.

Koda, Y., Takahashi, K., Kikuta, Y., Greulich, F., Toshima, H., and Ichihara, A., 1996, Similarities of the biological activities of coronatine and coronafacic acid to those of jasmonic acid, *Phytochemistry* **41:**93–96.

Lanham, P. G., McIlravey, K. I., and Perombelon, M. C. M., 1991, Production of the cell wall dissolving enzymes by *Erwinia carotovora* subsp. *atroseptica in vitro* at 27°C and 30.5°C, *J. Appl. Bacteriol.* **70:**20–24.

Leigh, J. A., and Coplin, D. L., 1992, Exopolysaccharides in plant–bacterial interactions, *Annu. Rev. Microbiol.* **46:**307–346.

Liang, L. Z., Sobiczewski, P., Paterson, J. M., and Jones, A. L., 1994, Variation in virulence, plasmid content, and genes for coronatine synthesis between *Pseudomonas syringae* pv. *morsprunorum* and *P. s. syringae* from *Prunus*, *Plant Dis.* **78:**389–392.

Liyanage, H., Palmer, D. A., Ullrich, M., and Bender, C. L., 1995a, Characterization and transcriptional analysis of the gene cluster for coronafacic acid, the polyketide component of the phytotoxin coronatine, *Appl. Environ. Microbiol.* **61:**3843–3848.

Liyanage, H., Penfold, C., Turner, J., and Bender, C. L., 1995b. Sequence, expression and transcriptional analysis of the coronafacate ligase-encoding gene required for coronatine biosynthesis by *Pseudomonas syringae*, *Gene* **153:**17–23.

Magasanik, B., 1995, Historical perspective, in *Two-Component Signal Transduction* (J. A. Hoch and T. J. Silhavy, eds.), pp. 1–5, American Society for Microbiology, Washington, D.C.

Mathews, D. E., and Durbin, R. D., 1990, Tagetitoxin inhibits RNA synthesis directed by RNA polymerases and *Escherichia coli*, *J. Biol. Chem.* **265:**493–498.

Mekalanos, J. J., 1992, Environmental signals controlling expression of virulence determinants in bacteria, *J. Bacteriol.* **174:**1–7.

Mills, S. D., Jasalavich, C. A., and Cooksey, D. A., 1993, A two-component regulatory system required for copper-inducible expression of the copper resistance operon of *Pseudomonas syringae*, *J. Bacteriol.* **175:**1656–1664.

Mitchell, R. E., 1982, Coronatine production by some phytopathogenic pseudomonads, *Physiol. Plant Pathol.* **20:**83–89.

Mitchell, R. E., 1985, Coronatine biosynthesis: incorporation of L-[U-$^{14}$C]isoleucine and L-[U-$^{14}$C]threonine into the 1-amido-1-carboxy-2-ethylcyclopropyl moiety, *Phytochemistry* **24:**247–249.

Mitchell, R. E., 1991, Implications of toxins in the ecology and evolution of plant pathogenic microorganisms: Bacteria, *Experientia* **47:**791–803.

Mitchell, R. E., Hale, C. N., and Shanks, J. C., 1983, Production of different pathogenic symptoms and different toxins by strains of *Pseudomonas syringae* pv. *tomato* not distinguishable to gel-immunodiffusion assay, *Physiol. Plant Pathol.* **23:**315–322.

Mitchell, R. E., Young, S. A., and Bender, C. L., 1994, Coronamic acid, an intermediate in coronatine biosynthesis by *Pseudomonas syringae*, *Phytochemistry* **35:**343–348.

Mitchell, R. E., Young, H., and Liddell, M. J., 1995, Isolation and structural characterization of 2-[1-oxo-2-cyclopenten-2-ylmethyl]-butanoic acid, a polyketide product of coronatine-producing *Pseudomonas* spp., *Tetrahedron Lett.* **36:**3237–3240.

Mittal, S. M., and Davis, K. R., 1995, Role of the phytotoxin coronatine in the infection of *Arabaidopsis thaliana* by *Pseudomonas syringae* pv. *tomato*, *Mol. Plant–Microbe Interact.* **8:**165–171.

Mo, Y.-Y., and Gross, D. C., 1991a, Expression *in vitro* and during plant pathogenesis of the *syB* gene required for syringomycin production by *Pseudomonas syringae* pv. *syringae*, *Mol. Plant–Microbe Interact.* **4:**28–36.

Mo, Y.-Y., and Gross, D. C., 1991b, Plant signal molecules activate the *syrB* gene, which is required for syringomycin production by *Pseudomonas syringae* pv. *syringae*, *J. Bacteriol.* **173:**5784–5792.

Morgan, M. K., and Chatterjee, A. K., 1988, Genetic organization and regulation of proteins associated with production of syringotoxin by *Pseudomonas syringae* pv. *syringae*, *J. Bacteriol.* **170:**5689–5697.

Mosqueda, G., Van den Broeck, G., Saucedo, O., Bailey, A., and Herrera-Estrella, L., 1990, Isolation and characterization of the gene from *Pseudomonas syringae* pv. *phaseolicola* encoding the phaseolotoxin-insensitive ornithine carbamoyltransferase, *Mol. Gen. Genet.* **222:**461–466.

Palmer, D. A., 1995, Ph.D. Dissertation, Oklahoma State University, Stillwater, Oklahoma.

Palmer, D. A., and Bender, C. L., 1993, Effects of environmental and nutritional factors on production of the polyketide phytotoxin coronatine by *Pseudomonas syringae* pv. *glycinea*, *Appl. Environ. Microbiol.* **59:**1619–1626.

Palmer, D. A., and Bender, C. L., 1995, Ultrastructure of tomato leaf tissue treated with the pseudomonad phytotoxin coronatine and comparison with methyl jasmonate, *Mol. Plant–Microbe Interact.* **8:**683–692.

Parkinson, J. S., 1995, Genetic approaches for signaling pathways and proteins, in *Two-Component Signal Transduction* (J. A. Hoch and T. J. Silhavy, eds.), pp. 9–23, American Society for Microbiology, Washington, D.C.

Parkinson, J. S., and Kofoid, E. C., 1992, Communication modules in bacterial signalling proteins, *Annu. Rev. Genet.* **26:**71–112.

Parry, R. J., Lin, M. T., Walker, A. E., and Mhaskar, S., 1991, Biosynthesis of coronatine: Investigations of the biosynthesis of coronamic acid, *J. Am. Chem. Soc.* **113:**1849–1850.

Parry, R. J., Mhaskar, S. V., Lin, M.-T., Walker, A. E., and Mafoti, R., 1994, Investigations of the biosynthesis of the phytotoxin coronatine, *Can. J. Chem.* **72:**86–99.

Parry, R. J., Jiralerspong, S., Mhaskar, S., Alemany, L., and Wilcott, R., 1996, Investigations of coronatine biosynthesis. Elucidation of the mode of incorporation of pyruvate into coronafacic acid, *J. Am. Chem. Soc.* **118:**703–704.

Patil, S. S., Kolattukudy, P. E., and Dimond, A. E., 1970, Inhibition of ornithine carbamoyl transferase from bean plants by the toxin of *Pseudomonas phaseolicola, Plant Physiol.* **46:**752–753.

Patil, S. S., Hayward, A. C., and Emmons, R., 1974, An ultraviolet-induced nontoxigenic mutant of *Pseudomonas phaseolicola* of altered pathogenicity, *Phytopathology* **64:**590–595.

Peet, R. C., Lindgren, P. B., Willis, D. K., and Panopoulos, N. J., 1986, Identification and cloning of genes involved in phaseolotoxin production by *Pseudomonas syringae* pv. *"phaseolicola," J. Bacteriol.* **166:**1096–1105.

Penfold, C. N., Bender, C. L., and Turner, J. G., 1996, Characterisation of genes involved in biosynthesis of coronafacic acid, the polyketide component of the phytotoxin coronatine, *Gene* **183:**167–173.

Quigley, N. G., Mo, Y.-Y., and Gross, D. C., 1993, SyrD is required for syringomycin production by *Pseudomonas syringae* pathovar *syringae* and is related to a family of ATP-binding secretion proteins, *Mol. Microbiol.* **9:**787–801.

Rahme, L. G., Mindrinos, M. N., and Panopoulos, N. J., 1992, Plant and environmental sensory signals control the expression of *hrp* genes in *Pseudomonas syringae* pv. *phaseolicola, J. Bacteriol.* **174:**3499–3507.

Rich, J. J., Hirano, S. S., and Willis, D. K., 1992, Pathovar-specific requirement for the *Pseudomonas syringae lemA* gene in disease lesion formation, *Appl. Environ. Microbiol.* **58:**1440–1446.

Rowley, K. B., Clements, D. E., Mandel, M., Humphreys, T., and Patil, S. S., 1993, Multiple copies of a DNA sequence from *Pseudomonas syringae* pathovar phaseolicola abolish thermoregulation of phaseolotoxin production, *Mol. Microbiol.* **8:**625–635.

Ryan, C., 1967, Quantitative determination of soluble cellular proteins by radial diffusion in agar gels containing antibodies, *Anal. Biochem.* **19:**434–440.

Sakai, R., Nishiyama, K., Ichihara, A., Shiraishi, K., and Sakamura, S., 1979, The relation between bacterial toxic action and plant growth regulation in *Recognition and Specificity in Plant Host–Parasite Interactions* (J. M. Daly and I. Uritani, eds.), pp. 165–179, University Park Press, Baltimore, Maryland.

Salmond, G. P. C., 1994, Secretion of extracellular virulence factors by plant pathogenic bacteria, *Annu. Rev. Phytopathol.* **32:**181–200.

Sato, M., 1988, *In planta* transfer of the gene(s) for virulence between isolates of *Pseudomonas syringae* pv. *atropurpurea, Ann. Phytopathol. Soc. Jpn.* **54:**20–24.

Sato, M., Nishiyama, K., and Shirata, A., 1983, Involvement of plasmid DNA in the productivity of coronatine by *Pseudomonas syringae* pv. *atropurpurea, Ann. Phytopathol. Soc. Jpn.* **49:**522–528.

Shumway, L. K., Rancour, J. M., and Ryan, C. A., 1970, Vacuolar protein bodies in tomato leaf cells and their relationship to storage of chymotrypsin inhibitor I protein, *Planta* **93:**1–14.

Shumway, L. K., Vie Yang, V., and Ryan, C. A., 1976, Evidence for the presence of proteinase inhibitor I in vacuolar protein bodies of plant cells, *Planta* **129:**161–165.

Staskawicz, B. J., and Panopoulos, N. J., 1979, A rapid and sensitive microbiological assay for phaseolotoxin, *Phytopathology* **69:**663–666.

Thomas, M. D., Langston-Unkefer, P. J., Uchytil, T. F., and Durbin, R. D., 1983, Inhibition of glutamine synthetase from pea by tabtoxinine-β-lactam, *Plant Physiol.* **71:**912–915.

Tobe, T., Yoshikawa, M., and Sasakawa, C., 1995, Thermoregulation of *virB* transcription in *Shigella flexneri* by sensing of changes in local DNA superhelicity, *J. Bacteriol.* **177:**1094–1097.

Uhl, M. A., and Miller, J. F., 1995, *Bordetella pertussis* BvgAS virulence control system, in *Two-Component Signal Transduction* (J. A. Hoch and T. J. Silhavy, eds.), pp. 333–349, American Society for Microbiology, Washington, D.C.

Ullrich, M., and Bender, C. L., 1994, The biosynthetic gene cluster for coronamic acid, an ethylcyclopropyl amino acid, contains genes homologous to amino acid activating enzymes and thioesterases, *J. Bacteriol.* **176:**7574–7586.

Ullrich, M., Bereswill, S., Völksch, B., Fritsche, W., and Geider, K., 1993, Molecular characterization of field isolates of *Pseudomonas syringae* pv. *glycinea* differing in coronatine production, *J. Gen. Microbiol.* **139:**1927–1937.

Ullrich, M., Guenzi, A. C., Mitchell, R. E., and Bender, C. L., 1994, Cloning and expression of genes required for coronamic acid (2-ethyl-1-aminocyclopropane 1-carboxylic acid), an intermediate in the biosynthesis of the phytotoxin coronatine, *Appl. Environ. Microbiol.* **60:**2890–2897.

Ullrich, M., Peñaloza-Vázquez, A., Bailey, A. M., and Bender, C. L., 1995, A modified two-component regulatory system is involved in temperature-dependent biosynthesis of the *Pseudomonas syringae* phytotoxin coronatine, *J. Bacteriol.* **177:**6160–6169.

Völksch, B., Bublitz, F., and Fritsche, W., 1989, Coronatine production by *Pseudomonas syringae* pathovars: Screening method and capacity of product formation. *J. Basic Microbiol.* **29:** 463–468.

Weiler, E. W., Kutchan, T. M., Gorba, T., Brodschelm, W., Niesel, U., and Bublitz, F., 1994, The *Pseudomonas* phytotoxin coronatine mimics octadecanoid signalling molecules of higher plants, *FEBS Lett.* **345:**9–13.

Willis, D. K., Rich, J. J., and Hrabak, E. M., 1991, *hrp* genes of phytopathogenic bacteria, *Mol. Plant–Microbe Interact.* **4:**132–138.

Willis, D. K., Rich, J. J., Kinscherf, T. G., and Kitten, T., 1994, Genetic regulation in plant pathogenic pseudomonads, in *Genetic Engineering*, Vol. 16 (J. K. Setlow, ed.), pp. 167–193, Plenum Press, New York.

Xiao, Y., Lu, Y., Heu, S., and Hutcheson, S. W., 1992, Organization and environmental regulation of the *Pseudomonas syringae* pv. *syringae* 61 *hrp* cluster, *J. Bacteriol.* **174:**1734–1741.

Xu, G.-W., and Gross, D. C., 1988a, Evaluation of the role of syringomycin in plant pathogenesis by using Tn*5* mutants of *Pseudomonas syringae* pv. *syringae* defective in syringomycin production, *Appl. Environ. Microbiol.* **54:**1345–1353.

Xu, G.-W., and Gross, D. C., 1988b, Physical and functional analysis of the *syrA* and *syrB* genes involved in syringomycin production by *Pseudomonas syringae* pv. *syringae*, *J. Bacteriol.* **170:**5680–5688.

Young, S. A., Park, S. K., Rodgers, C., Mitchell, R. E., and Bender, C. L., 1992, Physical and functional characterization of the gene cluster encoding the polyketide phytotoxin coronatine in *Pseudomonas syringae* pv. *glycinea*, *J. Bacteriol.* **174:**1837–1843.

Zhang, J.-H., Quigley, N. B., and Gross, D. C., 1995, Analysis of the *syrB* and *syrC* genes of *Pseudomonas syringae* pv. *syringae* indicates that syringomycin is synthesized by a thiotemplate mechanism, *J. Bacteriol.* **177:**4009–4020.

Zhang, Y., Rowley, K. B., and Patil, S. S., 1993, Genetic organization of a cluster of genes involved in the production of phaseolotoxin, a toxin produced by *Pseudomonas syringae* pv. *phaseolicola*, *J. Bacteriol.* **175:**6451–6458.

*Chapter 11*

# DNA Transfer from *Agrobacterium* to Plant Cells in Crown Gall Tumor Disease

Anath Das

## 1. INTRODUCTION

*Agrobacterium tumefaciens*, a Gram-negative soil bacterium, causes crown gall tumor disease on most plants (Hooykaas and Beijersbergen, 1994; Zambryski, 1992; Kado, 1991; Ream, 1989; Nester *et al.*, 1984). The bacterium infects plants at a wound site and forms tumors at the site of infection. Virulent strains of *Agrobacterium* contain a large tumor-inducing (Ti-) plasmid. Upon infection the bacterium donates a segment of the Ti-plasmid DNA, the transferred (T-) DNA, to the plant cell, where it is stably incorporated into the nuclear genome. Encoded within the T-DNA are genes for the biosynthesis of plant hormones auxin and cytokinin. Constitutive expression of these genes in a transformed plant results in an uncontrolled growth of cells, a cancerous phenotype. The related bacterium *Agrobacterium rhizogenes* causes the hairy-root disease that results from the stable transfer of the root-inducing (Ri-) plasmid DNA to plants. The

**Anath Das** Department of Biochemistry and Plant Molecular Genetics Institute, University of Minnesota, St. Paul, Minnesota 55108.

*Subcellular Biochemistry, Volume 29: Plant–Microbe Interactions*, edited by Biswas and Das. Plenum Press, New York, 1998.

process of DNA transfer from *Agrobacterium* to plants represents a naturally occurring gene transfer across a phylogenetic barrier.

The tumor tissue is unique in the ability to grow in a defined culture medium without exogenous addition of phytohormones. It produces a novel compound, an opine, a conjugate of a basic amino acid and an α-keto acid (Tempe and Goldmann, 1982). The bacterium can use the opines for growth. The Ti-plasmids differ in their repertoire of opine biosynthetic genes. This property has been used in the nomenclature of the Ti-plasmids, e.g., an octopine Ti-plasmid or a nopaline Ti-plasmid. The opine biosynthetic gene is encoded within the T-DNA. The T-DNA is 10–25 kb in size and can be a contiguous piece of DNA (the nopaline T-DNA) or two DNA segments separated by a nontransferred DNA (the octopine $T_L$- and $T_R$-DNA). A 24-base pair (bp) imperfect, direct repeat sequence, termed the border sequence, flanks the DNA to be transferred to the plant cell and defines the boundaries of the T-DNA. A second 24-bp sequence, the overdrive, is required for a high efficiency of transfer of the octopine Ti-plasmid $T_L$- and $T_R$-DNA (Peralta *et al.*, 1986).

## 2. FUNCTIONS ESSENTIAL FOR DNA TRANSFER

DNA transfer requires participation of both chromosomal and Ti-plasmid-encoded genes. Several chromosomal loci, *chvA–E, att, cel,* and *ros,* that facilitate attachment of bacteria to plant cells as well as affect other pathogenesis-related processes have been defined (Cooley *et al.*, 1991; Robertson *et al.*, 1988; Matthysse, 1987; Thomashow *et al.*, 1987; Douglas *et al.*, 1982, 1985; Close *et al.*, 1985; Deasey and Matthysse, 1984). The Ti-plasmid encodes two regions that are essential for crown gall tumor formation. One, the T-DNA, is required for the tumor phenotype of a transformed plant, while the second one, the virulence (*vir*) region, is essential for the processing, transfer, and integration of the T-DNA into plant cells. The T-DNA does not encode a gene essential for its own transfer. The dispensable nature of the genes has allowed the substitution of the T-DNA with a gene of interest, greatly facilitating the introduction of foreign genes into plants. The only essential elements in the T-DNA are the *cis*-acting border sequences (Simpson *et al.*, 1982; Yadav *et al.*, 1982; Zambryski *et al.*, 1982). The deletion of the right border sequence of the nopaline Ti-plasmid abolished DNA transfer. The tumor-forming ability of the mutant could be restored by the incorporation of a synthetic 24-bp border sequence. Complementation was orientation-dependent, indicating that DNA transfer occurs in a polar manner and proceeds from 5′ to 3′ direction (Shaw *et at.*, 1984; Wang *et al.*, 1984).

The *vir* region of the octopine Ti-plasmid pTiA6 encompasses about 35 kb of DNA and is composed of six major loci, *virA–E* and *virG* (Stachel and Nester,

**Table I**
**The Virulence Genes of the Octopine Ti-Plasmid**

| Locus | Open-reading frame | Role in DNA transfer |
|---|---|---|
| *virA* | 1 | Regulatory; environmental sensor |
| *virB* | 11 | Transport pore formation; VirB1 is not essential |
| *virC* | 2 | Host-range determinant; not essential |
| *virD* | 5 | T-DNA processing (VirD1 and VirD2) and transfer (VirD4); VirD3 and VirD5 are not required |
| *virE* | 2 | Transfer of VirE2 to plants (VirE1) and stabilization of T-strand DNA (VirE2) |
| *virF* | 1 | Not essential |
| *virG* | 1 | Regulatory; transcriptional activator |
| *virH* | 4 | Not essential |

1986). Mutations in *virA*, *virB*, *virD*, and *virG* are avirulent, while those in *virC* and *virE* have an attenuated phenotype (Stachel and Nester, 1986). Two other ancillary *vir* genes, *virF* and *virH*, are required for tumor formation under specific conditions (Melchers *et al.*, 1990; Kanemoto *et al.*, 1989). The nopaline Ti-plasmid *vir* region lacks *virF* and *virH*, but contains another ancillary gene, *tzs* (Akiyoshi *et al.*, 1985; Beaty *et al.*, 1986). The octopine Ti-plasmid *vir* region can encode 27 proteins, 17 of which are essential for DNA transfer (Table I).

## 3. EXPRESSION OF THE VIRULENCE GENES

Below I summarize our current understanding of *vir* gene expression (for reviews, see Winans, 1992; Das, 1994; Johnson and Das, 1997). In free-living bacteria *virA* and *virG* are constitutively expressed. Transcription of the *vir* genes is activated when the bacterium encounters a host plant. In response to a signal molecule(s) released by a wounded plant, *virA* and *virG* control expression of the *vir* genes (Stachel *et al.*, 1986a; Stachel and Zambryski, 1986). The induction of *vir* gene expression is sensitive to the cell growth conditions. A low pH (pH 5.3–6) of the growth medium and a temperature of ≤30°C are required for the inducible expression. Two phenolics from tobacco, acetosyringone and hydroxy-acetosyringone, were the first compounds identified as inducers of *vir* gene expression (Stachel *et al.*, 1985). Subsequently, other inducers have been isolated from both dicotyledonous and nonhost monocotyledonous plants. An analysis of the structural homologs of acetosyringone and studies on the response of isolates of *Agrobacterium* to different inducers indicate that a phenol is the basic molecule necessary for *vir* gene induction (Stachel *et al.*, 1985; Melchers *et al.*, 1989b; Lee *et al.*, 1995). A methoxy group at the R3 and/or R5 position increases the potency of the compound. Several functional groups, including —H, —CHO, —COOH, —$COCH_3$, and —CH═CH.COOH, can occupy the R1 position. A group of aldolases, viz., L-arabinose, D-xylose, D-lyxose, D-glucose, D-mannose, D-galactose, D-idose, and D-talose, greatly increase the sensitivity of the bacterium toward the inducer. This process is mediated by a chromosomal gene *chvE*, a homolog of the *Escherichia coli* sugar-binding proteins (Cangelosi *et al.*, 1990; M. Huang *et al.*, 1990; Shimoda *et al.*, 1990).

The two regulatory genes *virA* and *virG* are members of the two-component regulatory system (Leroux *et al.*, 1987; Winans *et al.*, 1986; Bourret *et al.*, 1991). In response to a signal the sensor VirA, an integral membrane protein, phosphorylates itself at a conserved histidine and transmits the signal to the regulator VirG through the phosphorylation of a conserved aspartic acid (Y. Huang *et al.*, 1990; Jin *et al.*, 1990a,c). Phosphorylated VirG functions as a transcriptional activator. A *cis*-acting regulatory element, the *vir* box, is essential for transcription activation at a *vir* gene promoter (Das *et al.*, 1986; Winans *et al.*, 1987; Das and Pazour, 1989). The *vir* box has the consensus sequence dPuPyTNCAATTGNAAPy, is found within 80 nucleotides upstream from the transcription initiation sites, and serves as the VirG binding site (Jin *et al.*, 1990b; Pazour and Das, 1990a,b).

VirA contains two membrane-spanning domains that are essential for function (Melchers *et al.*, 1989a; Pazour *et al.*, 1991). It consists of at least four domains: a periplasmic domain, a linker domain, a kinase domain, and a carboxy-terminal response regulator homologous domain (Leroux *et al.*, 1987; Melchers *et al.*, 1989a; Stock *et al.*, 1989; Chang and Winans, 1992). The periplasmic domain is essential for the interaction with ChvE, the sugar-binding protein (Cangelosi *et al.*, 1990). A deletion of this domain leads to the loss of sugar responsiveness and the low pH requirement, but not the inducibility of *vir* gene expression (Cangelosi *et al.*, 1990; Melchers *et al.*, 1989a). The linker domain is believed to interact with the inducer (Chang and Winans, 1992; Doty *et al.*, 1996). The kinase domain is conserved in the family of the sensor proteins and contains a conserved histidine that is the site of autophosphorylation. *In vitro* the cytoplasmic segment of VirA autophosphorylates at histidine 474 and transfers the phosphate to aspartic acid 52 of VirG. The C-terminal response regulator homologous domain is found only in a small subset of the sensor proteins. This domain of VirA functions as an autoinhibitory domain, since its deletion renders the mutant constitutive for *vir* gene expression (Chang and Winans, 1992; Gubba *et al.*, 1995).

VirG is the transcriptional activator that induces expression of all the *vir* genes. VirG, a cytoplasmic protein of 241 amino acid residues, is composed of two domains: the N-terminal receiver domain and the C-terminal output domain (Melchers *et al.*, 1986; Winans *et al.*, 1986; Powell *et al.*, 1987; Aoyama *et al.*, 1989; Pazour and Das, 1990b). The receiver domain is conserved in all members of the family, while the output domain is conserved only in a subset of the proteins. The latter allows the subdivision of these proteins into smaller subfamilies and suggests that the members of a subfamily may use a similar mechanism for function. The VirG subfamily of proteins includes the *E. coli* OmpR, PhoB, and SfrA (Melchers *et al.*, 1986; Winans *et al.*, 1986).

In response to an inducer, VirA is phosphorylated at histidine 474 and transfers the phosphate to aspartic acid 52 of VirG. *In vitro* phosphorylation of VirA and VirG has been demonstrated (Y. Huang *et al.*, 1990; Jin *et al.*, 1990a,c). Phosphorylation of the receiver domain is believed to induce a conformational change in the C-terminal output domain. The C-terminal domain of VirG encodes the DNA-binding function (Powell and Kado, 1990). VirG binds specifically to the *vir* box sequences of the Ti- and Ri-plasmid *vir* genes (Jin *et al.*, 1990b; Pazour and Das, 1990b; Tamamoto *et al.*, 1990). The *virC*, *virD*, and *virG* genes contain multiple *vir* boxes. In these genes VirG bound only to those boxes that are essential for gene expression *in vivo* (Pazour and Das, 1990b; Winans, 1990).

The isolation and characterization of *virG* mutants facilitated the analysis of the VirG functional domains. Three mutations, substitutions of asparagine 54 to aspartic acid (*virGN54D*), isoleucine 77 to valine (*virGI77V*), and isoleucine 106 to leucine (*virGI106L*), allow inducer-independent constitutive expression of a *vir* gene (Han *et al.*, 1992; Pazour *et al.*, 1992; Jin *et al.*, 1993; Scheeren-Groot *et al.*, 1994; Gubba *et al.*, 1995). While two of the mutants, *virGN54D* and *virGI106L*, require aspartic acid 52 for function, the third one, *virGI77V*, acts independently of aspartic acid 52 and was isolated as a second-site suppressor that restores function of the nonfunctional mutant *virGD52E* (Pazour *et al.*, 1992; Gubba *et al.*, 1995). The mutant *virGI77V* and two other mutants, *virGM13T* and *virGH15R*, are supersensitive to acetosyringone and are induced by a low concentration of the inducer (Han *et al.*, 1992; Scheeren-Groot *et al.*, 1994). No constitutive mutant that maps to the C-terminal domain of the protein was isolated in many attempts, suggesting that a single amino acid substitution in the output domain may not be sufficient for the activation of VirG.

Transcription activation by VirG requires phosphorylation. How phosphorylation modulates VirG is not known. One mode of action is that phosphorylated VirG promotes DNA-RNA polymerase closed complex formation. *In vitro* studies with the constitutive mutant protein VirGN54D support this model (P. Mallik and A. Das, unpublished results). A similar mechanism is used by the *E. coli* activators OmpR and PhoB (Tsung *et al.*, 1990; Makino *et al.*, 1993). NtrC, another VirG homolog, promotes transcription by facilitating the conversion of the closed promoter complex to an open complex (Ninfa *et al.*, 1987; Popham *et al.*, 1989).

Phosphorylation can also modulate DNA binding. For several VirG homologs the affinity for the cognate binding site is increased ~10-fold upon phosphorylation (Aiba *et al.*, 1989; Nakashima *et al.*, 1993; Roggiani and Dubuau, 1993). Studies with MalE–VirGN54D fusion protein suggest that a similar mechanism may be applicable to VirG as well (Han and Winans, 1994). Phosphorylation also affects the oligomerization of the regulators (Fiedler and Weiss, 1995). It is postulated that the output domain inhibits the N-terminal dimerization domain in the absence of an inducer. Phosphorylation of the receiver domain induces dimerization, activating a new function(s). Phosphorylation of NtrC and PhoB induces oligomerization. Oligomerization of NtrC activates its ATPase activity, but not DNA binding (Mettke *et al.*, 1995). Phosphorylation of OmpR induces a conformational change that results in an altered sensitivity to proteases (Kenney *et al.*, 1995).

## 4. PROCESSING OF THE T-DNA

Induction of the *vir* gene expression initiates a cascade of events leading to the integration of the T-DNA into the plant nuclear genome. Two proteins of the *virD* operon, VirD1 and VirD2, catalyze the first step in the processing of the T-DNA, nicking at the T-DNA borders (Yanofsky *et al.*, 1986; Jayaswal *et al.*, 1987). The site- and strand-specific nicks at the borders lead to the synthesis of a single-stranded (ss) T-strand DNA comprised of the bottom strand of the T-DNA (Stachel *et al.*, 1986b; Albright *et al.*, 1987). The T-strand DNA is an intermediate in DNA transfer (Tinland *et al.*, 1994; Yusibov *et al.*, 1994). Of the other VirD proteins, VirD4 is essential for DNA transfer, and VirD3 and VirD5 have no role in tumorigenesis (Stachel and Nester, 1986; Vogel and Das, 1992b; Koukolikova-Nicola *et al.*, 1993; Lin and Kado, 1993).

The synthesis of the T-strand DNA requires VirD1 and the N-terminal two-thirds of VirD2 (Yanofsky *et al.*, 1986). Mutational analysis of VirD1 showed that a short segment (residues 45–60) of the protein is important for the nicking activity (Vogel and Das, 1994). This region is conserved in the functionally analogous proteins of conjugatable bacterial plasmids and in the *K. pneumoniae* transcriptional activators NtrC and NifA (Porter *et al.*, 1987; Vogel and Das, 1994). In NtrC this segment functions as a DNA-binding motif (Contreras and Drummond, 1988). VirD2, in the presence of VirD1, catalyzes nicking at a T-DNA border. *In vitro* purified VirD1 and VirD2 catalyze site-specific nicking at a border indicating that no other protein is required for this reaction (Scheiffele *et al.*, 1995). Following nicking, VirD2 attaches covalently to the 5′ end of the T-strand DNA. The 5′-linked VirD2 is postulated to direct the transfer of the T-strand DNA in a 5′→3′ directional manner and to target it to the plant cell nucleus (Herrera-Estrella *et al.*, 1988, 1990; Ward and Barnes, 1988; Young and Nester, 1988; Howard *et al.*, 1992; Shurvinton *et al.*, 1992; Tinland *et al.*, 1992).

A tyrosine residue at position 29 of VirD2 participates in the formation of the protein–DNA complex (Vogel and Das, 1992a). Three other tyrosines at positions 68, 121, and 160 are required for optimal VirD2 activity. Tyrosine 121 lies near a Gly-x-Gly-x-x-Gly sequence (residues 111–116) that may function in mononucleotide binding. VirD2 and its functional homologs in conjugatable bacterial plasmids share a conserved 14-residue segment with the consensus sequence HxDxD(H/N)uHuHuuuN (invariant residues in capital letters; x, any amino acid; u, hydrophobic residue) (Ilyina and Koonin, 1992). Random mutagenesis of this domain of pTiA6 VirD2 (residues 126–139) indicated that all invariant residues except for asparagine 139 are important for function (Vogel *et al.*, 1995). The carboxylic acid side chain of aspartic acid 130 and three nonconserved residues at positions 129, 132, and 134 are also essential for function. This region is postulated to coordinate the essential cofactor magnesium by the two histidines at positions 133 and 135 (Ilyina and Koonin, 1992).

The N-terminal 286 residues of the pTiA6 VirD2 (a 424-residue polypeptide) are sufficient for nicking at the T-DNA borders. Except for the C-terminal 30 residues of the protein, the intervening sequence is not necessary for DNA transfer (Koukolikova-Nicola *et al.*, 1993). The C-terminal region encodes two important sequences: a bipartite nuclear localization sequence and the omega sequence. A foreign protein fused to the nuclear localization sequence targets the fusion protein into the plant cell nucleus (Howard *et al.*, 1992). The deletion of the omega sequence leads to a complete loss of the tumor-forming ability of the bacterium (Shurvinton *et al.*, 1992). The loss of function in the latter is not due to protein instability. The role of the C-terminal domain of VirD2 in T-DNA transfer is discussed in the chapter by Ream in this volume.

The single-stranded nature of the T-strand DNA poses an unusual problem in maintaining the integrity of the transformed DNA. The action of cellular nucleases will lead to the loss of information on the T-strand DNA. An effective way to safeguard against nucleases is to make the DNA inaccessible to these enzymes. This mechanism is used by *Agrobacterium* and the protective function is carried out by a protein of the *virE* operon, VirE2. VirE2 is a single-stranded DNA-binding protein that cooperatively binds DNA without sequence specificity (Gietl *et al.*, 1987; Das, 1988; Christie *et al.*, 1988; Citovsky *et al.*, 1988, 1989; Sen *et al.*, 1989). The protective role of VirE2 will require it to function in bacteria and plant cells. *In planta* function of VirE2 was demonstrated by Citovsky *et al.* (1992) using transgenic plants that express VirE2. That study found that an avirulent *virE* mutant can form tumors on the VirE2$^+$ transgenic tobacco plants. A second role of VirE2 is to target the T-strand DNA into the plant cell nucleus (Citovsky *et al.*, 1992). Like VirD2, VirE2 contains a bipartite nuclear localization sequence that when fused to a foreign gene targets the encoded fusion protein into the plant nucleus. The prevailing view on the T-DNA transfer is that a protein–DNA complex, the T-complex, composed of the T-strand DNA containing the 5′-linked VirD2 and coated with VirE2, is the transfer intermediate (reviewed in Zupan and Zambryski, 1995; Tinland, 1996). The transport of the large T-complex probably requires the nuclear localization functions of both VirD2 and VirE2 (Citovsky *et al.*, 1992).

An unusual property of VirE2 is the ability of a strain expressing *virE* and lacking T-DNA to complement a *virE* mutant in a tumor formation assay (Otten *et al.*, 1984). *In planta* complementation requires that the complementing strain harbors all non-T-DNA genes essential for tumor formation (Otten *et al.*, 1984; Christie *et al.*, 1988). These observations can be explained by a model that suggests that both T-strand DNA and VirE2 are independently exported to the plant cell (Binns *et al.*, 1995; Sundberg *et al.*, 1996). The two components form the T-complex in the transformed plant cell, not in the bacterium. Binns *et al.* (1995) studied the mechanism of inhibition of T-DNA transfer by the IncQ plasmid RSF1010. By using *in planta* complementation assays they observed that RSF1010 inhibits the ability of a strain to serve as a VirE2 donor, but only partially inhibits T-DNA transfer from a *virE2* mutant. The transfer of the RSF1010 R-DNA is dramatically increased by extracellular complementation with a *virE2*$^{+}$ strain, but not by intracellular overproduction of VirE2. In another study Sundberg *et al.* (1996) investigated the role of VirE1 in T-DNA transfer. A strain with a deletion in *virE1* can function as a T-DNA donor, but not as a VirE2 donor in an *in planta* complementation assay, indicating that both VirE2 and T-strand DNA are transferred to the plant cells, and VirE1 is essential for the transfer of VirE2, but not for that of T-strand DNA. That both DNA and protein are transported to plants during infection by *Agrobacterium* was first concluded by Regensburg-Tunik and Hooykaas (1993) in the course of their studies on *virF* which like *virE* can be complemented *in planta* (Otten *et al.*, 1985).

## 5. THE TRANSPORT PORE

The transfer of the T-strand DNA to the plant cells requires *virD4* and *virB* (Yoshioka *et al.*, 1994; Yusibov *et al.*, 1994). VirD4 and VirB proteins also catalyze DNA transfer between two *Agrobacterium* strains and from *Agrobacterium* to yeast (Beijersbergen *et al.*, 1992; Bundock *et al.*, 1995; Piers *et al.*, 1996). The C-terminus of VirD4, an inner membrane protein, is postulated to lie in the periplasmic space (Okamoto *et al.*, 1991). Our recent studies using *virD4–phoA* and *virD4–lacZ* fusions, however, indicate that the majority of the protein including the C-terminus lies in the cytoplasm (Das and Xie, 1997). The revised topology of VirD4 suggests that it probably interacts with a cytoplasmic protein. A very likely candidate is VirD2 linked to the T-strand DNA. Such an interaction can allow targeting of the T-strand DNA into the transport pore in preparation for its transfer. VirD4 contains type A and type B nucleotide-binding motifs that can provide energy essential for translocation. The type A motif of RP4 TraG, a VirD4 homolog, is essential for the DNA transfer function (Balzer *et al.*, 1994).

The *virB* operon encodes 11 proteins, most of which are membrane or membrane-associated proteins. Several VirB proteins contain a signal sequence and/or a significant stretch of hydrophobic residues that can function in membrane spanning (Thompson *et al.*, 1988; Ward *et al.*, 1988; Kuldau *et al.*, 1990; Shirasu *et al.*, 1990). These proteins function in a step(s) subsequent to the T-strand DNA synthesis (Stachel *et al.*, 1987). The prevailing view is that the VirB proteins form a transport pore through which the T-strand DNA exits the bacterium. The similarities between several VirB proteins and proteins essential for the conjugal transfer of bacterial plasmids and for the transport of the animal pathogen *Bordetella pertussis* toxin protein support this hypothesis (Lessl *et al.*, 1992a; Weiss *et al.*, 1993; Pohlman *et al.*, 1994). Biochemical and genetic studies on the VirB proteins are in agreement with the transport pore model. In fractionation studies all VirB proteins are found associated with the bacterial membranes (Fernandez *et al.*, 1996a; Finberg *et al.*, 1995; Stephens *et al.*, 1995; Jones *et al.*, 1994; Berger and Christie, 1993; Shirasu and Kado, 1993; Thorstenson *et al.*, 1993; Ward *et al.*, 1990b). However, genetic studies using fusions to the periplasmic marker protein *E. coli* alkaline phosphatase suggest that all, except VirB4 and VirB11, are membrane proteins (Beijersbergen *et al.*, 1994; Fernandez *et al.*, 1996b; A. Das, unpublished results). VirB4 and VirB11 are presumably associated with the cytoplasmic face of the inner membrane.

Ten of the 11 VirB proteins are essential for tumor formation (Berger and Christie, 1994). One, VirB1, is not essential, but is required for a high transformation efficiency. An analysis of VirB1 sequences shows that one domain has a limited homology with the active-site sequence of lysozyme. Sequence conservation is found around a glutamic acid (residue 53) and an aspartic acid (residue 93) that are the two catalytic residues in lysozyme. A recent molecular modeling study indicates that the tertiary structure of VirB1 resembles that of chicken egg white lysozyme (Mushegian *et al.*, 1996). The alteration of the glutamic acid 53 to glutamine, alanine, or threonine and that of aspartic acid 93 to asparagine or leucine almost completely abolished VirB1 function, indicating that these two residues are important for function (Mushegian *et al.*, 1996; Anderson, 1997). The structural similarity with lysozyme and the requirement of the putative catalytic-site residues for function strongly suggest that VirB1 is a glycosidase.

VirB2, a 121-amino acid preprotein, is processed at the N-terminus to a 74-amino acid polypeptide (Shirasu and Kado, 1993). VirB2 and VirB3 are homologous to the pilus proteins TraA and TraL of the F and other plasmids and are predicated to be involved in the assembly of a pilus. Both VirB4 and VirB11 possess ATP-hydrolyzing activity that presumably provides energy essential for the translocation of the T-complex and the assembly of the transport pore (Berger and Christie, 1993; Fullner *et al.*, 1994; Shirasu *et al.*, 1994; Christie *et al.*, 1989; Stephens *et al.*, 1995). Both proteins contain a mononucleotide-binding site consensus sequence Gly-x-x-Gly-x-Gly-Lys-Thr (Thompson *et al.*, 1988). Alteration of the conserved lysine residue in both proteins abolished the ability of the mutant to complement a deletion mutation in the respective gene, indicating that this domain in VirB4 and VirB11 is required for function (Berger and Christie, 1993; Fullner *et al.*, 1994; Stephens *et al.*, 1995). VirB4 is also required for the localization of VirB3 into the outer membrane (Jones *et al.*, 1994).

In addition to the transfer of T-DNA from *Agrobacterium* to plants, the VirB proteins and VirD4 catalyze the transfer of the IncQ plasmid RSF1010 DNA between two *Agrobacterium* strains and of T-DNA from *Agrobacterium* to yeast (Beijersbergen *et al.*, 1992; Bundock *et al.*, 1995; Piers *et al.*, 1995). Introduction of the plasmid RSF1010 into a virulent *Agrobacterium* strain inhibits T-DNA transfer to plants (Ward *et al.*, 1991). This inhibition is overcome by coexpression of three VirB proteins VirB9, VirB10, and VirB11, indicating that these proteins are limiting under specific conditions. The inhibition of T-DNA transfer by RSF1010 is best explained by the hypothesis that the R-DNA of RSF1010 has a higher affinity for the transport pore than does the T-strand DNA. Consequently, in the presence of RSF1010 the R-DNA saturates all the available pores, causing an inhibition of T-DNA transfer (Binns *et al.*, 1995). Overproduction of the three VirB proteins relieves the inhibition by promoting the formation of additional pores.

Little is known about the structure of the transport pore. The elucidation of the interactions between the VirB proteins is essential to address this question. The use of chemical cross-linkers led to the identification of a VirB10 multimer (Ward *et al.*, 1990b). Recently the first interaction between two VirB proteins, VirB7 and VirB9, has been identified (Anderson *et al.*, 1996; Spudich *et al.*, 1996). VirB7 and VirB9 form a disulfide-linked protein complex. VirB7, a 55-residue preprotein, is processed to a 41-residue mature peptide. VirB7 is a lipoprotein and is associated with the outer membrane (Fernandez *et al.*, 1996a). The N-terminal cysteine (Cys 15) is covalently attached to a lipid moiety, while the other cysteine (Cys 24) forms the disulfide linkage with Cys 262 of VirB9. VirB7 and VirB9 form homodimers in the absence of each other and form heterotetramer when both proteins are present. The tetramer results from the dimerization of a VirB7–VirB9 heterodimer (Anderson *et al.*, 1996). VirB7 is also essential for the stability of VirB9 and several VirB proteins (Berger and Christie, 1994; Fernandez *et al.*, 1996b).

Analyses of *virB–phoA* fusions indicate that VirB5 and VirB9 are primarily located in the periplasm (Beijersbergen *et al.*, 1994; Das and Xie, 1997). However, no VirB protein was localized to the periplasm in biochemical fractionation studies. A plausible reason is that the interactions between the various VirB proteins in the pore complex do not allow a clean partitioning of proteins in

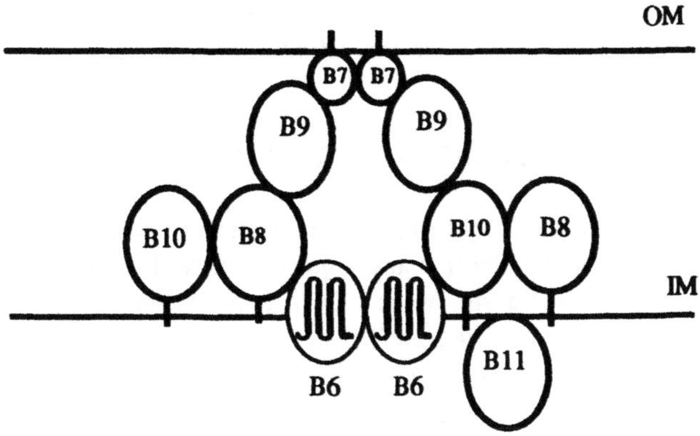


**FIGURE 1.** A hypothetical model for the T-DNA transport pore (see text for details). VirB proteins are depicted as Bx. IM, Inner membrane; OM, outer membrane.

cellular fractionation studies. The presence of internal hydrophobic segment in VirB3, VirB8, and VirB10 suggests that these proteins are anchored to the inner membrane and the C-terminal ends of the proteins lie in the periplasmic space. The *phoA* fusion studies are consistent with this hypothesis (Beijersbergen *et al.*, 1994; Das and Xie, 1997). The association of VirB7 with the outer membrane and the interaction between VirB7 and VirB9 suggest that the VirB7–VirB9 complex can serve as an anchor at the outer membrane. On the basis of the formation of the VirB7–VirB9 complex and the assignment of the VirB proteins into various cellular locations I propose the following model for the T-DNA transport pore (Fig. 1).

VirB6 has multiple membrane-spanning domains (Beijersbergen *et al.*, 1994; A. Das, unpublished results) and I propose that it is one component of a two-component ATP-binding cassette (ABC) transporter. VirB4 or VirB11, the ATPase, constitutes the second component of the transporter (for a review on the ABC transporter, see Fath and Kolter, 1993). Since VirB4 is proposed to be required for the translocation of VirB3 (Jones *et al.*, 1994), VirB11 is a likely candidate for the nucleotide-binding protein component. The VirB6–VirB4/11 transporter will function primarily in the transport of the T-strand DNA across the bacterial inner membrane. VirB6 and two of the proteins anchored to the inner membrane, VirB8 and VirB10, constitute the inner membrane subcomplex. Because the transporters are postulated to contain 12 transmembrane helices (Henderson, 1993), oligomerization of the VirB6 complex will be required for the formation of the functional pore. The subcomplex in the outer membrane consists of VirB7 and VirB9. The identification of a dimer of VirB7–VirB9 complex in cross-linking studies suggests that an oligomer of this complex forms at the outer membrane (Anderson *et al.*, 1996). The two subcomplexes interact to form the pore structure. Future studies on protein–protein interactions among the VirB proteins will test the validity of the model.

## 6. T-DNA TRANSFER AND OTHER TRANSFER SYSTEMS

The processes of T-DNA transfer to plants and of conjugal transfer of bacterial plasmids are similar in many respects. The VirB proteins and VirD4, in a manner similar to the transfer of the T-DNA to plants, support conjugal transfer of plasmids between two *A. tumefaciens* strains (Beijersbergen *et al.*, 1992). The intermediate in both processes is a ssDNA, although in bacterial conjugation the existence of the ssDNA intermediate is transient. Similarities are also found in the initiation of the DNA transfer processes. Functional homologs of VirD1 and VirD2, the processing enzymes, carry out initial processing of the bacterial plasmids. In all cases two proteins—e.g., TraJ, TraI in pRP4; NikA, NikB in pR64; and MobC, MobA in pRSF1010—participate in processing (reviewed in Lanka and Wilkins, 1995; Lessl and Lanka, 1994). In some cases, a third protein, e.g., TraH in pRP4 and MobB in pRSF1010, augment processing but is not essential (Pansegrau *et al.*, 1990). There is homology between the T-DNA border sequence and the RP4 *oriT* nick site, the sites of action of the 'nicases' (Waters *et al.*, 1991; Pansegran and Lanka, 1991). *In vitro* with a single-stranded DNA as a substrate, VirD2 nicks both the T-DNA border and the RP4 *oriT* DNA, but RP4 TraI nicks only at *oriT* DNA (Pansegrau *et al.*, 1993b). In all cases known, following nicking the nicase becomes covalently attached to the 5′ end of the nicked DNA. A tyrosine residue near the N-terminus, viz., Y29 in VirD2 and Y22 in RP4 TraI, participates in the covalent protein–DNA complex formation (Vogel and Das, 1992a; Pansegrau *et al.*, 1993a).

The VirD1 homologs, like VirD1, are small polypeptides (10–16 kDa) and share a loosely conserved 15-residue segment that presumably participates in DNA binding (Porter *et al.*, 1987; Vogel and Das, 1994). The VirD1 homologs in the conjugal plasmids bind at a specific site near the *oriT* and direct the nicase to the target site (Ziegelin *et al.*, 1989). In the *Agrobacterium* Ti- and Ri-plasmids, no *cis*-acting sequence except for the border sequence is either conserved or essential for DNA transfer. Therefore, if the mechanism of VirD1 action is similar to its Tra counterpart, the VirD1-binding site must reside within the border sequence. *In vitro* sequence-specific DNA binding, however, has not been observed with either VirD1 or VirD2. The VirD2 homologs are quite variable in size (40–180 kDa) mainly because these proteins encode an additional function(s). The nicking activity of VirD2 is encoded within the N-terminal two-thirds, while the C-terminal domain encodes a nuclear localization sequence and an essential motif of unknown function (Yanofsky *et al.*, 1986; Howard *et al.*, 1992; Shurvinton *et al.*, 1992). The F-plasmid TraI encodes a helicase activity (reviewed in Lanka and Wilkins, 1995). The transfer of the T-strand DNA to plants requires VirD4 (Yoshioka *et al.*, 1994; Yosibov *et al.*, 1994). Homologs of VirD4, e.g., RP4 TraG, R388 TrwB, CloDF13 MobB, and the F-plasmid TraD, are found in the conjugal plasmid transfer systems and are essential transfer proteins (Lessl *et al.*, 1992b).

DNA sequence analysis of the transfer regions of the bacterial conjugal plasmids and the *ptl* operon of *Bordetella pertussis* showed that several proteins encoded by these operons are homologous to the *Agrobacterium* VirB proteins (Lessl *et al.*, 1992a; Weiss *et al.*, 1993; Pohlman *et al.*, 1994). Conservation is observed in both individual protein sequences and the organization of the genes. The conservation in the primary sequence of the different gene products suggests functional similarity. In the three systems, the *A. tumefaciens virB* operon is essential for T-DNA transfer to plants; *tra* regions are essential for conjugal transfer of plasmids between two *E. coli* strains and the *ptl* operon is essential for the secretion of pertussis toxin, the causative agent of whooping cough. The nature of the products transported by these systems suggests the existence of a common macromolecule export machinery that different bacteria have adapted to carry out specialized functions.

ACKNOWLEDGMENTS. I thank my colleagues for sending preprints and sharing unpublished results prior to publication. Work in my laboratory is supported by a U.S. Public Health Service grant from the National Institutes of Health and by an American Cancer Society Faculty Research Award.

## 7. REFERENCES

Aiba, H., Nakasai, F., Mizushima, S., and Mizuno, T., 1989, Phosphorylation of a bacterial activator protein, OmpR, by a protein kinase, EnvZ, results in stimulation of its DNA-binding ability, *J. Biochem.* **106:**5–7.

Akiyoshi, D., Regier, D., and Gordon, M., 1985, Cloning and nucleotide sequences of the *tzs* gene from *Agrobacterium tumefaciens* strain T37, *Nucleic Acids Res.* **13:**2773–2788.

Albright, L. M., Yanofsky, M. F., Leroux, B., Ma, D. Q., and Nester, E. W., 1987, Processing of the T-DNA of *Agrobacterium tumefaciens* generates border nicks and linear, single-stranded T-DNA, *J. Bacteriol.* **169:**1046–1055.

Anderson, L. B., 1997, Characterization of the DNA transfer proteins of *Agrobacterium tumefaciens*, Ph.D. diss., University of Minnesota, Minneapolis, MN.

Anderson, L. B., Hertzel, A. V., and Das, A., 1996, *Agrobacterium tumefaciens* VirB7 and VirB9 form a disulfide-linked protein complex, *Proc. Natl. Acad. Sci. USA* **93:**8889–8894.

Aoyama, T., Hirayama, T., Tamamoto, S., and Oka, A., 1989, Putative start codon TTG for the regulatory protein VirG of the hairy-root-inducing plasmid pRiA4, *Gene* **78:**173–178.

Balzer, D., Pansegrau, W., and Lanka, E., 1994, Essential motifs of relaxase (TraI) and TraG proteins involved in conjugative transfer of plasmid RP4, *J. Bacteriol.* **176:**4285–4295.

Beaty, J., Powell, G., Lica, L., Regier, D., MacDonald, E., Hommes, N., and Morris, R., 1986, *tzs*, a nopaline Ti plasmid gene from *Agrobacterium tumefaciens* associated with *trans*-zeatin biosynthesis, *Mol. Gen. Genet.* **203:**274–280.

Beijersbergen, A., Dulk-Ras, A. D., Schilperoort, R. A., and Hooykaas, P. J., 1992, Conjugative transfer by the virulence system of *Agrobacterium tumefaciens*, *Science* **256:**1324–1327.

Beijersbergen, A., Smith, S., and Hooykaas, P. J., 1994, Localization and topology of VirB proteins of *Agrobacterium tumefaciens*, *Plasmid* **32:**212–218.

Berger, B. R., and Christie, P. J., 1993, The *Agrobacterium tumefaciens virB4* gene product is an essential virulence protein requiring an intact nucleoside triphosphate-binding domain, *J. Bacteriol.* **175:**1723–1734.

Berger, B., and Christie, P., 1994, Genetic complementation analysis of the *Agrobacterium tumefaciens virB* operon: *virB2* through *virB11* are essential virulence genes, *J. Bacteriol.* **176:**3646–3660.

Binns, A., Beaupre, C., and Dale, E., 1995, Inhibition of VirB-mediated transfer of diverse substrate from *Agrobacterium tumefaciens* by the IncQ plasmid RSF1010, *J. Bacteriol.* **177:**4890–4899.

Bourret, R., Borkovich, K., and Simon, M., 1991, Signal transduction pathways involving protein phosphorylation in prokaryotes, *Annu. Rev. Biochem.* **60:**401–441.

Bundock, P., den Dulk-Ras, A., Beijersbergen, A., and Hooykaas, P. J., 1995, Transkingdom T-DNA transfer from *Agrobacterium tumefaciens* to *Saccharomyces cerevisiae*, *EMBO J.* **14:**3206–3214.

Cangelosi, G., Ankenbauer, R., and Nester, E., 1990, Sugars induce the *Agrobacterium* virulence genes through a periplasmic binding protein and a transmembrane signal protein, *Proc. Natl. Acad. Sci. USA* **87:**6708–6712.

Chang, C.-H., and Winans, S., 1992, Functional roles assigned to the periplasmic, linker, and receiver domains of the *Agrobacterium tumefaciens* VirA protein, *J. Bacteriol.* **174:**7033–7039.

Christie, P. J., Ward, J. E., Winans, S., and Nester, E. W., 1988, The *Agrobacterium tumefaciens virE2* gene product is a single stranded DNA binding protein that associates with T-strands, *J. Bacteriol.* **170:**2659–2667.

Christie, P. J., Ward, J. E. J., Gordon, M. P., and Nester, E. W., 1989, A gene required for transfer of T-DNA to plants encodes an ATPase with autophosphorylating activity, *Proc. Natl. Acad. Sci. USA* **86:**9677–9681.

Citovsky, V., De Vos, G., and Zambryski, P., 1988, Single stranded DNA binding protein encoded by the *virE* locus of *Agrobacterium tumefaciens*, *Science* **240:**501–504.

Citovsky, V., Wong, M. L., and Zambryski, P., 1989, Cooperative interaction of *Agrobacterium* VirE2 protein with single-stranded DNA: Implications for the T-DNA transfer process, *Proc. Natl. Acad. Sci. USA* **86:**1193–1197.

Citovsky, V., Zupan, J., Warnick, D., and Zambryski, P., 1992, Nuclear localization of *Agrobacterium* VirE2 protein in plant cells, *Science* **256:**1802–1805.

Close, T. J., Tait, R. C., and Kado, C. I., 1985, Regulation of Ti plasmid virulence genes by a chromosomal locus of *Agrobacterium tumefaciens*, *J. Bacteriol.* **164:**774–781.

Contreras, A., and Drummond, M., 1988, The effect on the function of the transcriptional activator NtrC from *Klebsiella pneumoniae* of mutations in the DNA-recognition helix, *Nucleic Acids Res.* **16:**4025–4039.

Cooley, M. B., D'Souza, M. R., and Kado, C. I., 1991, *virC* and *virD* operons of the *Agrobacterium* Ti plasmid are regulated by the *ros* chromosomal gene: Analysis of the cloned *ros* gene, *J. Bacteriol.* **173:**2608–2616.

Das, A., 1988, *Agrobacterium tumefaciens virE* operon encodes a single-stranded DNA-binding protein, *Proc. Natl. Acad. Sci. USA* **85:**2909–2913.

Das, A., 1994, Regulation of *Agrobacterium tumefaciens* virulence gene expression, in *Molecular Mechanisms of Bacterial Virulence* (C. Kado and J. Crosa, eds.), pp. 477–489, Kluwer Dordrecht, The Netherlands.

Das, A., and Pazour, G., 1989, Delineation of the regulatory region of *Agrobacterium tumefaciens virB* operon, *Nucleic Acids Res.* **17:**4541–4550.

Das, A., and Xie, Y.-H., 1997, Construction of transposon Tn*3phoA*: Its application in defining the membrane topology of the *Agrobacterium tumefaciens* DNA transfer proteins, submitted.

Das, A., Stachel, S., Ebert, P., Allenza, P., Montoya, A., and Nester, E., 1986, Promoters of *Agrobacterium tumefaciens* Ti-plasmid virulence genes, *Nucleic Acids Res.* **14:**1355–1364.

Deasey, M., and Matthysse, A., 1984, Interactions of wild-type and a cellulose-minus mutant of *Agrobacterium tumefaciens* with tobacco mesophyll and tobacco tissue culture cells, *Phytopathology* **74:**991–994.

Doty, S. L., Yu, M. C., Lundin, J. I., Heath, J. D., and Nester, E. W., 1996, Mutational analysis of the input domain of the VirA protein of *Agrobacterium tumefaciens, J. Bacteriol.* **178:**961–970.

Douglas, C., Halperin, W., and Nester, E., 1982, *Agrobacterium tumefaciens* mutants affected in attachment to plant cells, *J. Bacteriol.* **152:**1265–1275.

Douglas, C. J., Staneloni, R. J., Rubin, R. A., and Nester, E. W., 1985, Identification and genetic analysis of an *Agrobacterium tumefaciens* chromosomal virulence region, *J. Bacteriol.* **161:**850–860.

Fath, M., and Kolter, R., 1993, ABC transporters: Bacterial exporters, *Microbiol. Rev.* **57:**995–1017.

Fernandez, D., Dang, T. A., Spudich, G., Zhou, X.-R., Berger, B., and Christie, P., 1996a, The *Agrobacterium tumefaciens virB7* gene product, a proposed component of the T-complex transport apparatus, is a membrane-associated lipoprotein exposed at the periplasmic surface, *J. Bacteriol.* **178:**3156–3167.

Fernandez, D., Spudich, G., Zhou, X.-R., and Christie, P., 1996b, The *Agrobacterium tumefaciens* VirB7 lipoprotein is required for stabilization of VirB proteins during assembly of the T-complex transport apparatus, *J. Bacteriol.* **178:**3168–3176.

Fiedler, U., and Weiss, V., 1995, A common switch in activation of the response regulators NtrC and PhoB: Phosphorylation induces dimerization of the receiver modules, *EMBO J.* **14:**3696–3705.

Finberg, K., Muth, T., Young, S., Maaarken, J., Heitritter, S., Binns, A. N., and Banta, L., 1995, Interactions of VirB9, -10 and -11 with the membrane fraction of *Agrobacterium tumefaciens:* Solubility studies provide evidence for tight associations, *J. Bacteriol.* **177:**4881–4889.

Fullner, K., Stephens, K., and Nester, E., 1994, An essential virulence protein of *Agrobacterium tumefaciens,* VirB4, requires an intact mononucleotide binding domain to function in transfer of T-DNA, *Mol. Gen. Genet.* **245:**704–715.

Gietl, C., Koukolikova-Nicola, Z., and Hohn, B., 1987, Mobilization of T-DNA from *Agrobacterium* to plant cells involves a protein that binds single-stranded DNA, *Proc. Natl. Acad. Sci. USA* **84:**9006–9010.

Gubba, S., Xie, Y.-H., and Das, A., 1995, Regulation of *Agrobacterium tumefaciens* virulence gene expression: Isolation of a mutation that restores *virGD52E* function, *Mol. Plant–Microbe Interact.* **8:**788–791.

Han, D., and Winans, S., 1994, A mutation in the receiver domain of the *Agrobacterium tumefaciens* transcriptional regulator VirG increases its affinity for operator DNA, *Mol. Microbiol.* **12:**23–30.

Han, D. C., Chen, C., Chen, Y., and Winans, S. C., 1992, Altered-function mutations of the transcriptional regulatory gene *virG* of *Agrobacterium tumefaciens, J. Bacteriol.* **174:**7040–7043.

Henderson, P., 1993, The 12-transmembrane helix transporters, *Curr. Opin. Cell Biol.* **5:**708–721.

Herrera-Estrella, A., Chen, Z., Van Montagu, M., and Wang, K., 1988, VirD proteins of *Agrobacterium tumefaciens* are required for the formation of a covalent DNA protein complex at the 5′ terminus of T-strand molecules, *EMBO J.* **7:**4055–4062.

Herrera-Estrella, A., Van Montagu, M., and Wang, K., 1990, A bacterial peptide acting as a plant nuclear targeting signal: The amino-terminal portion of *Agrobacterium* VirD2 protein directs a β-galactosidase fusion protein into tobacco nuclei, *Proc. Natl. Acad. Sci. USA* **87:**9534–9537.

Hooykaas, P. J., and Beijersbergen, A. G., 1994, The virulence system of *Agrobacterium tumefaciens, Annu. Rev. Phytopathol.* **32:**157–179.

Howard, A. E., Zupan, J. R., and Zambryski, P., 1992, The VirD2 protein of *Agrobacterium tumefaciens* contains a C-terminal bipartite nuclear localization signal: Implications for nuclear uptake of DNA in plant cells, *Cell* **68:**109–118.

Huang, M., Cangelosi, G., Halperin, W., and Nester, E., 1990, A chromosomal *Agrobacterium tumefaciens* gene required for effective plant signal transduction, *J. Bacteriol.* **172:**1814–1822.

Huang, Y., Morel, P., Powell, B., and Kado, C., 1990, VirA, a coregulator of Ti-specified virulence genes, is phosphorylated *in vitro, J. Bacteriol.* **172:**1142–1144.

Ilyina, T., and Koonin, E., 1992, Conserved sequence motifs in the initiator proteins for rolling circle DNA replication encoded by diverse replicons from eubacteria, eucaryotes and archaebacteria, *Nucleic Acids Res.* **20:**3279–3285.

Jayaswal, R., Veluthambi, K., Gelvin, S., and Slightom, J., 1987, Double-stranded cleavage of T-DNA and generation of single-stranded T-DNA molecules in *Escherichia coli* by a *virD*-encoded border-specific endonuclease from *Agrobacterium tumefaciens, J. Bacteriol.* **169:**5035–5045.

Jin, S., Prusti, R., Roitsch, T., Ankenbauer, R., and Nester, E., 1990a, Phosphorylation of the VirG protein of *Agrobacterium tumefaciens* by the autophosphorylated VirA protein: Essential role in the biological activity of VirG, *J. Bacteriol.* **172:**4945–4950.

Jin, S., Roitsch, T., Christie, P., and Nester, E., 1990b, The regulatory VirG protein specifically binds to a *cis*-acting regulatory sequence involved in transcriptional activation of *Agrobacterium tumefaciens* virulence genes, *J. Bacteriol.* **172:**531–537.

Jin, S., Roitsch, T., Ankenbauer, R., Gordon, M., and Nester, E., 1990c, The VirA protein of *Agrobacterium tumefaciens* is autophosphorylated and is essential for *vir* gene regulation, *J. Bacteriol.* **172:**525–530.

Jin, S. G., Song, Y. N., Pan, S. Q., and Nester, E. W., 1993, Characterization of a *virG* mutation that confers constitutive virulence gene expression in *Agrobacterium, Mol. Microbiol.* **7:**555–562.

Johnson, T., and Das, A., 1997, Organization and regulation of expression of the *Agrobacterium* virulence genes, in *The Rhizobiaceae* (H. Spaink, Kondorosi P., and P. Hooykaas, eds.), Kluwer, Dordrecht, The Netherlands.

Jones, A. L., Shirasu, K., and Kado, C., 1994, The product of the *virB4* gene of *Agrobacterium tumefaciens* promotes accumulation of VirB3 protein, *J. Bacteriol.* **176:**5255–5261.

Kado, C., 1991, Molecular mechanisms of crown gall tumorigenesis, *Crit. Rev. Plant Sci.* **10:**1–32.

Kanemoto, R. H., Powell, A. T., Akiyoshi, D. E., Regier, D. A., Kerstetter, R. A., Nester, E. W., Hawes, M. C., and Gordon, M. P., 1989, Nucleotide sequence and analysis of the plant-inducible locus *pinF* from *Agrobacterium tumefaciens, J. Bacteriol.* **171:**2506–2512.

Kenney, L., Bauer, M., and Shilhavy, T., 1995, Phosphorylation-dependent conformational changes in OmpR, an osmoregulatory DNA-binding protein of *Escherichia coli, Proc. Natl. Acad. Sci. USA* **92:**8866–8870.

Koukolikova-Nicola, Z., Raineri, D., Stephens, K., Ramos, C., Tinland, B., Nester, E. W., and Hohn, B., 1993, Genetic analysis of the *virD* operon of *Agrobacterium tumefaciens:* A search for functions involved in transport of T-DNA into the plant cell nucleus and in T-DNA integration. *J. Bacteriol.* **175:**723–731.

Kuldau, G. A., De Vos, G., Owen, J., McCaffrey, G., and Zambryski, P., 1990, The *virB* operon of *Agrobacterium tumefaciens* pTiC58 encodes eleven open reading frames, *Mol. Gen. Genet.* **221:**256–266.

Lanka, E., and Wilkins, B., 1995, DNA processing reactions in bacterial conjugation, *Annu. Rev. Biochem.* **64:**141–169.

Lee, Y.-W., Jin, S., Sim, W.-S., and Nester, E. W., 1995, Genetic evidence for direct sensing of phenolic compounds by the VirA protein of *Agrobacterium tumefaciens, Proc. Natl. Acad. Sci. USA* **92:**12245–12249.

Leroux, B., Yanofsky, M., Winans, S., Ward, J., Ziegler, S., and Nester, E., 1987, Characterization of the *virA* locus of *Agrobacterium tumefaciens:* A transcriptional regulator and host range determinant, *EMBO J.* **6:**849–856.

Lessl, M., and Lanka, E., 1994, Common mechanism in bacterial conjugation and Ti-mediated T-DNA transfer to plant cells, *Cell* **77:**321–324.

Lessl, M., Balzer, D., Pansegrau, W., and Lanka, E., 1992a, Sequence similarities between the RP4 Tra2 and the Ti VirB region strongly support the conjugation model for T-DNA transfer, *J. Biol. Chem.* **267:**20471–20480.

Lessl, M., Pansegrau, W., and Lanka, E., 1992b, Relationship of DNA-transfer-systems: Essential transfer factors of plasmid RP4, Ti and F share common sequences, *Nucleic Acids Res.* **20:**6099–6100.

Lin, T., and Kado, C. I., 1993, The *virD4* gene is required for virulence while *virD3* and *orf5* are not required for virulence of *Agrobacterium tumefaciens, Mol. Microbiol.* **9:**803–812.

Makino, K., Anemura, M., Kim, S.-K., Nakata, A., and Shinagawa, H., 1993, Role of $\sigma^{70}$ subunit of RNA polymerase in transcription activation by activator protein PhoB in *Escherichia coli, Genes Dev.* **7:**149–160.

Matthysse, A., 1987, Characterization of nonattaching mutants of *Agrobacterium tumefaciens, J. Bacteriol.* **169:**313–323.

Melchers, L., Thompson, D., Idler, K., Schilperoort, R., and Hooykaas, P., 1986, Nucleotide sequence of the virulence gene *virG* of the *Agrobacterium tumefaciens* octopine Ti plasmid: Significant homology between *virG* and the regulatory genes *ompR, phoB,* and *dye* of *E. coli, Nucleic Acids Res.* **14:**9933–9940.

Melchers, L. S., Regensburg-Tuink, T., Bourret, R., Sedee, J., Schilperoort, R., and Hooykaas, P., 1989a, Membrane topology and functional analysis of the sensory protein VirA of *Agrobacterium tumefaciens, EMBO J.* **8:**1919–1925.

Melchers, L., Regensburg-Tuink, A., Schilperoort, R., and Hooykaas, P., 1989b, Specificity of signal molecules in the activation of *Agrobacterium* virulence gene expression, *Mol. Microbiol.* **3:**969–977.

Melchers, L., Maroney, M. J., Dulk-Ras, A., Thompson, D. S., van Vuuren, H., Schilperoort, R. A., and Hooykaas, P., 1990, Octopine and nopaline strains of *Agrobacterium tumefaciens* differ in virulence: Molecular characterization of the *virF* locus, *Plant Mol. Biol.* **14:**249–259.

Mettke, I., Fiedler, U., and Weiss, V., 1995, Mechanism of activation of a response regulator: Interaction of NtrC-P dimers induces ATPase activity, *J. Bacteriol.* **177:**5056–5061.

Mushegian, A. R., Fullner, K. J., Koonin, E. V., and Nester, E. W., 1996, A family of lysozyme-like virulence factors in bacterial pathogens of plants and animals, *Proc. Natl. Acad. Sci. USA* **93:**7321–7326.

Nakashima, K., Sugiura, A., Kanamaru, K., and Mizuno, T., 1993, Signal transduction between the two regulatory components involved in the regulation of the *kdpABC* operon in *Escherichia coli:* Phosphorylation-dependent functioning of the positive regulator, KdpE, *Mol. Microbiol.* **7:**109–116.

Nester, E. W., Gordon, M. P., Amasino, R. M., and Yanofsky, M. Y., 1984, Crown gall: A molecular and physiological analysis, *Annu. Rev. Plant. Physiol.* **35:**387–413.

Ninfa, A. J., Reitzer, L. J., and Magasanik, B., 1987, Initiation of transcription at the bacterial *glnAp2* promoter by purified *E. coli* components is facilitated by enhancers, *Cell* **50:**1039–1046.

Okamoto, S., Toyoda-Yamamoto, A., Ito, K., Takebe, I., and Machida, Y., 1991, Localization and orientation of the VirD4 protein of *Agrobacterium tumefaciens* in the cell membrane, *Mol. Gen. Genet.* **228**:24–32.

Otten, L., DeGreve, H., Leemans, J., Hain, R., Hooykaas, P., and Schell, J., 1984, Restoration of virulence of *vir* region mutants of *Agrobacterium tumefaciens* strain B653 by coinfection with normal and mutant *Agrobacterium* strains, *Mol. Gen. Genet.* **175**:159–163.

Otten, L., Piotrowiak, G., Hooykaas, P., Dubois, M., Szegedi, E., and Schell, J., 1985, Identification of an *Agrobacterium tumefaciens* pTiB6S3 vir region fragment that enhances the virulence of pTiC58, *Mol. Gen. Genet.* **199**:189–193.

Pansegrau, W., and Lanka, E., 1991, Common sequence motifs in DNA relaxases and nick regions from a variety of DNA transfer systems, *Nucleic Acid Res.* **19**:3455.

Pansegrau, W., Blazer, D., Kruft, V., Lurz, R., and Lanka, E., 1990, *In vitro* assembly of relaxosomes at the transfer origin of plasmid RP4, *Proc. Natl. Acad. Sci. USA* **87**:6555–6559.

Pansegrau, W., Schroder, W., and Lanka, E., 1993a, Relaxase (TraI) of IncP-alpha plasmid RP4 catalyzes a site-specific cleaving–joining reaction of single-stranded DNA, *Proc. Natl. Acad. Sci. USA* **90**:2925–2929.

Pansegrau, W., Schoumacher, F., Hohn, B., and Lanka, E., 1993b, Site-specific cleavage and joining of single-stranded DNA by VirD2 protein of *Agrobacterium tumefaciens* Ti plasmids: Analogy to bacterial conjugation, *Proc. Natl. Acad. Sci. USA* **90**:11538–11542.

Pazour, G. J., and Das, A., 1990a, Characterization of the VirG binding site of *Agrobacterium tumefaciens, Nucleic Acids Res.* **18**:6909–6913.

Pazour, G. J., and Das, A., 1990b, *virG*, an *Agrobacterium tumefaciens* transcriptional activator, initiates translation at a UUG codon and is a sequence-specific DNA-binding protein, *J. Bacteriol.* **172**:1241–1249.

Pazour, G. J., Ta, C. N., and Das, A., 1991, Mutants of *Agrobacterium tumefaciens* with elevated *vir* gene expression, *Proc. Natl. Acad. Sci. USA* **88**:6941–6945.

Pazour, G. J., Ta, C. N., and Das, A., 1992, Constitutive mutations of *Agrobacterium tumefaciens* transcriptional activator *virG, J. Bacteriol.* **174**:4169–4174.

Peralta, E. G., Hellmiss, R., and Ream, W., 1986, Overdrive, a T-DNA transmission enhancer on the *Agrobacterium tumefaciens* tumor-inducing plasmid, *EMBO J.* **5**:1137–1142.

Piers, K. L., Heath, J. D., Liang, X., Stephens, K. M., and Nester, E. W., 1996, *Agrobacterium tumefaciens*-mediated transformation of yeast, *Proc. Natl. Acad. Sci. USA* **93**:1613–1618.

Pohlman, R., Genetti, H., and Winans, S., 1994, Common ancestry between IncN conjugal transfer genes and macromolecular export systems of plant and animal pathogens, *Mol. Microbiol.* **14**:655–668.

Popham, D., Szeto, D., Keener, J., and Kustu, S., 1989, Function of a bacterial activator protein that binds to transcriptional enhancers, *Science* **243**:629–635.

Porter, S. G., Yanofsky, M. F., and Nester, E. W., 1987, Molecular characterization of the *virD* operon from *Agrobacterium tumefaciens, Nucleic Acids Res.* **15**:7503–7517.

Powell, B., and Kado, C., 1990, Specific binding of VirG to the *vir* box requires a C-terminal domain and exhibits a minimum concentration threshold, *Mol. Microbiol.* **4**:1–8.

Powell, B., Powell, G., Morris, R., Rogowsky, P., and Kado, C., 1987, Nucleotide sequence of the *virG* locus of the *Agrobacterium tumefaciens* plasmid pTiC58, *Mol. Microbiol.* **1**:309–316.

Ream, W., 1989, *Agrobacterium tumefaciens* and interkingdom genetic exchange, *Annu. Rev. Phytopathol.* **27**:583–618.

Regensburg-Tuink, A., and Hooykaas, P. J., 1993, Transgenic *N. glauca* plants expressing bacterial virulence gene *virF* are converted into hosts for nopaline strains of *Agrobacterium tumefaciens, Nature* **363**:69–71.

Robertson, J., Holiday, T., and Matthysse, A., 1988, Mapping of *Agrobacterium tumefaciens* chromosomal genes affecting cellulose synthesis and bacterial attachment to host cells, *J. Bacteriol.* **170:**1408–1411.

Roggiani, M., and Dubneau, D., 1993, ComA, a phosphorylated response regulator protein of *Bacillus subtilis,* binds to the promoter region of *srfA, J. Bacteriol.* **175:**3181–3187.

Scheeren-Groot, E., Rodenburg, K., Dulk-Ras, A., Turk, S., and Hooykaas, P., 1994, Mutational analysis of the transcriptional activator VirG of *Agrobacterium tumefaciens, J. Bacteriol.* **176:**6418–6426.

Scheiffele, P., Pansegrau, W., and Lanka, E., 1995, Initiation of *Agrobacterium tumefaciens* T-DNA processing: Purified proteins VirD1 and VirD2 catalyze site- and strand-specific cleavage of superhelical T-border DNA *in vitro. J. Biol. Chem.* **270:**1269–1276.

Sen, P., Pazour, G. J., Anderson, D., and Das, A., 1989, Cooperative binding of *Agrobacterium tumefaciens* VirE2 protein to single-stranded DNA, *J. Bacteriol.* **171:**2573–2580.

Shaw, C. H., Watson, M. D., Carter, G. H., and Shaw, C. H., 1984, The right hand copy of the nopaline Ti plasmid 25 bp repeat is required for tumor formation, *Nucleic Acids Res.* **12:**6031–6041.

Shimoda, N., Toyoda-Yamamoto, A., Nagamine, J., Usami, S., Katayama, M., Sakagami, Y., and Machida, Y., 1990, Control of expression of *Agrobacterium vir* genes by synergistic actions of phenolic signal molecules and monosaccharides, *Proc. Natl. Acad. Sci. USA* **87:**6684–6688.

Shirasu, K., and Kado, C., 1993, Membrane localization of the Ti plasmid VirB proteins involved in biosynthesis of a pilin-like conjugative structure on *Agrobacterium tumefaciens, FEMS Microbiol. Lett.* **111:**287–294.

Shirasu, K., Morel, P., and Kado, C., 1990, Characterization of the *virB* operon of *Agrobacterium tumefaciens* Ti plasmid: Nucleotide sequence and protein analysis, *Mol. Microbiol.* **4:**1153–1163.

Shirasu, K., Koukolikova-Nicola, Z., Hohn, B., and Kado, C., 1994, An inner-membrane associated virulence protein essential for T-DNA transfer from *Agrobacterium tumefaciens* to plants exhibits ATPase activity and similarity to conjugative transfer genes, *Mol. Microbiol.* **11:**581–588.

Shurvinton, C. E., Hodges, L., and Ream, W., 1992, A nuclear localization signal and the C-terminal omega sequence in the *Agrobacterium tumefaciens* VirD2 endonuclease are important for tumor formation, *Proc. Natl. Acad. Sci. USA* **89:**11837–11841.

Simpson, R. B., O'Hara, P. J., Kwok, W., Montoya, A., Lichtenstein, C., Gordon, M. P., and Nester, E. W., 1982, DNA from the A6/S2 crown gall tumor contains scrambled Ti plasmid sequences near its junction with the plant DNA, *Cell* **29:**1005–1014.

Spudich, G., Fernandez, D., Zhou, X.-R., and Christie, P., 1996, Intermolecular disulfide bonds stabilize VirB7 homodimers and VirB7/VirB9 heterodimers during biogenesis of the *Agrobacterium tumefaciens* T-complex transport pore, *Proc. Natl. Acad. Sci. USA* **93:**7512–7517.

Stachel, S., and Nester, E., 1986, The genetic and transcriptional organization of the *vir* region of the A6 Ti plasmid of *Agrobacterium tumefaciens, EMBO J.* **5:**1445–1454.

Stachel, S., and Zambryski, P., 1986, *virA* and *virG* control the plant-induced activation of the T-DNA transfer process of *A. tumefaciens, Cell* **46:**325–333.

Stachel, S., Messens, E., Van Montagu, M., and Zambryski, P., 1985, Identification of the signal molecules produced by wounded plant cells that activate T-DNA transfer in *Agrobacterium tumefaciens, Nature* **318:**624–629.

Stachel, S. E., Nester, E. W., and Zambryski, P. C., 1986a, A plant cell factor induces *Agrobacterium tumefaciens vir* gene expression, *Proc. Natl. Acad. Sci. USA* **83:**379–383.

Stachel, S., Timmerman, B., and Zambryski, P., 1986b, Generation of single-stranded T-DNA molecules during the initial stages of T-DNA transfer from *Agrobacterium tumefaciens* to plant cells, *Nature* **322:**706–712.

Stachel, S. E., Timmerman, B., and Zambryski, P., 1987, Activation of *Agrobacterium tumefaciens vir* gene expression generates multiple single-stranded T-strand molecules from the pTiA6 T-region: Requirement for 5′ *virD* gene products, *EMBO J.* **6**:857–863.

Stephens, K., Roush, C., and Nester, E., 1995, *Agrobacterium tumefaciens* VirB11 protein requires a consensus nucleotide binding site for function in virulence, *J. Bacteriol.* **177**:27–36.

Stock, J., Ninfa, A., and Stock, A., 1989, Protein phosphorylation and regulation of adaptive responses in bacteria, *Microbiol. Rev.* **53**:450–490.

Sundberg, C., Meek, L., Carroll, K., Das, A., and Ream, W., 1996, VirE1 protein mediates export of VirE2 single-stranded DNA binding protein from *Agrobacterium tumefaciens* to plant cells, *J. Bacteriol.* **178**:1207–11212.

Tamamoto, S., Aoyama, T., Takanami, M., and Oka, A., 1990, Binding of the regulatory protein VirG to the phased signal sequences upstream from virulence genes on the hairy-root-inducing plasmid, *J. Mol. Biol.* **215**:537–547.

Tempe, J., and Goldmann, A., 1982, Occurrence and biosynthesis of opines, in *Molecular Biology of Plant Tumors* (G. Kahl and J. Schell, eds.), pp. 428–449, Academic Press, New York.

Thomashow, M. F., Karlinsey, J. E., Marks, J. R., and Hurlbert, R. E., 1987, Identification of a new virulence locus in *Agrobacterium tumefaciens* that affects polysaccharide composition and plant cell attachment, *J. Bacteriol.* **169**:3209–3216.

Thompson, D. V., Melchers, L. S., Idler, K. B., Schilperoort, R. A., and Hooykaas, P. J., 1988, Analysis of the complete nucleotide sequence of the *Agrobacterium tumefaciens virB* operon, *Nucleic Acids Res.* **16**:4621–4636.

Thorstenson, Y. R., Kuldau, G. A., and Zambryski, P. C., 1993, Subcellular localization of seven VirB proteins of *Agrobacterium tumefaciens*—Implications for the formation of a T-DNA transport structure, *J. Bacteriol.* **175**:5233–5241.

Tinland, B., 1996, The integration of T-DNA into plant genomes, **1**:178–184.

Tinland, B., Koukolikova-Nicola, Z., Hall, M. N., and Hohn, B., 1992, The T-DNA linked protein VirD2 contains two distinct functional nuclear localization regions, *Proc. Natl. Acad. Sci. USA* **89**:7442–7446.

Tinland, B., Hohn, B., and Puchta, H., 1994, *Agrobacterium tumefaciens* transfers single-stranded transferred DNA (T-DNA) into the plant cell nucleus, *Proc. Natl. Acad. Sci. USA* **91**:8000–8004.

Tsung, K., Brissette, R., and Inouye, M., 1990, Enhancement of RNA polymerase binding to promoters by a transcriptional activator, OmpR, in *Escherichia coli:* Its positive and negative effects on transcription, *Proc. Natl. Acad. Sci. USA* **87**:5940–5944.

Vogel, A. M., and Das, A., 1992a, Mutational analysis of *Agrobacterium tumefaciens virD2:* Tyrosine-29 is essential for endonuclease activity, *J. Bacteriol.* **174**:303–308.

Vogel, A. M., and Das, A., 1992b, The *Agrobacterium tumefaciens virD3* gene is not essential for tumorigenicity on plants, *J. Bacteriol.* **174**:5161–5164.

Vogel, A., and Das, A., 1994, Mutational analysis of *Agrobacterium tumefaciens* pTiA6 *virD1:* Identification of functionally important residues, *Mol. Microbiol.* **12**:811–817.

Vogel, A., Yoon, J., and Das, A., 1995, Mutational analysis of a conserved motif of *Agrobacterium tumefaciens* VirD2, *Nucleic Acids Res.* **23**:4087–4091.

Wang, K., Herrera-Estrella, L., Van Montagu, M., and Zambryski, P., 1984, Right 25 bp terminus sequences of the nopaline T-DNA is essential for and determines direction of DNA transfer from *Agrobacterium* to the plant genome, *Cell* **38**:455–462.

Ward, E., and Barnes, W., 1988, VirD2 protein of *Agrobacterium tumefaciens* very tightly linked to the 5′ end of T-strand DNA, *Science* **242**:927–930.

Ward, J. E., Akiyoshi, D. E., Regier, D., Datta, A., Gordon, M. P., and Nester, E., 1988, Characterization of the *virB* operon from *Agrobacterium tumefaciens* Ti plasmid, *J. Biol. Chem.* **263**:5804–5814.

Ward, J. E., Akiyoshi, D. E., Regier, D., Datta, A., and Gordon, M. P., 1990a, Correction: Characterization of the *virB* operon from *Agrobacterium tumefaciens* Ti plasmid, *J. Biol. Chem.* **265:**4768.

Ward, J. E., Dale, E. M., Nester, E. W., and Binns, A. N., 1990b, Identification of a *virB10* protein aggregate in the inner membrane of *Agrobacterium tumefaciens*, *J. Bacteriol.* **172:**5200–5210.

Ward, J. E., Dale, E. M., and Binns, A. N., 1991, Activity of the *Agrobacterium* T-DNA transfer machinery is affected by *virB* gene products, *Proc. Natl. Acad. Sci. USA* **88:**9350–9354.

Waters, V. L., Hirata, K. H., Pansegrau, W., Lanka, E., and Guiney, D. G., 1991, Sequence identity in the nick regions of IncP plasmid origins and T-DNA borders of *Agrobacterium* Ti plasmids, *Proc. Natl. Acad. Sci. USA* **88:**1456–1460.

Weiss, A. A., Johnson, F. D., and Burns, D. L., 1993, Molecular characterization of an operon required for pertussis toxin secretion, *Proc. Natl. Acad. Sci. USA* **90:**2970–2974.

Winans, S., 1990, Transcriptional induction of an *Agrobacterium* regulatory gene at tandem promoters by plant-released phenolic compounds, phosphate starvation, and acidic growth media, *J. Bacteriol.* **172:**2433–2438.

Winans, S., 1992, Two-way chemical signaling in *Agrobacterium*–plant interactions, *Microbiol. Rev.* **56:**12–31.

Winans, S., Ebert, P., Stachel, S., Gordon, M., and Nester, E., 1986, A gene essential for *Agrobacterium virulence* is homologous to a family of positive regulatory loci, *Proc. Natl. Acad. Sci. USA* **83:**8278–8282.

Winans, S., Jin, S., Komari, T., Johnson, K., and Nester, E., 1987, The role of virulence regulatory loci in determining *Agrobacterium* host range, in *Plant Molecular Biology* (D. von Wettstein and N.-H. Chua, eds.), pp. 573–582, Plenum Press, New York.

Yadav, N. S. J., V., Bennett, D. R., Barnes, W. M., and Chilton, M. D., 1982, Short direct repeats flank the T-DNA on a nopaline Ti plasmid, *Proc. Natl. Acad. Sci. USA* **79:**6322–6326.

Yanofsky, M. F., Porter, S. G., Young, C., Albright, L. M., Gordon, M. P., and Nester, E. W., 1986, The *virD* operon of *Agrobacterium tumefaciens* encodes a site-specific endonuclease, *Cell* **47:**471–477.

Yusibov, V., Steck, T., Gupta, V., and Gelvin, S., 1994, Association of single-stranded transferred DNA from *Agrobacterium tumefaciens* with tobacco cells, *Proc. Natl. Acad. Sci. USA* **91:**2994–2998.

Yoshioka, Y., Takahashi, Y., Matsumoto, S., Kojima, S., Matsuoka, K., Nakamura, K., Ohshima, K., Okada, N., and Machida, Y., 1994, Mechanisms of T-DNA transfer and integration into plant chromosomes: Role of *virB*, *virD4* and *virE2* and a short interspersed repetitive element (SINE) from tobacco in *Molecular Mechanisms of Bacterial Virulence* (C. Kado and J. Crosa, eds.), pp. 231–248, Kluwer, Dordrecht, The Netherlands.

Young, C., and Nester, E. W., 1988, Association of the VirD2 protein with the 5′ end of T strands in *Agrobacterium tumefaciens*, *J. Bacteriol.* **170:**3367–3374.

Yusibov, V., Steck, T., Gupta, V., and Gelvin, S., 1994, Association of single-stranded transferred DNA from *Agrobacterium tumefaciens* with tobacco cells, *Proc. Natl. Acad. Sci. USA* **91:**2994–2998.

Zambryski, P., 1992, Chronicles from the *Agrobacterium*–plant cell DNA transfer story, *Annu. Rev. Plant Physiol. Plant Mol. Biol.* **43:**465–490.

Zambryski, P., Depicker, A., Kruger, K., and Goodman, H., 1982, Tumor induction by *Agrobacterium tumefaciens:* Analysis of the boundaries of the T-DNA, *J. Mol. Appl. Genet.* **1:**361–370.

Ziegelin, G., Furste, J. P., and Lanka, E., 1989, TraJ protein of plasmid RP4 binds to a 19-base pair invert sequence repetition within the transfer origin, *J. Biol. Chem.* **264:**11989–11994.

Zupan, J., and Zambryski, P., 1995, Transfer of T-DNA from *Agrobacterium* to the plant cell, *Plant Physiol.* **107:**1041–1047.

*Chapter 12*

# Import of *Agrobacterium tumefaciens* Virulence Proteins and Transferred DNA into Plant Cell Nuclei

Walt Ream

## 1. INTRODUCTION

*Agrobacterium tumefaciens* causes crown gall tumors on many dicotyledonous plant species when the bacteria infect wounded tissue (DeCleene and DeLey, 1976). This bacterium harbors a tumor-inducing (Ti) plasmid, where genes essential for tumorigenesis are located (Watson *et al.*, 1975; Van Larebeke *et al.*, 1974). The transferred DNA (T-DNA) portion of the Ti plasmid enters plant cells and integrates into nuclear DNA (Chilton *et al.*, 1977, 1980; Willmitzer *et al.*, 1980). Tumorous growth results from expression of three T-DNA genes that encode biosynthetic enzymes for two plant growth hormones, auxin (indole acetic acid) and cytokinin (isopentenyl adenosine) for reviews see Ream, 1989; Winans, 1992; Zambryski, 1992). There are many excellent reviews of *Agrobacterium tumefaciens* biology (Zambryski, 1992; Winans, 1992; Hooykaas and Beijersbergen, 1994; Citovsky, and Zambryski, 1993; Dreiseikelmann, 1994; Ream, 1989; Binns and Thomashow, 1988), and the most recent presents an

---

**Walt Ream** Molecular and Cellular Biology Program and Department of Microbiology, Oregon State University, Corvallis, Oregon 97331.

*Subcellular Biochemistry, Volume 29: Plant–Microbe Interactions*, edited by Biswas and Das. Plenum Press, New York, 1998.

outstanding treatment of nuclear localization of T-DNA (Sheng and Citovsky, 1996), the topic of this chapter.

### 1.1. Regulation of Virulence Genes

Virulence (*vir*) genes necessary for T-DNA transmission (transfer and integration) lie elsewhere on the Ti plasmid (Stachel and Nester, 1986; Garfinkel and Nester, 1980). VirA, a signal receptor/kinase protein located in the bacterial membrane, phosphorylates VirG, a transcriptional activator, in response to phenolic compounds released by wounded plant cells (Stachel *et al.*, 1985; for reviews see Winans, 1992; Hooykaas and Beijersbergen, 1994). Phosphorylated VirG protein activates transcription of its own gene and other *vir* operons (Winans, 1992).

### 1.2. Protein Export Channel

Export of T-DNA and proteins (VirD2, VirE2, and VirF) from *A. tumefaciens* into plant cells depends on membrane-associated proteins encoded by the *virB* operon, which contains 11 genes (for reviews see Zambryski, 1992; Hooykaas and Beijersbergen, 1994), and the *virD4* gene (Jayaswal *et al.*, 1987; Okamoto *et al.*, 1991). The VirB proteins are similar in amino acid sequence to pertussis toxin liberation (Ptl) proteins of *Bordetella pertussis*, which mediate export of pertussis toxin (Covacci and Rappuoli, 1993; Weiss *et al.*, 1993), and to proteins that mediate conjugal transfer of IncPα plasmid RP4 (Trb proteins) (Lessl *et al.*, 1992) and IncN plasmid pKM101 (Tra proteins) (Pohlman *et al.*, 1994). VirD4 protein is similar to TraG, another protein required for conjugal transfer of RP4 (Lessl *et al.*, 1992). Thus, proteins essential for crown gall tumorigenesis appear to permit export of specific Vir proteins and T-DNA into plant cells using pathways similar to those that operate in other bacteria (see Table I).

### 1.3. Origin of Transfer

Direct repeats, known as border sequences, define the T-DNA ends; T-DNA transfer requires, in *cis*, the right-hand 25-base-pair border sequence, while deletions that remove it abolish tumorigenesis (Peralta and Ream, 1985; Wang *et al.*, 1984; Shaw *et al.*, 1984). Loss of a nearby sequence, designated *overdrive*, reduces tumorigenesis several hundredfold (Peralta *et al.*, 1986); the *virC* operon, which is also necessary for full virulence, encodes a protein (VirCl) that binds *overdrive* (Toro *et al.*, 1989). T-DNA transfer begins when an endonuclease comprised of VirD1 and VirD2 nicks at a specific site within the right-hand border sequence (Wang *et al.*, 1987; Yanofsky *et al.*, 1986) and the VirD2 subunit attaches to the 5′ end of the nicked strand (Ward and Barnes, 1988; Young and Nester, 1988; Herrera-Estrella *et al.*, 1988; Howard *et al.*, 1989; Durrenberger *et al.*, 1989). Displacement, in a 5′-to-3′ direction, of the bottom (nicked) strand of the T-DNA produces linear VirD2-bound single-stranded DNAs called T strands (Stachel *et al.*, 1986; Albright *et al.*, 1987; Jayaswal *et al.*, 1987), which the bacteria export into plant cells (Stachel and Zambryski, 1986; Yusibov *et al.*, 1994; Tinland *et al.*, 1994).

### 1.4. Secretion of a Single-Stranded DNA-Binding Protein

*A. tumefaciens* also exports VirE2, a cooperative single-stranded DNA-binding (SSB) protein (Citovsky *et al.*, 1988, 1989; Gietl *et al.*, 1987; Das, 1988; Christie *et al.*, 1988; Sen *et al.*, 1989), into plant cells via a *vir*-dependent pathway. However, transfer of VirE2 and the T strand–VirD2 complex can occur independently. A *virE2* mutant can transfer T strands into plant cells, and a *virE*$^+$ strain that lacks T-DNA can export VirE2 protein into plant cells; thus, inoculation of two such nonpathogenic strains onto the same plant wound restores tumorigenesis (Otten *et al.*, 1984). This phenomenon, called extracellular complementation, requires functional VirB proteins, VirD4, and loci (*chvA*, *chvB*, and *exoC*) necessary for binding of *A. tumefaciens* to plant cells (Christie *et al.*, 1988). The latter observation indicates that the *virE*$^+$ strains probably export VirE2 directly into plant cells, rather than into the *virE2* mutant.

### 1.5. VirE2 and T Strands Are Exported to Plant Cells Separately

Additional evidence shows that VirE2 and T strands can be exported separately to plant cells. First, export of VirE2, but not of T strands, requires VirE1 (Sundberg *et al.*, 1996). Second, mobilization of the IncQ plasmid RSF1010 from *A. tumefaciens* into plant cells through the VirB pore (Buchanan-Wollaston *et al.*, 1987, 1988) interferes with T-DNA transmission and abolishes tumorigenesis (Ward *et al.*, 1991) by preventing transfer of VirE2 into plant cells (Binns *et al.*, 1991) by preventing transfer of VirE2 into plant cells (Binns *et al.*, 1995). The presence of RSF1010 in *A. tumefaciens* reduces, but does not eliminate transfer of T-DNA to plant cells (Binns *et al.*, 1995). Thus, transfer of T strands can occur even though export of VirE2 is blocked (Binns *et al.*, 1995; Sundberg *et al.*, 1996).

### 1.6. VirE2 Is Required Only in Plant Cells

The VirE2 SSB protein plays an important role inside plant cells, but not inside *A. tumefaciens*. The T strands accumulate to wild-type levels in *virE2* mutants (Stachel *et al.*, 1987; Veluthambi *et al.*, 1988), indicating that VirE2 does not stabilize T strands inside bacterial cells. In addition, *virE2* mutants can transfer T strands into plant cells (Yusibov *et al.*, 1994), although at an unknown efficiency, proving that VirE2 is not essential for export of T strands. Transgenic tobacco plants that produce VirE2 are susceptible to transformation by *virE*

**Table I**
**Properties of Ti Plasmid-Encoded Virulence Proteins[a]**

| | | | | | Homologs | | | | |
|---|---|---|---|---|---|---|---|---|---|
| Protein | AA | $M_r$ | pI | Location | BPtl | RP4 Trb | pKM101 Tra | Other | Function/properties |
| VirA | 829 | 91.8 | 7.6 | IM & PP; HP domain | — | — | — | NtrB | Signal transduction; protein kinase; responds to plant phenolics; phosphorylates VirG & VirA |
| VirB1 | 239 | 26.0 | 6.4 | IM & OM; signal | — | — | L | — | Processed to smaller form & exported; not essential for transfer; lysozyme-like protein; putative glycosidase |
| VirB2 | 121 | 12.4 | 9.7 | IM & OM; signal | A | C | M | TraA of F | Processed to smaller form & exported; F-pilin homolog |
| VirB3 | 108 | 11.8 | 9.8 | IM & OM; HP domain | B | D | A | TraL of F | Needs VirB4 to move from IM to OM; pilus assembly? |
| VirB4 | 789 | 87.4 | 7.9 | IM | C | E | B | FliI; TraC of F | ATPase; needed to move VirB3 to OM; homolog of flagellar protein export and pilus assembly proteins |
| VirB5 | 220 | 23.5 | 5.1 | IM; signal | — | F | C | — | Homolog of protein (Trc) that acts extracellularly; putative pilus component |
| VirB6 | 295 | 31.7 | 4.4 | Trans; HP domain | D | L | D | — | Has memebrane-spanning domains |
| VirB7 | 55 | 5.9 | 8.2 | OM; signal | I | — | N | ColA, ColE1, ColE3, CloDF3 | Forms disulfide bond with and stabilizes VirB9; has lipid attachment site |
| VirB8 | 230 | 25.4 | 8.9 | IM; HP domain | E | — | E | — | May affect accumulation of VirB9 |
| VirB9 | 293 | 32.2 | 8.7 | IM & OM; signal | F | — | O | — | Forms high*Mr*, aggregate disulfide bond with VirB7; aggregation of VirB10 (in separate complex) requires VirB9 |
| VirB10 | 377 | 40.7 | 6.5 | IM & PP; HP Domain | G | I | F | — | Forms aggregates (separate from VirB9) that require VirB9 for assembly; has membrane-spanning domain |

| | | | | | | | | | |
|---|---|---|---|---|---|---|---|---|---|
| VirB11 | 343 | 38.0 | 5.7 | Cyto & IM | H | B | G | ComG *B. subt.* | ATPase; protein kinase; autophosphorylated |
| VirC1 | 231 | 25.7 | 4.9 | — | — | — | — | — | Binds *overdrive* sequence at origin of T-DNA transfer; needed for full virulence |
| VirC2 | 202 | 22.7 | 9.7 | — | — | — | — | — | Needed for full virulence |
| VirD1 | 147 | 16.2 | 8.9 | — | — | TraJ | — | — | Subunit of border-specific endonuclease (relaxosome); may help VirD2 bind border |
| VirD2 | 424 | 47.6 | 8.3 | Plant nucleus | — | TraI | — | — | Border-specific endonuclease; covalent bond to T strand; pilot protein; ligase activity |
| VirD3 | 201 | 21.4 | 11.6 | — | — | TraH | — | — | Not important for virulence; poorly conserved; may stabilize relaxosome |
| VirD4 | 656 | 72.5 | 7.1 | IM & PP; signal | — | TraG | — | TraD of F | ATPase; may link relaxosome to membrane pore |
| VirE1 | 65 | 7.2 | 4.9 | — | — | — | — | — | Needed for export of VirE2, but not for DNA transfer |
| VirE2 | 533 | 60.6 | 6.9 | Plant nucleus | — | — | — | — | Cooperative ssDNA-binding protein; protects T strands; can be supplied *in planta*; may have integrase function |
| VirF | 202 | 22.4 | 7.7 | Plant | — | — | — | — | Host-range factor; not found on nopaline-type pTi; can be supplied *in planta* |
| VirG | 267 | 30.0 | 8.1 | Cytoplasm | — | — | — | NtrC | Transcriptional activator of *vir* operons; phosphorylated by VirA; binds *vir* box |
| VirH1 | 422 | 47.5 | 6.1 | — | — | — | — | P450 | Cytochrome P-450 enzyme; may dextoxify plant defense compounds |
| VirH2 | 407 | 45.9 | 6.2 | — | — | — | — | P450 | Cytochrome P-450 enzyme; may detoxify plant defense compounds |

[a] Abbreviations: AA, amino acids; $M_r$, molecular weight (deduced from nucleotide sequence); BPtl, *Bordetella pertussis* toxin liberation operon; RP4 Trb, transfer (Tra2 or Trb) operon of plasmid RP4; IM, inner membrane; OM, outer membrane; PP, periplasm; signal, protein contains signal peptide sequence for protein export; HP domain, protein contains a hydrophobic transmembrane domain; cyto, cytoplasm; trans, transmembrane; *B. subt.*, *Bacillus subtilis*.

mutant *A. tumefaciens* (Citovsky *et al.*, 1992), indicating that VirE2 is necessary inside plant cells.

## 1.7. Nuclear Localization Signals

VirE2 and VirD2 both contain plant nuclear localization signals (NLSs) (Herrera-Estrella *et al.*, 1990; Howard *et al.*, 1992; Citovsky *et al.*, 1992; Shurvinton *et al.*, 1992; Tinland *et al.*, 1992); inside plant cells these signals appear to target both proteins and the T-DNA to the nucleus (Tinland *et al.*, 1995; Rossi *et al.*, 1996; Zupan *et al.*, 1996) where the T-DNA integrates into the plant genome. The NLSs in VirD2 and VirE2 are similar to the type found in nucleoplasmin from the amphibian *Xenopus laevis* (Robbins *et al.*, 1991). Proteins are directed to the nucleus by NLS sequences and enter by ATP-dependent active transport through nuclear pores (Silver, 1991). Typical NLSs are short regions rich in basic amino acids (Silver, 1991). Many NLSs, including those in VirE2, VirD2, and nucleoplasmin, are bipartite sequences that contain two interdependent basic domains, both needed for full activity (Robbins *et al.*, 1991; Howard *et al.*, 1992; Citovsky *et al.*, 1992) (Fig. 1). NLS-binding proteins recognize NLSs and direct NLS-containing proteins to nuclear pores where transport into nuclei occurs (Silver, 1991). The NLS in VirD2 plays an important role

| | | | |
|---|---|---|---|
| *A. tumefaciens* octopine-type | VirD2 | KRPRDRHDGELGGRKRAR | (396-413) |
| *A. tumefaciens* nopaline-type | VirD2 | KRPREDDDGEPSERKRER | (417-434) |
| *Agrobacterium rhizogenes* | VirD2 | KRPRVEDDGEPSERKRAR | (406-423) |
| *A. tumefaciens* octopine-type | VirE2 | KLRPEDRYVQTERYGRR | (205-221) |
| *A. tumefaciens* nopaline-type | VirE2 | KLRPEDRYIQTEKYGRR | (228-244) |
| *A. tumefaciens* octopine-type | VirE2 | KRRYGGETEIKLKSK | (273-287) |
| *A. tumefaciens* nopaline-type | VirE2 | KTKYGSDTEIKLKSK | (296-310) |
| *Xenopus laevis* nucleoplasmin | | KRPAATKKAGQAKKKKLD | (155-172) |

**FIGURE 1.** NLS from *A. tumefaciens*, *A. rhizogenes*, and *X. laevis*. Boldface denotes basic residues presumed important for nuclear targeting. Numbers in parentheses indicate positions of amino acids shown. Capital letters represent the amino acid sequence using the standard single-letter code.

in tumorigenesis, presumably by facilitating nuclear entry of the VirD2–T strand complex (Rossi *et al.*, 1996). However, deletion of this NLS does not abolish tumorigenesis (Shurvinton *et al.*, 1992), suggesting that VirE2, which contains two bipartite NLSs (Citovsky *et al.*, 1992), also may target T strands to plant nuclei. Indeed, purified VirE2 mediates nuclear uptake of fluorescently labeled single-stranded DNA in plant cells (Zupan *et al.*, 1996). A T strand 15 kb long can bind 500 molecules of VirE2 (Citovsky *et al.*, 1988), but only one of VirD2. It seems likely that the NLSs in VirE2 play an important role in nuclear transport of T strands.

## 2. THE ROLE OF VirD2 IN NUCLEAR LOCALIZATION AND INTEGRATION OF T-DNA

### 2.1. VirD2 and Nuclear Localization

#### 2.1.1. VirD2 as a Pilot Protein

One of the proteins transported into plant cells, the VirD2 endonuclease subunit, is attached to T-strand DNA (Ward and Barnes, 19988; Young and Nester, 1988; Herrera-Estrella *et al.*, 1988; Howard *et al.*, 1989; Durrenberger *et al.*, 1989); thus, T-DNA is transferred to plant cells through a protein secretion channel (the VirB pore) by virtue of its attachment to VirD2. The attachment of VirD2 to the 5′ end of T strands is through a phosphodiester bond with a conserved tyrosine residue at position 29 (Vogel and Das, 1992). The first 213 amino acid residues of VirD2 are highly conserved (85% identity), whereas the region from position 214 to the carboxy terminus (residue 424) contains only a few short stretches of conserved residues, primarily the penultimate 27 amino acids. The T-DNA border-specific endonuclease activity lies entirely within the first 228 residues of VirD2 (Ward and Barnes, 1988); however, the conserved residues at the carboxy terminus play a very important role in tumorigenesis (Shurvinton *et al.*, 1992; Rossi *et al.*, 1993). Because the carboxy terminus of VirD2 is not needed for normal T-strand production, it must act during later stages of T-DNA transmission.

#### 2.1.2. The VirD2 NLS

The conserved amino acids at the carboxy terminus of VirD2 perform two distinct functions. A bipartite nuclear localization signal occupies residues 396–413 (Fig. 1). This sequence, when fused to either β-glucuronidase or β-galactosidase, is necessary and sufficient to target these fusion proteins to plant cell nuclei (Howard *et al.*, 1992; Tinland *et al.*, 1992). A tobacco etch virus NLS, which differs considerably in primary sequence, can substitute for the VirD2

NLS in promoting tumorigenesis, demonstrating that NLS activity, rather than a particular amino acid sequence, is the important feature of this region of VirD2 (Shurvinton *et al.*, 1992). Mutations that destroy this NLS reduce, but do not abolish, tumorigenesis (Shurvinton *et al.*, 1992; Rossi *et al.*, 1993), indicating that although the VirD2 NLS plays a role in T-DNA transmission, it is not essential. Thus, other proteins apparently participate in targeting T strands to the nuclei of host cells; VirE2, which also contains NLS sequences, is a prime candidate for this role.

### 2.1.3. The Omega Sequence of VirD2

A second sequence at the carboxy terminus of VirD2 (DGRGG; residues 419–423), known as the omega domain, is very important for tumorigenesis (Shurvinton *et al.*, 1992) and appears necessary for efficient integration of T-DNA into plant DNA (Narasimhulu *et al.*, 1996). An *A. tumefaciens* strain with a mutant *virD2* gene that lacks the omega sequence (but contains a wild-type NLS in VirD2) delivers T-DNA to plant nuclei with nearly normal proficiency, but stable inheritance (integration) of T-DNA is severely reduced (Narasimhulu *et al.*, 1996). Despite its proximity to the NLS, the omega sequence does not contribute to nuclear localization activity (S. B. Gelvin and W. Ream, unpublished data).

### 2.1.4. A Second NLS in VirD2

Two groups have reported an NLS in the amino-terminal portion of VirD2 adjacent to the active site in the endonuclease domain (Herrera-Estrella *et al.*, 1990; Tinland *et al.*, 1992). This NLS is comprised of a single cluster of basic amino acids (RKGK or RKGR; residues 32–35); β-galactosidase fusion proteins containing this NLS accumulate in plant nuclei upon expression in plant cells (Herrera-Estrella *et al.*, 1990; Tinland *et al.*, 1992). However, this NLS does not play a role in nuclear localization of T strands. A substitution mutation that changes one lysine to a glutamic acid (RKGK to REGK) does not diminish tumorigenesis (Shurvinton *et al.*, 1992). Furthermore, just two amino acids separate this NLS from the tyrosine residue (position 29) that likely forms a phosphodiester bond with the 5′ end of the T-strand (Vogel and Das, 1992). Thus, the attached T-strand DNA probably hinders access to the adjacent NLS (Tinland *et al.*, 1992).

### 2.1.5. An Alternate Pilot Protein

A broad-host-range mobilizable bacterial plasmid (RSF1010), without a T-DNA border sequence, can transfer from *A. tumefaciens* through the VirB pore into plant cells and integrate into the host genome (Buchanan-Wollaston *et al.*, 1987, 1988). In addition to certain *vir* genes, this transfer requires the RSF1010 plasmid's origin of transfer (*oriT*) sequence and mobilization (*mob*) proteins that create a strand- and site-specific nick within *oriT* (Buchanan-Wollaston *et al.*, 1987). The RSF1010 MobA protein (Derbyshire *et al.*, 1987; Scholz *et al.*, 1989) is the equivalent of VirD2 and attaches to the nicked DNA strand. However, MobA lacks a putative NLS: it does not contain a cluster of four amino acids comprised of at least three basic residues (lysine or arginine). Nevertheless, MobA-linked T strands can enter plant nuclei at an easily detectable frequency (Buchanan-Wollaston *et al.*, 1987). This confirms that an NLS sequence on the pilot protein is not crucial for nuclear import of T strands.

## 2.2. VirD2 Participates in T-DNA Integration

### 2.2.1. Analysis of Integrated T-DNAs

The left and right ends of the T-DNA are joined to plant DNA via different mechanisms. Sequences of plant-to-T-DNA junctions indicate that the right-hand end of the integrated T-DNA very often corresponds exactly to the point at which the 5′ end of the T strand attached to VirD2; in contrast, the point at which the left end of the T-DNA joins to plant DNA often varies by hundreds of bases (Gheysen *et al.*, 1987, 1991; Matsumoto *et al.*, 1990; Mayerhofer *et al.*, 1991). This suggests not only that VirD2 pilots the 5′ end of the T-strand into the nucleus, but VirD2 also appears to protect the 5′ end of the T-strand from nuclease attack (Jasper *et al.*, 1994) and to ligate this end of the T-strand to plant DNA (Pansegrau *et al.*, 1993; Jasper *et al.*, 1994). In support of this hypothesis, conversion of an arginine residue (at position 129) to a glycine not only reduces the efficiency of T-DNA transmission, but also results in T-DNA structures with aberrant right-hand ends (Tinland *et al.*, 1995). Thus, proper joining of the VirD2-bound end of a T strand to plant DNA requires wild-type VirD2 activity.

### 2.2.2. Ligase Activity of VirD2

Although cleavage of T-DNA border sequences and production of T strands *in vivo* requires both VirD1 and VirD2 (Veluthambi *et al.*, 1987), *in vitro*, purified VirD2 protein cleaves single-stranded oligonucleotides that contain the bottom strand of the 25-base T-DNA border sequence (Jasper *et al.*, 1994; Pansegrau *et al.*, 1993). This *in vitro* activity appears identical to VirD2-promoted cleavage *in vivo* because the endonuclease cuts between the third and fourth bases (from the left end of the border sequence), and VirD2 protein remains bound to the 5′ end of the cleaved oligonucleotide (Jasper *et al.*, 1994; Pansegrau *et al.*, 1993). This VirD2–DNA complex has ligase activity. Approximately 50% of the oligonucleotide substrate is cleaved and bound by VirD2, and increased quantities of VirD2 or longer incubation times do not change this equilibrium between nicked and intact molecules (Jasper *et al.*, 1994). This suggests that VirD2 can ligate the nicked DNA molecules to reform the original substrate. Another study demonstrated that VirD2 can nick a border sequence oligonucleotide and join the VirD2-bound portion to another oligonucleotide in a sequence-specific manner (Pansegrau *et al.*, 1993). Apparently, this same ligase activity joins the 5′ end of T strands to plant DNA.

## 3. VirE2: T-STRAND PROTECTION, NUCLEAR LOCALIZATION, AND INTEGRATION

### 3.1. VirE2 Protects T Strands in Plant Cells

#### 3.1.1. T-Strand Accumulation in Plant Cells

Two recent studies demonstrate that VirE2 protects T strands from degradation by nucleases inside plant cells. Yusibov *et al.* (1994) measured accumulation of T strands in tobacco cells infected with either wild-type or *virE2* mutant *A. tumefaciens*. The *virE2* mutation drastically reduces the amount of T-strand DNA recovered from the cytoplasm of infected plant cells (Yusibov *et al.*, 1994), even though this mutation does not affect levels of T strands in the bacterial cells (Stachel *et al.*, 1987; Veluthambi *et al.*, 1988). This suggests that T strands are more susceptible to nucleolytic degradation in the absence of VirE2.

#### 3.1.2. Truncated T-DNA Structures

Recent observations by Rossi *et al.* (1996) also indicate that VirE2 protects T-strand DNA from degradation inside plant cells. As in several other studies, these authors found no evidence for degradation of T strands inside *virE2* mutant *A. tumefaciens* (Rossi *et al.*, 1996). Although T-DNA transmission and tumorigenesis is severely reduced by mutations that abolish *virE2* function, some transformation of plant cells does occur (Garfinkel and Nester, 1980; Stachel and Nester, 1986). In the absence of VirE2, integrated T-DNAs are often truncated at their left-hand ends (Rossi *et al.*, 1996), confirming the importance of VirE2 for protection of T strands from nuclease attack in plant cells.

### 3.2. VirE2 and Nuclear Localization

#### 3.2.1. VirE2 Contains Two Bipartite Nuclear Localization Signals

VirE2 contains two NLSs in the central region of the protein (Fig. 1) (Citovsky *et al.*, 1992). Although each VirE2 NLS is capable of directing a β-glucuronidase fusion protein to the nucleus of a plant cell, both are necessary for maximum accumulation in the nucleus (Citovsky *et al.*, 1992). Genetic testing of the importance of these NLSs for T-DNA transmission has not been possible because the NLS domains lie in regions of VirE2 important for binding to single-stranded DNA (ssDNA) and for cooperative interaction between VirE2 molecules (Citovsky *et al.*, 1992; Dombek and Ream, 1997). Thus, mutations in *virE2* that abolish the activity of one of these NLSs either prevent DNA binding (in the case of the NLS nearest the carboxy terminus) or eliminate cooperativity (Citovsky *et al.*, 1992; Dombek and Ream, 1997).

### 3.2.2. VirE2 Can Target ssDNA to Plant Nuclei

The NLSs in VirE2 appear to remain accessible for nuclear targeting when the protein is bound to DNA, despite the involvement of these regions in protein–DNA and protein–protein interactions. Fluorescently labeled ssDNA, when coated with VirE2, accumulates in nuclei upon microinjection of the complex into plant cells (Zupan *et al.*, 1996). This nuclear import is inhibited by wheat germ agglutinin or a nonhydrolyzable GTP analog, which block nuclear localization of proteins in animal cells (Zupan *et al.*, 1996). Thus, VirE2 retains its NLS-dependent nuclear localization capability even when bound to ssDNA.

### 3.2.3. VirE2 Changes the Structure of ssDNA

Electron micrographs of VirE2–ssDNA complexes indicate that this nucleoprotein complex forms a very thin extended filament (Citovsky *et al.*, 1989). This physical alteration may facilitate passage of the complex through the nuclear pore (Citovsky *et al.*, 1989), and some have suggested that it may obviate the need for the VirE2 NLSs to remain exposed when the protein is bound to ssDNA (Bravo Angel *et al.*, 1996; Rossi *et al.*, 1996).

### 3.2.4. Contradictory Observations

Two studies may argue against a role for VirE2 in nuclear localization of T strands. Rossi *et al.* (1996) found that deletion of *virE2* from an *A. tumefaciens* strain that lacks the carboxy-terminal NLS in VirD2 did not reduce T-DNA transmission, suggesting that VirE2 does not play a role in NLS-dependent localization of T strands to the nuclear pore, which is the initial step of nuclear import. Instead, they suggest that VirE2 may assist with the second stage of nuclear transport, the ATP-dependent translocation of T strands through the pore, by unfolding the ssDNA into an extended filament (Rossi *et al.*, 1996). However, plant cells possess a Vir-independent mechanism for nuclear import of T strands (see Section 3.2.5), and nuclear targeting via this pathway may mask the effect of eliminating the putative VirE2-dependent pathway.

The second observation that raises questions about the role of VirE2 in nuclear import of T strands comes from the proponents of VirE2-dependent nuclear localization. Citovsky *et al.* (1992) created transgenic tobacco that expresses VirE2, which accumulates in the nuclei of these plant cells. These plants fully restore pathogenicity to a *virE* mutant that does not induce tumors on wild-type tobacco (Citovsky *et al.*, 1992). If most of the VirE2 protein resides in the nucleus of these plants at the time of infection, how can it play a role in nuclear import of T strands? Conceivably, the cytoplasm may contain enough VirE2 to target T strands to the nucleus efficiently, even though a great percentage of the VirE2 is already in the nucleus. Alternatively, VirE2 may contain NLSs for the sole purpose of targeting the protein, without DNA, to the nucleus, where it may protect T strands from degradation and perhaps play a direct role in T-DNA integration (see Section 3.3).

#### 3.2.5. Alternate Pathways

Plant cells must have an alternate means, independent of the NLSs in both VirD2 and VirE2, to transport T strands into the nucleus. This was established by Rossi *et al.* (1996), who showed that a strain that lacks both the VirD2 NLS and VirE2 can transfer and integrate T-DNA at 7.7% of the wild-type efficiency. Thus, nuclear localization mediated by Vir proteins is not essential for T-DNA transmission.

### 3.3. VirE2 May Participate in T-DNA Integration

#### 3.3.1. The First Evidence for a VirE2 Integrase

Gardner and Knauf (1986) devised an extremely sensitive method to detect T-DNA transfer to plant cells even when the incoming T-DNA fails to integrate into the host genome. They constructed a T-DNA containing tandem copies of the potato spindle tuber viroid genome; the viroid can initiate autonomous replication from a single molecule of transferred T-DNA, spread systematically from a single host cell, and cause obvious viroid symptoms throughout the entire plant. Viroid replication occurs upon transfer of T-DNA into plant cells; it presumably does not depend on T-DNA integration. Appropriate host plants infected by a *virE* mutant that harbors viroid sequences in the T-DNA exhibit viroid symptoms, but do not develop crown gall tumors, whereas wild-type strains induce both tumors and viroid symptoms (Gardner and Knauf, 1986). These results suggest that a *virE* mutant can deliver T-DNA to the nucleus, but that, without VirE2, the T-DNA does not integrate stably. However, we know that plant cells infected by a *virE2* mutant contain significantly less T-strand DNA than those infected by a wild-type strain (Yusibov *et al.*, 1994). Thus, detectable tumorigenesis, the result of transfer and *integration* of intact T-DNA, may require more T-DNA molecules than a *virE2* mutant can deliver, whereas viroid replication may require a lower threshold amount of T-DNA.

### 3.3.2. VirE2 Is Important for Normal T-DNA Integration

The rare T-DNAs integrated after infection of a plant by a *virE2* mutant have aberrant structures (Rossi *et al.*, 1996). These T-DNAs are truncated at their left (3′) ends, which indicates that VirE2 protects T strands from nucleolytic degradation (Rossi *et al.*, 1996). In addition, most T-DNAs that are truncated at the left-hand plant-to-T-DNA junction also have aberrant junctions between their right ends and plant DNA (Rossi *et al.*, 1996). This observation suggests a direct role for VirE2 in T-DNA integration, beyond simple protection of T strands from nuclease activity. In the absence of VirE2, T-DNA integration apparently occurs via a VirD2-independent pathway. Hence, aberrant right-hand junctions may result from VirD2-independent integration events.

Rossi *et al.* (1996) argue that VirE2 does not influence the efficiency of T-DNA integration. However, plant cells infected by a *virE2* mutant accumulate much less T-strand DNA than cells infected by a wild-type strain (Yusibov *et al.*, 1994), and plant cells must possess a VirE2-independent T-DNA integration pathway, which is responsible for the rare tumors induced by *virE2* mutants (Garfinkel and Nester, 1980; Stachel and Nester, 1986). This alternate T-DNA integration pathway may be sufficient to process efficiently a small amount of T-strand DNA, such as that present during infection by a *virE2* mutant.

### 3.3.3. VirE2 May Promote Synapsis with Host Chromosomes

VirE2 SSB protein may resemble RecA protein of *Escherichia coli*, which acquires the ability to bind double-stranded DNA only after binding heterologous or homologous single-stranded DNA (Radding, 1991). Such an activity might mediate synapsis between incoming T strands and double-stranded host DNA. Indeed, VirE2 protein can promote, at extremely low frequency, homologous genetic exchange in *E. coli* that lack the *recA* gene (W. Ream, unpublished data). Thus, VirE2 may do more than simply escort T strands through the nuclear pore and protect them from nucleases. VirE2 may play an active role in T-DNA integration.

## 4. VirF: ANOTHER PROTEIN SECRETED INTO PLANT CELLS

### VirF Is Required Only in Plant Cells

VirF resembles VirE2 in that it, too, can be transported into plant cells through the VirB pore in the absence of T-DNA, and mutations in *virF* can be complemented extracellularly (Van Slogteren *et al.*, 1984), in the same manner as mutations in *virE2* (see Section 1.4). However, VirF is important only on some species, and the gene is absent in nopaline-type strains (Regensburg-Tuink, 1993). In contrast to VirE2, export of VirF does not depend on VirE1 (P. Hooykaas, personal communication). Transgenic plants that express VirF restore the virulence of *virF*-mutant *A. tumefaciens* strains, indicating that, like VirE2, VirF functions in plant cells, but not in *A. tumefaciens* (Regensburg-Tuink, 1993). VirF lacks an obvious NLS (a four-amino acid cluster containing at least three basic residues), although this protein is small enough (22 kDa) that it may not require an NLS to enter the nucleus. Because the function and subcellular localization of VirF in plants is unknown, I will not discuss it further.

## 5. A NEW MODEL FOR T-DNA TRANSMISSION

### 5.1. *Caveat Emptor*

The following model represents my personal view of T-DNA transfer and integration. Although it is based on the information available at present, some aspects of the model are speculation on my part. This model is intended to stimulate discussion, thought, and future experimentation.

### 5.2. Protein Export

*A. tumefaciens* exports at least three Vir proteins (VirD2, VirE2, and VirF) to plant cells through a pore formed by VirB proteins (and VirD4), which resemble the *B. pertussis* Ptl toxin protein export apparatus.

### 5.3. T-DNA Export

T-DNA export is simply export of a protein (VirD2) which happens to have DNA attached. Thus, T strands are transferred to plant cells through a protein secretion channel (the VirB pore) solely by virtue of their attachment to the VirD2 endonuclease.

### 5.4. Site of Action of VirE2

VirE2 is necessary only in plant cells.

### 5.5. VirE2 Export

The VirE2 SSB protein need not, and very likely does not, bind T-strand DNA until both have been transported independently through the VirB pore into plant cells. The T-strand transfer probably resembles conjugal transfer of the F plasmid: the plasmid-encoded SSB protein is not transferred to the recipient (Rees and Wilkins, 1990).

### 5.6. Interaction between the Origin of Transfer and the VirB Pore

VirCl, which binds the *overdrive* sequence within the origin of T-DNA transfer, may interact (directly or through VirC2) with membrane proteins in the VirB/VirD4 pore, thereby bringing the right-hand border sequence near to the pore. This might permit immediate export of VirD2 (and the attached T strand) when the VirD1/VirD2 endonuclease nicks the right-hand border sequence. Thus, T strands may leave the bacterial cell without contacting the bacterial cytoplasm, which may explain why VirE2 does not bind T-strand DNA in *A. tumefaciens*.

### 5.7. T-Strand Protection

VirD2 and VirE2 each protect T strands from nuclease attack, but they do so in different ways.

### 5.8. Nuclear Localization of T Strands

The NLS near the C-terminus of VirD2 plays a minor role in nuclear targeting of T strands, whereas VirE2 appears much more important for nuclear localization.

### 5.9. Alternate Pathways of Nuclear Import

Some T strands enter plant nuclei even without VirE2 or the VirD2 NLS.

### 5.10. T-DNA Integration

VirD2 appears to ligate the 5′ end of T strands to plant DNA. VirE2 probably also participates directly in T-DNA integration. Aberrant T-DNA integration may occur at a low frequency in the absence of VirE2 and VirD2.

## 6. FUTURE QUESTIONS

Several important questions will continue to occupy those interested in *Agrobacterium tumefaciens* T-DNA transfer. What is the molecular structure of the VirB pore, and how does it function? What will X-ray crystallography tell us about the structure of VirE2–DNA and VirD2–DNA complexes? What role does VirD1 play in cleavage of T-DNA borders?

Future studies must also investigate the plant proteins that interact with T-DNA and Vir proteins. For example, what plant proteins are necessary for nuclear localization of Vir proteins and T-DNA, and how many NLS-binding proteins are there? What plant proteins are involved in T-DNA integration?

I will conclude with perhaps the most important question. How can we modify *A. tumefaciens* to permit efficient gene replacement in plants via homologous recombination?

## 7. REFERENCES


Albright, L. M., Yanofsky, M. F., Leroux, B., Ma, D., and Nester, E. W., 1987, Processing of the T-DNA of *Agrobacterium tumefaciens* generates border nicks and linear, single-stranded T-DNA, *J. Bacteriol.* **169:**1046–1055.

Binns, A. N., and Thomashow, 1988, Cell biology of *Agrobacterium* infection and transformation of plants, *Annu. Rev. Microbiol.* **42:**575–606.

Binns, A. N., Beaupre, C. E., and Dale, E. M., 1995, Inhibition of VirB-mediated transfer of diverse substrates from *Agrobacterium tumefaciens* by the IncQ plasmid RSF1010, *J. Bacteriol.* **177:**4890–4899.

Bravo Angel, A. M., Tinland, B., and Hohn, B., 1996, The role of virulence proteins in nuclear targeting and integration of T-DNA, in *Crown Gall: Advances in Understanding Interkingdom Genetic Exchange* (W. Ream and S. B. Gelvin, eds.), pp. 30–43, APS Press, St. Paul, Minnesota.

Buchanan-Wollaston, V., Passiatore, J. E., and Cannon, F., 1987, The *mob* and *oriT* mobilization functions of a bacterial plasmid promote its transfer to plants, *Nature* **328:**172–175.

Buchanan-Wollaston, V., Passiatore, J. E., and Cannon, F., 1988, The effect of *vir* mutations on plasmid transfer into plants, in *Molecular Genetics of plant–Microbe Interactions* (R. Palachios and D. P. S. Verma, eds.), pp. 281–282, APS Press, Saint Paul, Minnesota.

Chilton, M. D., Drummond, M. H., Merlo, D. J., Sciaky, D., Montoya, A. L., Gordon, M. P., and Nester, E. W., 1977, Stable incorporation of plasmid DNA into higher plant cells: The molecular basis of crown gall tumorigenesis, *Cell* **11:**263–271.

Chilton, M. D., Saiki, R. K., Yadav, N., Gordon, M. P., and Quetier, F., 1980, T-DNA from *Agrobacterium* Ti plasmid is in the nuclear DNA fraction of crown gall tumor cells, *Proc. Natl. Acad. Sci. USA* **77:**4060–4064.

Christie, P. J., Ward, J. E., Winans, S. C., and Nester, E. W., 1988, The *Agrobacterium tumefaciens virE2* gene product is a single-stranded-DNA-binding protein that associates with T-DNA, *J. Bacteriol.* **170:**2659–2667.

Citovsky, V., and Zambryski, P., 1993, Transport of nucleic acids through membrane channels: Snaking through small holes, *Annu. Rev. Microbiol.* **47:**167–197.

Citovsky, V., De Vos, G., and Zambryski, P., 1988, Single-stranded DNA binding protein encoded by the *virE* locus of *Agrobacterium tumefaciens*, *Science* **240:**501–504.

Citovsky, V., Wong, M. L., and Zambryski, P., 1989, Cooperative interaction of *Agrobacterium* VirE2 protein with single-stranded DNA: Implications for the T-DNA transfer process, *Proc. Natl. Acad. Sci. USA* **86:**1193–1197.

Citovsky, V., Zupan, J., Warnick, D., and Zambryski, P., 1992, Nuclear localization of *Agrobacterium* VirE2 protein in plant cells, *Science* **256:**1802–1805.

Covacci, A., and Rappuoli, R., 1993, Pertussis toxin export requires accessory genes located downstream from the pertussis toxin operon, *Mol. Microbiol.* **8**:429–434.

Das, A., 1988, *Agrobacterium tumefaciens virE* operon encodes a single-stranded DNA-binding protein, *Proc. Natl. Acad. Sci. USA* **85**:2909–2913.

DeCleene, M., and DeLey, J., 1976, The host range of crown gall, *Bot. Rev.* **42**:389–466.

Derbyshire, K. M., Hatfull, G., and Willetts, N., 1987, Mobilization of the non-conjugative plasmid RSF1010: A genetic and DNA sequence analysis of the mobilization region, *Mol. Gen. Genet.* **206**:161–168.

Dombek, P., and Ream, W., 1997, Functional domains of *Agrobacterium tumefaciens* single-stranded DNA-binding protein VirE2, *J. Bacteriol.* **179**:1165–1173.

Dreiseikelmann, B., 1994, Translocation of DNA across bacterial membranes, *Microbiol. Rev.* **58**:293–316.

Durrenberger, F., Crameri, A., Hohn, B., and Koukolikova-Nicola, Z., 1989, Covalently bound VirD2 protein of *Agrobacterium tumefaciens* protects the T-DNA from exonucleolytic degradation, *Proc. Natl. Acad. Sci. USA* **86**:9154–9158.

Gardner, R. C., and Knauf, V. C., 1986, Transfer of *Agrobacterium* DNA to plants requires a T-DNA border but not the *virE* locus, *Science* **231**:725–727.

Garfinkel, D. J., and Nester, E. W., 1980, *Agrobacterium tumefaciens* mutants affected in crown gall tumorigenesis and octopine catabolism, *J. Bacteriol.* **144**:732–743.

Gheysen, G., Van Montagu, M., and Zambryski, P., 1987, Integration of *Agrobacterium tumefaciens* transfer DNA (T-DNA) involves rearrangements of target plant DNA sequences, *Proc. Natl. Acad. Sci. USA* **84**:6169–6173.

Gheysen, G., Villarroel, R., and Van Montagu, M., 1991, Illegitimate recombination in plants: A model for T-DNA integration, *Genes Dev.* **5**:287–297.

Gietl, C., Koukolikova-Nicola, Z., and Hohn, B., 1987, Mobilization of T-DNA from *Agrobacterium* to plant cells involves a protein that binds single-stranded DNA, *Proc. Natl. Acad. Sci. USA* **84**:9006–9010.

Herrera-Estrella, A., Chen, Z., Van Montagu, M., and Wang, K., 1988, VirD proteins of *Agrobacterium tumefaciens* are required for the formation of a covalent DNA–protein complex at the 5′ terminus of T-strand molecules, *EMBO J.* **7**:4055–4062.

Herrera-Estrella, A., Van Montagu, M., and Wang, K., 1990, A bacterial peptide acting as a plant nuclear targeting signal: The amino-terminal portion of *Agrobacterium* VirD2 protein directs a beta-galactosidase fusion protein into tobacco nuclei, *Proc. Natl. Acad. Sci. USA* **87**:9534–9537.

Hooykaas, P. J. J., and Beijersbergen, G. M., 1994, The virulence system of *Agrobacterium tumefaciens*, *Annu. Rev. Phytopathol.* **32**:157–179.

Howard, E. A., Winsor, B. A., De Vos, G., and Zambryski, P., 1989, Activation of the T-DNA transfer process in *Agrobacterium* results in the generation of a T-strand-protein complex: Tight association of VirD2 with the 5′ ends of T-strands, *Proc. Natl. Acad. Sci. USA* **86**:4017–4021.

Howard, E. A., Zupan, J. R., Citovsky, V., and Zambryski, P. C., 1992, The VirD2 protein of *Agrobacterium tumefaciens* contains a C-terminal bipartite nuclear localization signal: Implications for nuclear uptake of DNA in plant cells, *Cell* **68**:109–118.

Jasper, F., Koncz, C., Schell, J., and Steinbiss, H. H., 1994, *Agrobacterium* T-strand production *in vitro:* Sequence-specific cleavage and 5′ protection of single-stranded DNA templates by purified VirD2 protein, *Proc. Natl. Acad. Sci. USA* **91**:694–698.

Jayaswal, R. K., Veluthambi, K., Gelvin, S. B., and Slightom, J. L., 1987, Double-stranded T-DNA cleavage and the generation of single-stranded T-DNA molecules in *E. coli* by a *virD* encoded border specific endonuclease from *Agrobacterium tumefaciens*, *J. Bacteriol.* **169**:5035–5045.

Lessl, M., Balzer, D., Pansegrau, W., and Lanka, E., 1992, Sequence similarities between the RP4

Tra2 and the Ti VirB region strongly support the conjugation model for T-DNA transfer, *J. Biol. Chem.* **267**:20471–20480.

Matsumoto, S., Ito, Y., Hosoi, T., Takahashi, Y., and Machida, Y., 1990, Integration of *Agrobacterium* T-DNA into a tobacco chromosome: Possible involvement of DNA homology between T-DNA and plant DNA, *Mol. Gen. Genet.* **224**:309–316.

Mayerhofer, R., Koncz-Kalman, Z., Nawrath, C., Bakkeren, G., Crameri, A., Angelis, K., Redei, G. P., Schell, J., Hohn, B., and Koncz, C., 1991, T-DNA integration: A mode of illegitimate recombination in plants, *EMBO J.* **10**:697–704.

Narasimhulu, S. B., Deng, X., Sarria, R., and Gelvin, S. B., 1996, Early transcription of *Agrobacterium* T-DNA genes in tobacco and maize, *Plant Cell* **8**:873–886.

Okamoto, S., Toyoda-Yamamoto, A., Ito, K., Takebe, I., and Machida, Y., 1991, Localization and orientation of the VirD4 protein of *Agrobacterium tumefaciens* in the cell membrane, *Mol. Gen. Genet.* **228**:24–32.

Otten, L., DeGreve, H., Leemans, J., Hain, R., Hooykaas, P. J. J., and Schell, J., 1984, Restoration of virulence of *vir* region mutants of *Agrobacterium tumefaciens* strain B653 by coinfection with normal and mutant *Agrobacterium* strains, *Mol. Gen. Genet.* **175**:159–163.

Pansegrau, W., Schoumacher, F., Hohn, B., and Lanka, E., 1993, Site-specific cleavage and joining of single-stranded DNA by VirD2 protein of *Agrobacterium tumefaciens* Ti plasmids: Analogy to bacterial conjugation, *Proc. Natl. Acad. Sci. USA* **90**:11538–11542.

Peralta, E. G., and Ream, L. W., 1985, T-DNA border sequences required for crown gall tumorigenesis, *Proc. Natl. Acad. Sci. USA* **82**:5112–5116.

Peralta, E. G., Hellmiss, R., and Ream, W., 1986, *Overdrive*, a T-DNA transmission enhancer on the *A. tumefaciens* tumour-inducing plasmid, *EMBO J.* **5**:1137–1142.

Pohlman, R. F., Genetti, H. D., and Winans, S. C., 1994, Common ancestry between IncN conjugal transfer genes and macromolecular export systems of plant and animal pathogens, *Mol. Microbiol.* **14**:655–668.

Radding, C. M., 1991, Helical interactions in a homologous pairing and strand exchange driven by RecA protein, *J. Biol. Chem.* **266**:5355–5358.

Ream, W., 1989, *Agrobacterium tumefaciens* and interkingdom genetic exchange, *Annu. Rev. Phytopathol.* **27**:583–618.

Rees, C. E. D., and Wilkins, B. M., 1990, Protein transfer into the recipient cell during bacterial conjugation: Studies with F and RP4, *Mol. Microbiol.* **4**:1199–1205.

Regensburg-Tuink, A. J. G., 1993, Transgenic *N. glauca* plants expressing bacterial virulence gene *virF* are converted into hosts for nopaline strains of *A. tumefaciens*, *Nature* **363**:69–71.

Robbins, J., Dilworth, S. M., Laskey, R. A., and Dingwall, C., 1991, Two interdependent basic domains in nucleoplasmin nuclear targeting sequence: Identification of a class of bipartite nuclear targeting sequence, *Cell* **64**:615–623.

Rossi, L., Hohn, B., and Tinland, B., 1993, The VirD2 protein of *Agrobacterium tumefaciens* carries nuclear localization signals important for transfer of T-DNA to plants, *Mol. Gen. Genet.* **239**:345–353.

Rossi, L., Hohn, B., and Tinland, B., 1996, Integration of complete transferred DNA units is dependent on the activity of virulence E2 protein of *Agrobacterium tumefaciens*, *Proc. Natl. Acad. Sci. USA* **93**:126–130.

Scholz, P., Haring, V., Wittmann-Liebold, B., Ashman, K., Bagdasarian, M., and Scherzinger, E., 1989, Complete nucleotide sequence and gene organization of the broad-host-range plasmid RSF1010, *Gene* **75**:271–288.

Sen, P., Pazour, G. J., Anderson, D., and Das, A., 1989, Cooperative binding of the VirE2 protein to single-stranded DNA, *J. Bacteriol.* **171**:2573–2580.

Shaw, C. H., Watson, M., and Carter, G., 1984, The right hand copy of the nopaline Ti plasmid 25 bp repeat is required for tumour formation, *Nucleic Acids Res.* **12**:6031–6041.

Sheng, J., and Citovsky, V., 1996, *Agrobacterium*–plant cell DNA transport: Have virulence proteins, will travel, *Plant Cell* **8:**1699–1710.

Shurvinton, C. E., Hodges, L., and Ream, W., 1992, A nuclear localization signal and the C-terminal omega sequence in the *Agrobacterium tumefaciens* VirD2 endonuclease are important for tumor formation, *Proc. Natl. Acad. Sci. USA* **89:** 11837–11841.

Silver, P. A., 1991, How proteins enter the nucleus, *Cell* **64:**489–497.

Stachel, S. E., and Nester, E. W., 1986, The genetic and transcriptional organization of the *vir* region of the A6 Ti plasmid of *Agrobacterium tumefaciens*, *EMBO J.* **5:**1445–1454.

Stachel, S. E., and Zambryski, P., 1986, *Agrobacterium tumefaciens* and the susceptible plant cell: A novel adaptation of extracellular recognition and DNA conjugation, *Cell* **47:**155–157.

Stachel, S. E., Messens, E., Van Montagu, M., and Zambryski, P., 1985, Identification of the signal molecules produced by wounded plant cells that activate T-DNA transfer in *Agrobacterium tumefaciens*, *Nature* **318:**624–629.

Stachel, S. E., Timmerman, B., and Zambryski, P., 1986, Generation of single-stranded T-DNA molecules during the initial stages of T-DNA transfer from *Agrobacterium tumefaciens* to plant cells, *Nature* **322:**706–712.

Stachel, S. E., Timmerman, B., and Zambryski, P., 1987, Activation of *Agrobacterium tumefaciens vir* gene expression generates multiple single-stranded T-strand molecules from the pTiA6 T-region: Requirement of 5′ *virD* gene products, *EMBO J.* **6:**857–863.

Sundberg, C., Meek, L., Carroll, K., Das, A., and Ream, W., 1996, VirEl protein mediates export of the single-stranded DNA-binding protein VirE2 from *Agrobacterium tumefaciens* into plant cells, *J. Bacteriol.* **178:**1207–1212.

Tinland, B., Koukolikova-Nicola, Z., Hall, M. N., and Hohn, B., 1992, The T-DNA-linked VirD2 protein contains two distinct functional nuclear localization signals, *Proc. Natl. Acad. Sci. USA* **89:**7442–7446.

Tinland, B., Hohn, B., and Puchta, H., 1994, *Agrobacterium tumefaciens* transfers single-stranded transferred DNA (T-DNA) into the plant cell nucleus, *Proc. Natl. Acad. Sci. USA* **91:**8000–8004.

Tinland, B., Schoumacher, F., Gloeckler, V., Bravo-Angel, A., and Hohn, B., 1995, The *Agrobacterium tumefaciens* virulence D2 protein is responsible for precise integration of T-DNA into the plant genome, *EMBO J.* **14:**3585–3595.

Toro, N., Datta, A., Carmi, O. A., Young, C., Prusti, R. K., and Nester, E. W., 1989, The *Agrobacterium tumefaciens virC1* gene products binds to Overdrive, a T-DNA transfer enhancer, *J. Bacteriol.* **171:**6845–6849.

Van Larebeke, N., Engler, G., Holsters, M., Van den Elsacker, S., Zaenen, I., Schilperoort, R. A., and Schell, J., 1974, Large plasmid in *Agrobacterium tumefaciens* essential for crown gall inducing activity, *Nature* **252:**169–170.

Van Slogteren, G. M. S., Hooykaas, P. J. J., and Schilperoort, R. A., 1984, Tumor formation on plants by mixtures of attenuated *Agrobacterium tumefaciens* T-DNA mutants, *Plant Mol. Biol.* **3:**337–344.

Veluthambi, K., Jayaswal, R. K., and Gelvin, S. B., 1987, Virulence genes A, G, and D mediate the double-stranded border cleavage of T-DNA from the *Agrobacterium* Ti plasmid, *Proc. Natl. Acad. Sci. USA* **84:**1881–1885.

Veluthambi, K., Ream, W., and Gelvin, S. B., 1988, Virulence genes, borders, and overdrive generate single-stranded T-DNA molecules from the A6 Ti plasmid of *Agrobacterium tumefaciens*, *J. Bacteriol.* **170:**1523–1532.

Vogel, A. M., and Das, A., 1992, Mutational analysis of *Agrobacterium tumefaciens virD2:* Tyrosine 29 is essential for endonuclease activity, *J. Bacteriol.* **174:**303–308.

Wang, K., Herrera-Estrella, L., Van Montagu, M., and Zambryski, P., 1984, Right 25 bp terminus sequence of the nopaline T-DNA is essential for and determines direction of DNA transfer from *Agrobacterium* to the plant genome, *Cell* **38:**455–462.

Wang, K., Stachel, S. E., Timmerman, B., Van Montagu, M., and Zambryski, P., 1987, Site-specific nick in the T-DNA border sequence as a result of *Agrobacterium vir* gene expression, *Science* **235**:587–591.

Ward, E. R., and Barnes, W. M., 1988, VirD2 protein of *Agrobacterium tumefaciens* very tightly linked to the 5′ end of T-strand DNA, *Science* **242**:927–930.

Ward, J. E., Dale, E. M., and Binns, A. N., 1991, Activity of the *Agrobacterium* T-DNA transfer machinery is affected by *virB* gene products, *Proc. Natl. Acad. Sci. USA* **88**:9350–9354.

Watson, B., Currier, T. C., Gordon, M. P., Chilton, M. D., and Nester, E. W., 1975, Plasmid required for virulence of *Agrobacterium tumefaciens*, *J. Bacteriol.* **123**:255–264.

Weiss, A. A., Johnson, F. D., and Burns, D. L., 1993, Molecular characterization of an operon required for pertussis toxin secretion, *Proc. Natl. Acad. Sci. USA* **90**:2970–2974.

Willmitzer, L., Debeuckeleer, M., Lemmers, M., Van Montagu, M., and Schell, J., 1980, DNA from Ti plasmid present in nucleus and absent from plastids of crown gall plant cells, *Nature* **287**:359–361.

Winans, S. C., 1992, Two-way chemical signaling in *Agrobacterium*–plant interactions, *Microbiol. Rev.* **56**:12–31.

Yanofsky, M. F., Porter, S. G., Young, C., Albright, L. M., Gordon, M. P., and Nester, E. W., 1986, The *virD* operon of *Agrobacterium tumefaciens* encodes a site-specific endonuclease, *Cell* **47**:471–477.

Young, C., and Nester, E. W., 1988, Association of the VirD2 protein with the 5′ end of T strands in *Agrobacterium tumefaciens*, *J. Bacteriol.* **170**:3367–3374.

Yusibov, V. M., Steck, T. R., Gupta, V., and Gelvin, S. B., 1994, Association of single-stranded transferred DNA from *Agrobacterium tumefaciens* with tobacco cells, *Proc. Natl. Acad. Sci. USA* **91**:2994–2998.

Zambryski, P. C., 1992, Chronicles from the *Agrobacterium*–plant cell DNA transfer story, *Annu. Rev. Plant Physiol. Plant Mol. Biol.* **43**:465–490.

Zupan, J. R., Citovsky, V., and Zambryski, P., 1996, *Agrobacterium* VirE2 protein mediates nuclear uptake of single-stranded DNA in plant cells, *Proc. Natl. Acad. Sci. USA* **93**:2392–2397.

*Chapter 13*

# Oligosaccharide Elicitors in Host–Pathogen Interactions

## Generation, Perception, and Signal Transduction

François Côté, Kyung-Sik Ham, Michael G. Hahn, and Carl W. Bergmann

## 1. INTRODUCTION

Studies of plant–microorganism interactions yielded the first evidence that oligosaccharides could serve as biological signals. Much of this research focused on the synthesis and accumulation of antimicrobial phytoalexins in response to microbial attack. Phytoalexin synthesis and accumulation are observed not only after microbial infection, but also after treatment of plant tissue with cell-free extracts of microbial origin. The active components in these extracts are com-

**François Côté and Michael G. Hahn** Complex Carbohydrate Research Center and Department of Botany, University of Georgia, Athens, Georgia 30602-4712. **Kyung-Sik Ham** Complex Carbohydrate Research Center and Department of Biochemistry and Molecular Biology, University of Georgia, Athens, Georgia 30602-4712; and Department of Food Engineering, Mopko National University, Chonnam 534-729, Korea. **Carl W. Bergmann** Complex Carbohydrate Research Center and Department of Biochemistry and Molecular Biology, University of Georgia, Athens, Georgia 30602-4712.

*Subcellular Biochemistry, Volume 29: Plant–Microbe Interactions*, edited by Biswas and Das. Plenum Press, New York, 1998.

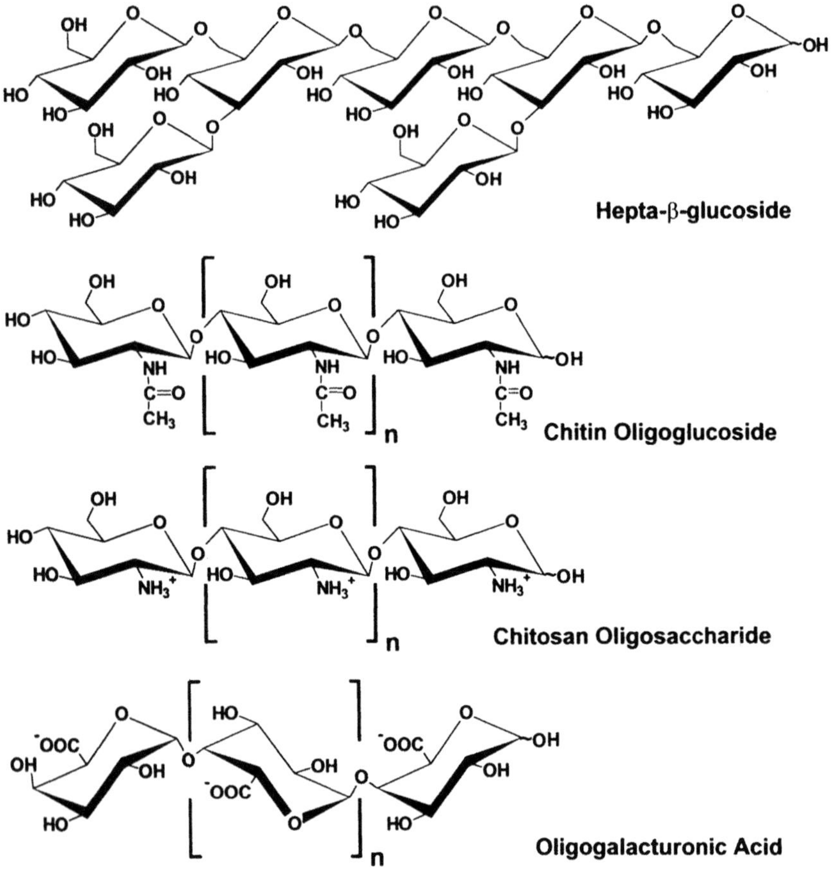


**FIGURE 1.** Structures of oligosaccharide elicitors involved in plant–pathogen interactions.

monly referred to as "elicitors." The term "elicitor" was originally used to refer to molecules and other stimuli that induce the synthesis and accumulation of phytoalexins in plant cells (Keen, 1975), but is now commonly used for molecules that stimulate any plant defense mechanism (Dixon and Lamb, 1990; Hahlbrock and Scheel, 1987; Dixon, 1986; Ebel, 1986). Recent reviews (Côté and Hahn, 1994; Ebel and Cosio, 1994; Hahn *et al.*, 1989a; Ryan, 1988; Darvill and Albersheim, 1984) provide an overview of the structures and the activities of diverse elicitors.

The purpose of this chapter is to review the biological activities, the generation, the perception by the plant cell, and the signal processing for four oligosaccharides thought to be involved in host–pathogen interactions. Two of these oligosaccharides originate from polysaccharides present in the fungal wall, and the third from plant primary cell wall polysaccharides (see Fig. 1). Almost all of the investigations of oligosaccharide elicitors described below utilized *in vitro* preparations, often made using harsh chemical treatments that are unlikely to be encountered *in planta*. Legitimate questions can thus be raised as to whether these oligosaccharides actually are present *in vivo*, how they are released from larger complex carbohydrates, how that release is regulated, and whether the oligosaccharides can interact with their apparent target tissue to effect a biological response. Several recent studies have begun to address these questions and it is within the scope of this review to summarize our knowledge in this area. Therefore, we will focus on those oligosaccharides that have been shown to be released by enzymes present in host–pathogen interactions, and that have been purified to apparent homogeneity and whose structures have been determined. There are a number of studies, which will not be discussed here, that suggest the involvement of additional oligosaccharides as signals in plants, although the oligosaccharides have not been purified and characterized (for review see Aldington and Fry, 1993). Other biologically active oligosaccharides that are active in regulating plant growth and development (for reviews see Fry *et al.*, 1993; Darvill *et al.*, 1992) are not considered. The lipo-chitooligosaccharides will also not be considered here, as they are synthesized by bacteria in symbiotic interaction with plants and therefore, they are not truly involved in host–pathogen interactions, although some of the regulatory mechanisms are common (Spaink and Lugtenberg, 1994).

The purification of a complex carbohydrate to homogeneity often requires elaborate methodology due to the potential structural complexity of these molecules. Therefore, some attention will be given to the purification and characterization procedures developed for each of the oligosaccharide elicitors discussed in this review. Finally, we will also provide an overview of the current understanding of signal transduction pathways activated by these elicitors. In that respect, this review updates a previously published overview (Côté and Hahn, 1994) of our knowledge in this very rapidly expanding field.

## 2. BIOLOGICAL ACTIVITIES AND PURIFICATION

### 2.1. Oligosaccharins Derived from Fungal Wall Polysaccharides

#### 2.1.1. Glucan-Derived Oligosaccharins

Elicitor-active glucans were first detected in the culture filtrates of *Phytophthora sojae*, a phytopathogenic oomycete (Ayers *et al.*, 1976a), and were later also isolated from a commercially available yeast extract (Hahn and Albersheim, 1978). Glucan elicitors have been shown to induce phytoalexins in several legumes, including soybean (Sharp *et al.*, 1984c; Ayers *et al.*, 1976b), chickpea (Gunia *et al.*, 1991), bean (Cline *et al.*, 1978), alfalfa (Kobayashi *et al.*, 1995; Kobayashi *et al.*, 1993), and pea (Bhandal and Paxton, 1991), and also in the solanaceous species potato (Cline *et al.*, 1978) and sweet pepper (Bhandal and Paxton, 1991). These elicitor preparations consisted primarily of glucans containing 3-, 6-, and 3,6-linked β-glucosyl residues (Hahn and Albersheim, 1978; Ayers *et al.*, 1976a), a composition very similar to the β-glucans that are important structural components of various mycelial walls (Bartnicki-Garcia, 1968). Thus, subsequent research focused on elicitors released from mycelial walls.

Partial hydrolysis of the mycelial walls of *P. sojae* using either hot water (Ayers *et al.*, 1976b) or 2 N trifluoroacetic acid (Sharp *et al.*, 1984c) releases a complex mixture of oligo- and polysaccharides. This mixture contains β-glucan fragments ranging in size from short oligomers up to polysaccharides (MW > 100,000) (Sharp *et al.*, 1984c; Ayers *et al.*, 1976c). Oligoglucans in this mixture elicit phytoalexin accumulation (Sharp *et al.*, 1984c) and production of hydroxyproline-rich glycoproteins (Roby *et al.*, 1985) in soybean, as well as activate glycine-rich protein gene expression (Brady *et al.*, 1993) and induce resistance to viruses (Rouhier *et al.*, 1995; Kopp *et al.*, 1989) in tobacco. The structural motifs of the glucans involved in the elicitation of different defense responses are not identical (Rouhier *et al.*, 1995). There may be other, as yet unidentified, activities in the oligoglucan mixture as well. The mycelial wall hydrolysate also contains a large number of biologically inactive molecules and noncarbohydrate elicitors. For example, a proteinaceous component of the crude *P. sojae* elicitor preparation elicits phytoalexin synthesis in suspension-cultured parsley cells (Parker *et al.*, 1988, 1991). This protein elicitor is not active in soybean (Parker *et al.*, 1988). The following sections focus on the glucan elicitors of phytoalexin accumulation, and describe the purification of a hepta-β-glucoside elicitor as well as those structure–activity studies that define essential structural elements of this elicitor.

Fractionation of the mixture of oligoglucosides generated by partial acid hydrolysis of mycelial walls of *P. sojae* on a size-exclusion column revealed that elicitor-active oligosaccharides are present in all fractions containing oligomers of degree of polymerization (DP) ≥ 6 (Sharp *et al.*, 1984c). The heptamer-enriched fraction was further fractionated on a series of normal- and reversed-phase high-performance liquid chromatography (HPLC) columns (Sharp *et al.*, 1984c). It was estimated that the original heptaglucoside mixture contained >100 structurally distinct heptaglucosides, based on the number of peaks observed in the various chromatographic steps. Homogeneous preparations of the aldehyde-reduced forms (i.e., the hexa-β-glucosyl glucitols) of one elicitor-active hepta-β-glucoside (Fig. 1) and of seven other elicitor-inactive hepta-β-glucosides were obtained in amounts sufficient to determine their structures (Sharp *et al.*, 1984b). The structure of the elicitor-active hepta-β-glucoside

(Sharp *et al.*, 1984a) was subsequently confirmed by its chemical synthesis (Nicolaou *et al.*, 1997; Verduyn *et al.*, 1993; Lorentzen *et al.*, 1991; Fügedi *et al.*, 1987; Ossowski *et al.*, 1984).

The ability of the chemically synthesized, unreduced hepta-β-glucoside elicitor to induce phytoalexin accumulation in soybean cotyledons is identical to that of the corresponding hexa-β-glucosyl glucitol purified from fungal wall hydrolysates (Sharp *et al.*, 1984a). Both are active at concentrations of ~10 nM, making them some of the most active elicitors of phytoalexin accumulation yet observed. The seven other hexa-β-glucosyl glucitols that were purified from the partial hydrolysates of fungal cell walls had no elicitor activity over the concentration range (≤ 400 μM) tested (Sharp *et al.*, 1984c). These results provided the first evidence that specific structural features are required for an oligo-β-glucoside to be an effective elicitor of phytoalexin accumulation.

Detailed structure–activity studies (Cheong *et al.*, 1991, 1993) using synthetic oligosaccharides structurally related to the hepta-β-glucoside elicitor (Hong and Ogawa, 1990; Birberg *et al.*, 1989; Fügedi *et al.*, 1987; Ossowski *et al.*, 1984; N. Hong, T. Ogawa, R. Verduyn, and J. van Boom, unpublished results) demonstrate that the three terminal nonreducing glucosyl residues are essential for elicitor activity. The reducing terminal glucosyl residue, on the other hand, is not important for activity and can be modified without significant impact on the activity of the elicitor (Cheong *et al.*, 1991). Coupling of aromatic groups to biologically inactive oligoglucosides does not endow those oligosaccharides with phytoalexin elicitor activity. Thus, the radioiodinated form of the tyramine-coupled hepta-β-glucoside elicitor can be prepared for use as a labeled ligand to search for the presence of elicitor-specific, high-affinity binding sites in soybean membranes (see Section 4 on signal transduction).

### 2.1.2. Chitin- and Chitosan-Derived Oligosaccharins

Oligosaccharide fragments derived from chitin, a linear polysaccharide composed of 4-linked β-*N*-acetylglucosaminyl residues, and chitosan, its de-*N*-acetylated derivative (see Fig. 1), have been shown to elicit defense responses in various plants. Chitin and chitosan polysaccharides are structural components of the cell walls of many fungi, many of which are plant pathogens (Ruiz-Herrera, 1991; Farkas, 1979; Bartnicki-Garcia, 1968).

**2.1.2a. Purification.** Chitosan- and chitin-derived oligomers can be generated from commercially available chitosan derived from the shells of crustaceans. Depending on the source, 80–90% of the glucosaminyl residues are de-*N*-acetylated (Kendra *et al.*, 1989); thus, chitin is generally obtained by re-*N*-acetylation of chitosan (Aiba, 1994a; Molano *et al.*, 1977). Oligomers of either chitin or chitosan can be generated with relatively low yield by partial acid hydrolysis of the corresponding polymer, and the products fractionated on high-resolution size-exclusion columns. Pure chitin oligomers are separated from partially *N*-acetylated fragments by cation-exchange chromatography. The fragments are then further purified by HPLC on amine-modified silica columns (Felix *et al.*, 1993). Alternatively, *N*-acetyl-chitooligosaccharides can be prepared by re-*N*-acetylation of the corresponding chitosanoligosaccharides (Aiba, 1994b; Yamada *et al.*, 1993). Chitin- and chitosan-derived oligosaccharides have also been purified from fungal spores and mycelial walls (Kendra *et al.*, 1989). Chitosan and chitin oligomers can also be generated by enzymatic treatment of the corresponding polymer with chitosanases (Li *et al.*, 1995) or chitinases (Aiba, 1994b). Chitin and chitosan fragments of DP 3–6 are commercially available (Baureithel *et al.*, 1994; Kuchitsu *et al.*, 1993).

**2.1.2b. Induction of Plant Defense Responses by Oligochitins and Oligochitosans.** Heterogeneous preparations of chitosan and chitosan-derived fragments elicit the accumulation of phytoalexins in pea pods (Walker-Simmons *et al.*, 1983; Hadwiger and Beckman, 1980) and in suspension-cultured soybean (Köhle *et al.*, 1984) and parsley (Conrath *et al.*, 1989) cells. Other plant defenses induced by chitosan-derived oligosaccharides include the accumulation of defense-related proteinase inhibitors in tomato and potato leaves (Peña-Cortes *et al.*, 1988; Walker-Simmons *et al.*, 1983, 1984; Walker-Simmons and Ryan, 1984) and the synthesis of callose, a β-1,3-glucan, in suspension-cultured parsley (Conrath *et al.*, 1989), tomato (Grosskopf *et al.*, 1991), and *Catharanthus roseus* (Kauss *et al.*, 1989) cells. Oligosaccharide fragments of chitosan and chitin have both been shown to induce defense-related lignification of the walls of suspension-cultured slash pine (*Pinus elliottii*) cells (Lesney, 1989, 1990). The deposition of both callose and lignin is thought to function in plant defense by strengthening the plant cell wall. Chitosan has also been reported to induce, in mechanically wounded leaves, protection from subsequent infection with fungi (Ride and Barber, 1987) or viruses (Pospieszny *et al.*, 1991; Pospieszny and Atabekov, 1989), though the mechanisms underlying this induced resistance are unknown. Pretreatment of plants with chitosan improves the ability of cucumber (El Ghaouth *et al.*, 1994a), tomato (Benhamou and Lafontaine, 1995; Benhamou *et al.*, 1994), and pepper (El Ghaouth *et al.*, 1994b) to resist subsequent fungal infection. The observed effects in the two latter plants range from induction of cell wall appositions to fungal cell disorganization. Finally, chitosan elicits free radical formation that is thought to lead to cell death, cell wall thickening, and chitinase secretion in suspension-cultured rice cell (Masuta *et al.*, 1991).

The structural requirements for bioactive chitosan- and chitin-derived oligosaccharides have only recently been investigated in detail. In most studies, the fragments used have been heterogeneous in size and often in structure (for example, in the extent and pattern of *N*-acetylation). The oligosaccharides generally must have a DP $>$ 4 to induce a biological response, but beyond that requirement, it is not possible to generalize about structural features essential for activity. In large part, the structural elements required for activity depend on the plant and the defense response being studied. For example, de-*N*-acetylated and partially *N*-acetylated β-1,4-oligoglucosamines with DP > 4 are both effective elicitors of proteinase inhibitors in tomato and potato leaves (Walker-Simmons *et al.*, 1983, 1984). In contrast, chitosan-derived, but not chitin-derived, oligosaccharides with DP > 7 elicit pisatin accumulation in pea pods (Kendra *et al.*, 1989; Kendra and Hadwiger, 1984; Hadwiger and Beckman, 1980). Stereocontrolled chemical synthesis of chitosan oligomers confirmed these results and indicated that the octasaccharide is the most active (Hadwiger *et al.*, 1994). This elicitor also inhibits growth of the pathogenic fungus *Fusarium solani f.sp. phaseoli* (Hadwiger *et al.*, 1994). These acetylation and size requirements for pisatin accumulation seem to be different from those necessary for the induction of less active antimicrobial flavonoids (Akiyama *et al.*, 1994). Maximal induction of pisatin was observed with partially deacetylated oligomers or with chitosan with a degree of acetylation of 56% (Akiyama *et al.*, 1995; Kobayashi *et al.*, 1994). Partial *N*-acetylation or chemical fragmentation reduces the ability of chitosan to elicit callose formation in *C. roseus* cells (Kauss *et al.*, 1989). In other plants, the chitosan oligosaccharides are inactive. Thus, in suspension-cultured rice cells, chitosan and its oligomers have little or no activity, while homogeneous preparations of chitin oligomers with a minimum DP of 4 induce diterpene phytoalexin biosynthesis and accumulation, oligomers of DP ≥ 6 being the most active (Yamada *et al.*, 1993; Ren and West, 1992). The chitin heptamer has been shown to induce the rapid and transient expression of two novel genes with no known role or function (Minami *et al.*, 1996). Chitosan-derived oligosaccharides also do not induce lignification in wounded wheat leaves, while the chitosan and chitin polysaccharides and chitin-derived oligosaccharides with DP > 4 are effective elicitors of this response (Barber *et al.*, 1989).

## 2.2. Oligosaccharins Derived from Plant Cell Wall Polysaccharides

### 2.2.1. Pectin-Derived Oligosaccharins

The family of pectic polysaccharides constitutes a major fraction of the cell walls of dicotyledons and nongraminaceous monocots as well as a minor yet important part of the walls of the graminacea. Pectic polysaccharides include rhammogalacturonans I and II (RG-I and RG-II) and the methyl-esterified and free acid homogalacturonans (O'Neill *et al.*, 1990; Morris, 1986). Partial depolymerization of homogalacturonan, the pectic polysaccharide composed of 1,4-linked α-D-galactosyluronic acid residues (Fig. 1), generates oligogalacturonides that exhibit various regulatory effects in plants including the elicitation of defense responses, the regulation of growth and development, and the induction of rapid responses at the cell surface (Ryan and Farmer, 1991; Roberts, 1990; Albersheim and Darvill, 1985). These effects are extensively reviewed elsewhere (Côté and Hahn, 1994; Darvill *et al.*, 1992).

**2.2.1a. Purification.** The biological activities of oligogalacturonides have often been determined using heterogeneous mixtures of oligosaccharides containing both active and inactive molecules. The bioactive molecules were not, in all cases, purified to homogeneity and shown to be oligogalacturonides. However, the active molecules must contain several contiguous 1,4-linked α-D-galactosyluronic acids since they are inactivated by treatment with homogeneous endo-polygalacturonases (EPGs). Chemically synthesized homogalacturonides (Nakahara and Ogawa, 1989) are required to obtain unequivocal evidence for bioactivity, although homogeneous and well-characterized oligogalacturonides generated from cell walls can be used in their absence since they are more readily available and the chemical synthetic process is particularly difficult in this case (Spiro *et al.*, 1993). Homogeneous oligogalacturonides are essential for the identification of oligogalacturonide-specific receptors and for determining the effects of oligogalacturonides on the physicochemical properties of membranes.

The preparations of oligogalacturonides used for biological studies have been generated by various methods from different source materials and purified to varying degrees. The heterogeneous nature of many of the utilized oligogalacturonide preparations complicates the interpretation of the biological activity studies. Oligogalacturonides can be released from homogalacturonans present in plant primary cell walls either by partial acid hydrolysis of the walls (Nothnagel *et al.*, 1983; Hahn *et al.*, 1981), or by treating the walls with pectic-degrading enzymes such as EPG (Favaron *et al.*, 1992; De Lorenzo *et al.*, 1990; Cervone *et al.*, 1989a; Moloshok and Ryan, 1989; Baldwin and Biggs, 1988; Robertsen, 1987; Davis *et al.*, 1984; Jin and West, 1984) or endo-pectate lyase (Forrest and Lyon, 1990; Hashimoto *et al.*, 1989; Moloshok and Ryan, 1989; Robertsen, 1989; Davis *et al.* 1986b,c). The released oligomers are sometimes deesterified using a mild alkali treatment in order to reduce the complexity of the mixture and facilitate purification. Deesterification is also required in order for EPG digestion to be used to establish the oligogalacturonide nature of the active molecules; EPG will only cleave a deesterified substrate. A more readily available starting material for the preparation of bioactive oligogalacturonides is polygalacturonic acid (PGA) obtainable from commercial sources. While cell wall homogalacturonans are methyl-esterified to varying degrees, PGA is a (essentially) methyl deesterified pectin. Similar, if not identical, active oligogalacturonides may be released from cell walls or PGA by chemical or enzymatic fragmentation (Spiro *et al.*, 1983; Davis *et al.*, 1986c).

The mixtures of oligogalacturonides generated by fragmentation of either plant cell walls or polygalacturonic acid are heterogeneous in size, and include galacturonic acid and oligomers with DP between 2 and 20. The released oligogalacturonides in such mixtures are usually separated by low-pressure anion-exchange chromatography (Robertsen, 1987; Jin and West, 1984; Nothnagel *et al.*, 1983; Hahn *et al.*, 1981) into fractions enriched for specific sizes of oligogalacturonides. Such size-fractionated oligomers have been used in many of the bioactivity studies. However, evidence has shown that oligogalacturonides, particularly those of DP > 10, isolated by low-pressure anion-exchange chromatography, contain significant amounts (~30%) of modified oligogalacturonides (Spiro *et al.*, 1993; Davis *et al.*, 1986b). For example, a fraction containing predominantly a tridecagalacturonide was further analyzed by high-performance anion exchange chromatography in combination with pulsed amperometric detection (HPAEC-PAD) and resolved into seven peaks, three of which had galactaric acid-containing oligogalacturonides (Spiro *et al.*, 1993). Galactaric acid, the C1-oxidized derivative of galacturonic acid, may be formed during commercial processing of PGA or it may be a natural product formed by the action of a plant galacturonic acid oxidase (Pressey, 1993). The HPAEC-purified, apparently homogeneous, tridecagalacturonide is an active elicitor (Spiro *et al.*, 1993). None of the galactaric acid-containing oligogalacturonides elicit phytoalexin accumulation in soybean tis ue (A. Koller, J.-J. Cheong, and M. O'Neill, unpublished results), although relatively high concentrations activate the *in vitro* anionic peroxidase-catalyzed oxidation of indole-3-acetic acid (Pressey, 1991). These results emphasize the importance of obtaining homogeneous oligogalacturonides in order to avoid uncertainty in the attribution of biological activity to homo-oligogalacturonides.

**2.2.1b. Elicitation of Plant Defense Responses by Oligogalacturonides.** The presence of endogenous molecules (elicitors) capable of inducing defense responses in plants was first observed in experiments in which the placement of frozen-thawed bean stem segments in contact with healthy stem segments resulted in the accumulation of phytoalexins in the healthy stem segments (Hargreaves and Bailey, 1978). Subsequently, elicitor-active material was found to be present in extracts of autoclaved bean (*Phaseolus vulgaris*) hypocotyls (Hargreaves and Selby, 1978) and isolated soybean cell walls (Hahn *et al.*, 1981). Elicitor-active molecules present in these extracts were later identified as linear oligomers of 1,4-linked α-D-galactosyluronic acid residues (Nothnagel *et al.*, 1983; Hahn *et al.*, 1981).

The nature of the plant defense response(s) induced by oligogalacturonides depends on the plant being studied. Thus, oligogalacturonides have been shown to induce the accumulation of phytoalexins in soybean (Davis *et al.*, 1986b,c; Nothnagel *et al.*, 1983; Hahn *et al.*, 1981), castor bean (Jin and West, 1984), bean (Tepper and Anderson, 1990; Dixon *et al.*, 1989), pea (Walker-Simmons *et al.*, 1984), and parsley (Davis and Hahlbrock, 1987). Oligogalacturonides also induce the accumulation of antimicrobial shikonins in suspension-cultured *Lithospermum erythrorhizon* cells (Tani *et al.*, 1992). In soybean and parsley, oligogalacturonide and oligo-β-glucoside elicitors (see above) act synergistically, i.e., the concentrations required to elicit phytoalexins when both elicitors are present are less than the concentration required for each elicitor to elicit phytoalexins individually (Davis and Hahlbrock, 1987; Davis *et al.*, 1986a). This synergistic effect has been observed with a synthetic dodecagalacturonide, but not with the decagalacturonide (Hahn *et al.*, 1989b; Nakahara and Ogawa, 1987, 1989). Other oligogalacturonide-induced defense responses include the induction of glycosylhydrolases (β-glucanase, chitinase, lysozyme) in parsley (Davis and Hahlbrock, 1987) and tobacco (Broekaert and Peumans, 1988), the increased deposition of lignin in cucumber (Robertsen, 1986) and castor bean (Bruce and West, 1989), and the accumulation of proteinase inhibitors in tomato (Farmer *et al.*, 1991; Bishop *et al.*, 1984).

The size range of oligogalacturonides that activate defense responses is usually quite narrow. For example, oligogalacturonides with DP between 10 and 15 are generally required to elicit most of the plant defense responses listed above. Although the reason for the frequently observed size dependence of the response to oligogalacturonides is not known, this requirement suggests that oligogalacturonides need 10 or more galactosyluronic acid residues in order to assume a solution conformation that is biologically active. Evidence that oligogalacturonides undergo a conformational transition at DP $> 10$ has been obtained from binding studies with a monoclonal antibody that preferentially binds oligogalacturonides of DP $\geq 10$ (Liners *et al.*, 1992). Di- and trigalacturonides induce proteinase inhibitor production in tomato (Moloshok *et al.*, 1992), suggesting that in some tissues there may be a different size or conformational requirement for biological activity.

Whether or not esterification is important for the biological activity of oligogalacturonides has been partially assessed with somewhat contradictory results. A crude mixture of oligosaccharides released from the cell wall by EPG is active in a tobacco thin-cell layer flower induction assay both before and after deesterification (Marfà *et al.*, 1991). Retreatment of the deesterified cell wall-derived oligosaccharides with EPG destroys the flower-inducing activity, showing that the active component contains 1,4-linked α-D-galactosyluronic acid residues. A crude preparation of rhamnogalacturonan II containing oligogalacturonides elicits soybean phytoalexin accumulation whether or not it has been deesterified (Hahn *et al.*, 1981). However, casbene synthetase induction in castor bean by oligomers of galacturonic acid of DP 12–15 was significantly decreased by *in vitro* methyl-esterification of the carboxylate groups (Jin and West, 1984). The elicitor activity was restored by deesterification of the esterified oligogalacturonides.

## 3. GENERATION OF OLIGOSACCHARIDE ELICITORS AND ITS REGULATION

Some evidence has been obtained that bioactive oligoglucosides, oligochitin, and oligogalacturonides can be generated *in vivo*. Enzymatic hydrolysis offers the most likely biological mechanism for the generation of oligosaccharide elicitors. Plants and pathogens are known to produce a number of hydrolytic enzymes that could be involved in this process. For example, plant β-1,3-glucanases and chitinases exist as families of isoforms whose members are expressed either constitutively or induced following pathogen infection or elicitor treatment (Bol *et al.*, 1990). These enzymes are mostly localized in either the cell wall, the extracellular space, or the vacuole (Mauch and Staehelin, 1989; Boller, 1987). The induction of glucanase and chitinase has often been observed to be coordinated (Boller, 1987). Inducible glucanases, chitinases, and chitosanases are thought to function in plant–pathogen interactions directly by inhibiting fungal growth (Grenier and Asselin, 1990; Kendra *et al.*, 1989; Mauch *et al.*, 1988; Schlumbaum *et al.*, 1986; Hadwiger and Line, 1981). Alternatively, any of these enzymes are capable of generating biologically active glucans, oligochitins, and oligochitosans from the mycelial walls of an invading pathogen at the plant–pathogen interface (Ren and West, 1992; Barber *et al.*, 1989; Kendra *et al.*, 1989), thereby triggering plant defenses.

Other examples of hydrolytic enzymes possibly involved in the generation of oligosaccharins during a plant–pathogen interaction are the microbial pectic-degrading enzymes, in particular, endo-polygalacturonase and endo-pectate lyase. These enzymes are among the first to be secreted by pathogens when cultured on plant cell walls (Mankarios and Friend, 1980; Jones *et al.*, 1972). In addition, a number of different pectic-modifying enzymes exist in uninfected plant tissues; such enzymes have the potential to generate bioactive oligogalacturonides *in planta* by the deesterification of pectin and cleavage of the resulting deesterified homogalacturonans. Preliminary studies have shown that oligogalacturonides are present in the growth medium of suspension-cultured sycamore maple cells (V. Marfà, P. Albersheim, and A. Darvill, unpublished results); their presence in intact plants remains to be investigated.

*In vitro*, the action of the glycosylhydrolases ultimately results in the rapid and total fragmentation of the plant and fungal cell wall polysaccharides to short, biologically inactive oligomers. Therefore, regulatory mechanisms must exist *in planta* in order for the action of wall polysaccharide-degrading enzymes to result in the production of longer, biologically active oligosaccharides. For some lyases, the cell wall pH is several units below the pH optimum of the enzyme and could therefore serve as one regulatory mechanism. The involvement of specific inhibitors of glycosylhydrolases has been presented as another possible regulatory mechanism, and is discussed below.

### 3.1. Generation of β-Glucan Elicitors by Plant Enzymes

The demonstration that bioactive glucan-derived oligosaccharides are effectively generated at the plant–pathogen interface by hydrolytic enzymes would provide evidence that these fragments do have a role *in vivo*. Except for one report (Xu and Mendgen, 1994), there is no clear demonstration that glucan elicitors are present in plants infected by fungi and play a role in the plant–pathogen interaction. In this report, linear and branched β-1,3-glucans were immunodetected in coated vesicles and in clathrin-coated pits around the penetration site of the fungal pathogen *Uromyces vignae* in bean (Xu and Mendgen, 1994). However, the composition and any eliciting ability of the immunodetected β-glucans remain to be established. A similar study of the interaction between *P. sojae* and soybean using the same antibodies failed to show the presence of similar vesicles (Enkerli *et al.*, 1997).

The biological significance of the β-glucan elicitors is supported by the presence *in planta* of enzymes that produce these elicitors from fungal mycelial walls. Several proteins with β-1,3-glucanase activity have been purified from soybean and shown to release elicitor-active fragments from the mycelial walls of *P. sojae* (Ham *et al.*, 1991; Yoshikawa *et al.*, 1990). The expression of these glucanases appears to be differentially regulated. One glucanase has been cloned and its RNA is induced by exogenous application of ethylene (Takeuchi *et al.*, 1990b). Two other β-1,3-glucanases are induced by both pathogen infection and the abiotic elicitor, mercuric chloride (Ham *et al.*, 1991, 1997). The soybean endo-β-1,3-glucanases appear to be quantitatively predominant and perhaps the only enzyme activities capable of generating elicitor-active fragments from fungal mycelial walls (Ham *et al.*, 1991; Takeuchi *et al.*, 1990a). These enzymes do not destroy the activity of glucan elicitors even after long incubation, nor after addition of fresh enzymes (Keen *et al.*, 1993); K.-S. Ham, P. Albersheim, and A. G. Darvill, unpublished results).

The activity of the glucanase-released elicitors is heat-stable (Ham *et al.*, 1991), but is destroyed by soybean exo-β-1,3-glucanase (Okinaka *et al.*, 1995) or by *P. sojae* exo-β-1,3-glucanase (K.-S. Ham, P. Albersheim, and A. G. Darvill, unpublished results). These observations indicate that the chemical nature of the elicitors is carbohydrate; this has been confirmed by composition analysis (Ham *et al.*, 1996). The structure of the elicitors generated by a soybean endo-β-1,3-glucanase appears to consist of β-1,6-linked glucan backbone chains with frequent side branches composed of one or two β-1,3-linked glucose residues (Okinaka *et al.*, 1995). The elicitor-active carbohydrates are probably not major components of the partially purified fractions used for structural analysis. Therefore, the structural characteristics of the oligoglucosides determined in these studies may not reflect the structure of the elicitors. In addition, relatively high molecular weight carbohydrates, the structures of which are almost impossible to determine, were used for the structural characterization (Okinaka *et al.*, 1995).

Purification of the smallest elicitor generated by soybean enzymes has been attempted (Ham *et al.*, 1996). The elicitor-active hepta-β-glucoside, which was isolated from acid hydrolyzates of *P. sojae* mycelial walls, was not detected in the enzyme hydrolysates. The smallest elicitor-active oligoglucoside generated appears to be an octamer or more likely a nonamer. Enzymatically released elicitor-active carbohydrates of various sizes were used to determine whether these elicitors have the ability to compete with the elicitor-active hepta-β-glucoside for the elicitor-binding site on soybean membranes (see Section 4 on signal transduction). All of the tested enzyme-released glucan fragments compete effectively for the binding site. This suggests that at least some of the enzyme-generated oligoglucoside elicitors share those structural elements of the hepta-β-glucoside elicitor that are required for biological activity and detection by the putative receptor (Ham *et al.*, 1996). Thus, there are enzymes *in planta* that generate, from fungal cell walls, β-glucan elicitors with structural similarity to the hepta-β-glucoside elicitor.

### 3.2. Other Proteins that Might Be Involved in the Regulation of β-Glucan Elicitor Production

Glucanase-released β-glucan elicitors can be degraded *in planta*. An exo-β-1,3-glucanase was purified from the walls of suspension-cultured soybean cells and has the ability to degrade β-glucan elicitors that were prepared from acid hydrolysates of *P. sojae* mycelial walls (Cline and Albersheim, 1981). This enzyme can also destroy β-glucan elicitors generated from *P. sojae* mycelial walls by a soybean endo-β-1,3-glucanase (Okinaka *et al.*, 1995). This exo-glucanase has been hypothesized to play a role in limiting the presence of elicitor activity to the tissues immediately surrounding an infection site. Phytoalexin accumulation induced by elicitors must be prevented where it is not required since phytoalexins are toxic to plant as well as pathogen cells (Giannini *et al.*, 1991, 1995; Spessard *et al.*, 1994).

Fungal pathogens also produce a variety of β-glucanases (Pitson *et al.*, 1993). These enzymes are believed to be involved in fungal carbon metabolism or in morphogenetic events such as cell wall growth and extension by plasticizing mycelial walls (Pitson *et al.*, 1993). Some of these enzymes could also play a role in plant–pathogen interactions. For example, elicitor-active β-glucans are released during the germination of *P. sojae* zoospores in the absence of plant host endo-β-1,3-glucanases (Waldmüller *et al.*, 1992). These elicitors might be generated by the fungal endo-β-1,3-glucanases during normal developmental processes and plants may take advantage of such glucans as a signal to induce their defense responses at the initial stages of fungal infection. Fungal pathogens also produce exo-β-1,3-glucanases. An exo-β-1,3-glucanase was isolated from *P. sojae* (K.-S. Ham, P. Albersheim, and A. G. Darvill, unpublished results). This enzyme has the ability to degrade β-glucan phytoalexin elicitors generated from *P. sojae* mycelial walls by acid hydrolysis or by the action of plant endo-β-1,3-glucanases. Thus, fungi may have evolved this mechanism to inactivate plant defense-inducing signals, thereby enhancing their pathogenic potential.

Plant endo-β-1,3-glucanases have the ability to inhibit fungal growth in culture and to generate elicitors by hydrolyzing the polysaccharides of the fungal mycelial walls. As plants evolved endo-β-1,3-glucanases to fight off fungi, fungi may have evolved a mechanism to counter plant endo-β-1,3-glucanases. Indeed, a glucanase inhibitor protein (GIP) was detected in the bean pathogen *Colletotrichum lindemuthianum* (Albersheim and Valent, 1974). More recently, a GIP (GIP-1) was purified to homogeneity from the culture filtrate of the soybean pathogen *P. sojae* (Ham *et al.*, 1997). GIP-1 inhibits only one of the two soybean endo-β-1,3-glucanases (Ham *et al.*, 1997). This observation demonstrates that there is a specificity in the ability of GIP-1 to inhibit individual members of the endo-β-1,3-glucanase family of the host. GIP-1 does not inhibit the fungal endo- or exo-β-1,3-glucanase or a tobacco pathogenesis-related endo-β-1,3-glucanase. In addition, *P. sojae* produces at least one other GIP that has different specificity than GIP-1 (Ham *et al.*, 1997). In the presence of GIP-1, the ability of soybean endo-β-1,3-glucanase to generate oligoglucoside elicitors from *P. sojae* walls is significantly reduced *in vitro* (K.-S. Ham, P. Albersheim, and A. G. Darvill, unpublished results). Thus, GIP may have an important role in the regulation of the β-glucan elicitor production at the plant–fungal pathogen interface.

### 3.3. Chitinases and Chitosanases

Chitinases have been found in all plant species that have been tested, and nearly all plant chitinases isolated are endo-chitinases. Plant chitinases are one of the class of pathogenesis-related proteins that are induced by pathogen infection (Stintzi *et al.*, 1993). Consequently, chitinases have been suggested to be involved in defense against fungal pathogens. There are numerous reports supporting this involvement by demonstrating that plant chitinases can inhibit fungal growth in culture (for review see Boller, 1993; Collinge *et al.* 1993; Stintzi *et al.*, 1993; Flach *et al.*, 1992). The antifungal action of plant chitinases is characterized by swelling of the hyphal tips followed by their lysis (Arlorio *et al.*, 1992; Mauch *et al.*, 1988). Furthermore, transgenic overexpression of chitinase genes in plants increases their resistance to fungal pathogens (Grison *et al.*, 1996; Punja and Raharjo, 1996; Jach *et al.*, 1995; Lin *et al.*, 1995; Zhu *et al.*, 1994; Vierheilig *et al.*, 1993; Broglie *et al.*, 1991).

In addition to the ability of chitinases to inhibit fungal growth in culture, another mechanism by which chitinases are involved in defense against fungi *in planta* has been suggested: plant chitinases generate chitin oligosaccharide elicitors from fungal mycelial walls much in the same way as plant endo-β-1,3-glucanases generate β-glucan elicitors from fungal walls. Several isoforms of chitinases purified from wheat leaves have the ability to generate chitooligosaccharides that elicit lignification on wheat leaves (Ride and Barber, 1990). These observations indicate that plant tissues have the ability to hydrolyze the chitin polymer and generate elicitor-active oligosaccharides from it.

Chitinases can potentially also have the opposite role of limiting the extent of an oligochitin-induced defense response in plant tissues. Incubation of chitin polysaccharides with a purified chitinase initially generates large elicitor-active oligosaccharides, while prolonged incubation leads to the production of the elictor-inactive dimer (Ride and Barber, 1990). In addition, plant tissues contain enzymes that are capable of hydrolyzing the elicitor-active chitooligosaccharides. This is exemplified by the presence of several isoforms of *N*-acetyl-β-D-hexosaminidase in plants (Jordan and Barber, 1995; Bouquelet and Spik, 1978). Some of these isoforms show preferential activity toward large, elicitor-active chitin oligosaccharides, while others are active toward smaller, inactive oligosaccharides. The complete degradation of chitin polymers and the associated production of monomeric *N*-acetylglucosamine is probably achieved by the combined action of chitinases and *N*-acetyl-β-D-hexosaminidase (Jordan and Barber, 1995). Thus, plants seem to have mechanisms to generate and degrade elicitor-active chitooligosaccharides.

In return, fungi may have evolved mechanisms to protect themselves from plant chitinases. Fungi produce chitin deacetylase, which converts chitin to chitosan, another fungal wall polymer which is resistant to chitinase (Ride and Barber, 1990). In the interaction between the bean rust fungus *Uromyces viciae-fabae* and bean, chitin deacetylase activity increases dramatically upon fungal penetration through the stomata (Deising and Siegrist, 1995). The chitin present in the fungal infection structure disappears as chitin deacetylase activity increases. One may speculate that chitin deacetylase modifies the fungal wall and make it resistant to degradation by plant chitinases. If this is the case, the conversion of chitin to chitosan by chitin deacetylase would lead to the reduced production of elicitor-active chitooligosaccharides. This conclusion is challenged by the induction of plant chitosanases (e.g., barley, tomato, cucumber, and spruce) in response to stresses, including pathogen infection and abiotic elicitors (Sharma *et al.*, 1993; Grenier and Asselin, 1990). Moreover, chitosan oligosaccharides have been shown to induce defense responses in several plants, and elicitor-active fragments larger than heptamer have been demonstrated to be generated *in vivo* from fungal walls during the interaction between plants and fungi. These fragments were also shown to be released by the action of a mixture of purified plant β-1,3-glucanase and chitinase (Kendra *et al.*, 1989). These results demonstrate that plants indeed have mechanisms to generate elicitor-active chitosan oligosaccharides. However, it is not clear how chitosan oligosaccharides are generated *in vivo*. Although experimental evidence is lacking, chitosanases are likely to be a significant class of enzyme for the generation of elicitor-active chitosan oligosaccharides *in planta*.

Exo-chitinases could constitute another fungal mechanism for inactivation of oligochitin elicitors. An exo-chitinase from the bean pathogen *Colletotrichum lindemuthianum* hydrolyzes those chitooligosaccharides which elicit alkalinization of bean suspension cell medium (R. Clay, personal communication).

It has been hypothesized that fungal pathogens produce inhibitor proteins against plant chitinases, analogous to the inhibitor proteins produced against plant endo-β-1,3-glucanases (Ham *et al.*, 1997). Thus far, there is no evidence of the presence of such inhibitors. However, this possibility is supported by the following observations. Like plant endo-β-1,3-glucanases, plant chitinases exhibit inconsistency in their antifungal activity, since some fungi are resistant to chitinases while others are sensitive; all tested fungi had chitin-containing mycelial walls (Ham *et al.*, 1997; Stintzi *et al.*, 1993). Second, some phytopathogenic fungi whose growth is inhibited by plant chitinases and endo-β-1,3-glucanases have the ability to overcome the effect of these hydrolyses over a period of hours, suggesting that these fungi have the capacity to adapt to the enzymes (Ludwig and Boller, 1990). Finally, endo-β-1,3-glucanases and chitinases share several common characteristics: they both function in defense by degrading fungal wall polysaccharides (Stintzi *et al.*, 1993), both generate elicitor-active oligosaccharide fragments from their substrates (Ham *et al.*, 1991; Takeuchi *et al.*, 1990b; Kendra *et al.*, 1989), they degrade the mycelial wall polysaccharides more effectively when acting together (Mauch *et al.*, 1988), and their induction is often coordinately regulated (Ohashi and Ohshima, 1992; Ward *et al.*, 1991a,b; Vögeli *et al.*, 1988). Thus, it would be advantageous for the fungi to produce an inhibitor protein against plant chitinases at the same time that the glucanase inhibitor proteins are produced. Such an inhibitor protein could be a component in control of the production of chitooligosaccharide elicitors.

### 3.4. Endo-Polygalacturonase and Polygalacturonase Inhibitor Protein

Plant pathogens produce a number of glycosidases, including pectin lyases, pectate lyases, endo-polygalacturonases (EPGs), and exo-polygalacturonases, which degrade homogalacturonans. The potential role of EPGs and pectate lyases in generating elicitor-active oligogalacturonides (reviewed earlier in this chapter) has been recognized for some time (Côté and Hahn, 1994; Darvill *et al.*, 1992; Davis *et al.*, 1984, 1986b). *In vitro*, these enzymes have been demonstrated to transiently produce midsize oligogalacturonides (DP 9–16) which are then further degraded to (in most cases inactive) mono-, di-, and trimer. However, if the hydrolytic activity of these enzymes is sufficiently slowed, the lifespan of the midsized, elicitor-active oligogalacturonides is increased dramatically (Cervone *et al.*, 1989a,b).

Pectate lyases have pH optima several pH units above the pH of the cell wall, and thus *in vivo* their rate of action is slowed by several orders of magnitude. There is also at least one report in the literature of the presence *in planta* of an inhibitor protein for pectin lyase (Bugbee, 1993). The EPGs of fungal pathogens, on the other hand, have pH optima at approximately the pH of the cell wall, and therefore the primary mechanism of regulation of homogalacturonan hydrolysis is a result of their interaction with, and inhibition by, plant cell wall glycoproteins. These glycoproteins, known as PGIPs (polygalacturonase inhibitor proteins), are present in the walls of all dicotyledonous plants examined as well as in some monocots (Benen *et al.*, 1996). One important consequence of the action of PGIP might be to raise the local concentration of oligogalacturonides at the site of an infection sufficiently high to trigger plant defense responses (Cervone *et al.*, 1993). PGIPs were first recognized over two decades ago (Anderson and Albersheim, 1972) in *Phaseolus vulgaris*. A sizable amount of data, some of it conflicting, has been generated on the abilities of PGIPs from various plant species to inhibit fungal EPGs. PGIPs have been reported to show competitive (Abu-Goukh *et al.*, 1983), noncompetitive (Johnston *et al.*, 1993; Lafitte *et al.*, 1984), and partial-competitive (Yao *et al.*, 1995) modes of action, depending on the species of origin. In general, PGIPs inhibit only fungal EPGs, although there are reports of inhibition of a maize pollen exopolygalacturonase by *Phaseolus vulgaris* PGIP (Cervone *et al.*, 1990) and of a *Venturia inequalis* exopolygalacturonase by apple PGIP (Müller and Gessler, 1993).

PGIPs do not show equal abilities to inhibit EPGs from different fungi (Cervone *et al.*, 1996; Bergmann *et al.*, 1995; Stotz *et al.*, 1994a). For example, the PGIP isolated from bean inhibits the EPGs from all fungi investigated, while those of pear and tomato are far more limited in their inhibitory capabilities (Stotz *et al.*, 1994a). There are significant differences in the amino acid sequence for these three PGIPs (Stotz *et al.*, 1994b) which may account for the differences in inhibitory specificity. However, the differing inhibitory capabilities of various PGIPs may also be due in some degree to glycosylation of either PGIP or EPG or both, as a correlation appears to exist between increased levels of glycosylation and greater inhibitory specificity for pear, tomato, and bean PGIPs. Further, PGIP isolated from cultured *P. vulgaris* cv. *cannelino* cells, which is more heavily glycosylated than the PGIP isolated from intact bean tissues, has only 10% of the inhibitory activity against *Fusarium moniliforne* EPG expressed in *Saccharomyces cereviseae* (also heavily glycosylated) as it does against the native *F. moniliforne* EPG (Bergmann *et al.*, 1995). While this latter example does not represent a set of conditions which would be seen *in vivo*, it serves to indicate that glycosylation of either polypeptide can affect the inhibition of EPGs by PGIPs. Recent work provides another, intriguing explanation for the differences in inhibitory specificity of PGIPs (Cervone *et al.*, 1996). A small family of PGIP genes appears to be present in the *P. vulgaris* genome (Frediani *et al.*, 1993). A single PGIP gene was sequenced (Toubart *et al.*, 1992) and has been expressed in tomato under the control of the cauliflower mosaic virus (CaMV) 35S promoter. The PGIP isolated from the transgenic tomato has an increased inhibitory specificity to EPGs (interestingly, the specificity appears to be the same as that of native tomato PGIP). The argument has been made that PGIP, as isolated from *P. vulgaris,* is actually a collection of different PGIPs, each with its own specificity and expressed under different control mechanisms (Cervone *et al.*, 1996). This argument is worthy of further careful investigation.

PGIP has been shown to fall into the class of proteins whose structure is based on a leucine-rich repeat motif (LPR protein) (De Lorenzo *et al.*, 1994; Stotz *et al.*, 1994b). This is a motif seen in prokaryotic and eukaryotic proteins that are involved in interactions with other macromolecules (Kobe and Deisenhofer, 1995). Whether this LRR motif, which is also present in several gene products involved in resistance to pathogens (Jones, 1996; Dangl, 1995; Staskawicz *et al.*, 1995), is an essential element of the interaction of PGIP with EPG as part of a defense mechanism is yet to be conclusively demonstrated. PGIP increases at the site of fungal infection (Bergmann *et al.*, 1994), as does the mRNA encoding PGIP (Nuss *et al.*, 1996). In addition, PGIP mRNA levels increase after treatment with elicitors such as oligogalacturonides, fungal β-glucans, and salicylic acid, and after wounding (Bergmann *et al.*, 1994). Western blot analysis using an antibody specific to PGIP demonstrated increased expression in salicylic acid-treated and wounded tissues. Moreover, an immunocytological study using the same antibody in bean infected with *Colletotrichum lindemuthianum* has shown that PGIP accumulates preferentially in cells adjacent to the infection site (Bergmann *et al.*, 1994). It has been suggested that soluble PGIP, the production of which is induced by the presence of oligogalacturonides, may bind to a receptor in the membrane, forming the trigger for the hypersensitive response (Cervone *et al.*, 1994).

Our laboratory has been studying the mechanism of fungal EPGs as well as their interaction with PGIPs to test whether specific EPG–PGIP combinations yield a higher proportion of elicitor-active oligogalactronides, which may, in turn, correlate with increased resistance to invading fungal pathogens. Using HPAEC-PAD, our studies have shown that many (or perhaps most) EPGs act through a "mixed endo/exo" or "multiple attack" mechanism, rather than through a true "endo" mechanism (Hahn *et al.*, 1989a, 1992). This dual-action mechanism had been noticed for a few isolated EPGs (Benen *et al.*, 1996; English *et al.*, 1972), but its prevalence was unexpected. For EPGs employing such a mechanism, the levels of elicitor-active oligogalacturonides produced, even in the presence of PGIP, is at most 10% of what would be expected for an inhibited classic EPG. Thus, any interpretation of induction of defense responses based on the release of elicitor-active oligogalacturonides must take this line of evidence into account.

## 4. SIGNAL TRANSDUCTION

Cells that make up an organism have complex and diverse mechanisms for perceiving and responding to stimuli originating not only from within the organism, but also from the external environment. The process of cellular signaling can be divided into three steps: signal perception, usually by cellular receptors that specifically recognize the signal (e.g., Hollenberg, 1991; Hahn, 1989), signal transduction and amplification [transmission of the signal, either directly or indirectly to its site(s) of action within the cell], and signal translation (conversion of the signal into a specific cellular response such as the activation of specific genes).

The investigation of the structures and activities of oligosaccharide elicitors has yielded important insights into how plant cells respond to external signals and has provided useful model systems for molecular studies on signal perception, signal transduction, and gene regulation in plants (for recent reviews, see Hahn, 1996; Côté and Hahn, 1994; De Wit, 1992; Scheel and Parker, 1990; Lamb *et al.*, 1989; Dixon, 1986). Thus, binding sites for oligoglucoside and oligochitin elicitors have recently been identified and are discussed in more detail below. The study of oligogalacturonide elicitor perception is more problematic, mainly due to the difficulty of generating a biologically active derivative of oligogalacturonides that can be used for binding studies. As a consequence, the translation of oligogalacturonide signals has been studied more than their perception.

### 4.1. Oligoglucoside Signal Transduction

The specificity of the response of soybean tissue to oligoglucoside elicitors of phytoalexin accumulation (Cheong *et al.*, 1991; Sharp *et al.*, 1984c), discussed above, suggests that a specific receptor for the hepta-β-glucoside elicitor exists in soybean cells. Several studies utilizing heterogeneous mixtures of mycelial glucan fragments indicated that binding sites for glucan fragments exist in soybean membranes (Yoshikawa and Sugimoto, 1993; Cosio *et al.*, 1988; Schmidt and Ebel, 1987; Yoshikawa *et al.*, 1983; Peters *et al.*, 1978). In particular, binding studies carried out with partially purified elicitor-active glucans from *P. sojae* mycelial walls demonstrated the presence of high-affinity glucan-binding sites on soybean root plasma membranes (Schmidt and Ebel, 1987) and protoplasts from suspension-cultured soybean cells (Cosio *et al.*, 1988). Lower-affinity glucan-binding sites have also been identified in soybean membranes when a radiolabeled preparation of glucanase-released glucan fragments (Yoshikawa and Sugimoto, 1993) and radiolabeled laminarin, a β-1,3-glucan (Yoshikawa *et al.*, 1983) were used. Since the glucan preparations utilized in these binding studies were not homogeneous, it is difficult to determine whether the observed binding was specific for the biologically active oligoglucosides.

Subsequent research with homogenous oligoglucosides has substantiated the hypothesis that soybean cells have membrane-localized glucan elicitor-binding sites. These later binding studies utilized as the labeled ligand the hepta-β-glucoside elicitor coupled to iodinatable (with $^{125}I$) functional groups (Cheong and Hahn, 1991; Cosio *et al.*, 1990b). The hepta-β-glucoside elicitor-binding sites are present in membranes prepared from every major organ of young soybean plants (Cheong and Hahn, 1991). The hepta-β-glucoside elicitor-binding sites are predominantly found in plasma membranes (Cheong *et al.*, 1993; Schmidt and Ebel, 1987). Binding of the radiolabeled ($^{125}I$) hepta-β-glucoside elicitor to the root membranes is saturable over a concentration range of 0.1–5 nM, which is somewhat lower than the range of concentrations (6–200 nM) required to saturate the bioassay for phytoalexin accumulation (Cheong *et al.*, 1991; Sharp *et al.*, 1984c). Ligand saturation experiments indicate that the root membranes possess only a single class of high-affinity hepta-β-glucoside-binding sites (apparent $K_d \approx 1$ nM) of relatively low abundance ($B_{max} \approx 1$ pmol/mg protein) (Cheong and Hahn, 1991; Cosio *et al.*, 1990b). Binding of the active hepta-β-glucoside to the membrane preparation is reversible, indicating that the elicitor does not become covalently attached to the binding protein(s) (Frey *et al.*, 1993; Cheong and Hahn, 1991). The membrane-localized, elicitor-binding proteins (EBPs) exhibit a high degree of specificity with respect to the oligoglucosides that they bind. More importantly, the ability of an oligoglucoside to bind to soybean root membranes correlates with its ability to induce phytoalexin accumulation (Cheong *et al.*, 1993; Cheong and Hahn, 1991). The results of the structure–activity (Cheong *et al.*, 1991) and ligand-binding (Cheong *et al.*, 1993; Cheong and Hahn, 1991) studies demonstrate that those structural elements of the hepta-β-glucoside required to elicit phytoalexin synthesis are also essential for efficient binding of the elicitor to its putative receptor; these data provide strong evidence that the binding proteins are physiological receptors for the hepta-β-glucoside elicitor.

Cyclic β-glucans produced by the symbiotic bacterium *Bradyrhizobium japonicum* have been reported to elicit phytoalexin accumulation in soybean cotyledons (Miller *et al.*, 1994). However, this report is contradicted by more recent findings indicating that these oligosaccharides act as antagonists in the interaction between the hepta-β-glucoside elicitor and its binding proteins (Mithöfer *et al.*, 1996a). The bacterial cyclic (1→3), (1→6)-β-glucans do not elicit phytoalexin accumulation in soybean at concentrations up to 1 mg/ml. However, they competitively inhibit the binding of radiolabeled hepta-β-glucoside elicitor to soybean membranes at concentrations only one order of magnitude higher than required for fungal β-glucans. Cyclic glucans also compete with fungal β-glucans in soybean bioassays, inhibiting the accumulation of phytoalexin in a dose-dependent manner. In contrast, bacterial cyclic (1→2)-β-glucans have no antagonistic activity in either bio- or binding assays (Mithöfer *et al.*, 1996a; Miller *et al.*, 1994). The discrepancies between the different studies have yet to be resolved, and the relevance of these results for plant–microbe interactions have yet to be understood. However, the first identification of an antagonist against the hepta-β-glucoside merits further attention because of the utility of such a molecule as a tool in pharmacological studies.

Solubilization of fully functional EBPs from soybean root microsomal membranes has been achieved with the aid of several nonionic and zwitteronic detergents (Cheong *et al.*, 1993; Cosio *et al.*, 1990a). The *n*-dodecanoylsucrose-solubilized EBPs retain their high affinity for the hepta-β-glucoside elicitor, having an apparent $K_d$ = 1.8 nM (Cheong *et al.*, 1993). The apparent $K_d$ determined for ZW 3-12-solubilized binding proteins is very similar (Cheong *et al.*, 1993; Frey *et al.*, 1993; Cosio *et al.*, 1992). Equally significant, the solubilized EBPs also retain the specificity for elicitor-active oligoglucosides characteristic of the membrane-localized proteins, regardless of which detergent is used for solubilization (Cheong *et al.*, 1993).

The size of the solubilized EBPs has been characterized in an effort to determine the molecular complexity of the elicitor-binding species. Chromatography of dodecanoylsucrose-solubilized soybean root membrane proteins on Superose-6 and Superose-12 columns revealed that the EBPs behave as molecular species having a molecular mass of >300 kDa, while 90% of the membrane proteins eluted at positions suggesting molecular masses of 50–200 kDa (Cheong *et al.*, 1993; Côté *et al.*, 1997). The gel permeation data were confirmed by ultrafiltration studies with 100- and 300-kDa molecular weight cutoff (MWCO) membranes wherein over 90% of the elicitor-binding activity did not pass through the membranes. Recently, the analysis of EBPs by blue native gel electrophoresis revealed a single diffuse band migrating as a 240-kDa polypeptide (Mithöfer *et al.*, 1996b). Finally, solubilized EBPs can be precipitated using 50% ammonium sulfate, another indication of the large size of the binding protein complex (Côté *et al.*, 1997).

Efforts to purify the EBPs from the mixture of proteins solubilized from soybean root membranes have primarily relied on affinity chromatography using a matrix carrying either an immobilized mixture of elicitor-active glucan fragments (Umemoto *et al.*, 1997; Frey *et al.*, 1993; Cosio *et al.*, 1992) or immobilized hepta-β-glucoside elicitor (Côté *et al.*, 1997). The EBPs eluted from the ligand-affinity columns retain the same affinity and specificity for the hepta-β-glucoside elicitor as that observed for the membrane-localized and crude detergent-solubilized proteins (Côté *et al.*, 1997). The eluted fraction contains several polypeptides as visualized on polyacrylamide gels (Frey *et al.*, 1993; Cosio *et al.*, 1992).

It is not yet known which of the proteins present in the affinity-purified fractions are essential for elicitor-binding activity. Photoaffinity-labeling experiments carried out on affinity column eluates suggest that three polypeptides (~75, 100, 150 kDa) carry an elicitor-binding domain (Frey *et al.*, 1993; Cosio *et al.*, 1992); the 150-kDa polypeptide appears to be a dimer of the predominant 75-kDa protein (Mithöfer *et al.*, 1996b). However, the specificity of the photolabeling of these three polypeptides has not been defined rigorously. The cloning of a candidate cDNA for the 75-kDa polypeptide present in affinity-purified EBP fractions has been reported independently by two laboratories (Fliegmann *et al.*, 1997; Umemoto *et al.*, 1997). The reported amino acid sequence for this polypeptide, which has no homolog in various sequence databases, has no obvious hydrophobic transmembrane domain and no signal peptide. These characteristics are very puzzling, since the EBPs behave as integral membrane proteins in plant extracts (Cheong *et al.*, 1993; Cosio *et al.*, 1990a). Expression of the cloned cDNA in bacteria yielded a polypeptide that was reported to be able to bind an elicitor-active heterogeneous preparation of oligoglucosides (Umemoto *et al.*, 1997), but this result could not be independently verified (Fliegmann *et al.*, 1997). The affinity and specificity of the binding site on the 75-kDa polypeptide for a homogeneous glucan elicitor such as the hepta-β-glucoside elicitor has not been determined.

Possible mechanisms for the transmission of the glucan elicitor signal within plant cells have been examined in several studies. For example, neither cAMP (Hahn and Grisebach, 1983) nor phosphoinositides (Strasser *et al.*, 1986), two important secondary messengers in animal signal transduction, appear to be involved in glucan elicitor signal transduction. On the other hand, glucan-induced phytoalexin accumulation in suspension-cultured soybean cells is inhibited when the extracellular calcium concentration is reduced (Stäb and Ebel, 1987). Further, a calcium ionophore was shown to induce phytoalexin accumulation (Stäb and Ebel, 1987), while selected calcium channel blockers inhibit elicitor-induced increases in chalcone synthase activity in soybean cells (Ebel *et al.*, 1993). Thus, calcium ion fluxes appear to have a role in the glucan elicitor signaling pathway. Chlorine channel antagonists such as anthracene 9-carboxylate, flufenamic acid, and 5-nitro-2-(3-phenylpropylamino)-benzoic acid (NPPB) at low concentrations have been reported to block the stimulating activity of fungal β-glucans on chalcone synthase, an enzyme of the phenylpropanoid pathway involved in phytoalexin synthesis in soybean (Ebel *et al.*, 1995). Vanadate, an inhibitor of plasma membrane ATPase, mimics the effect of a cell wall elicitor from *Phytophthora cambivora* on suspension-cultured peanut cells at concentrations matching those required for inhibition of the plasma membrane ATPase. Such results led those authors to propose that changes in the ATPase-associated membrane potentials are intermediates in the signal cascade leading to phytoalexin accumulation (Steffens *et al.*, 1989). Finally, an oxidative burst has been hypothesized to play a role in the transduction of the β-glucan signal inside the plant cell. *P. sojae* β-glucans cause accumulation of $H_2O_2$ in soybean suspension-cultured cells (Levine *et al.*, 1994). This oxidative burst is observed in concomitance with an increase in RNA transcripts amounts for glutathione *S*-transferase (GST), phenylammonia lyase (PAL), and chalcone synthase (CHS). Effectors of oxidative burst such as 2,5-dihydroxycinnamic acid methyl ester, which acts as a peroxide molecular trap, and diphenylene iodonium, a NADPH oxidase inhibitor, inhibit the elicitor-mediated $H_2O_2$ accumulation, but not the accumulation of chalcone synthase transcripts. These results suggest that the oxidative burst and phytoalexin accumulation are two distinct responses activated in plant cells by glucan elicitors.

The role of protein phosphorylation in the glucan elicitor-stimulated signaling pathway has been investigated in several studies. Rapid changes in the pattern of protein phosphorylation were observed after treating suspension-cultured soybean cells with a heterogeneous glucan elicitor preparation (Grab *et al.*, 1989), but the roles, if any, of the phosphoproteins in the signaling pathway leading to phytoalexin accumulation remain unknown. A protein kinase inhibitor, K-252a, when in the presence of pure hepta-β-glucoside or partially purified β-glucan elicitors, synergistically induces chalcone synthase activity in soybean cell cultures (Ebel *et al.*, 1993). This result has been contradicted by a more recent study showing that K-252A and staurosporine, another kinase inhibitor, inhibit glucan elicitation of defense responses (Levine *et al.*, 1994). Interestingly, staurosporine inhibits the glucan-elicited oxidative burst and GST RNA transcript accumulation, but not the chalcone synthase transcript accumulation. Several protein phosphatase inhibitors, including okadaic acid, were recently found to induce the production of daidzein, an isoflavone precursor of the soybean phytoalexins, in soybean cotyledons (MacKintosh *et al.*, 1994). Moreover, okadaic acid stimulates the expression of phenylalanine-ammonia lyase in suspension-cultured soybean cells, but this effect is blocked by the addition of K-252a (MacKintosh *et al.*, 1994), contrasting with the synergistic stimulation of soybean chalcone synthase activity by this protein kinase inhibitor and β-glucans (Ebel *et al.*, 1993).

The effect of the various signal translation inhibitors or phytoalexin accumulation has not been assessed directly; instead their effect on the induction of early enzymes in the biosynthetic pathway and on the accumulation of precursors was measured. The inducibility of the early enzymes in the phenylpropanoid pathway in response to various stimuli is a matter of debate (Dhawale *et al.*, 1989; Bonhoff *et al.*, 1986). Precursors of glyceollin, such as the isoflavone daidzein, are present in soybean tissue as glycosyl conjugates (Graham *et al.*, 1990); soybean cotyledons contain particularly high and constant levels of these conjugates (Graham, 1991). Following an incompatible pathogen infection, the daidzein conjugates are rapidly hydrolyzed; subsequently, high levels of glyceollin accumulate (Graham *et al.*, 1990). In a compatible interaction, conjugate hydrolysis is retarded, and only low levels of glyceollin accumulate (Graham *et al.*, 1990). A spatial study of the elicitation of phytoalexin accumulation in soybean cotyledons has shown that glyceollin and conjugates of the isoflavonoid precursors accumulate to high levels in cells adjacent to the elicitor application point; in more distant cell layers, only the precursor conjugates accumulate (Graham and Graham, 1991). Abiotic elicitors, unlike biotic elicitors such as oligoglucosides, induce glyceollin accumulation and a decrease in the amounts of the daidzein conjugates. Such abiotic elicitors may act by inducing the release of daidzein from a preexisting pool (Graham and Graham, 1991). Thus, okadaic acid-induced production of daidzein (MacKintosh *et al.*, 1994) may be interpreted as showing that the phosphatase inhibitor is itself an abiotic elicitor in soybean cells. In conclusion, although evidence has been presented suggesting that protein phosphorylation–dephosphorylation plays a role in the elicitor signal transduction, it is not clear at which step of the signaling cascade or the biosynthetic pathway the phosphorylated proteins act.

The results of most studies to date on glucan elicitor signal transmission within plant cells must be interpreted cautiously since they have been carried out with heterogeneous glucan preparations, thereby making it difficult to relate the observed effects unambiguously to the glucan elicitors themselves. Clearly, investigations of the signaling pathway downstream of the signal perception merit reexamination utilizing the homogeneous hepta-β-glucoside elicitor.

### 4.2. Oligochitin and Oligochitosan Signal Transduction

The nature of chitin oligosaccharides and the specificity of the plant response upon their recognition suggest the presence of specific receptors. Binding sites satisfying some of the criteria for bona fide receptors have been identified in rice and tomato (Baureithel *et al.*, 1994; Shibuya *et al.*, 1993). A binding site for *N*-acetylchitooctaose (chitin octamer) with high affinity (apparent $K_d \approx 5.4$ nM) and low abundance ($B_{max} \approx 0.3$ pmol/mg protein) is present in membrane preparations of suspension-cultured rice cells (Shibuya *et al.*, 1993). The observed binding is reversible and the binding specificity has been investigated (Shibuya *et al.*, 1996a). The abilities of chitooligosaccharides to compete for the chitooctaose-binding site correlate with their biological activity and are dependent on their degree of polymerization, the heptamer and octamer competing most effectively and the trimer being the least effective. The de-*N*-acetylated heptamer did not compete for the binding site even at concentrations four orders of magnitude higher than were required for the *N*-acetylchitooctaoside elicitor (Shibuya *et al.*, 1993).

The presence of a high-affinity oligochitin elicitor-binding site in tomato cell membranes and on the surface of whole cells was demonstrated using a radioactive derivative of the chitin pentamer active in the tomato cell-culture alkalinization assay (Baureithel *et al.*, 1994). The tomato membrane-localized binding site has an apparent $K_d$ of 23 nM and is of low abundance ($B_{max} \approx 2.45$ pmol/mg protein) (Baureithel *et al.*, 1994). Ligand specificity studies demonstrated that oligomers of DP 4 and 5 are effective competitors, while smaller oligomers are significantly less effective. Thus, there is a direct correlation between the ability of a chitin oligomer to bind to tomato membranes and its ability to induce alkalinization of the medium of suspension-cultured tomato cells (Baureithel *et al.*, 1994; Felix *et al.*, 1993). A Nod factor isolated from *Rhizobium leguminosarum* is able to compete for the oligochitin-binding site in tomato membrane preparations at concentrations similar to those required for the chitin pentamer; such results are consistent with the observation that some Nod factors stimulate alkalinization of the medium of suspension-cultured tomato cells (Baureithel *et al.*, 1994; Staehelin *et al.*, 1994).

The rice and the tomato binding sites have both been solubilized from membranes using nonionic detergents (Shibuya *et al.*, 1996b; Baureithel and Boller, 1995). Photoaffinity-labeling experiments have been carried out in rice, and results suggest that a 75-kDa polypeptide carries the oligochitin-binding site (Ito *et al.*, 1997). Attempts to purify the chitin elicitor-binding proteins using ligand affinity chromatography have yielded a single protein of 70 kDa as determined by sodium dodecyl sulfate–polyacrylamide gel electrophoresis (Shibuya *et al.*, 1996b).

Chitin oligosaccharides have been shown to induce rapid responses in plant cells that may be part of the signal cascade leading to the induction of plant defense responses. For example, the elicitation of phytoalexin production in rice cell cultures by chitin oligomers of DP $\geq$ 6 at subnanomolar concentrations (Ren and West, 1992) is accompanied by transient membrane depolarization (Kuchitsu *et al.*, 1993), while shorter oligomers induced neither response. The oligochitin-induced membrane depolarization is rapid (within 1 min), suggesting that a change in membrane potential is one of the initial events of the signal transduction pathway leading to the synthesis of phytoalexins. Phytoalexin elicitation is also accompanied by a transient generation of reactive oxygen species (Kuchitsu *et al.*, 1995), an increase in jasmonic acid levels (Nojiri *et al.*, 1996), and a rapid and transient gene expression (Minami *et al.*, 1996). Jasmonic acid (JA) has been suggested as an intermediary in this signal transduction pathway because it mimics by itself the chitin-elicited phytoalexin accumulation, and because Ibuprofen, an inhibitor of JA biosynthesis, inhibits both the chitin-stimulated accumulation of JA and phytoalexins in rice cells (Nojiri *et al.*, 1996). Chitin oligomers of DP $\geq$ 4 also stimulate alkalinization of the medium of suspension-cultured tomato cells at subnanomolar concentrations (Felix *et al.*, 1993). More than a 1000-fold higher concentration of the chitin trimer was required to induce an equivalent increase in the extracellular pH (Felix *et al.*, 1993). The reducing end of the chitin fragments could be modified without any loss of activity as long as four *N*-acetylglucosaminyl residues in their pyranose form remained intact (Baureithel *et al.*, 1994). The chitosan pentamer showed no activity in tomato, even in the micromolar range. The extracellular alkalinization of tomato cell cultures by chitin oligomers is also accompanied by changes in the pattern of protein phosphorylation (Felix *et al.*, 1993). The activity of oligochitins on tomato cells is only transient, and it is followed by a refractory state where the cells are no longer responsive to subsequent signals. Moreover, oligochitins and a lipo-chitooligosaccharride nodulation factor from *Rhizobium leguminosarum* desensitize tomato cells to each other's action. These results suggest the operation of an endocytic mechanism to remove both signal and receptor from the cell surface. However, these rapid responses of cells to oligochitins have yet to be directly tied to a specific signaling pathway triggered by these oligosaccharide elicitors. In oat, where the chitin pentamer elicits phytoalexin accumulation, $Ca^{2+}$ has been shown to be involved (Ishihara *et al.*, 1996). The calcium ionophore A23187 mimics the chitopentaose elicitation by strongly inducing phytoalexin accumulation, while calcium channel and calcium movement inhibitors repressed it.

The cellular signaling pathway triggered by oligochitosan elicitors has not been studied as well as that for oligochitin. Treatment of suspension-cultured soybean cells with chitosan induces electrolyte fluxes across the plasma membrane (Waldmann *et al.*, 1988; Young and Kauss, 1983; Young *et al.*, 1982) that were proposed to result in the activation of $Ca^{2+}$-dependent callose synthase (Köhle *et al.*, 1985). In contrast, chitosan-induced defense responses in pea pods are not correlated with membrane leakage and $Ca^{2+}$ flux (Kendra and Hadwiger, 1987a,b). Based on studies of the correlation between the DP of the chitosan oligomers, the degree of *N*-acetylation of these oligosaccharides, and the abilities of oligochitosans to elicit callose synthesis, it has been suggested that the oligochitosan elicitors interact primarily with regularly spaced negative charges on the plant plasma membrane rather than with a discrete receptor (Kauss *et al.*, 1989). Evidence has also been presented suggesting that protein phosphorylation is not involved in the chitosan-induced synthesis of callose in tomato cells (Grosskopf *et al.*, 1991).

### 4.3. Oligogalacturonide Signal Transduction

Information about the molecules responsible for initial recognition of the active oligogalacturonides is notably lacking. One problematic characteristic of oligogalacturonides is their tendency to form intermolecular complexes in solution, due in part to their polyanionic nature. Several characteristics of oligogalacturonides impede the utilization of common biochemical approaches for the identification of binding sites for this ligand. Moreover, the synthesis of derivatives of biologically active oligogalacturonides suitable for the preparation of radiolabeled ligands for use in ligand-binding studies is more complicated than for the other oligosaccharins described in previous sections. This is due to the occurrence of side reactions involving the carboxylic acid groups present in oligogalacturonides. Nevertheless, homogeneous tyraminylated and biotinylated derivatives of the tridecagalacturonide have been prepared (Ridley *et al.*, 1997; Reymond *et al.*, 1996; Spiro *et al.*, 1996). The preparation of a tyrosine hydrazone of the dodecagalacturonide has also been described (Horn *et al.*, 1989), although the purity and structure of the derivative were not documented. Saturable binding of the radioiodinated form of the latter derivative to intact soybean cells has been reported (Low *et al.*, 1993), and evidence suggesting uptake of the bound oligogalacturonide via receptor-mediated endocytosis has been presented (Horn *et al.*, 1989). A fluorescein derivative of a tetradecagalacturonide-enriched fraction was prepared and binding of this derivative to the surface of soybean protoplasts was visualized using a novel silver-enhanced immunogold technique (Diekmann *et al.*, 1994). The observed binding is competed for by a large excess of unlabeled heterogeneous oligogalacturonides, although only 50% competition was achieved and high standard deviations for the data were observed (Diekmann *et al.*, 1994).

The most recent report of identification of a protein binding to oligogalacturonides concerns a 34-kDa membrane protein whose *in vitro* phosphorylation is enhanced by size-specific (DP 14 and 15) oligogalacturonides (Reymond *et al.*, 1996). The 34-kDa protein is associated with purified plasma membranes isolated from potato and tomato leaves (Farmer *et al.*, 1989, 1991). The phosphorylation of the 34-kDa protein does not appear to be directly related to the induction of proteinase inhibitor accumulation in tomato, since oligogalacturonides with DP $< 14$ induce the production of these inhibitors, but do not enhance *in vitro* phosphorylation of the 34-kDa protein (Farmer *et al.*, 1991). Oligoguluronide preparations, although less effective than oligogalacturonides, induce proteinase inhibitor activity and *in vitro* phosphorylation of the 34-kDa protein in tomato (Farmer *et al.*, 1991). The 34-kDa protein, renamed remorin, is widely distributed in angiosperms, and has been cloned from potato (Reymond *et al.*, 1996). The sequence reveals a surprisingly hydrophilic nature for a protein that behaves like an integral membrane protein. A recombinant GST–remorin fusion protein has been expressed and shown to bind oligogalacturonides (DP 20) by ligand blotting using a biotinylated ligand. This binding is not displaced by biotin or by an unrelated biotinylated compound, nor by most polyanions or a series of uronides known not to enhance the phosphorylation of remorin. On the other hand, polygalacturonic acid, fractionated pectin, and the strong negatively charged polyanion heparin can displace the binding to varying degrees. Remorin shares with bacterial pectate lyase C (Yoder and Jurnak, 1995) a conserved distribution of amino acid residues thought to form a uronide-binding groove. These two proteins are otherwise very different. According to the authors, the overall characteristics of remorin are suggestive of a role in cell-to-cell signaling involving an association with plasmodesmata, a role similar to the one suggested for the viral movement proteins (Reymond *et al.*, 1996; Lucas and Gilbertson, 1994). Proof of such a role must still be obtained.

In no case has the specificity of the observed binding of the oligogalacturonides been rigorously demonstrated, raising questions about the significance of the findings with respect to oligogalacturonide signal transduction. These preliminary studies indicate, though, that the molecular tools are becoming more available to pursue the identification and characterization of oligogalacturonide-binding proteins.

The cellular signaling pathway(s) by which oligogalacturonides induce responses in plant cells remain unknown, although recent research has demonstrated that oligogalacturonides induce several rapid responses at the plant cell surface. The effects of oligogalacturonides on ion fluxes across the plasma membrane have been examined in several studies. For example, the membranes of tomato leaf mesophyll cells (Thain *et al.*, 1990) are depolarized within 5 min of exposure to relatively high concentrations (1 mg/ml) of oligogalacturonide mixtures. This effect is similar to the membrane depolarization caused by cyanide, suggesting that oligogalacturonides depolarize the electrical potential difference of the membrane by inhibiting the plasma membrane $H^+$-ATPase (Thain *et al.*, 1995). Lower concentrations (10 $\mu$g/ml) of size-specific oligogalacturonides (DP 12–15) induce within 5 min a transient stimulation of $K^+$ efflux, alkalinization of the medium, depolarization of the plasma membrane, and a decrease in the external $Ca^{2+}$ concentration in suspension-cultured tobacco cells (Mathieu *et al.*, 1991). This last effect of oligogalacturonides has also been observed in carrot protoplast culture medium (Mesiaen *et al.*, 1993). In the carrot system, oligogalacturonides induce the mobilization of cytosolic $Ca^{2+}$, the activation of various defense genes (such as PAL), and a decrease of the internal pH, only in the presence of calcium. These effects were blocked by calcium channel antagonists (Messiaen and Van Cutsem, 1994). In tobacco, prior depletion of calcium in the growth medium of suspension cells abolished the oligogalacturonide-stimulated $K^+$ efflux. Size-heterogeneous mixtures of oligomannuronides and oligoguluronides are approximately 400-fold less effective in inducing $K^+$ efflux, suggesting that the biological response to oligogalacturonides is structurally specific. Furthermore, treatment of tobacco cells with an oligoguluronide preparation did not result in a measurable decrease in the $Ca^{2+}$ concentration of the incubation medium. The decrease of $Ca^{2+}$ concentration in the incubation medium can be explained by either an increase in the $Ca^{2+}$ influx or by the formation of intermolecular complexes (dimers) between oligogalacturonides with a DP > 10 in the presence of $Ca^{2+}$ (Kohn, 1975, 1987; Powell *et al.*, 1982), thereby immobilizing the divalent cation in the medium or in the cell wall. The latter hypothesis has led to the suggestion that an interaction between oligogalacturonide–$Ca^{2+}$ complexes and plant cells induces the localized physiological and morphological responses (Ryan and Farmer, 1991; Farmer *et al.*, 1990), although no evidence for a specific binding site (receptor) for such complexes has been presented. Altogether, these observations provide evidence that size-specific oligogalacturonides at low concentrations can specifically induce rapid responses at the plant cell surface.

There is evidence that at least some plant cells become rapidly desensitized to oligogalacturonides. For example, tobacco cells treated with bioactive oligogalacturonides to induce $K^+$ efflux respond less intensively to a second treatment with the same fragments (Mathieu *et al.*, 1991). Moreover, oligogalacturonide-induced ethylene biosynthesis in cultured pear cells results in a time-dependent decrease in the ability of the cells to produce ethylene in response to a second elicitor treatment (Campbell and Labavitch, 1991). These desensitization effects may be explained by internalization of oligogalacturonide receptors from the plasma membrane (Horn *et al.*, 1989).

Suspension-cultured soybean cells release $H_2O_2$ within 5 min of exposure to heterogeneous oligogalacturonide mixtures (Legendre *et al.*, 1993a; Apostol *et al.*, 1989), and evidence suggesting the involvement of GTP-binding proteins (Legendre *et al.*, 1992) and phospholipase C (Legendre *et al.*, 1993b) in this signaling pathway has been presented. Recent data suggest that the oxidative burst induced by oligogalacturonides results from a separate signaling pathway from that leading to phytoalexin accumulation (D. Davis *et al.*, 1993). In tobacco cells, treatment with fungal pectin lyase elicits an oxidative burst like the one induced by oligogalacturonides. It is not clear if pectin lyase elicits by itself or releases a compound different from the oligogalacturonide elicitor, but it has been shown that the two elicitation events are associated with different transduction pathways (Rouet-Mayer *et al.*, 1997). Altogether, these data suggests a widespread role for oxidative burst in different signaling pathways.

As is the case with chitin oligosaccharide-induced phytoalexin accumulation in rice, jasmonic acid seems to be part of the signal transduction pathway associated with the wounding response in tomato (Doares *et al.*, 1995). The wound response is elicited by oligogalacturonides of DP 20 and chitosan oligosaccharides of DP 4–6. It has been suggested that systemin, oligogalacturonides, and chitosan oligomers ultimately all activate the octadecanoid signaling pathway, but are initially perceived by different receptors (Farmer and Ryan, 1992). Salicylic acid and sodium diethyldithiocarbamate (an active oxidant; DIECA), both inhibitors of the conversion of linolenic acid to jasmonic acid (jasmonic acid is biosynthesized via the octadecanoid pathway; Sembdner and Parthier, 1993), inhibit the elicitor-induced stress response (accumulation of proteinase inhibitor). Moreover, sodium *p*-chloromercuribenzenesulfonate (PCMBS), an inhibitor of apoplastic phloem loading, inhibits the induction of tomato proteinase inhibitor I by wounding and systemin (Doares *et al.*, 1995). PCMBS does not inhibit chitosan elicitation of the tomato proteinase inhibitor.

The rapid effects produced by oligogalacturonides at the plant cell surface have not yet been correlated with any of the known plant responses induced by oligogalacturonides. Thus, further investigation of the oligogalacturonide-induced membrane responses will be required in order to couple these responses to specific signaling pathways.

## 5. CONCLUDING REMARKS

The study of any physiological event mediated by receptor–effector interactions goes through several experimental stages. These stages can be defined as being the identification of the biological response and the development of an effective and reproducible biological assay for the response, the description and the characterization of signal molecules that trigger this response, and, finally, the understanding of the underlying mechanisms involved in the perception and the translation of the signals leading to the observed biological events. The results of extensive research in a number of laboratories over the past 20 years have clearly established that oligosaccharide elicitors are important signal molecules that play major roles in triggering plant defense responses during plant–pathogen interactions. The structural characterization of these molecules has been in some cases a major experimental challenge. However, knowledge of the structures of oligosaccharide elicitors and the biological responses to these signals has now progressed to the point where detailed studies of the modes of action of these elicitors can be undertaken, and the mechanisms by which they affect plant cellular physiology are beginning to be identified.

Oligosaccharides were first described as signals in plants. Due to the complexity of the plant cell wall, it is probable that more plant-derived oligosaccharins will be identified, possibly with novel activities. Equally complex, the cell surfaces of the many microorganisms that interact with plants could also provide a pool of oligosaccharide signals in addition to those reviewed here. It is becoming equally clear that oligosaccharides can act as signals in organisms other than plants. For example, oligosaccharides derived from plant cell walls have been recently found to affect the expression of specific genes in bacteria (Rong *et al.*, 1994), although the active component has not been purified and characterized. Bacterial complex carbohydrates such as lipopolysaccharides and K-antigens (a capsular polysaccharide or an exopolysaccharide) are also known to trigger or prevent specific and general defense responses in mammals (Jann and Jann, 1990; Raetz, 1990); similar bacterial components are known to play a role in plant–bacteria interactions, although the specific functions have not yet been determined (Benen *et al.*, 1996; González *et al.*, 1996; Reuhs *et al.*, 1993; Noel, 1992). Fungus-derived oligosaccharides can act as regulators of the immune responses in animals (Velupillai and Harn, 1994). Putative receptors for fungal β-glucans have been identified in mammals (Müller *et al.*, 1996; Thornton *et al.*, 1996; Czop and Austen, 1985). The identification of new activities of oligosaccharides in a broader range of organisms, together with extensive structure–activity and signal transduction studies, will undoubtedly broaden our understanding of the crucial role of these molecules as biological signals and yield valuable insights into how cells perceive and respond to stimuli from their environment (Ryan, 1994).

Progress in elucidating the mechanisms by which oligosaccharide elicitors exert their effects on plant cells has often been hampered by the heterogeneity of the oligosaccharide preparations used in many studies. Detailed biochemical investigations of the cellular signaling pathways triggered by oligosaccharide elicitors require the isolation and/or chemical synthesis of homogeneous oligosaccharides in order to assign unambiguously the observed effects to single inducer-stimulated pathways. Recent improvements in the techniques for the purification of oligosaccharides and contributions from synthetic organic chemists have made available homogeneous preparations of active oligosaccharides and structurally related, less active agonists in milligram quantities. The availability of these homogeneous oligosaccharides has been essential for the tentative identification of binding proteins (putative receptors) for the elicitors discussed in this review.

We have touched on the question of the biological relevance of oligosaccharides in host–pathogen interactions, particularly with regard to possible mechanisms for the generation of these elicitors *in planta*. We have also discussed possible mechanisms for the regulation of elicitor production. However, the ability of the released oligosaccharide elicitors to move freely in the apoplast and the vascular tissues to their target sites is another important consideration with regard to the biological relevance of these signals. In systemic responses, the target site can be localized in tissues or organs at a distance from the sites where the signal is generated. To date, only the movement of radiolabeled oligogalacturonides *in planta* has been examined. These studies have shown that oligogalacturonides can move freely in the vascular system, although modifications such as the shortening of the chain and esterification occurred (MacDougall *et al.*, 1992). However, oligogalacturonides applied to a wound were shown not to move to adjacent tissues (Baydoun and Fry, 1985).

At the cellular level, the ability of oligosaccharide elicitors to pass through the plant cell wall in order to interact with plasma membrane-localized receptors is of concern. For example, nonesterified homogalacturonans can aggregate via a calcium-mediated cooperative mechanism (Vreeland *et al.*, 1989; Grant *et al.*, 1973). This property of homogalacturonans has been used to localize pectin components in the plant cell wall using labeled oligogalacturonides of DP ~30 (Vreeland *et al.*, 1989). Furthermore, the binding of a monoclonal antibody which recognizes a calcium-induced conformation of pectin (Liners *et al.*, 1989) is inhibited by oligogalacturonides of DP 9 and 10, but only in the presence of $Ca^{2+}$ (Liners *et al.*, 1992). These results suggest that bioactive oligogalacturonides in the presence of $Ca^{2+}$ could be bound to partially deesterified stretches of pectin, thereby immobilizing the fragments in the wall matrix. The immobilization of oligochitosan in the cell wall and at the cell surface might occur through the interaction of the positive charges on this oligosaccharide with regularly spaced negative charges on galacturonans and on the plasma membrane. The differences between the activities of chitosan and chitin fragments might be attributable to the ability of oligochitosan, but not oligochitin, to participate in such ionic interactions with the cell wall. Until evidence is provided that specific plasma membrane-localized receptors exist for oligogalacturonides and oligochitosans, one cannot rule out that these oligosaccharide elicitors exert their biological effects by interactions within the cell wall.

A significant body of research has accumulated on binding proteins for two oligosaccharide elicitors, namely the chitooligosaccharide and the hepta-β-glucoside elicitors. The high affinity, saturability, and ligand specificity of these oligosaccharide-binding proteins strongly suggests that these are physiological receptors for the two oligosaccharide elicitors. Moreover, the low apparent $K_d$ values for these binding proteins indicate maximal occupancy of the putative receptor at low signal concentrations, thereby fully triggering the cellular transduction pathways induced by these elicitors. However, the purification and characterization of oligosaccharide elicitor-binding proteins and/or the isolation of the gene(s) encoding these proteins will be required to prove that the binding proteins function as physiological receptors. Evidence supporting a role for these protein(s) in the cellular signaling pathway leading to the biological response could be obtained by reconstitution of an oligosaccharin-responsive system using purified binding protein(s) or by functional expression of the genes encoding the binding protein(s). It will be of great interest to determine the specific functions of oligosaccharide elicitor receptors in the plant signal transduction pathways, and to see if new paradigms for signal perception and signal translation emerge from these studies.

ACKNOWLEDGMENTS. Our work on glucan elicitor-binding proteins and on the structures and functions of biologically active oligosaccharides is supported by grants from the National Science Foundation (MCB-9723685) and the Mizutani Foundation for Glycoscience to M. G. H., by a grant from the Department Energy (DE-

FG02-96ER20221) to Peter Albersheim, and in part by the Department of Energy-funded Center for Plant and Microbial Complex Carbohydrates (DE-FG09-93ER20097). We are grateful to P. Garegg (University of Stockholm, Sweden), T. Ogawa (RIKEN, Japan), J. van Boom (University of Leiden, The Netherlands), and their colleagues for their generous gifts of synthetic oligoglucosides.

## 6. REFERENCES


Abu-Goukh, A. A., Greve, L. C., and Labavitch, J. M., 1983, Purification and partial characterization of "Bartlett" pear fruit polygalacturonase inhibitors, *Physiol. Plant Pathol.* **23:**111–122.

Aiba, S., 1994a, Preparation of *N*-acetylchitooligosaccharides from lysozymatic hydrolysates of partially *N*-acetylated chitosans, *Carbohydr. Res.* **261:**297–306.

Aiba, S., 1994b, Preparation of *N*-acetylchitooligosaccharides by hydrolysis of chitosan with chitinase followed by *N*-acetylation, *Carbohydr. Res.* **265:**323–328.

Akiyama, K., Kawazu, K., and Kobayashi, A., 1994, Partially N-deacetylated chitin elicitor induces antimicrobial flavonoids in pea epicotyls, *Z. Naturforsch.* **49c:**811–818.

Akiyama, K., Kawazu, K., and Kobayashi, A., 1995, Partially N-deacetylated chitin oligomers (pentamer to heptamer) are potential elicitors for (+)-pisatin induction in pea epicotyls, *Z. Naturforsch.* **50c:**391–397.

Albersheim, P., and Darvill, A. G., 1985, Oligosaccharins, *Sci. Am.* **253:**58–64.

Albersheim, P., and Valent, B. S., 1974, Host–pathogen interactions. VII. Plant pathogens secrete proteins which inhibit enzymes of the host capable of attacking the pathogen, *Plant Physiol.* **53:**684–687.

Aldington, S., and Fry, S. C., 1993, Oligosaccharins, *Adv. Bot. Res.* **19:**1–101.

Anderson, A. J., and Albersheim, P., 1972, Host–pathogen interactions. V. Comparison of the abilities of proteins isolated from three varieties of *Phaseolus vulgaris* to inhibit the endo-polygalacturonases secreted by three races of *Colletotrichum lindemuthianum*, *Physiol. Plant Pathol.* **2:**339–346.

Apostol, I., Heinstein, P. F., and Low, P. S., 1989, Rapid stimulation of an oxidative burst during elicitation of cultured plant cells. Role in defense and signal transduction, *Plant Physiol.* **90:**109–116.

Arlorio, M., Ludwig, A., Boller, T., and Bonfante, P., 1992, Inhibition of fungal growth by plant chitinases and β-1,3-glucanases. A morphological study, *Protoplasma* **171:**34–43.

Ayers, A. R., Ebel, J., Finelli, F., Berger, N., and Albersheim, P., 1976a, Host–pathogen interactions. IX. Quantitative assays of elicitor activity and characterization of the elicitor present in the extracellular medium of cultures of *Phytophthora megasperma* var. *sojae*, *Plant Physiol.* **57:**751–759.

Ayers, A. R., Ebel, J., Valent, B., and Albersheim, P., 1976b, Host–pathogen interactions. X. Fractionation and biological activity of an elicitor isolated from the mycelial walls of *Phytophthora megasperma* var. *sojae*, *Plant Physiol.* **57:**760–765.

Ayers, A. R., Valent, B., Ebel, J., and Albersheim, P., 1976c, Host–pathogen interactions. XI. Composition and structure of wall-released elicitor fractions, *Plant Physiol.* **57:**766–774.

Baldwin, E. A., and Biggs, R. H., 1988, Cell-wall lysing enzymes and products of cell-wall digestion elicit ethylene in citrus, *Physiol. Plant* **73:**58–64.

Barber, M. S., Bertram, R. E., and Ride, J. P., 1989, Chitin oligosaccharides elicit lignification in wounded wheat leaves, *Physiol. Mol. Plant Pathol.* **34:**3–12.

Bartnicki-Garcia, S., 1968, Cell wall chemistry, morphogenesis, and taxonomy of fungi, *Annu. Rev. Microbiol.* **22:**87–108.

Baureithel, K., and Boller, T., 1995, Characterization and solubilization of a specific binding site for chitin fragments in suspension-cultured tomato cells and microsomal membranes, *J. Cell. Biochem. Suppl.* **19B:**151(Abstract).

Baureithel, K., Felix, G., and Boller, T., 1994, Specific, high affinity binding of chitin fragments to tomato cells and membranes. Competitive inhibition of binding by derivatives of chitin fragments and a nod factor of *Rhizobium, J. Biol. Chem.* **269:**17931–17938.

Baydoun, E. A.-H., and Fry, S. C., 1985, The immobility of pectic substances in injured tomato leaves and its bearing on the identity of the wound hormone, *Planta* **165:**269–276.

Benen, J. A. E., Kester, H. C. M., Parenicová, L., and Visser, J., 1996, Kinetics and mode of action of *Aspergillus niger* polygalacturonases, in *Pectins and Pectinases,* Vol. 14 (J. Visser and A. G. J. Voragen, eds.), pp. 221–230, Elsevier, Amsterdam, The Netherlands.

Benhamou, N., and Lafontaine, P. J., 1995, Ultrastructural and cytochemical characterization of elicitor-induced structural responses in tomato root tissues infected by *Fusarium oxysporum* f.sp. *radicis-lycopersici, Planta* **197:**89–102.

Benhamou, N., Lafontaine, P. J., and Nicole, M., 1994, Induction of systemic resistance to fusarium crown and root rot in tomato plants by seed treatment with chitosan, *Phytopathology* **84:**1432–1444.

Bergmann, C. W., Ito, Y., Singer, D., Albersheim, P., Darvill, A. G., Benhamou, N., Nuss, L., Salvi, G., Cervone, F., and De Lorenzo, G., 1994, Polygalacturonase-inhibiting protein accumulates in *Phaseolus vulgaris* L. in response to wounding, elicitors and fungal infection, *Plant J.* **5:**625–634.

Bergmann, C. W., Cook, B., Darvill, A. G., Albersheim, P., Bellincampi, D., and Caprari, C., 1995, The effect of glycosylation of endopolygalacturonases and polygalacturonase inhibiting proteins on the production of oligogalacturonides (Abstract), International Symposium on Pectins and Pectinases, Wageningen, The Netherlands.

Bhandal, I. S., and Paxton, J. D., 1991, Phytoalexin biosynthesis induced by the fungal glucan polytran L in soybean, pea, and sweet pepper tissues, *J. Agric. Food Chem.* **39:**2156–2157.

Birberg, W., Fügedi, P., Garegg, P. J., and Pilotti, Å., 1989, Syntheses of a heptasaccharide β-linked to an 8-methoxycarbonyl-oct-1-yl linking arm and of a decasaccharide with structures corresponding to the phytoelicitor active glucan of *Phytophthora megasperma* f.sp. *glycinea, J. Carbohydr. Chem.* **8:**47–57.

Bishop, P. D., Pearce, G., Bryant, J. E., and Ryan, C. A., 1984, Isolation and characterization of the proteinase inhibitor-inducing factor from tomato leaves. Identity and activity of poly- and oligogalacturonide fragments, *J. Biol. Chem.* **259:**13172–13177.

Bol, J. F., Linthorst, H. J. M., and Cornelissen, B. J. C., 1990, Plant pathogenesis-related proteins induced by virus infection, *Annu. Rev. Phytopathol.* **28:**113–138.

Boller, T., 1987, Hydrolytic enzymes in plant disease resistance, in *Plant–Microbe Interactions. Molecular and Genetic Perspectives,* Vol. 2 (T. Kosuge and E. W. Nester, eds.), pp. 385–413, Macmillan, New York.

Boller, T., 1993, Antimicrobial functions of the plant hydrolases, chitinase and β-1,3-glucanase, in *Mechanisms of Plant Defense Responses,* Vol. 2 (B. Fritig and M. Legrand, eds.), pp. 391–400, Kluwer, Dordrecht, The Netherlands.

Bonhoff, A., Loyal, R., Ebel, J., and Grisebach, H., 1986, Race:cultivar-specific induction of enzymes related to phytoalexin biosynthesis in soybean roots following infection with *Phytophthora megasperma* f. sp. *glycinea, Arch. Biochem. Biophys.* **246:**149–154.

Bouquelet, S., and Spik, G., 1978, Properties of four molecules forms of *N*-acetyl-β-D-hexosaminidase isolated from germinating seeds of fenugreek (*Trigonella foenumgraecum*), *Eur. J. Biochem.* **84:**551–559.

Brady, K. P., Darvill, A. G., and Albersheim, P., 1993, Activation of a tobacco glycine-rich protein gene by a fungal glucan preparation, *Plant J.* **4:**517–524.

Broekaert, W. F., and Peumans, W. J., 1988, Pectic polysaccharides elicit chitinase accumulation in tobacco, *Physiol. Plant.* **74**:740–744.

Broglie, K., Chet, I., Holliday, M., Cressman, R., Biddle, P., Knowlton, S., Mauvais, C. J., and Broglie, R., 19991, Transgenic plants with enhanced resistance to the fungal pathogen *Rhizoctonia solani, Science* **254**:1194–1197.

Bruce, R. J., and West, C. A., 1989, Elicitation of lignin biosynthesis and isoperoxidase activity by pectic fragments in suspension-cultures of castor bean, *Plant Physiol.* **91**:889–897.

Bugbee, W. M., 1993, A pectin lyase inhibitor protein from cell walls of sugar beet, *Phytopathology* **83**:63–68.

Campbell, A. D., and Labavitch, J. M., 1991, Induction and regulation of ethylene biosynthesis by pectic oligomers in cultured pear cells, *Plant Physiol.* **97**:699–705.

Cervone, F., De Lorenzo, G., Salvi, G., Bergmann, C., Hahn, M. G., Ito, Y., Darvill, A., and Albersheim, P., 1989a, Release of phytoalexin elicitor-active oligogalacturonides by microbial pectic enzymes, in *Signal Molecules in Plants and Plant-Microbe Interactions* NATO ASI Series, Vol. H36 (B. J. J. Lugtenberg, ed.), pp. 85–89, Springer-Verlag, Berlin.

Cervone, F., Hahn, M. G., De Lorenzo, G., Darvill, A., and Albersheim, P., 1989b, Host–pathogen interactions. XXXIII. A plant protein converts a fungal pathogenesis factor into an elicitor of plant defense responses, *Plant Physiol.* **90**:542–548.

Cervone, F., De Lorenzo, G., Pressey, R., Darvill, A. G., and Albersheim, P., 1990, Can *Phaseolus* PGIP inhibit pectic enzymes from microbes and plants? *Phytochemistry* **29**:447–449.

Cervone, F., De Lorenzo, G., Caprari, C., Clark, A. J., Desiderio, A., Devoto, A., Leckie, F., Nuss, L., Salvi, G. and Toubart, P., 1993, The interaction between fungal endopolygalacturonase and plant cell wall PGIP (polygalacturonase-inhibiting protein), in *Mechanisms of Plant Defense Responses*, Vol. 2 (B. Fritig and M. Legrand, eds.), pp. 64–67, Kluwer, Dordrecht, The Netherlands.

Cervone, F., De Lorenzo, G., Bellincampi, D., Caprari, C., Clark, A. J., Desiderio, A., Devoto, A., Leckie, F., Nuss, L., and Salvi, G., 1994, Accumulation of PGIP, a leucine-rich receptor-like protein, correlates with the hypersensitive response in race-cultivar interactions, in *Advances in Molecular Genetics of Plant–Microbe Interactions*, Vol. 3 (M. J. Daniels, J. A. Downie, and A. E. Osbourn, eds.), pp. 319–322, Kluwer, Dordrecht, The Netherlands.

Cervone, F., De Lorenzo, G., Aracri, B., Bellincampi, D., Caprari, C., Devoto, A., Leckie, F., Mattei, B., Nuss, L., and Salvi, G., 1996, The PGIP (polygalacturonase-inhibiting protein) family: Extracellular proteins specialized for recognition, in *Biology of Plant–Microbe Interactions* (G. Stacey, B. Mullin, and P. M. Gresshoff, eds.), pp. 93–98, International Society for Molecular Plant–Microbe Interactions, St. Paul, Minnesota.

Cheong, J.-J., and Hahn, M. G., 1991, A specific, high-affinity binding site for the hepta-β-glucoside elicitor exists in soybean membranes, *Plant Cell* **3**:137–147.

Cheong, J.-J., Birberg, W., Fügedi, P., Pilotti, Å., Garegg, P. J., Hong, N., Ogawa, T., and Hahn, M. G., 1991, Structure–activity relationships of oligo-β-glucoside elicitors of phytoalexin accumulation in soybean, *Plant Cell* **3**:127–136.

Cheong, J.-J., Alba, R., Côté, F., Enkerli, J., and Hahn, M. G., 1993, Solubilization of functional plasma membrane-localized hepta-β-glucoside elicitor binding proteins from soybean, *Plant Physiol.* **103**:1173–1182.

Cline, K., and Albersheim, P., 1981, Host–pathogen interactions. XVII. Hydrolysis of biologically active fungal glucans by enzymes isolated from soybean cells, *Plant Physiol.* **68**:221–228.

Cline, K., Wade, W., and Albersheim, P., 1978, Host–pathogen interactions. XV. Fungal glucans which elicit phytoalexin accumulation in soybean also elicit the accumulation of phytoalexins in other plants, *Plant Physiol.* **62**:918–921.

Collinge, D. B., Kragh, K. M., Mikkelsen, J. D., Nielsen, K. K., Rasmussen, U., and Vad, K., 1993, Plant chitinases, *Plant J.* **3**:31–40.

Conrath, U., Domard, A., and Kauss, H., 1989, Chitosan-elicited synthesis of callose and of coumarin derivatives in parsely cell suspension cultures, *Plant Cell Rep.* **8:**152–155.

Cosio, E. G., Pöpperl, H., Schmidt, W. E., and Ebel, J., 1988, High-affinity binding of fungal β-glucan fragments to soybean (*Glycine max* L.) microsomal fractions and protoplasts, *Eur. J. Biochem.* **175:**309–315.

Cosio, E. G., Frey, T., and Ebel, J., 1990a, Solubilization of soybean membrane binding sites for fungal β-glucans that elicit phytoalexin accumulation, *FEBS Lett.* **264:**235–238.

Cosio, E. G., Frey, T., Verduyn, R., Van Boom, J., and Ebel, J., 1990b, High-affinity binding of a synthetic heptaglucoside and fungal glucan phytoalexin elicitors to soybean membranes, *FEBS Lett.* **271:**223–226.

Cosio, E. G., Frey, T., and Ebel, J., 1992, Identification of a high-affinity binding protein for a hepta-β-glucoside phytoalexin elicitor in soybean, *Eur. J. Biochem.* **204:**1115–1123.

Côté, F., and Hahn, M. G., 1994, Oligosaccharins: structures and signal transduction, *Plant Mol. Biol.* **26:**1375–1411.

Côté, F., Cheong, J.-J., Chen, H., Dhugga, K. S., and Hahn, M. G., 1997, Proteins that co-purify with a glucan elicitor-binding protein complex from soybean react with antisera against callose synthase-associated polypeptides, submitted.

Czop, J. K., and Austen, K. F., 1985, A β-glucan inhibitable receptor on human monocytes: Its identity with the phagocytic receptor for particulate activators of the alternative complement pathway, *J. Immunol.* **134:**2588–2593.

Dangl, J. L., 1995, Pièce de résistance: Novel classes of plant disease resistance genes, *Cell* **80:**363–366.

Darvill, A. G., and Albersheim, P., 1984, Phytoalexins and their elicitors—A defense against microbial infection in plants, *Annu. Rev. Plant Physiol.* **35:**243–275.

Darvill, A., Augur, C., Bergmann, C., Carlson, R. W., Cheong, J.-J., Eberhard, S., Hahn, M. G., Lo, V.-M., Marfà, V., Meyer, B., Mohnen, D., O'Neill, M. A., Spiro, M. D., van Halbeek, H., York, W. S., and Albersheim, P., 1992, Oligosaccharins—Oligosaccharides that regulate growth, development and defense responses in plants, *Glycobiology* **2:**181–198.

Davis, D., Merida, J., Legendre, L., Low, P. S., and Heinstein, P., 1993, Independent elicitation of the oxidative burst and phytoalexin formation in cultured plant cells, *Phytochemistry* **32:**607–611.

Davis, K. R., and Hahlbrock, K., 1987, Induction of defense responses in cultured parsley cells by plant cell wall fragments, *Plant Physiol.* **85:**1286–1290.

Davis, K. R., Lyon, G. D., Darvill, A. G., and Albersheim, P., 1984, Host–pathogen interactions. XXV. Endopolygalacturonic acid lyase from *Erwinia carotovora* elicits phytoalexin accumulation by releasing plant cell wall fragments, *Plant Physiol.* **74:**52–60.

Davis, K. R., Darvill, A. G., and Albersheim, P., 1986a, Host–pathogen interactions. XXXI. Several biotic and abiotic elicitors act synergistically in the induction of phytoalexin accumulation in soybean, *Plant Mol. Biol.* **6:**23–32.

Davis, K. R., Darvill, A. G., Albersheim, P., and Dell, A., 1986b, Host–pathogen interactions. XXIX. Oligogalacturonides released from sodium polypectate by endopolygalacturonic acid lyase are elicitors of phytoalexins in soybean, *Plant Physiol.* **80:**568–577.

Davis, K. R., Darvill, A. G., Albersheim, P., and Dell, A., 1986c, Host–pathogen interactions. XXX. Characterization of elicitors of phytoalexin accumulation in soybean released from soybean cell walls by endopolygalacturonic acid lyase, *Z. Naturforsch.* **41c:**39–48.

Deising, H., and Siegrist, J., 1995, Chitin deacetylase activity of the rust *Uromyces viciae-fabae* is controlled by fungal morphogenesis, *FEMS Microbiol. Lett.* **127:**207–212.

De Lorenzo, G., Ito, Y., D'Ovidio, R., Cervone, F., Albersheim, P., and Darvill, A. G., 1990, Host–pathogen interactions. XXXVII. Abilities of the polygalacturonase-inhibiting proteins from four cultivars of *Phaseolus vulgaris* to inhibit the *endo*polygalacturonases from three races of *Colletotrichum lindemuthianum*, *Physiol. Mol. Plant Pathol.* **36:**421–435.

De Lorenzo, G., Cervone, F., Bellincampi, D., Caprari, C., Clark, A. J., Desiderio, A., Devoto, A., Forrest, R., Leckie, F., Nuss, L., and Salvi, G., 1994, Polygalacturonase, PGIP and oligogalacturonides in cell–cell communication, *Biochem. Soc. Trans.* **22:**394–397.

De Wit, P. J. G. M., 1992, Molecular characterization of gene-for-gene systems in plant–fungus interactions and the application of avirulence genes in control of plant pathogens, *Annu. Rev. Phytopathol.* **30:**391–418.

Dhawale, S., Souciet, G., and Kuhn, D. N., 1989, Increase of chalcone synthase mRNA in pathogen-inoculated soybeans with race-specific resistance is different in leaves and roots, *Plant Physiol.* **91:**911–916.

Diekmann, W., Herkt, B., Low, P. S., Nümberger, T., Scheel, D., Terschüren, C., and Robinson, D. G., 1994, Visualization of elicitor-binding loci at the plant cell surface, *Planta* **195:**126–137.

Dixon, R. A., 1986, The phytoalexin response: Elicitation, signalling and control of host gene expression, *Biol. Rev.* **61:**239–291.

Dixon, R. A., and Lamb, C. J., 1990, Molecular communication in interactions between plants and microbial pathogens, *Annu. Rev. Plant Physiol. Plant Mol. Biol.* **41:**339–367.

Dixon, R. A., Jennings, A. C., Davies, L. A., Gerrish, C., and Murphy, D. L., 1989, Elicitor active components from French bean hypocotyls, *Physiol. Mol. Plant Pathol.* **34:**99–115.

Doares, S. H., Syrovets, T., Weiler, E. W., and Ryan, C. A., 1995, Oligogalacturonides and chitosan activate plant defensive genes through the octadecanoid pathway, *Proc. Natl. Acad. Sci. USA* **92:**4095–4098.

Ebel, J., 1986, Phytoalexin synthesis: The biochemical analysis of the induction process, *Annu. Rev. Phytopathol.* **24:**235–264.

Ebel, J., and Cosio, E. G., 1994, Elicitors of plant defense responses, *Int. Rev. Cytol.* **148:**1–36.

Ebel, J., Cosio, E. G., Feger, M., Frey, T., Kissel, U., Reinold, S., and Waldmüller, T., 1993, Glucan elicitor-binding proteins and signal transduction in the activation of plant defence, in *Advances in Moleclar Genetics of Plant–Microbe Interactions*, Vol. 2 (E. W. Nester and D. P. S. Verma, eds.), pp. 477–484, Kluwer, Dordrecht, The Netherlands.

Ebel, J., Bhagwat, A. A., Cosio, E. G., Feger, M., Kissel, U., Mithöfer, A., and Waldmüller, T., 1995, Elicitor binding proteins and signal transduction in the activation of a phytoalexin defense response, *Can. J. Bot.* **73:**S506–S510.

El Ghaouth, A., Arul, J., Grenier, J., Benhamou, N., Asselin, A., and Bélanger, R., 1994a, Effect of chitosan on cucumber plants: Suppression of *Pythium aphanidermatum* and induction of defense reactions, *Phytopathology* **84:**313–320.

El Ghaouth, A., Arul, J., Wilson, C., and Benhamou, N., 1994b, Ultrastructural and cytochemical aspects of the effect of chitosan on decay of bell pepper fruit, *Physiol. Mol. Plant Pathol.* **44:**417–432.

English, P. D., Maglothin, A., Keegstra, K., and Albersheim, P., 1972, A cell wall-degrading endo-polygalacturonase secreted by *Colletotrichum lindemuthianum*, *Plant Physiol.* **49:**293–297.

Enkerli, K., Hahn, M. G., and Mims, C. W., 1997, Immunogold localization of callose and other plant cell wall components in soybean roots infected with the oomycete *Phytophthora sojae*, *Can. J. Bot.* **75:**1509–1517.

Farkas, V., 1979, Biosynthesis of cell walls of fungi, *Microbiol. Rev.* **43:**117–144.

Farmer, E. E., and Ryan, C. A., 1992, Octadecanoid precursors of jasmonic acid activate the synthesis of wound-inducible proteinase inhibitors, *Plant Cell* **4:**129–134.

Farmer, E. E., Pearce, G., and Ryan, C. A., 1989, *In vitro* phosphorylation of plant plasma membrane proteins in response to the proteinase inhibitor inducing factor, *Proc. Natl. Acad. Sci. USA* **86:**1539–1542.

Farmer, E. E., Moloshok, T. D., and Ryan, C. A., 1990, *In vitro* phosphorylation in response to oligouronide elicitors: Structural and biological relationships, *Curr. Top. Plant Biochem. Physiol.* **9:**249–258.

Farmer, E. E., Moloshok, T. D., Saxton, M. J., and Ryan, C. A., 1991, Oligosaccharide signaling in plants: Specificity of oligouronide-enhanced plasma membrane protein phosphorylation, *J. Biol. Chem.* **266:**3140–3145.

Favaron, F., Alghisi, P., and Marciano, P., 1992, Characterization of two *Sclerotinia sclerotiorum* polygalacturonases with different abilities to elicit glyceollin in soybean, *Plant Sci.* **83:**7–13.

Felix, G., Regenass, M., and Boller, T., 1993, Specific perception of subanomolar concentrations of chitin fragments by tomato cells: Induction of extracellular alkalinization, changes in protein phosphorylation, and establishment of a refractory state, *Plant J.* **4:**307–316.

Flach, J., Pilet, P.-E., and Jollès, P., 1992, What's new in chitinase research, *Experientia* **48:**701–716.

Fliegmann, J., Mithöfer, A., Antelo, L., Lottspeich, F., and Ebel, J., 1997, Molecular characterization of the putative receptor for a β-glucan elicitor in the interaction between soybean and its pathogen *Phytophthora sojae*, *Plant Physiol.* **114:**279 (abstract).

Forrest, R. S., and Lyon, G. D., 1990, Substrate degradation patterns of polygalacturonic acid lyase from *Erwinia carotovora* and *Bacillus polymyxa* and release of phytoalexin-eliciting oligosaccharides from potato cell walls, *J. Exp. Bot.* **41:**481–488.

Frediani, M., Cremonini, R., Salvi, G., Caprari, C., Desiderio, A., D'Ovidio, R., Cervone, F., and De Lorenzo, G., 1993, Cytological localization of the *PGIP* genes in the embryo suspensor cells of *Phaseolus vulgaris* L., *Theor. Appl. Genet.* **87:**369–373.

Frey, T., Cosio, E. G., and Ebel, J., 1993, Affinity purification and characterization of a binding protein for a hepta-β-glucoside phytoalexin elicitor in soybean, *Phytochemistry* **32:**543–550.

Fry, S. C., Aldington, S., Hetherington, P. R., and Aitken, J., 1993, Oligosaccharides as signals and substrates in the plant cell wall, *Plant Physiol.* **103:**1–5.

Fügedi, P., Birberg, W., Garegg, P. J., and Pilotti, A., 1987, Syntheses of a branched heptasaccharide having phytoalexin-elicitor activity, *Carbohydr. Res.* **164:**297–312.

Giannini, J. L., Holt, J. S., and Briskin, D. P., 1991, The effect of glyceollin on soybean (*Glycine max* L.) tonoplast and plasma membrane vesicles, *Plant Sci.* **74:**203–211.

Giannini, J. L., Nelson, M., and Spessard, G. O., 1995, The effect of rishitin on potato tonoplast vesicle and vacuole proton transport, *Phytochemistry* **40:**1655–1658.

González, J. E., Reuhs, B. L., and Walker, G C., 1996, Low molecular weight EPS II of *Rhizobium meliloti* allows nodule invasion in *Medicago sativa*, *Proc. Natl. Acad. Sci. USA* **93:**8636–8641.

Grab, D., Feger, M., and Ebel, J., 1989, An endogenous factor from soybean (*Glycine max* L.) cell cultures activates phosphorylation of a protein which is dephosphorylated *in vivo* in elicitor-challenged cells, *Planta* **179:**340–348.

Graham, T. L., 1991, Flavonoid and isoflavonoid distribution in developing soybean seedling tissues and in seed and root exudates, *Plant Physiol.* **95:**594–603.

Graham, T. L., and Graham, M. Y., 1991, Glyceollin elicitors induce major but distinctly different shifts in isoflavonoid metabolism in proximal and distal soybean cell populations, *Mol. Plant–Microbe Interact.* **4:**60–68.

Graham, T. L., Kim, J. E., and Graham, M. Y., 1990, Role of constitutive isoflavone conjugates in the accumulation of glyceollin in soybean infected with *Phytophthora megasperma*, *Mol. Plant–Microbe Interact.* **3:**157–166.

Grant, G. T., Morris, E. R., Rees, D. A., Smith, P. J. C., and Thom, D., 1973, Biological interactions between polysaccharides and divalent cations: The egg-box model, *FEBS Lett.* **32:**195–198.

Grenier, J., and Asselin, A., 1990, Some pathogenesis-related proteins are chitosanases with lytic activity against fungal spores, *Mol. Plant–Microbe Interact.* **3:**401–407.

Grison, R., Grezes-Besset, B., Schneider, M., Lucante, N., Olsen, L., Leguay, J.-J., and Toppan, A., 1996, Field tolerance to fungal pathogens of *Brassica napus* constitutively expressing a chimeric chitinase gene, *Bio/technology* **14:**643–646.

Grosskopf, D. G., Felix, G., and Boller, T., 1991, A yeast-derived glycopeptide elicitor and chitosan or digitonin differentially induce ethylene biosynthesis, phenylalanine ammonia-lyase and callose formation in suspension-cultured tomato cells, *J. Plant Physiol.* **138:**741–746.

Gunia, W., Hinderer, W., Wittkampf, U., and Barz, W., 1991, Elicitor induction of cytochrome P-450 monooxygenases in cell suspension cultures of chickpea (*Cicer arietinum* L.) and their involvement in pterocarpan phytoalexin biosynthesis, *Z. Naturforsch.* **46c:**58–66.

Hadwiger, L. A., and Beckman, J. M., 1980, Chitosan as a component of pea–*Fusarium solani* interactions, *Plant Physiol.* **66:**205–211.

Hadwiger, L. A., and Line, R. F., 1981, Hexosamine accumulations are associated with the terminated growth of *Puccinia striiformis* on wheat isolines, *Physiol. Mol. Plant Pathol.* **19:**249–255.

Hadwiger, L. A., Ogawa, T., and Kuyama, H., 1994, Chitosan polymer sizes effective in inducing phytoalexin accumulation and fungal suppression are verified with synthesized oligomers, *Mol. Plant–Microbe Interact.* **7:**531–533.

Hahlbrock, K., and Scheel, D., 1987, Biochemical responses of plants to pathogens, in *Innovative Approaches to Plant Disease Control* (I. Chet, ed.), pp. 229–254, Wiley, New York.

Hahn, M. G., 1989, Animal receptors—Examples of cellular signal perception molecules, in *Signal Molecules in Plants and Plant–Microbe Interactions*, H36 (B. J. J. Lugtenberg, ed.), pp. 1–26, Springer-Verlag, Berlin.

Hahn, M. G., 1996, Microbial elicitors and their receptors in plants, *Annu. Rev. Phytopathol.* **34:**387–412.

Hahn, M. G., and Albersheim, P., 1978, Host–pathogen interactions. XIV. Isolation and partial characterization of an elicitor form yeast extract, *Plant Physiol.* **62:**107–111.

Hahn, M. G., and Grisebach, H., 1983, Cyclic AMP is not involved as a second messenger in the response of soybean to infection by *Phytophthora megasperma* f. sp. *glycinea*, *Z. Naturforsch.* **38c:**578–582.

Hahn, M. G., Darvill, A. G., and Albersheim, P., 1981, Host–pathogen interactions. XIX. The endogenous elicitor, a fragment of a plant cell wall polysaccharide that elicits phytoalexin accumulation in soybeans, *Plant Physiol.* **68:**1161–1169.

Hahn, M. G., Bucheli, P., Cervone, F., Doares, S. H., O'Neill, R. A., Darvill, A., and Albersheim, P., 1989a, Roles of cell wall constituents in plant–pathogen interactions, in *Plant–Microbe Interactions. Molecular and Genetic Perspectives*, Vol. 3 (T. Kosuge and E. W. Nester, eds.), pp. 131–181, McGraw-Hill, New York.

Hahn, M. G., Cheong, J.-J., Birberg, W., Fügedi, P., Pilotti, Å., Garegg, P., Hong, N., Nakahara, Y., and Ogawa, T., 1989b, Elicitation of phytoalexins by synthetic oligoglucosides, synthetic oligogalacturonides, and their derivatives, in *Signal Molecules in Plants and Plant–Microbe Interactions*, H36 (B. J. J. Lugtenberg, ed.), pp. 91–97, Springer-Verlag, Berlin.

Hahn, M. G., Darvill, A., Albersheim, P., Bergmann, C., Cheong, J.-J., Koller, A., and Lò, V.-M., 1992, Preparation and characterization of oligosaccharide elicitors of phytoalexin accumulation, in *Molecular Plant Pathology. Volume II: A Practical Approach* (S. Gurr, M. McPherson, and D. J. Bowles, eds.), pp. 103–147, Oxford University Press, Oxford.

Ham, K.-S., Kauffmann, S., Albersheim, P., and Darvill, A. G., 1991, Host–pathogen interactions XXXIX. A soybean pathogenesis-related protein with β-1,3-glucanase activity releases phytoalexin elicitor-active heat-stable fragments from fungal walls, *Mol. Plant–Microbe Interact.* **4:**545–552.

Ham, K.-S., Côté, F., Hahn, M. G., Thurlby, T., Albersheim, P., and Darvill, A., 1996, Soybean β-1,3-glucanases generate, from the walls of *Phytophthora*, elicitors recognized by the putative receptor of the hepta-β-glucoside elicitor, *Plant Physiol.* **111:**170(Abstract).

Ham, K.-S., Darvill, A. G., and Albersheim, P., 1997, Fungal pathogens secrete an inhibitor protein that distinguishes isoforms of plant pathogenesis-related *endo*-β-1,3-glucanases, *Plant J.* **11:**169–179.

Hargreaves, J. A., and Bailey, J. A., 1978, Phytoalexin production by hypocotyls of *Phaseolus vulgaris* in response to constitutive metabolites released by damaged bean cells, *Physiol. Plant Pathol.* **13**:89–100.

Hargreaves, J. A., and Shelby, C., 1978, Phytoalexin formation in cell suspensions of *Phaseolus vulgaris* in response to an extract of bean hypocotyls, *Phytochemistry* **17**:1099–1102.

Hashimoto, H., Abe, Y., Horito, S., and Yoshimura, J., 1989, Synthesis of chitooligosaccharide derivatives by condensation polymerization, *J. Carbohydr. Chem.* **8**:307–311.

Hollenberg, M. D., 1991, Structure–activity relationships for transmembrane signaling: The receptor's turn, *FASEB J.* **5**:178–186.

Hong, N., and Ogawa, T., 1990, Stereocontrolled syntheses of phytoalexin elicitor-active β-D-glucohexaoside and β-D-glucononaoside, *Tetrahedron Lett.* **31**:3179–3182.

Horn, M. A., Heinstein, P. F., and Low, P. S., 1989, Receptor-mediated endocytosis in plant cells, *Plant Cell* **1**:1003–1009.

Ishihara, A., Miyagawa, H., Kuwahara, Y., Ueno, T.,, and Mayama, S., 1996, Involvement of $Ca^{2+}$ ion in phytoalexin induction in oats, *Plant Sci.* **115**:9–16.

Ito, Y., Kaku, H., and Shibuya, N., 1997, Identification of a high-affinity binding protein for *N*-acetylchitooligosaccharide elicitor in the plasma membrane of suspension-cultured rice cells by affinity labeling, *Plant J.* **12**:347–356.

Jach, G., Görnhardt, B., Munday, J., Logemann, J., Pinsdorf, E., Leah, R., Schell, J., and Maas, C., 1995, Enhanced quantitative resistance against fungal disease by combinatorial expression of different barley antifungal proteins in transgenic tobacco, *Plant J.* **8**:97–109.

Jann, B., and Jann, K., 1990, Structure and biosynthesis of the capsular antigens of *Escherichia coli, Curr. Top. Microbiol. Immunol.* **150**:19–42.

Jin, D. F., and West, C. A., 1984, Characteristics of galacturonic acid oligomers as elicitors of casbene synthetase activity in castor bean seedlings, *Plant Physiol.* **74**:989–992.

Johnston, D. J., Ramanathan, V., and Williamson, B., 1993, A protein from immature raspberry fruits which inhibits endopolygalacturonases from *Botrytis cinerea* and other micro-organisms, *J. Exp. Bot.* **44**:971–976.

Jones, J. D. G., 1996, Plant disease resistance genes: Structure, function and evolution, *Curr. Opin. Biotechnol.* **7**:155–160.

Jones, T. M., Anderson, A. J., and Albersheim, P., 1972, Host–pathogen interactions. IV. Studies on the polysaccharide-degrading enzymes secreted by *Fusarium oxysporum* f. sp. *lycopersici, Physiol. Plant Pathol.* **2**:153–166.

Jordan, N. D., and Barber, M. S., 1995, Multiple forms of what leaf *N*-acetyl-β-D-hexosaminidase, *Plant Sci.* **107**:41–48.

Kauss, H., Jeblick, W., and Domard, A., 1989, The degree of polymerization and N-acetylation of chitosan determine its ability to elicit callose formation in suspension cells and protoplasts of *Catharanthus roseus, Planta* **178**:385–392.

Keen, N. T., 1975, Specific elicitors of plant phytoalexin production: Determinants of race specificity in pathogens? *Science* **187**:74–75.

Keen, N. T., Yoshikawa, M., and Wang, M. C., 1983, Phytoalexin elicitor activity of carbohydrates from *Phytophthora megasperma* f. sp. *glycinea* and other sources, *Plant Physiol.* **71**:466–471.

Kendra, D. F., and Hadwiger, L. A., 1984, Characterization of the smallest chitosan oligomer that is maximally antifungal to *Fusarium solani* and elicits pisatin formation in *Pisum sativum, Exp. Mycol.* **8**:276–281.

Kendra, D. F., and Hadwiger, L. A., 1987a, Cell death and membrane leakage not associated with the induction of disease resistance in peas by chitosan or *Fusarium solani* f. sp. *phaseoli, Phytopathology* **77**:100–106.

Kendra, D. F., and Hadwiger, L. A., 1987b, Calcium and calmodulin may not regulate the disease resistance and pisatin formation responses of *Pisum sativum* to chitosan or *Fusarium solani*, *Physiol. Mol. Plant Pathol.* **31**:337–348.

Kendra, D. F., Christian, D., and Hadwiger, L. A., 1989, Chitosan oligomers from *Fusarium solani* pea interactions, chitinase/β-glucanase digestion of sporelings and from fungal wall chitin actively inhibit fungal growth and enhance disease resistance, *Physiol. Mol. Plant Pathol.* **35**:215–230.

Kobayashi, A., Tai, A., Kanzaki, H., and Kawazu, K., 1993, Elicitor-active oligosaccharides from algal laminaran stimulate the production of antifungal compounds in alfalfa, *Z. Naturforsch.* **48c**:575–579.

Kobayashi, A., Akiyama, K., and Kawazu, K., 1994, Partially N-deacetylated chitin fragments are strong elicitors for (+)-pisatin induction in epicotyls of pea, *Z. Naturforsch.* **49c**:302–308.

Kobayashi, A., Tai, A., and Kawazu, K., 1995, Structural elucidation of an elicitor-active oligosaccharide, LN-3, prepared from algal laminaran, *J. Carbohydr. Chem.* **14**:819–832.

Kobe, B., and Deisenhofer, J., 1995, A structural basis of the interactions between leucine-rich repeats and protein ligands, *Nature* **374**:183–186.

Köhle, H. Young, D. H., and Kauss, H., 1984, Physiological changes in suspension-cultured soybean cells elicited by treatment with chitosan, *Plant Sci. Lett.* **33**:221–230.

Köhle, H., Jeblick, W., Poten, F., Blashek, W., and Kauss, H., 1985, Chitosan-elicited callose synthesis in soybean cells as a $Ca^{2+}$-dependent process, *Plant Physiol.* **77**:544–551.

Kohn, R., 1975, Ion binding on polyuronates—alginate and pectin, *Pure Appl. Chem.* **42**:371–397.

Kohn, R., 1987, Binding of divalent cations to oligomeric fragments of pectin, *Carbohydr. Res.* **160**:343–353.

Kopp, M., Rouster, J., Fritig, B., Darvill, A., and Albersheim, P., 1989, Host–pathogen interactions. XXXII. A fungal preparation protects *Nicotianae* against infection by viruses, *Plant Physiol.* **90**:208–216.

Kuchitsu, K., Kikuyama, M., and Shibuya, N., 1993, *N*-Acetylchitooligosaccharides, biotic elictor for phytoalexin production, induce transient membrane depolarization in suspension-cultured rice cells, *Protoplasma* **174**:79–81.

Kuchitsu, K., Kosaka, H., Shiga, T., and Shibuya, N., 1995, EPR evidence for generation of hydroxyl radical triggered by *N*-acetylchitooligosaccharide elicitor and a protein phosphatase inhibitor in suspension-cultured rice cells, *Protoplasma* **188**:138–142.

Lafitte, C., Barthe, J. P., Montillet, J. L., and Touzé, A., 1984, Glycoprotein inhibitors of *Colletotrichum lindemuthianum* endopolygalacturonase in near isogenic lines of *Phaseolus vulgaris* resistant and susceptible to anthracnose, *Physiol. Plant Pathol.* **25**:39–53.

Lamb, C. J., Lawton, M. A., Dron, M., and Dixon, R. A., 1989, Signals and transduction mechanisms for activation of plant defenses against microbial attack, *Cell* **56**:215–224.

Legendre, L., Heinstein, P. F., and Low, P. S., 1992, Evidence for participation of GTP-binding proteins in elicitation of the rapid oxidative burst in cultured soybean cells, *J. Biol. Chem.* **267**:20140–20147.

Legendre, L., Rueter, S., Heinstein, P. F., and Low, P. S., 1993a, Characterization of the oligogalacturonide-induced oxidative burst in cultured soybean (*Glycine max*) cells, *Plant Physiol.* **102**:233–240.

Legendre, L., Yueh, Y. G., Crain, R., Haddock, N. Heinstein, P. F., and Low, P. S., 1993b, Phospholipase C activation during elicitation of the oxidative burst in cultured plant cells, *J. Biol. Chem.* **268**:24559–24563.

Lesney, M. S., 1989, Growth responses and lignin production in cell suspensions of *Pinus elliottii* 'elicited' by chitin, chitosan or mycelium of *Cronartium quercum* f.sp. *fusiforme*, *Plant Cell Tiss. Organ Cult.* **19**:23–31.

Lesney, M. S., 1990, Effect of 'elicitors' on extracellular peroxidase activity in suspension-cultured slash pine (*Pinus elliottii* Engelm.), *Plant Cell Tiss. Organ Cult.* **20**:173–175.

Levine, A., Tenhaken, R., Dixon, R., and Lamb, C., 1994, $H_2O_2$ from the oxidative burst orchestrates the plant hypersensitive disease resistance response, *Cell* **79:**583–593.

Li, T., Brzezinski, R,. and Beaulieu, C., 1995, Enzymatic production of chitosan oligomers, *Plant Physiol. Biochem.* **33:**599–603.

Lin, W., Anuratha, C. S., Datta, K., Potrykus, I., Muthukrishnan, S., and Datta, S. K., 1995, Genetic engineering of rice for resistance to sheath blight, *Bio/technology* **13:**686–691.

Liners, F., Letesson J.-J., Didembourg, C., and Van Cutsem, P., 1989, Monoclonal antibodies against pectin. Recognition of a conformation induced by calcium, *Plant Physiol.* **91:**14199–1424.

Liners, F., Thibault, F.-J., and Van Cutsem, P., 1992, Influence of the degree of polymerization of oligogalacturonates and of esterification pattern of pectin on their recognition by monoclonal antibodies, *Plant Physiol.* **99:**1099–1104.

Lorentzen, J. P., Helpap, B., and Lockhoff, O., 1991, Synthese eines elicitoraktiven Heptaglucansaccharides zur Untersuchung pflanzlicher Abwehrmechanismen, *Angew. Chem.* **103:**1731–1732.

Low, P. S., Legendre, L., Heinstein, P. F., and Horn, M. A., 1993, Comparison of elicitor and vitamin receptor-mediated endocytosis in cultured soybean cells, *J. Exp. Bot.* **44**(Suppl.):269–274.

Lucas, W. J., and Gilbertson, R. L., 1994, Plasmodesmata in relation to viral movement within leaf tissues, *Annu. Rev. Phytopathol.* **32:**387–411.

Ludwig, A., and Boller, T., 1990, A method for the study of fungal growth inhibition by plant proteins, *FEMS Microbiol. Lett.* **69:**61–66.

MacDougall, A. J., Rigby, N. M., Needs, P. W., and Selvendran, R. R., 1992, Movement and metabolism of oligogalacturonide elicitors in tomato shoots, *Planta* **188:**566–574.

MacKintosh, C., Lyon, G. D., and Mackintosh, R. W., 1994, Protein phosphatase inhibitors activate anti-fungal defence responses of soybean cotyledons and cell cultures, *Plant J.* **5:**137–147.

Mankarios, A. T., and Friend, J., 1980, Polysaccharide degrading enzymes of *Botrytis allii* and *Sclerotium cepivorum*. Enzyme production in culture and the effect of the enzymes on isolated onion cell walls, *Physiol. Plant Pathol.* **17:**93–104.

Marfà, V., Gollin, D. J., Eberhard, S., Mohnen, D., Darvill, A., and Albersheim, P., 1991, Oligogalacturonides are able to induce flowers to form on tobacco explants, *Plant J.* **1:**217–225.

Masuta, C., Van Den Bulcke, M., Bauw, G., Van Montagu, M., and Caplan, A. B., 1991, Differential effects of elicitors on the viability of rice suspension cells, *Plant Physiol.* **97:**619–629.

Mathieu, Y., Kurkdijan, A., Xia, H., Guern, J., Koller, A., Spiro, M., O'Neill, M., Albersheim, P., and Darvill, A., 1991, Membrane responses induced by oligogalacturonides in suspension-cultured tobacco cells, *Plant J.* **1:**333–343.

Mauch, F., and Staehelin, L. A., 1989, Functional implications of the subcellular localization of ethylene-induced chitinase and β-1,3-glucanase in bean leaves, *Plant Cell* **1:**447–457.

Mauch, F., Mauch-Mani, B., and Boller, T., 1988, Antifungal hydrolases in pea tissue. II. Inhibition of fungal growth by combinations of chitinase and β-1,3-glucanase, *Plant Physiol.* **88:**936–942.

Messiaen, J., and Van Cutsem, P., 1994, Pectic signal transduction in carrot cells: Membrane, cytosolic and nuclear responses induced by oligogalacturonides, *Plant Cell Physiol.* **35:**677–689.

Messiaen, J., Read, N. D., Van Cutsem, P., and Trewavas, A. J., 1993, Cell wall oligogalacturonides increase cytosolic free calcium in carrot protoplasts, *J. Cell Sci.* **104:**365–371.

Miller, K. J., Hadley, J. A., and Gustine, D. L., 1994, Cyclic β-1,6-1,3 glucans of *Bradyrhizobium japonicum* USDA 110 elicit isoflavonoid production in the soybean (*Glycine max*) host, *Plant Physiol.* **104:**917–923.

Minami, E., Kuchitsu, K., He, D. Y., Kouchi, H., Midoh, N., Ohtsuki, Y., and Shibuya, N., 1996, Two novel genes rapidly and transiently activated in suspension-cultured rice cells by treatment with *N*-acetylchitoheptaose, a biotic elicitor for phytoalexin production, *Plant Cell Physiol.* **37:**563–567.

Mithöfer, A., Bhagwat, A. A., Feger, M., and Ebel, J., 1996a, Suppression of fungal β-glucan-induced plant defence in soybean (*Glycine max* L.) by cyclic 1,3-1,6-β-glucans from the symbiont *Bradyrhizobium japonicum*, *Planta* **199:**270–275.

Mithöfer, A., Lottspeich, F., and Ebel, J., 1996b, One-step purification of the β-glucan elicitor-binding protein from soybean (*Glycine max* L.) roots and characterization of an anti-peptide antiserum, *FEBS Lett.* **381:**203–207.

Molano, J., Durán, A., and Cabib, E., 1977, A rapid and sensitive assay for chitinase using tritiated chitin, *Anal. Biochem.* **83:**648–656.

Moloshok, T., and Ryan, C. A., 1989, Di- and trigalacturonic acid and Delta$^{4,5}$-di- and Delta$^{4,5}$-trigalacturonic acids: Inducers of proteinase inhibitor genes in plants, *Meth. Enzymol.* **179:**566–569.

Moloshok, T., Pearce, G., and Ryan, C. A., 1992, Oligouronide signaling of proteinase inhibitor genes in plants: Structure–activity relationships of di- and trigalacturonic acids and their derivatives, *Arch. Biochem. Biophys.* **294:**731–734.

Morris, E. R., 1986, Molecular interactions in polysaccharide gelation, *Br. Polymer J.* **18:**14–21.

Müller, M., and Gessler, C., 1993, A protein from apple leaves inhibits pectinolytic activity of *Venturia inaequalis in vitro*, in *Mechanisms of Plant Defense Responses* (B. Fritig and M. Legrand, eds.), pp. 68–71, Kluwer, Dordrecht, The Netherlands.

Müller, A., Rice, P. J., Ensley, H. E., Coogan, P. S., Kalbfleisch, J. H. Kelley, J. L., Love, E. J., Portera, C. A., Ha, T., Browder, I. W., and Williams, D. L., 1996, Receptor binding and internalization of a water-soluble (1→3)-β-D-glucan biologic response modifier in two monocyte macrophage cell lines, *J. Immunol.* **156:**3418–3425.

Nakahara, Y., and Ogawa, T., 1987, Stereocontrolled, total synthesis of α-D-GalA-[(1/4)-α-D-GalA]$_8$-(1/4)-β-D-GalA-1/OPr, a synthetic model for phytoalexin elictor-active oligogalacturonic acids, *Carbohydr. Res.* **167:**c1–c7.

Nakahara, Y., and Ogawa, T., 1989, Total synthesis of galactododecaosiduronic acid, an endogenous phytoalexin elicitor isolated from soybean cell wall, *Tetrahedron Lett.* **30:**87–90.

Nicolaou, K. C., Winssinger, N., Pastor, J., and DeRoose, F., 1997, A general and highly efficient solid phase synthesis of oligosaccharides: Total synthesis of a heptasaccharide phytoalexin elicitor (HPE), *J. Am. Chem. Soc.* **119:**449–450.

Noel, K. D., 1992, Rhizobial polysaccharides required in symbioses with legumes, in *Molecular Signals in Plant–Microbe Communications* (D. P. S. Verma, ed.), pp. 341–357, CRC Press, Boca Raton, Florida.

Nojiri, H., Sugimori, M., Yamane, H., Nishimura, Y., Yamada, A., Shibuya, N., Kodama, O., Murofushi, N., and Omori, T., 1996, Involvement of jasmonic acid in elicitor-induced phytoalexin production in suspension-cultured rice cells, *Plant Physiol.* **110:**387–392.

Nothnagel, E. A., McNeil, M., Albersheim, P., and Dell, A., 1983, Host–pathogen interactions. XXII. A galacturonic acid oligosaccharide from plant cell walls elicits phytoalexins, *Plant Physiol.* **71:**916–926.

Nuss, L., Mahé, A., Clark, A. J., Grisvard, J., Dron, M., Cervone, F., and De Lorenzo, G., 1996, Differential accumulation of PGIP (polygalacturonase-inhibiting protein) mRNA in two near-isogenic lines of *Phaseolus vulgaris* L. upon infection with *Colletotrichum lindemuthianum*, *Physiol. Mol. Plant Pathol.* **48:**83–89.

Ohashi, Y., and Ohshima, M., 1992, Stress-induced expression of genes for pathogenesis-related proteins in plants, *Plant Cell Physiol.* **33:**819–826.

Okinaka, Y., Mimori, K., Takeo, K., Kitamura, S., Takeuchi, Y., Yamaoka, N., and Yoshikawa, M., 1995, A structural model for the mechanisms of elicitor release from fungal cell walls by plant β-1,3-endoglucanase, *Plant Physiol.* **109:**839–845.

O'Neill, M., Albersheim, P., and Darvill, A., 1990, The pectic polysaccharides of primary cell walls, in *Methods in Plant Biochemistry*, Vol. 2 (P. M. Dey, ed.), pp. 415–441, Academic Press, London.

Ossowski, P., Pilotti, Å, Garegg, P. J., and Lindberg, B., 1984, Synthesis of a glucoheptaose and a glucooctaose that elicit phytoalexin accumulation in soybean, *J. Biol. Chem.* **259:**11337–11340.

Parker, J. E., Hahlbrock, K., and Scheel, D., 1988, Different cell-wall components from *Phytophthora megasperma* f. sp. *glycinea* elicit phytoalexin production in soybean and parsley, *Planta* **176:**75–82.

Parker, J. E., Schulte, W., Hahlbrock, K., and Scheel, D., 1991, An extracellular glycoprotein from *Phytophthora megasperma* f. sp. *glycinea* elicits phytoalexin synthesis in cultured parsley cells and protoplasts, *Mol. Plant–Microbe Interact.* **4:**19–27.

Peña-Cortes, H., Sanchez-Serrano, J., Rocha-Sosa, M., and Willmitzer, L., 1988, Systemic induction of proteinase-inhibitor-II gene expression in potato plants by wounding, *Planta* **174:**84–89.

Peters, B. M., Cribbs, D. H., and Stelzig, D. A., 1978, Agglutination of plant protoplasts by fungal cell wall glucans, *Science* **201:**364–365.

Pitson, S. M., Seviour, R. J., and McDougall, B. M., 1993, Noncellulolytic fungal β-glucanases: Their physiology and regulation, *Enzyme Microb. Technol.* **15:**178–192.

Pospieszny, H., and Atabekov, J. G., 1989, Effect of chitosan on the hypersensitive reaction of bean to alfalfa mosaic virus, *Plant Sci.* **62:**29–31.

Pospieszny, H., Chirkov, S., and Atabekov, J., 1991, Induction of antiviral resistance in plants by chitosan, *Plant Sci.* **79:**63–68.

Powell, D. A., Morris, E. R., Gidley, M. J., and Rees, D. A., 1982, Conformations and interactions of pectins II. Influence of residue sequence on chain association in calcium pectate gels, *J. Mol. Biol.* **155:**517–531.

Pressey, R., 1991, Oxidized oligogalacturonides activate the oxidation of indoleacetic acid by peroxidase, *Plant Physiol.* **96:**1167–1170.

Pressey, R., 1993, Uronic acid oxidase in orange fruit and other plant tissues, *Phytochemistry* **32:**1375–1379.

Punja, Z. K., and Raharjo, S. H. T., 1996, Response of transgenic cucumber and carrot plants expressing different chitinase enzymes to inoculation with fungal pathogens, *Plant Dis.* **80:**999–1005.

Raetz, C. R. H., 1990, Biochemistry of endotoxins, *Annu. Rev. Biochem.* **59:**129–170.

Ren, Y.-Y., and West, C. A., 1992, Elicitation of diterpene biosynthesis in rice (*Oryza sativa* L.) by chitin, *Plant Physiol.* **99:**1169–1178.

Reuhs, B. L., Carlson, R. W., and Kim, J. S., 1993, *Rhizobium fredii* and *Rhizobium meliloti* produce 3-deoxy-D-*manno*-2-octulosonic acid-containing polysaccharides that are structurally analogous to group II K antigens (capsular polysaccharides) found in *Escherichia coli*, *J. Bacteriol.* **175:**3570–3580.

Reymond, P., Kunz, B., Paul-Pletzer, K., Grimm, R., Eckerskorn, C., and Farmer, E. E., 1996, Cloning of a cDNA encoding a plasma membrane-associated, uronide binding phosphoprotein with physical properties similar to viral movement proteins, *Plant Cell* **8:**2265–2276.

Ride, J. P., and Barber, M. S., 1987, The effects of various treatments on induced lignification and the resistance of wheat to fungi, *Physiol. Mol. Plant Pathol.* **31:**349–360.

Ride, J. P., and Barber, M. S., 1990, Purification and characterization of multiple forms of endochitinase from wheat leaves, *Plant Sci.* **71:**185–197.

Ridley, B. L., Spiro, M. D., Glushka, J., Albersheim, P., Darvill, A., and Mohnen, D., 1997, A method for biotin labeling of biologically active oligogalacturonides using a chemically stable hydrazide linkage, *Anal. Biochem.* **249:**10–19.

Roberts, K., 1990, Structures at the plant cell surface, *Curr. Opin. Cell Biol.* **2:**920–928.

Robertsen, B., 1986, Elicitors of the production of lignin-like compounds in cucumber hypocotyls, *Physiol. Mol. Plant Pathol.* **28:**137–148.

Robertsen, B., 1987, Endo-polygalacturonase from *Cladosporium cucumerinum* elicits lignification in cucumber hypocotyls, *Physiol. Mol. Plant Pathol.* **31:**361–374.

Robertsen, B., 1989, Pectate lyase from *Cladosporium cucumerinum*, purification, biochemical properties and ability to induce lignification in cucumber hypocotyls, *Mycol. Res.* **94:**595–602.

Roby, D., Toppan, A., and Esquerré-Tugayé, M.-T., 1985, Cell surfaces in plant–microorganism interactions V. Elicitors of fungal and of plant origin trigger the synthesis of ethylene and of cell wall hydroxyproline-rich glycoprotein in plants, *Plant Physiol.* **77:**700–704.

Rong, L., Carpita, N. C., Mort, A., and Gelvin, S. B., 1994, Soluble cell wall compounds from carrot roots induce the *picA* and *pgl* loci of *Agrobacterium tumefaciens*, *Mol. Plant–Microbe Interact.* **7:**6–14.

Rouet-Mayer, M.-A., Mathieu, Y., Cazalé, A.-C., Guern, J., and Laurière, C., 1997, Extracellular alkalinization and oxidative burst induced by fungal pectin lyase in tobacco cells are not due to the perception of oligogalacturonide fragments, *Plant Physiol. Biochem.* **35:**321–330.

Rouhier, P., Kopp, M., Begot, V., Bruneteau, M., and Fritig, B., 1995, Structural features of fungal β-D-glucans for the efficient inhibition of the initiation of virus infection on *Nicotiana tabacum*, *Phytochemistry* **39:**57–62.

Ruiz-Herrera, J., 1991, *Fungal Cell Wall: Structure, Synthesis, and Assembly*, CRC Press, Boca Raton, Florida.

Ryan, C. A., 1988, Oligosaccharides as recognition signals for the expression of defensive genes in plants, *Biochemistry* **27:**8879–8883.

Ryan, C. A., 1994, Oligosaccharide signals: From plant defense to parasite offense, *Proc. Natl. Acad. Sci. USA* **91:**1–2.

Ryan, C. A., and Farmer, E. E., 1991, Oligosaccharide signals in plants: A current assessment, *Annu. Rev. Plant Physiol. Plant Mol. Biol.* **42:**651–674.

Scheel, D., and Parker, J. E., 1990, Elicitor recognition and signal transduction in plant defense gene activation, *Z. Naturforsch.* **45c:**569–575.

Schlumbaum, A., Mauch, F., Vögeli, U., and Boller, T., 1986, Plant chitinases are potent inhibitors of fungal growth, *Nature* **324:**365–367.

Schmidt, W. E., and Ebel, J., 1987, Specific binding of a fungal glucan phytoalexin elicitor to membrane fractions from soybean *Glycine max*, *Proc. Natl. Sci. USA* **84:**4117–4121.

Sembdner, G., and Parthier, B., 1993, The biochemistry and the physiological and molecular actions of jasmonates, *Annu. Rev. Plant Physiol. Plant Mol. Biol.* **44:**569–589.

Sharma, P., Borja, D., Stougaard, P., and Lönneborg, A., 1993, PR-proteins accumulating in spruce roots infected with a pathogenic *Pythium* sp. isolate include chitinases, chitosanases and β-1,3-glucanases, *Physiol. Mol. Plant Pathol.* **43:**57–67.

Sharp, J. K., Albersheim, P., Ossowski, P., Pilotti, Å., Garegg, P. J., and Lindberg, B., 1984a, Comparison of the structures and elicitor activities of a synthetic and a mycelial-wall-derived hexa(β-D-glucopyranosyl)-D-glucitol, *J. Biol. Chem.* **259:**11341–11345.

Sharp, J. K., McNeil, M., and Albersheim, P., 1984b, The primary structures of one elicitor-active and seven elicitor-inactive hexa(β-D-glucopyranosyl)-D-glucitols isolated from the mycelial walls of *Phytophthora megasperma* f. sp. *glycinea*, *J. Biol. Chem.* **259:**11321–11336.

Sharp, J. K., Valent, B., and Albersheim, P., 1984c, Purification and partial characterization of a β-glucan fragment that elicits phytoalexin accumulation in soybean, *J. Biol. Chem.* **259:**11312–11320.

Shibuya, N., Kaku, H., Kuchitsu, K., and Maliarik, M. J., 1993, Identification of a novel high-affinity binding site for *N*-acetylchitooligosaccharide elicitor in the membrane fraction from suspension-cultured rice cells, *FEBS Lett.* **329**:75–78.

Shibuya, N., Ebisu, N., Kamada, Y., Kaku, H., Cohn, J., and Ito, Y., 1996a, Localization and binding characteristics of a high-affinity binding site for *N*-acetylchitooligosaccharide elicitor in the plasma membrane from suspension-cultured rice cells suggest a role as a receptor for the elicitor signal at the cell surface, *Plant Cell Physiol.* **37**:894–898.

Shibuya, N., Ito, Y., and Kaku, H., 1996b, Perception of oligochitin (*N*-acetylchitooligosaccharide) elicitor signal in rice, in *Biology of Plant–Microbe Interactions* (G. Stacey, B. Mullin, and P. M. Gresshoff, eds.), pp. 83–88, International Society for Molecular Plant–Microbe Interactions, St. Paul, Minnesota.

Spaink, H. P., and Lugtenberg, B. J. J., 1994, Role of rhizobial lipo-chitin oligosaccharide signal molecules in root nodule organogenesis, *Plant Mol. Biol.* **26**:1413–1422.

Spessard, G. O., Hanson, C., Halvorson, J. S., and Giannini, J. L., 1994, Effects of phaseollin on membrane leakage in red beet vacuoles and tonoplast vesicles, *Phytochemistry* **35**:43–47.

Spiro, M. D., Kates, K. A., Koller, A. L., O'Neill, M. A., Albersheim, P., and Darvill, A. G., 1993, Purification and characterization of biologically active 1,4-linked α-D-oligogalacturonides after partial digestion of polygalacturonic acid with endopolygalacturonase, *Carbohydr. Res.* **247**:9–20.

Spiro, M. D., Ridley, B. L., Glushka, J., Darvill, A. G., and Abersheim, P., 1996, Synthesis and characterization of tyramine-derivatized 1→4-linked α-D-oligogalacturonides, *Carbohydr. Res.* **290**:147–157.

Stäb, M. R., and Ebel, J., 1987, Effects of $Ca^{2+}$ on phytoalexin induction by fungal elicitor in soybean cells, *Arch. Biochem. Biophys.* **257**:416–423.

Staehelin, C., Granado, J., Müller, J., Wiemken, A., Mellor, R. B., Felix, G., Regenass, M., Broughton, W. J., and Boller, T., 1994, Perception of *Rhizobium* nodulation factors by tomato cells and inactivation by root chitinases, *Proc. Natl. Acad. Sci. USA* **91**:2196–2200.

Staskawicz, B. J., Ausubel, F. M., Baker, B. J., Ellis, J. G., and Jones, J. D. G., 1995, Molecular genetics of plant disease resistance, *Science* **268**:661–667.

Steffens, M., Ettl, F., Kranz, D., and Kindl, H., 1989, Vanadate mimics effects of fungal cell wall in eliciting gene activation in plant cell cultures, *Planta* **177**:160–168.

Stintzi, A., Heitz, T., Prasad, V., Wiedemann-Merdinoglu, S., Kauffmann, S., Geoffroy, P., Legrand, M., and Fritig, B., 1993, Plant 'pathogenesis-related' proteins and their role in defense against pathogens, *Biochimie* 75:687–706.

Stotz, H. U., Bergmann, C. W., Powell, A. L. T., Contos, J. J., Albersheim, P., Darvill, A. G., and Labavitch, J. M., 1994a, Structural and functional comparison of polygalacturonase inhibitor proteins from pear, tomato, and bean (Abstract), 7th International Symposium on Molecular Plant–Microbe Interactions, Edinburgh, U.K.

Stotz, H. U., Contos, J. J. A., Powell, A. L. T., Bennett, A. B., and Labavitch, J. M., 1994b, Structure and expression of an inhibitor of fungal polygalacturonases from tomato, *Plant Mol. Biol.* **25**:607–617.

Strasser, H., Hoffmann, C., Grisebach, H., and Matern, U., 1986, Are polyphosphoinositides involved in signal transduction of elicitor-induced phytoalexin synthesis in cultured plant cells? *Z. Naturforsch.* **41c**:717–724.

Takeuchi, Y., Yoshikawa, M., and Horino, O., 1990a, Immunological evidence that β-1,3-endoglucanase is the major elicitor-releasing factor in soybean, *Ann. Phytopath. Soc. Japan* **56**:523–531.

Takeuchi, Y., Yoshikawa, M., Takeba, G., Tanaka, K., Shibata, D., and Horino, O., 1990b, Molecular cloning and ethylene induction of mRNA encoding a phytoalexin elicitor-releasing factor, β-1,3-endoglucanase, in soybean, *Plant Physiol.* **93**:673–682.

Tani, M., Fukui, H., Shimomura, M., and Tabata, M., 1992, Structure of endogenous oligogalacturonides inducing shikonin biosynthesis in *Lithospermum* cell cultures, *Phytochemistry* **31**:2719–2723.

Tepper, C. S., and Anderson, A. J., 1990, Interactions between pectic fragments and extracellular components from the fungal pathogen *Colletotrichum lindemuthianum*, *Physiol. Mol. Plant Pathol.* **36**:147–158.

Thain, J. F., Doherty, H. M., Bowles, D. J., and Wildon, D. C., 1990, Oligosaccharides that induce proteinase inhibitor activity in tomato plants cause depolarization of tomato leaf cells, *Plant Cell Environ.* **13**:569–574.

Thain, J. F., Gubb, I. R., and Wildon, D. C., 1995, Depolarization of tomato leaf cells by oligogalacturonide elicitors, *Plant Cell Environ.* **18**:211–214.

Thornton, B. P., Vetvicka, V., Pitman, M., Goldman, R. C., and Ross, G. D., 1996, Analysis of the sugar specificity and molecular location of the β-glucan-binding lectin site of complement receptor type 3 (CD11b/CD18), *J. Immunol.* **156**:1235–1246.

Toubart, P., Desiderio, A., Salvi, G., Cervone, F., Daroda, L., De Lorenzo, G., Bergmann, C., Darvill, A. G., and Albersheim, P., 1992, Cloning and characterization of the gene encoding the endopolygalacturonase-inhibiting protein (PGIP) of *Phaseolus vulgaris* L., *Plant J.* **2**:367–373.

Umemoto, N., Kakitani, M., Iwamatsu, A., Yoshikawa, M., Yamaoka, N., and Ishida, I., 1997, The structure and function of a soybean β-glucan-elicitor-binding protein, *Proc. Natl. Acad. Sci. USA* **94**:1029–1034.

Velupillai, P., and Harn, D. A., 1994, Oligosaccharide-specific induction of interleukin 10 production by $B220^+$ cells from schistosome-infected mice: A mechanism for regulation of $CD4^+$ T-cell subsets, *Proc. Natl. Acad. Sci. USA* **91**:18–22.

Verduyn, R., Douwes, M., Van der Klein, P. A. M., Mösinger, E. M., Van der Marel, G. A., and van Boom, J. H., 1993, Synthesis of a methyl heptaglucoside: Analogue of the phytoalexin elicitor from *Phytophtora megasperma*, *Tetrahedron* **49**:7301–7316.

Vierheilig, H., Alt, M., Neuhaus, J.-M., Boller, T., and Wiemken, A., 1993, Colonization of transgenic *Nicotiana sylvestris* plants, expressing different forms of *Nicotiana tabacum* chitinase, by the root pathogen *Rhizoctonia solani* and by the mycorrhizal symbiont *Glomus mosseae*, *Mol. Plant–Microbe Interact.* **6**:261–264.

Vögeli, U., Meins, F., Jr., and Boller, T., 1988, Co-ordinated regulation of chitinase and β-1,3-glucanase in bean leaves, *Planta* **174**:364–372.

Vreeland, V., Morse, S. R., Robichaux, R. H., Miller, K. L., Hua, S.-S. T., and Laetsch, W. M., 1989, Pectate distribution and esterification in *Dubautia* leaves and soybean nodules, studied with a fluorescent hybridization probe, *Planta* **177**:435–446.

Waldmann, T., Jeblick, W., and Kauss, H., 1988, Induced net $Ca^{2+}$ uptake and callose biosynthesis in suspension-cultured plant cells, *Planta* **173**:88–95.

Waldmüller, T., Cosio, E. G., Grisebach, H., and Ebel, J., 1992, Release of highly elicitor-active glucans by germinating zoospores of *Phytophthora megasperma* f. sp. *glycinea*, *Planta* **188**:498–505.

Walker-Simmons, M., and Ryan, C. A., 1984, Proteinase inhibitor synthesis in tomato leaves. Induction by chitosan oligomers and chemically modified chitosan and chitin, *Plant Physiol.* **76**:787–790.

Walker-Simmons, M., Hadwiger, L., and Ryan, C. A., 1983, Chitosans and pectic polysaccharides both induce the accumulation of the antifungal phytoalexin pisatin in pea pods and antinutrient proteinase inhibitors in tomato leaves, *Biochem. Biophys. Res. Commun.* **110**:194–199.

Walker-Simmons, M., Jin, D., West, C. A., Hadwiger, L., and Ryan, C. A., 1984, Comparison of proteinase inhibitor-inducing activities and phytoalexin elicitor activities of a pure fungal endopolygalacturonase, pectic fragments, and chitosan, *Plant Physiol.* **76**:833–836.

Ward, E. R., Payne, G. B., Moyer, M. B., Williams, S. C., Dincher, S. S., Sharkey, K. C., Beck,

J. J., Taylor, H. T., Ahl-Goy, P., Meins, F., Jr., and Ryals, J. A., 1991a, Differential regulation of β-1,3-glucanse messenger RNAs in response to pathogen infection, *Plant Physiol.* **96:**390–397.

Ward, E. R., Uknes, S. J., Williams, S. C., Dincher, S. S., Widerhold, D. L., Alexander, D. C., Ahl-Goy, P., Métraux, J.-P., and Ryals, J. A., 1991b, Coordinate gene activity in response to agents that induce systemic acquired resistance, *Plant Cell* **3:**1085–1094.

Xu, H., and Mendgen, K., 1994, Endocytosis of 1,3-β-glucans by broad bean cells at the penetration site of the cowpea rust fungus (haploid stage), *Planta* **195:**282–290.

Yamada, A., Shibuya, N., Kodama, O., and Akatsuka, T., 1993, Induction of phytoalexin formation in suspension-cultured rice cells by *N*-acetylchitooligosaccharides, *Biosci. Biotech. Biochem.* **57:**405–409.

Yao, C. L., Conway, W. S., and Sams, C. E., 1995, Purification and characterization of a polygalacturonase-inhibiting protein from apple fruit, *Phytopathology* **85:**1373–1377.

Yoder, M. D., and Jurnak, F., 1995, The refined three-dimensional structure of pectate lyase C from *Erwinia chrysanthemi* at 2.2 Ångstrom resolution. Implications for an enzymatic mechanism, *Plant Physiol.* **107:**349–364.

Yoshikawa, M., and Sugimoto, K., 1993, A specific binding site on soybean membranes for a phytoalexin elicitor released from fungal cell walls by β-1,3-endoglucanase, *Plant Cell Physiol.* **34:**1229–1237.

Yoshikawa, M., Keen, N. T., and Wang, M.-C., 1983, A receptor on soybean membranes for a fungal elicitor of phytoalexin accumulation, *Plant Physiol.* **73:**497–506.

Yoshikawa, M., Takeuchi, Y., and Horino, O., 1990, A mechanism for ethylene-induced disease resistance in soybean: Enhanced synthesis of an elicitor-releasing factor, β-1,3-endoglucanase, *Physiol. Mol. Plant Pathol.* **37:**367–376.

Young, D. H., and Kauss, H., 1983, Release of calcium from suspension-cultured *Glycine max* cells by chitosan, other polycations, and polyamines in relation to effects on membrane permeability, *Plant Physiol.* **73:**698–702.

Young, D. H., Köhle, H., and Kauss, H., 1982, Effect of chitosan on membrane permeability of suspension-cultured *Glycine max* and *Phaseolus vulgaris* cells, *Plant Physiol.* **70:**1449–1454.

Zhu, Q., Maher, E. A., Masoud, S., Dixon, R. A., and Lamb, C. J., 1994, Enhanced protection against fungal attack by constitutive co-expression of chitinase and glucanase genes in transgenic tobacco, *Bio/technology* **12:**807–812.

# Index

GPSR Compliance

*The European Union's (EU) General Product Safety Regulation (GPSR) is a set of rules that requires consumer products to be safe and our obligations to ensure this.*

*If you have any concerns about our products, you can contact us on ProductSafety@springernature.com*

In case Publisher is established outside the EU, the EU authorized representative is:

Springer Nature Customer Service Center GmbH
Europaplatz 3
69115 Heidelberg, Germany

**Batch number: 10755284**

Printed by Printforce, the Netherlands